Psychology of Science

PSYCHOLOGY OF SCIENCE IMPLICIT AND EXPLICIT PROCESSES

2ND PURDUE SYMPOSIUM ON PSYCHOLOGICAL SCIENCES

Edited by
Robert W. Proctor
and
E. J. Capaldi

OXFORD
UNIVERSITY PRESS

OXFORD
UNIVERSITY PRESS

Oxford University Press, Inc., publishes works that further
Oxford University's objective of excellence
in research, scholarship, and education.

Oxford New York
Auckland Cape Town Dar es Salaam Hong Kong Karachi
Kuala Lumpur Madrid Melbourne Mexico City Nairobi
New Delhi Shanghai Taipei Toronto

With offices in
Argentina Austria Brazil Chile Czech Republic France Greece
Guatemala Hungary Italy Japan Poland Portugal Singapore
South Korea Switzerland Thailand Turkey Ukraine Vietnam

Published by Oxford University Press, Inc.
198 Madison Avenue, New York, New York 10016
www.oup.com

Library of Congress Cataloging-in-Publication Data

Purdue Symposium on Psychological Sciences (2nd : 2010 : Purdue University)
Psychology of science : implicit and explicit processes / edited by Robert W. Proctor and E.J. Capaldi.
p. cm.
Symposium held at Purdue Univ. in June 4–5, 2010.
Includes bibliographical references and index.
ISBN 978–0–19–975362–8 (hardback : alk. paper)
1. Reasoning–Congresses. 2. Science–Philosophy. 3. Science–Psychological aspects–Congresses.
I. Proctor, Robert W. II. Capaldi, E. John. III. Title.
Q175.32.R45P87 2010
501'.9–dc23
2011045711

9 8 7 6 5 4 3 2 1
Printed in the United States of America
on acid-free paper

Contents

vii Preface
ix Contributors

1 **Introduction** Implicit and Explicit Processes in the Psychology of Science, *Robert W. Proctor and E. J. Capaldi*

Part One Role of the Psychology of Science and Its Methods

13 1. The Psychology of Science Is Off and Running but Where Do We Go from Here?, *Gregory J. Feist*
40 2. Psychology of Science: Influence on the Philosophy of Science, *E. J. Capaldi and Robert W. Proctor*
57 3. Methodological Approaches to Scientific and Technological Thinking, *Michael E. Gorman*

Part Two Agency and Reasoning in the Psychology of Science

73 4. The Role of Psychology in an Agent-Centered Theory of Science, *Ronald N. Giere*
86 5. The Acting Person in Science Practice, *Lisa M. Osbeck and Nancy J. Nersessian*
112 6. Inference to the Best Explanation (IBE) and the Causal and Scientific Reasoning of Nonscientists, *Barbara Koslowski*
137 7. Detecting, Classifying, and Remediating: Children's Explicit and Implicit Misconceptions about Experimental Design, *Stephanie A. Siler and David Klahr*

Part Three Implicit and Explicit Processes in the Cognitive Psychology of Science

183 8. How to Define and Examine Implicit Processes?, *Jan De Houwer and Agnes Moors*
199 9. How Should We Understand Implicit and Explicit Processes in Scientific Thinking?, *Donelson E. Dulany*
228 10. The Interaction of Implicit versus Explicit Processing and Problem Difficulty in a Scientific Discovery Task, *Corinne Zimmerman and Jean E. Pretz*

Part Four Psychological Perspectives: Influence on Science

255 11. Scientists Are Human: Implicit Cognition and Researcher Conflict of Interest, *Anthony G. Greenwald*

267 12. Bias, Feminism, and the Psychology of Investigating Gender, *Alice H. Eagly*

289 13. The Theory Ladenness of the Mental Processes Used in the Scientific Enterprise: Evidence from Cognitive Psychology and the History of Science, *William F. Brewer*

335 14. The Practice of Psychological Science in Social-Personality Research: Are We Still a Science of Two Disciplines?, *Jessica L. Tracy, Richard W. Robins, and Jeffrey W. Sherman*

Part Five Scientific Creativity

363 15. Scientific Creativity as *Blind* Variation: Explicit and Implicit Procedures, Mechanisms, and Processes, *Dean Keith Simonton*

389 16. Creative Combination of Representations: Scientific Discovery and Technological Invention, *Paul Thagard*

406 17. On the Unreasonable Reasonableness of Mathematical Physics: A Cognitive View, *Ryan D. Tweney*

436 18. Digging into Implicit/Explicit States and Processes: The Case of Cognitive/Social Process Interaction in Scientific Groups, *Susannah B. F. Paletz and Christian D. Schunn*

Part Six Unconventional Perspectives on the Conduct of Science

459 19. Implicit Ontological Reasoning: The Problems of Dualism in Psychological Science, *Brent D. Slife, Jeffrey S. Reber, and James E. Faulconer*

479 20. Notre Trahison des Clercs: Implicit Aspirations—Explicit Exploitations, *P. A. Hancock*

499 Index

Preface

Psychology of science is a new field that attempts to describe the influence of psychological processes on the creation of science. Concepts and theories developed in various areas of psychology have implications for the development of understanding of scientific theory and practice.

Consider two broad approaches to the psychology of science. One emphasizes various means for examining the psychology of science, including such factors as its methodology, concepts, and so on. The second identifies explicit psychological factors that influence various aspects of science, including what we observe and what conclusions we come to. Some chapters in the book highlight both of these approaches, whereas other chapters stress one or the other. It is the case that psychological factors in the creation of science have been considered to some extent both in the philosophy and psychology of science. However, these factors are emphasized in this book to a much greater extent than in any previous treatment. We see the emphasis on psychological factors in the creation of science contained in this book as a significant novel and major contribution to the psychology of science. Of course, the credit for this goes to the individuals who contributed the chapters in this book.

Most prior work in the psychology of science has focused on the role of explicit processes and reasoning, which include all of logic, mathematics, and, more generally, conscious processes. Contemporary research in the field of psychology, however, has underscored the importance of implicit processes, and performance of many tasks seemingly involves a complex relationship between implicit and explicit processes. The goal of this book is to examine the roles played by both implicit and explicit processes in the empirical and theoretical aspects of science. Thus, another novel contribution of this book is identifying the roles of implicit and explicit processes together in science.

The book includes chapters from leaders in the emerging discipline of the psychology of science and from various other individuals who are experts on the roles of implicit and explicit processes in thinking. These chapters, taken collectively, cover a wide variety of current issues that are central to the psychology of science. It is our hope that the issues covered in this book will further the interests of many psychologists to develop the implications of psychology for better understanding science.

This book is the outgrowth of the Second Purdue Symposium on Psychological Sciences held at Purdue University on June 4 and 5, 2010. The financial support for the symposium and book was provided by the Bradley fund of the Department of Psychological Sciences. We wish to thank two individuals who encouraged our development of the project and allowed us access to the Bradley fund, the current department head, Chris Agnew, and the former head, Howard Weiss. We also thank Erica Wilson of the Conference Division at Purdue University for coordinating all of the details of the symposium.

Robert W. Proctor
E. J. Capaldi

Contributors

William F. Brewer
Department of Psychology
University of Illinois
Champaign-Urbana, IL

E. J. Capaldi
Department of Psychological Sciences
Purdue University
West Lafayette, IN

Jan De Houwer
Department of Experimental-Clinical and Health Psychology
Ghent University
Ghent, Belgium

Donelson E. Dulany
Department of Psychology
University of Illinois
Champaign-Urbana, IL

Alice H. Eagly
Department of Psychology
Northwestern University
Evanston, IL

James E. Faulconer
Department of Philosophy
Brigham Young University
Provo, UT

Gregory J. Feist
Department of Psychology
San Jose State University
San Jose, CA

Ronald N. Giere
Center for Philosophy of Science
University of Minnesota
Minneapolis, MN

Michael E. Gorman
Science, Technology and Society
University of Virginia
Charlottesville, VA

Anthony G. Greenwald
Department of Psychology
University of Washington
Seattle, WA

P. A. Hancock
Department of Psychology
University of Central Florida
Orlando, FL

David Klahr
Department of Psychology
Carnegie Mellon University
Pittsburgh, PA

Barbara Koslowski
Department of Human Development
Cornell University
Ithaca, NY

Agnes Moors
Department of Experimental-Clinical and Health Psychology
Ghent University
Ghent, Belgium

Nancy J. Nersessian
School of Interactive Computing
Georgia Institute of Technology
Atlanta, GA

Lisa M. Osbeck
Department of Psychology
University of West Georgia
Carrollton, GA

Susannah B. F. Paletz
Learning Research and Development Center
University of Pittsburgh
Pittsburgh, PA

Jean E. Pretz
Department of Psychology
Elizabethtown College
Elizabethtown, PA

Robert W. Proctor
Department of Psychological Sciences
Purdue University
West Lafayette, IN

Jeffrey S. Reber
Psychology Department
University of West Georgia
Carrollton, GA

Richard W. Robins
Department of Psychology
University of California, Davis
Davis, CA

Christian D. Schunn
Learning Research and Development Center
University of Pittsburgh
Pittsburgh, PA

Jeffrey W. Sherman
Department of Psychology
University of California, Davis
Davis, CA

Stephanie A. Siler
Carnegie Mellon University
Department of Psychology
Pittsburgh, PA

Dean Keith Simonton
Department of Psychology
University of California, Davis
Davis, CA

Brent D. Slife
Department of Psychology
Brigham Young University
Provo, UT

Paul Thagard
Department of Philosophy
University of Waterloo
Waterloo, Canada

Jessica L. Tracy
Department of Psychology
University of British Columbia
Vancouver, Canada

Ryan D. Tweney
Department of Psychology
Bowling Green State University
Bowling Green, OH

Corinne Zimmerman
Department of Psychology
Illinois State University
Normal, IL

Psychology of Science

Introduction

Implicit and Explicit Processes in the Psychology of Science

Robert W. Proctor and E. J. Capaldi

The psychology of science is a growing field that uses the methods, data, and theory of psychology to better understand the practice of science. In recent years, psychology has come increasingly to examine implicit as well as explicit processes in decision making and thought. This book is the first to examine the roles of both implicit and explicit processes in the creation of science. The major purpose of this Introduction to the book is to acquaint the reader with some of the major issues in the field. As such, it provides an overview that should be of value to readers in understanding the relations between the individual chapters. The major headings for this Introduction correspond to the section headings in the remainder of the book.

ROLE OF THE PSYCHOLOGY OF SCIENCE AND ITS METHODS

In the formal sense, the psychology of science is new and novel. As Feist (Chapter 1) indicates, it is reasonable to regard 2006 as the year in which the field was formally established. However, as his chapter and our later one (Chapter 2) highlight, informally, various ideas in psychology historically have had a great influence on views of science since well before 2006. Why is the psychology of science needed? Feist answers the question as follows:

> Compared to other studies of science—with the exception of history—psychology of science focuses on the development of the individual scientist in the context of his or her social environment and group. Moreover, psychology of science is unique in the studies of science to focus on influences such as intelligence, motivation, personality, and development over the lifespan of scientific interest, thought, ability, and achievement. (ms, p. 14)

We strongly agree with Feist's answer, for which the most central point is his emphasis on the focus on the individual scientist within the social context of science and society. This focus on the individual differentiates the psychology of science from the sociology of science (Freeman, 2004), which sees the group as primary. Many chapters in this book emphasize the role of the individual scientist in various scientific activities.

Feist's answer to the question of why the psychology of science is needed also suggests that not only cognitive and social psychology, but also various other areas of psychology, including life-span development, personality, clinical, and neuroscience, have much to contribute to the psychology of science. Feist points out that early work on the psychology of science and scientists dates to the latter part of the 1800s. In Chapter 2, we emphasize that various developments in psychology have had a profound effect on developments in the philosophy of science from its beginning. Our chapter illustrates the influence of four different areas of psychology (psychophysics, Gestalt approaches to perception, behaviorism, and cognitive psychology) on the development of the philosophy of science from its inception to the present day. As we show, this influence is much wider and deeper than typically recognized. For example, Gestalt psychology greatly influenced the work of Rudolph Carnap, Norwood Hanson, and Thomas Kuhn, among other philosophers of science (see also Brewer, Chapter 13).

Gorman (Chapter 3) surveys the variety of methods that have been, and can be, employed in the psychology of science, most of which are illustrated in other chapters of this book. These methods include analyzing the content of the notebooks and other materials of scientists (Gorman, Chapter 3; Tweney, Chapter 17), isolating factors common to many scientific and technological innovations (Thagard, Chapter 16), observing scientists at work in laboratory settings (Osbeck & Nersessian, Chapter 5), employing correlational methods to understand psychological factors in the creation of science (Tracy, Robins, & Sherman, Chapter 14), and employing experimental methods to examine under controlled conditions the thought processes and procedures of nonscientists and scientists (e.g., Zimmerman & Pretz, Chapter 10). More generally, how the thought processes of scientists, ranging from simple biases to their more fundamental theoretical assumptions, influence many aspects of scientific practice, such as observation and memory, among others, can be illuminated by utilizing the concepts and data from many different areas of psychology (Eagly, Chapter 12; Brewer, Chapter 13).

Both Gorman (Chapter 3) and Feist (Chapter 1) suggest, in considerable detail, what they see as necessary steps to be taken by a novel field, like the psychology of science, at the outset of its development to ensure its growth and survival. Of the six factors that Gorman mentions, we cite two, which we consider to be among the most important. First, the field must demonstrate the usefulness of established methods as well as novel ones. Second, the field must prove that it can solve problems and provide unique insights that elude other approaches, something considered in many of the book's chapters. In his chapter, Feist cites a number of practical efforts that must occur as well in order for a new field to be successful. These range from exciting teaching to obtaining research grants and establishing degree programs. An important perspective supplied by Feist is that the psychology of science is unique in that it employs all of the disciplines of psychology for a common purpose, that of contributing to the development and understanding of science.

AGENCY AND REASONING IN THE PSYCHOLOGY OF SCIENCE

Giere (Chapter 4) and Osbeck and Nersessian (Chapter 5) develop arguments for an agent-centered approach to science, which emphasizes the individual scientist as agent acting within the social context. Giere focuses initially on distributed cognition, which is concerned with the methods and instruments used in the production of knowledge, as well as the expertise of various individuals concerned with a particular problem in science. Giere goes beyond distributed cognition, though, by considering the individual conscious efforts of the scientist, or agent, who employs the various aspects of distributed cognition. One way to understand Giere is that he favors an individual agent approach to distributed cognition.

Giere situates his overall review with respect to five approaches to science. In the logical approach, such as that employed by logical positivists, the focus is on the logical structure of theories and on inductive logic, which provides a logical relation between observation and theory. The methodological approach, such as that used by Karl Popper, makes discoveries by employing scientific methods. The historical approach, initiated by Kuhn (1962), treats scientists as natural agents to whom empirical psychological categories apply. In the social approach, as represented by the sociology of science, emphasis is on agreement among the relevant community of scientists. Finally, in the cognitive approach, the one of most interest to Giere, the emphasis is on the representation of events and models of the world. Giere says that his overall goal is not to justify science but to understand it, which is a goal of the psychology of science in general.

In agreement with Giere, Osbeck, and Nersessian (Chapter 5) also focus on distributed cognition and the acting person in the construction of science. Their approach emphasizes that problem solving is accomplished by individuals in conjunction with the environment, suggesting that cognition is distributed in the sense that brain and environment constitute a single complex system and that problem solving would not be possible in the absence of the environment, including the social environment. Osbeck and Nersessian report new empirical work in their chapter that is similar in many respects to earlier work by Dunbar (1995), in that they observe scientists in their work setting and interview them in some detail, considering social, emotional, and cognitive aspects of the observed scientists.

Koslowski (Chapter 6) describes several factors that enter into sound psychological reasoning. She notes that examination of evidence in isolation is not very useful for the construction of valid scientific theory. Koslowski begins by highlighting inference to the best explanation (IBE) as a significant approach to science, a view that we endorse (Capaldi & Proctor, 2008; Proctor & Capaldi, 2006). IBE consists of choosing a particular explanation over its competitors because it provides a better causal account of the data. Koslowski goes on to argue that science consists of more than just applying formal rules to scientific problems. She indicates that the identification of a causal event requires that background information be taken into account, suggesting that data do not become evidence until we have an explanation of them. She reports an empirical study that demonstrates that evidential relevance is theory dependent.

Koslowski also says that whether information qualifies as evidence depends on the alternatives contrasted with a given explanation. The explanation of some phenomenon affects the information for which people search as relevant. Background information has an important role in generating relevant alternative interpretations, and it is useful in distinguishing between strong and weak anomalies. Koslowski reports interesting empirical data that merely thinking about an issue can result in background information becoming explicitly relevant. Importantly, she demonstrates that in evaluating a proposition, information diagnostic of the proposition is more valuable than information that is not diagnostic. Koslowski describes some circumstances under which people are more prone to employ diagnostic information. Additionally, she demonstrates that when a proposition is embedded in a network, undermining the network undermines the embedded belief. Finally, she identifies certain things that can reduce our ability to treat background information and explanation as interdependent.

Siler and Klahr (Chapter 7) show how prior beliefs, both implicit and explicit, may influence children's approaches to and understanding of the design of experiments. They point out that students interpret instructions as to how to design experiments as being domain specific rather than domain general. Siler and Klahr describe common misconceptions about the goal of their instructions, how these misconceptions lead to delays or failures in learning, and, importantly, they suggest instructional procedures that can be used to correct these misconceptions. They show that the prior beliefs of students may distort both the understanding of how to design an experiment and its purpose in identifying causal factors. In examining these issues, Siler and Klahr employ a procedure they call The Control of Variables Strategy, which consists of the necessary steps for designing an experiment that does not have confounds.

IMPLICIT AND EXPLICIT PROCESSES IN THE COGNITIVE PSYCHOLOGY OF SCIENCE

A major accomplishment of several of the chapters in this volume is to specify factors constituent of implicit and explicit processes. De Houwer and Moors (Chapter 8) are primarily concerned with clearly defining implicit processes. They supply a number of useful characteristics for determining when a process is explicit. In our view, anyone attempting to deal with implicit processes could usefully consult the De Houwer and Moors chapter. The same could be said of Dulany (Chapter 9), who questions the standard cognitive view of implicit reasoning. Other chapters in the book are concerned, not so much with defining implicit and explicit processes, but with employing these processes in empirical settings.

De Houwer and Moors (Chapter 8) equate the term "implicit" with "automatic," noting, for example, that "uncontrolled" is the most common feature attributed to automatic processes. Because different features of automaticity do not necessarily co-occur, they recommend specifying the automatic features one has in mind when investigating implicit processing. Such specification is necessary in order to compare results from different experiments. De Houwer and Moors indicate that these features should be defined not in mental terms but in functional terms, that is, in terms of observable elements in the environment. They see the major benefits of their approach as focusing on empirical phenomena as defining implicit processes, with mental processes being secondary.

Dulany's (Chapter 9) treatment, which differs from that of De Houwer and Moors, contrasts what he calls the standard cognitive theory of the explicit as conscious and the implicit as unconscious with his own mentalistic theory. Dulany argues that the standard view owes its acceptance to a variety of factors that act together in its support. The standard view, according to Dulany, is given more credence than it deserves. He describes his own mentalistic view of implicit and explicit, which is orthogonal to the unconscious versus conscious distinction, and supplies evidence for this view. Essentially, Dulany is applying an alternative approach to the standard one in considering implicit and explicit processes.

Zimmerman and Pretz (Chapter 10) take an experimental approach to the problem of implicit and explicit processing in problem solving. Their procedure is to produce implicit and explicit strategies by varying instructions: Explicit is defined by them as instructions to find a rule to solve the problem and implicit as not providing participants with an instruction to look for a rule. They state, "This is a notoriously difficult issue..., and our approach cannot guarantee that participants in the implicit condition were processing the problems implicitly" (ms, p. 239). We suggest that, this being the case, it may be best to interpret their experimental results in terms of whether or not participants were instructed to search for a rule. Zimmerman and Pretz used a balance of forces task in which participants had to provide weights to one side of a fulcrum to make it balance with the other side. One of their findings was that in difficult forms of this problem, noninstructed subjects performed better than instructed subjects. Another finding was that when participants learned the rule, they were much faster and much more correct. A major finding was that when subjects learned the rule on the basis of being provided a simple task, they responded more poorly than uninstructed subjects when shifted to a difficult task. An interesting feature of their chapter is that they applied their results to real-world scientific problems. For example, on the basis of their data, they suggested that when a practicing scientist is confronted with a difficult problem that evades understanding, what she or he must do is to shift from an explicit attempt to solve the problem to an implicit attempt.

PSYCHOLOGICAL PERSPECTIVES: INFLUENCE ON SCIENCE

Many chapters of the book demonstrate the influence of a person's perspective on psychological processes ranging from perception to theorizing. By perspective, we mean processes as basic as biases and attitudes, and as complex as hypotheses and theories.

Greenwald (Chapter 11) expresses concern about how an individual's bias may influence his or her scientific judgment. He identifies bias in the form of conflict of interest in four professional areas: (1) scientists who make decisions pertaining to publication or funding, (2) reviewers for journal articles and grants, (3) expert witnesses who provide testimony in court cases, and (4) researcher workers in the conduct of their research. Greenwald offers advice for minimizing the influence of conflict of interest in science. Furthermore, he suggests that an ethical code to correct bias might be appropriate. One suggestion for inclusion in this code is that reviewers having theoretical differences with a particular paper should decline to review it or, alternately, make clear their theoretical predilections. It is our view that problems from conflict of interest arising from biases of individual scientists may be ameliorated to some extent by the activities of the scientific group (Ziman, 1998).

Eagly (Chapter 12) points out that correspondence inference bias, in which the actions of an individual are seen as corresponding to what the actor is thinking rather than to external circumstances, has an influence on science. She is indicating that our normal reasoning processes emphasize what may be going on in the minds of other people to a greater extent than the contextual factors operating in the environment. One interpretation of this is that the thinking of the individual before us may be more evident than the contextual environmental factors, which may have occurred at some other time and in some other place. Another type of bias Eagly emphasizes is ingroup bias, in which one tends to look favorably on the activities of one's own group. She illustrates the influence of these biases on gender differences. For example, she has conducted meta-analyses showing that males tend to cite males and females tend to cite females. A similar observation is made by Brewer (Chapter 13), who notes, "Writers from microbiology and immunology included twice as many infectious disease articles as did psychiatrists, whereas psychiatrists included twice as many psychiatric articles as did the writers from microbiology and immunology" (ms, p. 324). Notice also that Kuhn (1962), in his monumental book, discussed ingroup bias in relation to nationality. Clearly, the effect of ingroup bias on science can arise from many sources, and its influence deserves to be recognized more than it has been.

Brewer (Chapter 13) emphasizes the influence of theory on various aspects of scientific activity ranging from perception to attention to memory to thinking to communication. He draws initially on the work of Hanson and of Kuhn. Although they emphasized perception and observation, Brewer goes far beyond them in emphasizing the other processes outlined above. He presents considerable evidence from a variety of sources that many types of perspectives influence all aspects of the activities in which scientists engage. The range of topics covered is so broad that many who may agree with Brewer to begin with may be surprised at how widespread the influence of theory is on so many scientific processes. Brewer's chapter should bring to the attention of many scientists the influence of theorizing on various forms of scientific activity.

Tracy, Robins, and Sherman (Chapter 14) demonstrate that one's methodological perspective influences the kind of scientific activity one engages in. Their chapter, joining in the work of Cronbach (1957), is concerned with two different ways of thinking about science, the correlational approach and the experimental approach. Using questionnaires sent to both correlational and experimental psychological scientists, they demonstrate that despite the methodological differences between these individuals, they often study the same problems, and some individuals bridge the divide between correlational and experimental approaches. One difference between the two groups is that personality psychologists (who take a correlational approach) tend to emphasize external validity, whereas social psychologists (who take an experimental approach) are more concerned with internal validity. Still another difference between the two groups is that the former tends to approach research using more commonsense ideas, whereas the latter group tends to approach research from a theoretical perspective. Tracy et al. provide an example of how the two groups differ in terms of how each studies aggression. The personality psychologists are primarily interested in stable individual differences in aggressive tendencies, whereas the social psychologists are interested in the role of situational factors in producing aggression.

SCIENTIFIC CREATIVITY

Simonton (Chapter 15) and Thagard (Chapter 16) examine various factors that lead to creative thinking both in general and in science. While their approaches are somewhat different, both emphasize that diversity of ideas is essential to the creation of novel scientific theories. Simonton, drawing on the influential work of Donald Campbell (1960), suggests that novel scientific ideas result from the blind variation and selective retention of ideas. Blind variation is the basis of novel suggestions, whereas selective retention is the basis for determining utility. He suggests that creative ideas should not only be new and useful, but also nonobvious. An important criterion for a novel idea is that it be both surprising and valuable.

Thagard describes how the combination of mental representations gives rise to novel ideas in both science and technology, emphasizing that all creativity results from such combination, described as the combinatorial conjecture. His chapter provides a thorough evaluation of his conjecture by determining whether it applies to 100 important cases of scientific discovery and 100 important cases of technological invention. This examination provides support for the combinatorial conjecture. In addition, no counterexamples were found. He emphasizes that human thinking, including perception, inference, learning, language, and action, involves the use of representations, with both visual and verbal representations being the most important for creative scientific and technological thinking.

Thagard considers the relative importance of theory and observation in science, examining 100 examples to determine whether the advances were observational, instrumental, or theoretical. His results show that 70 of the 100 discoveries were theoretical, with the remaining 30 being observational, some employing unaided observation (Mendel) and others employing instruments (Galileo and the telescope). Thagard also was concerned with the extent to which previous technological advances led to novel scientific discoveries, concluding that 37 of the 100 discoveries depended in important ways on technological advances. Thagard and

Simonton disagree about the role of blind variation in creativity, Thagard rejecting it and Simonton, as indicated, accepting it.

Tweney (Chapter 17) deals with the role of mathematics in science, particularly physics. He indicates that the computational and derivational advantages of the mathematical approach in science are central, but he suggests that there is an important role for representations. Tweney goes on to describe the nature of mathematical representation and how it works in physics. Importantly, he suggests that mathematical representations can aid and extend visual representations.

Paletz and Schunn (Chapter 18) discuss the seldom considered and neglected issue of how social and cognitive variables interact to influence scientific thinking and behavior. These authors are also concerned with creativity in science. They distinguish between divergent thinking, which leads to hypotheses being generated, and convergent thinking, which leads to testing hypotheses. Their distinction here corresponds in some respects to one commonly made in the philosophy of science, that between the context of discovery and the context of justification. Paletz and Schunn suggest that both divergent and convergent thinking may involve both implicit and explicit factors. The theory they present is tested by studying the Mars exploratory Rover team of 100 members. Video cameras were moved around a large room in which the scientists worked and carried out conversations among subgroups of various sizes. The transcripts of these sessions were examined for the presence of both analogies and conflicts. One of their findings was that within-domain analogies tended to increase the occurrence of conflict. However, within-discipline and outside-discipline analogies did not increase occurrence of conflict. The chapter is very valuable for detailing in specific ways how social variables and cognitive variables interact in a group in the pursuit of problem solving.

UNCONVENTIONAL PERSPECTIVES ON THE CONDUCT OF SCIENCE

The chapters in the last section of the book take relatively unconventional views of science. According to Slife, Reber, and Faulconer (Chapter 19), conventional scientists implicitly accept a distinction between theory and data. They argue that this distinction is not appropriate and suggest that many of the things that scientists see as objective are really value laden. For example, Slife et al. indicate that self-esteem, which many scientists think of as objective, is instead a subjective value. Slife et al. believe that scientific understanding requires combining theory and data in such a way that the result is what many might think of as problematic. For example, in evaluating an automobile, in addition to various objective facts (e.g., miles per gallon), they suggest that we might include such subjective variables as its aesthetic appeal, much as the magazine *Consumer Reports* does in its evaluations of products. Slife et al.'s belief that there is not a sharp distinction between theory and data is a view shared by other authors in the book, namely Brewer, Eagly, and Koslowski. However, they see the distinction as valuable, even if a bit fuzzy.

Although Hancock's chapter (Chapter 20) differs from that of Slife et al. substantively, it is in agreement in proposing that implicit factors not normally considered by scientists influence their reconstructions. In Hancock's formulation, scientific theories inevitably involve moral considerations. Hancock suggests that scientists think that science as an enterprise is good overall, a view with which he disagrees. He submits instead that science often has implications that are deleterious to humans; for example, attacks by drone airplanes may kill and injure innocent people as well as intended targets. Hancock asserts that we need a moral component in science that directs that technologies arising from science should be put to acceptable moral purposes and should avoid untoward, evil purposes. To accomplish this, he recommends the formation of a superagency of scientists who would endorse the good purposes of some scientific advance and prohibit the immoral purposes to which that advance might be put.

Hancock is not alone in this regard. For example, in this chapter he employs quotes from Noam Chomsky (2010) that suggest a similar conclusion. A similar view is also put forward by Sam Harris (2010) in a recently published popular book. Hancock's view advises that moral considerations cannot be divorced from the conduct of science, and he exhorts scientists to explicitly recognize and act on this. Notice that Hancock's concern is voiced not only by other scientists such as Chomsky and Harris, but also by literary figures such as Mary Shelley, who in 1818 created the story of *Frankenstein* (2009), which suggests that science had evil components that were not in the best interests of humanity.

CONCLUDING REMARKS

The chapters in this volume demonstrate that the time has arrived for psychology to exercise a more coordinated effort to influence the practice and understanding of science. For many years, the attempt to understand science consisted in consulting one's intuitions. However, as illustrated by the chapters in this book, intuition is not always reliable and empirical evidence is required. To some extent the need for empirical evidence has been acknowledged in the philosophy of science, with the turn since Kuhn to use of historical analyses, but this need is accepted within the psychology of science to a much greater extent. It is impossible within the psychology of science to even begin to consider how to create useful science in the absence of empirical information provided by multiple research methods.

As discussed in Chapters 1 and 2, the formalization of the psychology of science is in contrast to earlier developments in which psychology exercised an influence on the philosophy of science, but one that was uncoordinated and informal. The contributors to this volume are in agreement that the psychology of science is a viable enterprise. Any disagreements that exist among the contributors do not undermine the general proposition that psychology has much to offer to better the practice and understanding of science.

A unique emphasis of this book, consistent with contemporary approaches to psychology itself, is that both implicit and explicit processes must be considered in the psychology of science. Various chapters in this book emphasize that these processes include both cognitive and social factors. It is our view that linking the psychology of science more closely to psychology than in the past will result in many novel and important developments in the pursuit and conduct of science. In addition, while psychology can improve the psychology of science, we believe that the influence of these two fields is reciprocal and that studying the psychology of science will improve the understanding and application of psychology.

REFERENCES

Campbell, D. T. (1960). Blind variation and selective retention in creative thought as in other knowledge processes. *Psychological Review, 67,* 380–400.

Capaldi, E. J., & Proctor, R. W. (2008). Are theories to be evaluated in isolation or relative to alternatives? An abductive view. *American Journal of Psychology, 121,* 617–641.

Chomsky, N. (2010). *Hopes and prospects.* Chicago: Haymarket Books.

Cronbach, L. J. (1957). The two disciplines of scientific psychology. *American Psychologist, 12,* 671–684.

Dunbar, K. (1995). How scientists really reason: Scientific reasoning in real-world laboratories. In R. J. Sternberg & J. E. Davidson (Eds.), *The nature of insight* (pp. 365–395). Cambridge, MA: MIT Press.

Freeman, L. C. (2004). *The development of social network analysis: A study in the sociology of science.* Vancouver, BC: Empirical Press.

Harris, S. (2010). *The moral landscape: How science can determine human values.* New York: Free Press.

Kuhn, T. S. (1962). *The structure of scientific revolutions.* Chicago: University of Chicago Press.

Proctor, R. W., & Capaldi, E. J. (2006). *Why science matters: Understanding the methods of psychological research.* Malden, MA: Blackwell.

Shelley, M. (2009). *Frankenstein or the modern Prometheus: The 1818 text.* Oxford: Oxford University Press.

Ziman, J. (1998). What is science? In E. D. Klemke, R. Hollinger, & D. W. Rudge (Eds.), *Introductory readings in the philosophy of science* (pp. 48–53). Amherst, NY: Prometheus.

PART ONE
ROLE OF THE PSYCHOLOGY OF SCIENCE AND ITS METHODS

1 The Psychology of Science Is Off and Running but Where Do We Go from Here?

Gregory J. Feist

THE PSYCHOLOGY OF SCIENCE

Science exists in no species other than humans, if by science we mean explicitly and systematically testing hypotheses. Indeed, even if we expand our definition of science to include all of the implicit forms of science, theory construction, and hypothesis testing we see in children and nonscientific adults, no other species performs such tasks in the way we do. Science, in the formal sense of the word, to be sure, is not a "natural" state for humans. Many cultures historically and even some currently have no formal science institutions or individuals who make a living doing science. What, therefore, is the nature of science, and who does it? The first question is more the domain of philosophers, whereas the second one more the domain of psychologists.

Even more rare and recent are the studies of science, sometimes known as the metasciences. These include philosophy of science, history of science, sociology of science, and sometimes anthropology of science. Yet until recently, missing from these studies of science has been a formal psychology of science. But in the last 10 years that has begun to change. This chapter will explore and describe the nature of these changes. Among the primary questions I will address are: What is the psychology of science? How has it developed? And where is it going?

An obvious question arises, however: Why another science of science? Don't history, philosophy, sociology, and even anthropology do a good enough job in exploring the various dimensions of science? My answer is no, because they at best only touch on the psychological dimension and at worst are hostilely opposed to it.

A full-fledged field is needed, and we are a good way toward having it. This chapter describes that journey.

Psychology can offer not only a psychological perspective to science studies, but also a methodological and theoretical one. Compared to other studies of science—with the exception of history—psychology of science focuses on the development of the individual scientist in the context of his or her social environment and group. Moreover, psychology of science is unique in the studies of science to focus on influences such as intelligence, motivation, personality, and development over the lifespan of scientific interest, thought, ability, and achievement (Feist, 2006c; Gholson, Shadish, Neimeyer, & Houts, 1989; Simonton, 2009).

THE PSYCHOLOGY OF SCIENCE DEFINED

In order to define the psychology of science, I must first define each term separately. Psychology is the scientific study of human thought and behavior (Feist & Rosenberg, 2012). Of course, it is not quite that simple, but this provides a good working definition. If psychology is the study of human thought and behavior, then the psychology of science is simply the study of scientific thought and behavior, both narrowly and broadly defined.

Narrowly defined, science refers to thought and behavior of professional scientists, technologists, inventors, and mathematicians. Broadly defined, science refers to thought and behavior of any person or people (present or past) of any age (from infants to the elderly) engaged in theory construction, learning scientific or mathematical concepts, model building, hypothesis testing, scientific reasoning, problem finding or solving, or creating or working on technology. In short, science narrowly defined is explicit science, whereas science broadly defined is implicit science (Feist, 2006c).

SUBDISCIPLINES OF THE PSYCHOLOGY OF SCIENCE

If truth be told, psychologies of science exist rather than "the" psychology of science. Just as the parent discipline of psychology is broken down into subdisciplines, the psychology of science can be divided along the same lines, namely developmental, cognitive, personality, social, clinical, and neuroscience. Interestingly, the subdisciplines of clinical and neuroscience are so new that they were not broadly discussed in my 2006 book (Feist, 2006c). In what follows, I outline the parameters of each of the subdisciplines. I focus more on the developments in the formation of the field rather than elaborate on a review of the literature. For more detailed reviews of the literature see Feist (2006c).

DEVELOPMENTAL PSYCHOLOGY OF SCIENCE

Developmental psychologists examine thought and behavior over time. Developmental psychologists of science, therefore, study scientific thought and behavior over time. If scientific thinking involves developing and testing mental models of how the world works—the physical, biological, and social worlds—and coordinating these models with evidence, then infants, children, adolescents, and adults can all be the object of study in a developmental psychology of science. Indeed, one of the most fruitful applications of developmental psychology to science comes from defining science broadly. By so doing, we see that children as well as adults are "folk" physicists, biologists, and psychologists, among other things. They develop and modify their implicit theories of the world given their most recent experiences and new evidence. To be sure, developmentalists have explored the similarities as well as differences between children, adults, and scientists in reasoning, theory construction, and openness to evidence contrary to their prior beliefs (e.g., Brewer & Samarapungavan, 1991; Gopnik, Meltzoff, & Kuhl, 1999; Klaczynski & Robinson, 2000; Klahr, 2000; Koerber & Sodian, 2009; Koslowski, 1996; Kuhn & Pearsall, 2000).

COGNITIVE PSYCHOLOGY OF SCIENCE

Science is a cognitive act, to paraphrase Ryan Tweney. At its core science involves theory formation, hypothesis testing, theory change, problem solving, reasoning, and creativity, each of which deserves the label "cognitive." Cognitive psychologists of science have investigated each one of these topics. It is safe to say that cognitive psychology of science is the most elaborate and deeply developed of the psychologies of science. Among the major topics researched in the cognitive psychology of science are model building, conceptual change, causal maps, use of analogy, metaphor, and imagery in scientific problem solving and creativity (e.g., Gorman, 1992; Gruber, 1981; Miller, 1996; Nersessian, 2002; Shrager & Langley, 1990; Tweney, Doherty, & Mynatt, 1981).

Developmental and cognitive psychologists of science share in common the opportunity to examine both implicit and explicit modes of thought in relation to scientific reasoning. Implicit scientific thought and reasoning, as expressed in children, adolescents, and even nonscientist adults is one of the great wells of opportunity for psychologists of science. By continuing to unpack implicit processes such as scientific reasoning, theory construction, and folk domains of scientific knowledge, developmental and cognitive psychologists can shed important light on the mechanisms of how implicit processes influence how scientific thinking and behavior develop or become thwarted over the lifespan. One general principle of cognitive development is that with development and expertise, cognitive processes move from being primarily implicit to more and more explicit (Karmiloff-Smith, 1992). Cognitive psychologists can also compare and contrast implicit and explicit thought processes in novices and experts.

PERSONALITY PSYCHOLOGY OF SCIENCE

Personality psychology focuses on the study of individual differences and the consistency of behavior over time and across situations. The major focus of personality psychology is on the unique traits that individuals have, such as extraversion, openness, anxiety, and conscientiousness. Traits lower behavioral thresholds, making some behaviors more likely than others (Feist, 2008). One assumption in the personality psychology of science, therefore, is that particular traits make interest in and talent for science more or less likely. Fifty years of research support this assumption (e.g., Cattell & Drevdahl, 1955; Feist, 1993, 1998, 2006a, 2006d; Gough, 1987; Helson, 1971; Simonton, 2009; Terman, 1954).

SOCIAL PSYCHOLOGY OF SCIENCE

Social psychology is the study of how real or imagined others influence our thought and behavior. Scientific thought and behavior is very much influenced by others, real, imagined, or from the past. As Newton famously said: His achievements exist because he stood on the shoulders of giants. Science is more and more an enterprise that exists in teams and large groups and groups behave differently than individuals. For example, social psychologists have begun to examine how scientific groups cooperate with, compete against, and develop controversies between each other (Schulze, 1990; Schulze & Meischner, 2008), as well as the nature of collaborations (Liberman & Wolf, 1998). Moreover, scientific knowledge is created, evaluated, and disseminated in a social context (Hemlin, 2009; Kruglanski, 1994; Shadish, 1989; Shadish & Fuller, 1994; Shadish, Fuller, & Gorman, 1994; Westrum, 1989).

CLINICAL PSYCHOLOGY OF SCIENCE

Although the largest domain of study in its parent discipline, clinical psychology of science is perhaps the smallest area of investigation in scientific thought and behavior. Clinical psychology investigates disordered thought and behavior, understanding that disordered involves deviant, dysfunctional, distressing, and sometimes dangerous thoughts and behaviors (American Psychiatric Association, 2000). The mental health status of scientists is a relatively unexplored area partly due to the fact that as a group, scientists are relatively free of mental disorders (Ludwig, 1995). Yet certain nondebilitating disorders, such as high-functioning autism (Aspergers), appear to exist in scientists, engineers, and mathematicians at higher levels than the population (Baron-Cohen et al., 1998; Baron-Cohen, Wheelwright, Skinner, Martin, & Clubley, 2001; Baron-Cohen, Wheelwright, Stone, & Rutherford, 1999).

NEUROSCIENTIFIC PSYCHOLOGY OF SCIENCE

Neuroscience involves examining the neurological substrates of thought and behavior, often using brain-imaging technology. Until recently, examining the brain regions involved in scientific thinking was nearly impossible, but with the advances in imaging new investigations into brain activity and thought have flourished. Some of these investigations have turned their attention to scientific thinking and brain activity, in particular causal reasoning and reasoning about consistent and inconsistent evidence (e.g., Dunbar, Fugelsang, & Stein, 2007; Fugelsang & Dunbar, 2004; Fugelsang, Stein, Green, & Dunbar, 2004; Fugelsang, Roser, Corballis, Gazzaniga, & Dunbar, 2005; Roser, Fugelsang, Dunbar, Corballis, & Gazzaniga, 2005). Moreover, non-imaging research has demonstrated associations between the frontal lobe executive functioning and scientific and mathematical reasoning (Bull & Scerif, 2001; Fuster, 2002; Kwon & Lawson, 2000; Menon et al., 2000 ; Rourke & Conway, 1997).

THE ESTABLISHMENT OF SCIENTIFIC DISCIPLINES

To understand where the psychology of science came from and where it might be going, we have to understand how disciplines develop in science. Important lessons can be learned from analyzing how other studies of science began and flourished.

The most important lesson for the psychology of science comes from knowing the general stages that any scientific discipline goes through in its path toward maturity. Guiding my discussion of the development of each study of science, I borrow from and modify Nicholas Mullins's (1973) stage model of theory or network development as well as Joseph Matarazzo's (1987) criteria for a new psychological field.

> In stage 1, "Isolation," a leading scholar or two discovers and/or uncovers a new problem in isolation, with the founding intellectual figures setting the stage. There is no social organization in terms of training centers, conferences, or societies. Late in stage 1 and early in stage 2, a core group of scholars may be working in the field, but doing so implicitly rather than explicitly, not yet labeling themselves as members of the field.
>
> In stage 2, "Identification" happens when a wider circle of scholars become convinced by the founding figures that the problem at hand is an important one and more and more researchers start to explicitly identify themselves with the field. Semiregular meetings are organized and the first training-research centers may form. Such training centers are usually highly centralized around an intellectual leader, whose students have begun to have a major impact on the field. A leading journal becomes necessary as the outlet for the increased level of productivity of the field.
>
> In the final stage 3, "Institutionalization," the field becomes well established and institutionalized. Meetings become annual conferences because societies have now formed with their own social structure and hierarchy. Often multiple societies, some of them international, become necessary. Training centers proliferate and become less centralized, and at least one journal is now required for the expanding productivity of the field. Indeed, splinter movements, with different foci or agendas, may form and either break away or stay on the edge of the central field.

Similar to this three-stage model of disciplinary development, but taking less of a stage perspective, Matarazzo (1987) argued that a main criteria for a new subfield in psychology is that it have its own association or society, journals, postdoctoral training programs, and recognition from other scholars.

THE DEVELOPMENT OF OTHER STUDIES OF SCIENCE

In this section, I briefly summarize where the other studies of science are in this three-stage model of disciplinary development. I focus on the major three disciplines of philosophy, history, and sociology. I also compare and contrast the similarities and differences each of these other metasciences have to the psychology of science.

PHILOSOPHY OF SCIENCE

Although philosophy of knowledge (i.e., epistemology) was a central theme in ancient Greek philosophy, the field of the philosophy of science is a much more recent development. As an established scholarly discipline, it has its origins in the late 19th century and early 20th century (Feist, 2006c). Let me highlight briefly one or two major events.

In Table 1.1 I lay out the major events in the three distinct phases in the establishment of philosophy of science as a viable discipline. In short, the Isolation stage of development for the philosophy of science occurred in the 1840s with works by William Whewell and continued until the 1890s with publications by Mill, Comte, Jevons, Mach, and Pearson among others. The Identification stage was instigated around the turn of the last century with the First International Congress of Philosophy devoting a session to philosophy of science and Henri Poincaré's *La science et l'hypothese* in 1908. Probably the most well-known event of this Identification stage was the formation of the Vienna Circle from the 1910s until the 1930s. When the Vienna Circle broke up in the 1930s, philosophy of science had been well established for a generation, with formal societies, conferences, and journals. Philosophy of science rose to prominence in the 1950s and 1960s with the widely influential works of Karl Popper and Thomas Kuhn.

Historically, the primary method used by philosophers of science was rational argumentation often with a focus on the epistemological nature of science and its demarcation from nonscience. Moreover, philosophers of science were most interested in science as it should be rather than science as it is. That is, they didn't argue for or examine how scientists actually thought and reasoned, but rather how they ideally should if they were to do real science. As Popper made famous with his falsifiability criterion, scientists proposed falsifiable and testable hypotheses and nonscientists did not.

By the 1980s and 1990s, the logical analysis of science approach started to change. Although still not empirical, philosophers of science moved toward a more empirical approach with natural epistemology. This change was summed up best by David Hull, a major proponent of natural epistemology: "Throughout most of its history, philosophy has been defined in such a way that evidence cannot possibly bear on it. Increasingly, however, those of us who are philosophers by training have begun to interpret philosophy in such a way that evidence does bear on the views that we express. Testing strictly scientific hypothesis is very difficult. Testing claims about science is even more difficult, but test them we must" (Hull, 2001, p. 7).

HISTORY OF SCIENCE

History documents and contextualizes the major events and trends that occur over time. The history of science therefore documents, describes, and explains the

Table 1.1 The three stages of field development for philosophy of science

Stage of development	*Decades*	*Major representative event*
Isolation: Articles, books	1840s–1890s	• Publication of William Whewell's *The Philosophy of the Inductive Sciences, 1840, and On the Philosophy of Discovery,* 1856 • Auguste Comte publishes *Discours sur l'ensemble du positivisme* (*A General View of Positivism*), 1848 • William Jevons publishes *The Principles of Science,* 1874 • Ernst Mach publishes *Die Mechanik in ihrer Entwicklung Historisch-Kritisch Dargestellt* (*The Science of Mechanics*), 1883 • Karl Pearson publishes *Grammar of Science,* 1892
Identification: conferences, centers	1900s–1950s	• First International Congress of Philosophy, with a section devoted to philosophy of science, 1900 • Henri Poincaré publishes *La science et l'hypothèse,* 1908 • Vienna Circle (Neurath, Mach, Schlick, Feigl, Gödel, Carnap, et al.) and logical positivism, 1910s to 1930s • Hans Reichenbach publishes *Philosophie der Raum-Zeit-Lehrer,* 1928 • Karl Popper publishes *Logik der Forschung* (*Logic of Scientific Discovery*), 1935 • Philosophy of Science Group (later becomes British Society for the Philosophy of Science) has first congress, 1935 • First International Congress for the Unity of Science, 1935 • First International Congress of the Philosophy of Science, 1949 • Minnesota Center for the Philosophy of Science founded by Herbert Feigl, 1953
Institutionalization: journals, societies, degrees	1930s–1950s	• *Philosophy of Science* (journal) and Philosophy of Science Association, 1934 • *British Journal for Philosophy of Science,* early 1950s • British Society for Philosophy of Science

Adapted from Feist 2006c

developments of science, from its origins to its most contemporary forms of expression. It is not hard to understand that of all the disciplines that examine science, history would be the oldest. In fact, as George Sarton documents, the earliest known treatises on the history of science go back to ancient Greece and Syria and have been parts of official national academies of science since the 1700s (Sarton, 1952). For the history of science, the Isolation stage began in the 1700s and lasted through the end of the 1800s (see Table 1.2).

The Identification stage consists primarily of conferences, centers, and departments forming, and for the history of science these developments started in earnest right around the turn of the century and up through World War I. The organizational (and intellectual) leader was clearly George Sarton. He organized the first journal , many of the national and international congresses, and formed the leading department in the field at Harvard. By 1952 there were five universities that granted the PhD in the history of science: London, Harvard, Cornell, Columbia, and Wisconsin (Sarton, 1952).

The third stage of a field's development, Institutionalization, occurred in the history of science in earnest in the first decades of the 20th century, culminating in 1912 with the first international journal (*Isis*) devoted to the history of science. Moreover, the need for organization grew steadily, and by the turn of the century most European countries and one Asian country (Japan) had formed societies for the history of science. As Sarton's *A Guide to the History of Science* made clear, by the late 1940s there were already dozens of journals and many organizations throughout Europe and the United States devoted to the history of science; the history of a particular science; or the history of medicine, mathematics, or technology. Currently, there are at least 16 societies in the history of science and at least 57 different journals currently being published on the topics of the history of science, technology, medicine, or mathematics.

Explicated first in the 19th century, the primary methodology used by historians of science is known simply as "historical method." The essence of this method is to critically examine the most reliable and primary documents possible when writing any historical account (Howell & Prevenier, 2001). The more primary and closer the source is to the event the more reliable it is. Modern historians do talk about "hypotheses" and "evidence" but to historians these are logical rather than empirical constructs (McCullagh, 1984). That is, hypotheses are not testable statements but rather explanations based on historical evidence. Indeed, historians are not only nonempirical but sometimes explicitly antiempirical (Sulloway, 1996, in preparation; cf., Simonton, 1995).

SOCIOLOGY OF SCIENCE

To a large extent, the sociology of science developed out of the philosophy and history of science and the sociology of knowledge. Perhaps the first major intellectual contribution to the field (Isolation) would have to be William Ogburn's *Social Change* from 1922. Ogburn (1886–1959) was more generally interested in social change, but scientific and technological change played a key role: technological invention is

Table 1.2 The three stages of field development for history of science

Stage of development	*Decades*	*Major representative event*
Isolation: articles, books	To 1890s	• Some histories of science originated as early as ancient Greece and Syria, and many scientific societies in Europe had their own history from the 1700s on[a] • Jean Etienne Montucla publishes *Historie des mathématiques*, 1758 • Baden Powell publishes *Historical View of the Progress of the Physical and Mathematical Sciences from Earliest Ages to the Present Time*, 1834 • Publication of William Whewell's *The History of the Inductive Sciences*, 1837 • Royal Academy of Bavaria begins publication of its *Geschichte der Wissenschaften in Deutschland*, 1864 • William Youmans publishes first history of American science, *Pioneers of Science in America*, 1896
Identification: conferences, centers	1900–1930	• First congresses of the history of science, as a subsection of the International Congress of Philosophy, 1900 and 1904 • First International Congress for the History of Science, 1929 • History of Science Society (U.S.) starts holding annual conferences, 1924
Institutionalization: journals, societies, degrees	1910–1930s	• Institutes, museums, and libraries of science and the history of science open all over the world, especially in the 1920s and 1930s[a] • *Isis* (journal) first published, 1912, by George Sarton • History of Science Society forms, 1924 (3,000 members in 2003) • Académie Internationale d'Historie des Sciences forms, 1928 • Society for the History of Natural History, 1936 • British Society for the History of Science (BSHS) forms, mid-1940s • Currently dozens of different societies and more than 100 journals devoted to the history of science, technology, medicine, or mathematics[b]

Notes: [a] see Sarton 1952.

[b] see Sarton 1952 for extensive documentation of history of science up to 1950; for more current information, see Web sites in text or links on the BSHS Web site.

Adapted from Feist 2006c

the precursor of cultural change, which in turn is followed by social disorganization, which in turn is followed by social and cultural adjustment (see Table 1.3).

But it was not until the mid-1930s that the founding intellectual works really began to appear: In 1935 the undisputed intellectual father of the sociology of science, Robert K. Merton, finished his PhD dissertation entitled "Science, Technology, and Society in Seventeenth-Century England" (Feist, 2006c). Two years later he published a chapter in *Sociology and Social Research* entitled "Civilization and Culture," in which he laid bare the need to include knowledge (especially scientific) as a focus of sociological investigation. Moreover, he argued that "civilization" (that is, theoretical knowledge) was more accumulative than "culture" (values and norms), but progress in the former was not linear. Other figures who laid the foundation for the sociology of science were Merton's mentor Pitirim Sorokin (1889–1968) and John Bernal (1901–1971).

The Identification stage for the sociology of science, or science and technology studies (STS) as it came to be called, lasted from the 1960s to the mid-1970s, with a highlight being the first established division of the International Sociology Association forming in 1966. Merton and his students Harriet Zuckerman, Stephen Cole, and Jonathan Cole played key roles in moving the sociology of science into its third stage (Institutionalization) in the 1970s. The first journal, *Science Studies* (later renamed *Social Studies of Science*), was founded in 1971 and the "Society for the Social Studies of Science" (4S) was formed in 1975.

In contrast to philosophy and history of science, sociology of science is or was mostly an empirical science. From the 1930s until the 1970s sociologists of science collected large-scale data sets and tested hypotheses statistically (Cole & Cole, 1973; Merton, 1973; Zuckerman, 1977).

With the advent of the social constructivist movement in the late 1970s, however, things started to change. The mistrust of science and the scientific method held sway and the methods moved away from traditional hypothesis testing and statistical inference. The field became more focused on qualitative data analysis at best and it became antiscience at worst—to use Jonathan Cole's phrase, it became "voodoo sociology" (Cole, 1996; Feist, 2006c). The argument, in a nutshell, is that science has little direct connection to nature and the world but consists mostly of socially constructed arguments made by powerful groups. Currently, much of sociology of science uses quasi-anthropological methods of interviewing and studying scientists as if they were from a different culture.

The methods of the psychology of science share some things in common with each of the other main studies of science, but at their core are unique in their investigations of the thought and behavior of scientists and/or people who are thinking and behaving scientifically. The most notable difference is that psychology is the only approach that brings true experimental design to the study of scientific thought. Participants can be randomly assigned to an experimental or control group and some variable (e.g., scientific problem) can be manipulated. Moreover, psychological approaches to science are more likely than each of the other studies of science to test explicit hypotheses (experimental, quasi-experimental, correlational, or historical in nature) through empirical techniques. In short, psychology of science studies science using the scientific method and does not treat scientists as a foreign culture.

Table 1.3 The three stages of field development for sociology of science

Stage of development	*Decades*	*Major representative event*
Isolation: articles, books	1920s–1950s	• William Ogburn publishes *Social Change*, 1922 • R. K. Merton's PhD thesis "Science, Technology, and Society in Seventeenth Century England," 1935 • Karl Mannheim establishes the field of sociology of knowledge with publication of *Ideology and Utopia*, 1936 • Pitirim Sorokin publishes *Social and Cultural Dynamics*, 1936 • Merton publishes "Civilization and Culture" in *Sociology and Social Research*, 1936, and "The Sociology of Knowledge" in *Isis*, 1937 • John Bernal publishes *The Social Function of Science*, 1939 • Merton lays out the paradigm for the field in his chapter "Sociology of Knowledge," 1945, in *Twentieth Century Sociology* • Bernard Barber publishes *Science and the Social Order*, 1952
Identification: conferences, centers	1960s–1970s	• Division for Sociology of Science and Technology in the International Sociological Association forms, 1966 • National Association for Science, Technology, and Society (NASTS) forms, 1988
Institutionalization: journals, societies, degrees	1960s–1970s	• Three of Merton's students at Columbia University (Stephen and Jonathon Cole and Harriett Zuckerman) receive their PhDs between 1965 and 1969; each becomes a major figure in sociology of science • *Science Studies* (journal) is founded in 1971 by Roy MacLeod and David Edge, later is renamed *Social Studies of Science* • Society for the Social Studies of Science (4S) forms in 1975 and currently has about 1,000 members

Adapted from Feist 2006c.

In terms of their development as disciplines, however, the three other major metasciences are fully ensconced in all three stages and each has been fully established for at least 40 years. The psychology of science, on the other hand, is in the early stages of Institutionalization (see Table 1.4).

Only in the last decade have we seen a society for the psychology of science formed, regular conferences held, and a journal established. Based on these criteria, I would argue that the psychology of science has recently moved out of the identification stage and is now in the early stages of institutionalization. As recently as 2006, I was arguing for the formation of a society, conference, and journal and now we have each of these. But how did we get here?

A DEEPER LOOK AT THE HISTORY OF THE PSYCHOLOGY OF SCIENCE

As the British statesman Edmund Burke once wrote: "Those who do not know history are destined to repeat it." The fact is that the history of the psychology of science has been a struggle for existence. If we wish to move away from struggle and toward a comfortable existence, then we need to learn lessons from the other disciplines that successfully have made the transition from fledgling field to fully established scientific discipline. The history, philosophy, and sociology of science are just such established disciplines.

The French historian Alphonse de Candolle published a book in 1873 entitled *Histoire des sciences et des savants depuis deux siècles.* Candolle's book analyzed the history of 200 scientific men who lived between the late 1600s and 1800s. Not only did the book inspire Francis Galton to publish what is often regarded as the first work in the psychology of science, *English Men of Science: Their Nature and Nurture,* it also touched off one of the earliest examples of the now classic "nature-nurture" debate in psychology (Fancher, 1983). Indeed, Candolle's 1873 book was a response to Galton's earlier work, *Hereditary Genius,* and was an attempt to show how environmental forces, more than genetics, created scientific eminence (Fancher, 1983). Because it took more of a cultural/sociological perspective of eminence, Candolle's work more appropriately belongs to the sociology of science than the psychology of science. Galton's *English Men of Science* was the first ever to use the survey method in questioning scientists about their personal histories, attitudes, and beliefs about their interests in science. Here are Galton's own words:

> The intent of this book is to supply what may be termed a Natural History of the English Men of Science of the present day. It will describe their earliest antecedents, including their hereditary influences, the inborn qualities of their mind and body, the causes that first induced them to pursue science, the education they received and their opinions on its merits. (Galton, 1874/1895, p. 1)

Unfortunately, there was very little follow-up to Candolle's and Galton's work for nearly 50 years. Some philosophers occasionally touched on psychological factors in scientific thought and creativity. For example, in the mid-1930s, a French

Table 1.4 The three stages of field development for psychology of science

Stage of development	*Decades*	*Major representative event*
Isolation: articles, books	To 1980s	• Francis Galton publishes *English Men of Science*, 1874 • S. S. Stevens publishes first attempts at identifying psychology as a metascience, 1936, 1939 • Mieczyslaw Choynowski becomes president of the circle for the science of science and he founds the Institute for the Science of Sciences in 1948 in Cracow, Poland • Mikhail Grigoryevich Yaroshevsky founds the "Department of Psychology of Scientific Creative Works" at the Institute of the History of Science and Technology at the Russian Academy of Science, 1960 • Abraham Maslow publishes the first book entitled *Psychology of Science*, 1966 • Rudolf Fisch publishes first bibliography on the psychology of science, with more than 300 references, 1977 • Michael Mahoney publishes "Psychology of the Scientist" in *Social Studies of Science*, 1979 • Dean K. Simonton publishes *Scientific Genius: A Psychology of Science*, 1988 • "Memphis Group" (Gholson, Shadish, Neimeyer, and Houts) edits *Psychology of Science*, with chapters by numerous psychologists reviewing literatures in many different topics, 1989 • Review of field is published by Gregory Feist and Michael Gorman, 1998

(continued)

Table 1.4 (Continued)

Stage of development	*Decades*	*Major representative event*
Identification: conferences, centers	1980s–2006	• Conference "The Psychology of Science," Memphis State University, 1986 • Gregory Feist receives one of the first PhDs granted on the topic at University of California, Berkeley, 1991 • Conferences on "The Cognitive Basis of Science" held at Rutgers University (1999) and University of Sheffield (2000); a book by same name is published in 2002 • Feist publishes *The Psychology of Science and the Origins of the Scientific Mind,* 2006 • First Biennial Conference on the Psychology of Science, September 2006, Zacatecas Mexico; ISPST conferences also held in Berlin, Germany 2008 and Berkeley, California, in 2010
Institutionalization: journals, societies, degrees	2006–present	• International Society for the Psychology of Science and Technology (ISPST) is formed in Zacatecas, Mexico, September 2006; Gregory Feist elected founding president • *Journal of Psychology of Science & Technology,* first published in 2008 (Feist is founding editor; Gorman, Liberman, and Simonton are associate editors) • ISPST awards its first David C. Gooding Prize for Best Student Paper to Jordan Lippman, August 2010

Adapted from Feist 2006c.

philosopher, Gaston Bachelard, took a psychoanalytic approach to understanding the scientific mind in *Le nouvel esprit scientifique (The Scientific Mind; 1934)* and *La formation de l'esprit scientifique (The Formation of the Scientific Mind; 1938)*. Moreover, Bachelard's concept of "epistemological rupture" apparently influenced Kuhn's thinking and more specifically Kuhn's idea of "paradigm shift" (Gaston Bachelard, n.d.). A few years later, the American S. S. Stevens (1939) wrote an article entitled "Psychology and the science of science" in which he was the first to coin the phrase "psychology of science."

Between the 1940s and 1970s, Eastern bloc countries were more active in the psychology of science than Western countries. Mieczyslaw Choynowski (1948a, 1948b), for example, was president of the circle for the science of science and he founded the Institute for the Science of Sciences in 1948 in Cracow, Poland. The institute consisted of departments for all the major areas of science studies, including history, sociology, pedagogy/education, and psychology. The department of psychology of science was described as "the study of different kinds of scholars, the psychology of research work, the hygiene of intellectual work, the changes of specialty by well-known scientists, the correlation between scientific interests and abilities, etc." (Choynowski, 1948b, p. 4) His main work on the psychology of science and other studies of science was a monograph entitled *Life of Science* (Choynowski, 1948a; cf. Walentynowicz, 1975).

A similar movement was also begun by Mikhail Grigoryevich Yaroshevsky (1915–2000) in Russia in 1960 (Volodarskaya, 2005). He guided the Department of Psychology of Scientific Creative Works at the Institute of the History of Science and Technology at the Russian Academy of Science. Yaroshevsky's group defined the psychology of science as "the study of the psychological peculiarity of people whose activity produces new knowledge; it is also the study of the psychological regularity of the development and functioning of scientific collectives" (Volodarskaya, 2005, p. 478).

During the decade of the 1950s there was an upswing in research and theory on the psychology of creativity, including scientific creativity. Especially post-Sputnik, anything that helped foster an interest in science was encouraged and relatively well funded. Noteworthy among the efforts from the 1950s and early 1960s were Anne Roe's *The Making of a Scientist* (1952), Bernice Eiduson's *Scientists: Their Psychological World* (1962), and Calvin Taylor and Frank Barron's *Scientific Creativity* (1963). Also in the early 1950s, Lewis Terman (1954) published an important monograph on the psychological qualities of scientists. By the mid-1960s, Abraham Maslow published the very first book with psychology of science in the title (Maslow, 1966).

Between the late 1960s and mid-1980s, however, there was very little systematic work done by psychologists on scientific thought, reasoning, or behavior in Western countries. Starting in the mid-1960s, however, one university was about to begin a very productive and impressive string of scholars who would go on to conduct many groundbreaking studies in the psychology of science. That school was Carnegie Mellon and its founding scholar was Herb Simon. Simon published one of the explicit pieces in the psychology of science, a chapter on scientific discovery and the psychology of scientific problem solving (Simon, 1966). Just as important for the psychology of science, Simon produced a graduate student, David Klahr, who himself would go on to produce a string of one of the more impressive lists of graduate students and postdoctoral fellows in the psychology of science for the 1990s and 2000s: Kevin Dunbar, Jeffrey Shrager, Amy Masnick, and Chris Schunn. Indeed, we could dub this Simon lineage the "Carnegie Mellon School."

But by the mid- to late 1980s, the field really stood at the precipice of being a full-fledged discipline. Or so it appeared. The 1986 conference held in Memphis and organized by William Shadish, Barry Gholson, Robert Neimeyer, and Arthur Houts was a good beginning (see Figure 1.1). Indeed, the books that followed from that conference (Gholson et al., 1989; Simonton, 1988) were important beginnings in moving toward an established discipline. However, no society, regular conferences, or journal sprouted up afterward. In the 1990s Ron Westrum at Eastern Michigan University started a newsletter ("Social Psychology of Science Newsletter"), but even that lasted but a few years. These starts, however, did inspire me while I was working on my dissertation on scientific creativity at the University of California, Berkeley. They planted in me the idea that psychology of science should be an independent discipline of study.

Starting in the 1990s a new and exciting avenue in the psychology of science was opened up by developmentalists. In fact, developmental work in the psychology of science owes its existence to the seminal research and theory of Jean Piaget on the development of children's understanding of the physical, biological, and numeric worlds (Inhelder & Piaget, 1958; Piaget, 1952). Starting in the late 1980s, Deanna Kuhn and her students blazed the trail in examining how scientific reasoning develops in children and adolescents (Kuhn, 1989, 1993; Kuhn, Amsel, & O'Loughlin, 1988; Kuhn & Pearsall, 2000). Others soon followed (e.g., Brewer & Samarapungavan, 1991; Fay, Klahr, & Dunbar, 1990; Koslowski, 1996; Klaczynski & Narasimham, 1998).

Figure 1.1 First ever Psychology of Science Conference: Memphis, Tennessee, April 1986
Front Row (L-R): Cecilia Heyes, Ryan Tweney, Robert Neimeyer, Will Shadish, Dean Simonton
Center Row (L-R): Howard Gruber, Arthur Miller, Dedre Gentner, Jeffrey Berman, Michael Mahoney, Art Houts
Back Row (L-R): Donald Campbell, William McGuire, Peter Barker, Barry Gholson, Arthur Graesser, Ron Westrum, Marc De Mey

One work of primary importance in this movement was the publication of *The Scientist in the Crib: Minds, Brains and How Children Learn* by Gopnik et al. (1999). One general argument they put forth in the book is that in the relationship between children and scientists, it is best to think of scientists as big children rather than children as little scientists.

Starting with his participation in the Memphis meetings, Dean Simonton continued to publish many important works in the 1990s and 2000s on the psychology of science, the most influential of which was his book *Scientific Genius: A Psychology of Science* (1988; see also Simonton, 1989, 1991, 1992, 1995, 1999, 2000, 2004, 2009). Most of these works focused on the cognitive, personality, and motivational forces behind scientific creativity and scientific genius.

After completing in 1991 my dissertation entitled *The Psychology of Science: Personality, Cognitive, Motivational and Working Styles of Eminent and Less Eminent Scientists,* I published a few papers on the psychology of science (Feist, 1993, 1994, 1997). The most interesting finding was that blind raters (from tape recordings) perceived the most eminent scientists as arrogant and hostile (Feist, 1993). From this work, I became acquainted with Mike Gorman and his work in the psychology of science. Initially, we organized symposia on the psychology of science at the annual conference for the American Psychological Association (APA) in 1992 and for the Society for the Social Studies of Science (4S) in 1995 and from that we collaborated on the first review article of the published literature (Feist & Gorman, 1998). That article, in turn, inspired me to work up a book-length review of the literature, which I began working on in 2001. In the early stages of working on the book, I got sidetracked by a fascinating set of evolutionary questions, namely how and when did humans as a species become capable of doing science. The book, therefore, has two major sections, each of which is evident from the book's title: *The Psychology of Science and the Origins of the Scientific Mind* (Feist, 2006c).

In the book and a subsequent article, I lamented the fact that the discipline still had no society, no conferences, and no journal (Feist, 2006b, 2006c). Little did I know at the time, however, that within two years, all three would exist. Sure enough, soon after my book appeared, Sofia Liberman and Javier Rayas organized the second-ever conference on the psychology of science in September 2006 in Zacatecas Mexico (see Figure 1.2). It was a very small, but very productive conference. Attendees at this conference went on to establish the International Society for the Psychology of Science and Technology (ISPST) and elected me as its first president.[1] The society, too, started off small with only about 40 members in its first year, 2007. By 2008 the society had signed a contract to publish the first peer-reviewed journal, the *Journal of Psychology of Science and Technology* with Springer Publishing.[2] Conferences are now biennial, with the second one taking place in Berlin, Germany, in 2008 and the third one in Berkeley, California, in 2010. The society is still small (roughly 100 active members), but if the Berkeley conference is any indication, the field is alive with exciting ideas and programs of research.

[1] To learn more about the society, go to http://www.ispstonline.org.

[2] The journal is no longer published by Springer Publishing, but is transitioning to be an online, open-access peer-reviewed journal.

Figure 1.2 Sitting, L to R: Verona Seuffert, Ron Westrum, Javie Soto, Carl Martin Allwood, Sofia Liberman, Gregory Feist, Michael Gorman, Sven Hemlin
Standing, L to R: Gregory Trafton, Lisa Osbeck, Annedore Schulze, Barbara Koslowski, K. Bernardo Wolf, Laura Narvarro, Javier Rayas

THE FUTURE OF THE PSYCHOLOGY OF SCIENCE

The psychology of science has made some important first steps toward becoming an established discipline. But they are merely that; first steps. Our long-term survival is not yet guaranteed. To me and other leaders of the movement, it is clear that there is as much latent as actual interest in the psychology of science. In other words, there are many psychologists and psychologically oriented scholars who are doing psychology of science but only implicitly. Their identification with the field is not yet explicit. They may not even know that the field and society exist.

So the question we need to be asking ourselves at this point is "What can we do to ensure the next generation wants to carry the tradition forward and that scholars start to identify themselves as psychologists of science?" (Feist & Gorman, 2009). Any field, young or old, needs to excite the interest of young scholars and in this sense the psychology of science is no different from any other discipline and indeed it stands to gain from learning how the other studies of science—philosophy, history, and psychology—made this transition. As Gorman and I editorialized, there are at least three major activities the current generation of psychologists of science need to do in order to assure the next generation inherits an active and healthy discipline: teach exciting courses, carry out and publish exciting research, and reward student excellence (Feist & Gorman, 2009). The first two concern getting the next generation into the field, whereas the last concerns keeping them in the field.

TEACH EXCITING COURSES IN THE PSYCHOLOGY OF SCIENCE

As academics we teach "old" knowledge in the classroom and create "new" knowledge in the lab. Ideally, good teachers are also good researchers. Granted some of us tend to be more research oriented and others more teaching oriented. Sometimes this happens by choice and other times because the institutions we are in dictate it. But an ideal in higher education is that those who create knowledge should also pass on that knowledge to the next generation by teaching as well as publishing.

One principle of teaching is great teachers excite. Typically, they excite students by being excited themselves. As Richard Feynman apparently once said: "Teaching is like sex: If you're not enjoying it, you're not doing it right!" As instructors, we may sometimes forget to focus on enjoyment. Good teaching is exciting and enjoyable for both teacher and student. A second principle is "steal good ideas from others." In contrast to research and business, education is one domain in life where stealing from (really simply being inspired by) others is both acceptable and even encouraged. Teachers should be constantly open to new ideas in the classroom and talk to their colleagues and other instructors. When they hear about something other teachers do that sounds interesting they should not be shy about trying it. No one owns patents or copyrights on good teaching techniques. The more widely disseminated good teaching is, the better.

In the spirit of sharing good teaching ideas, we recently culled from ISPST members examples of syllabi for courses on the psychology of science or some related course. We received a number of syllabi, many of which we have posted on the ISPST Web site (http://www.ispstonline.org). Here are some examples we received:

- David Klahr teaches a freshman seminar entitled "The Psychology of Scientific Thinking" in which he examines four major questions: (1) What does it mean to do science? (2) How do people think, solve problems, reason, and make decisions? (3) How do scientists think when they are doing science? and (4) What are some of the more effective ways to teach science?
- Ryan Tweney has taught a course whose readings revolve around seeing, intervening, representing, and explaining the world as well as science in culture and society.
- Sofia Liberman teaches a graduate-level course that has two major phases: first, knowledge of the field is acquired through systematic reading of the literature and second, the participants identify problems and raise questions that need empirical attention. The idea is to start new lines of research.
- Kevin Dunbar makes use of an innovative technique in his course "Science, Education, and the Scientific Mind." In addition to a term paper, students can produce and turn in a four-to-six minute movie. As he writes in his syllabus: "Instead of handing in an essay, you can hand in a movie. The movie would be similar to the essay, except here it is presenting information visually. For the movie you need to pick a science education topic and film a class, a lab, and or a teacher, a professor, students etc. You will edit everything, give a voice over, and output the final product to the computer."
- Barbara Koslowski's course outlines and examines three aspects of scientific reasoning: deciding whether one event causes another; deciding whether an explanation seems credible; and deciding whether, and if so how, an explanation should be refined or rejected.

- Mike Gorman has taught a course on scientific and technological thinking to freshman honors engineering students, culminating in an interactive simulation of social-psychological factors in the development of new nanoscience and nanotechnology.

These are just a few of the interesting and creative topics and techniques some of us are using to excite student interest in the psychology of science. I can only hope that others will be inspired to incorporate some of these ideas and techniques. If you are not currently teaching a course in the psychology of science, perhaps these course syllabi and ideas will inspire you to propose such a course to your department.

CARRY OUT AND PUBLISH EXCITING RESEARCH IN THE PSYCHOLOGY OF SCIENCE

In the end, teaching depends on research. In science, if a field does not advance, there won't be anything to teach. The purpose behind the journal after all is to not only publish the best and most interesting research in the psychology of science, but to encourage it to happen. Research needs not only an outlet for publication once it is completed, but also support and facilities to be carried out.

If we succeed as teachers and researchers in inspiring young people to join us and commit themselves to a career as psychologists of science, then they need to be trained and supported. At the Berlin conference in July 2008, Mike Gorman gave voice to the need to develop a collaborative training center so that aspiring PhDs can be trained. Instead of a single graduate program, his idea was to create a graduate program distributed among multiple institutions. Only a few programs might be able to generate an actual degree; distance learning technologies have evolved to the point where psychology of science students at one institution could have learning, research, and mentoring experience with faculty at other institutions. In that way, we could leverage the depth and breadth of the field—and one or two institutions could step forward to offer a degree without having to hire a half-dozen faculty, or provide all the research space. A distributed institutional network also increases the base of contacts a student can take advantage of when looking for a job.

Relatedly, psychologists of science need to obtain funding for their research. Currently, the atmosphere at the federal level is increasingly cognizant of and receptive to psychology of science research. This is particularly true at the National Science Foundation. For example, as Gorman (2009) wrote, the National Science Foundation (NSF) has a "Science of Science and Innovation Policy" (SciSIP) program that funds research geared toward developing knowledge, theories, and human capital in science of science studies. There is no reason why psychology cannot be one of the major players in these grant proposals.

In addition, one of the core leaders of the psychology of science movement, Michael Gorman, recently served a two-year term as one of three Program Directors in Science, Technology & Society (STS) at the NSF. This program "considers proposals that examine historical, philosophical, and sociological questions that arise in connection with science, engineering, and technology, and their respective interactions with society" (Science, Technology, and Society, STS, n.d.). Although the psychology of science is not explicitly mentioned on the STS NSF Web site, Gorman was the first NSF program director who identified explicitly as a psychologist of science and therefore actively encouraged proposals from psychologists.

REWARD STUDENT EXCELLENCE IN THE PSYCHOLOGY OF SCIENCE

The other incentive that came out of the Berlin conference was that young aspiring psychologists of science not only need support for their research, but also need to be reinforced for it. Prizes, especially those that involve money and a guaranteed publication, are rewards and reinforcers for scholars, young and old. The idea, therefore, was floated to have the society support a "Best Student Paper Award." The first such award, named in honor of the late David C. Gooding, was bestowed to Jordan Lippman at the 2010 ISPST conference in Berkeley. Of course, the only reward and reinforcement need not be in the form of a prize. Praise and encouragement by senior faculty of high-quality research and thinking can go a long way.

APPLICATION OF CUTTING-EDGE PSYCHOLOGICAL METHODS TO SCIENTIFIC THOUGHT AND BEHAVIOR

Neuroscience, behavioral genetic, and longitudinal growth modeling research methods need to become more common if we are to really understand the psychology of the scientist and scientific thinking and behavior. Brain research has brought a revolution to the study of human thought and behavior in general and yet scientific thought and behavior have mostly been left out of the revolution. The work by Jonathan Fugelsang and Kevin Dunbar has been a notable exception to this trend (Dunbar et al., 2007; Fugelsang & Dunbar, 2004).

In addition, as Simonton's (2008) recent article on the genetics of scientific talent makes clear, there is also a lack of behavioral genetic evidence on scientific thought and talent. With more and more research using the twin-adoption paradigm to tease apart heritable and nonheritable contributions to thought and behavior, one can hope and assume that this method will be applied more frequently in the future to scientific thought and behavior to study the relative contribution of nature and nurture to scientific talent, creativity, and achievement.

Work begun by Julian Stanley in the 1970s and continued by David Lubinski and Camilla Benbow on a 35-year longitudinal investigation into mathematical and scientific talent is a benchmark for the kind of research that is needed to fully understand how scientific interest, aptitude, talent, and achievement evolve from childhood to adolescence and into young, middle, and late adulthood (Park, Lubinski, & Benbow, 2008; Wai, Lubinski, & Benbow, 2005). This investigation notwithstanding, there is an obvious lack of longitudinal studies focusing on the development of scientific interest, reasoning, talent, and achievement. Relatively new and sophisticated statistical techniques such as structural modeling and growth curve modeling can answer longitudinal and causal questions in ways that older statistical techniques could not. Therefore, the field would benefit from more researchers in the future focusing on how the nature and nurture of scientific thought and behavior develop and grow, from childhood to adulthood. Developmental psychologists of science are in a unique position to provide such important programs of research.

Finally, the psychology of science is a unique discipline within psychological science. It is one of the few subdisciplines that actually integrate all of the major subdisciplines in psychology, namely cognitive, developmental, and social psychology; biological-neuroscience; personality; and even clinical psychology. The need for psychological science to develop a unifying perspective is as urgent now as it has ever been (Staats, 1999). Not that psychology of science will unify all of psychology, but in fact it offers the possibility of demonstrating how one topic—scientific thought and behavior—requires each and every major subdiscipline within the field of psychology. Indeed, Division 1 of the American Psychological Association (APA), which is devoted to general psychology, has been quite receptive to the psychology of science. In 2006, an entire special issue of Division 1's flagship journal, *Review of General Psychology*, was devoted to the psychology of science. The following year, the division awarded the William James Book Award to my book *The Psychology of Science and the Origins of the Scientific Mind*. And more recently, president-elect of the division, Dean Simonton, has argued that he would uphold psychology of science as an exemplar of general psychology during his upcoming presidency.

In conclusion, the psychology of science has begun its journey toward being an established and contributing member to the studies of science. To be sure, it is still young and finding its way. Yet with the theoretical and methodological tools available to psychologists of science and our ability to teach, conduct cutting-edge research, and recognize and reward a next generation of psychologists of science, the future of the field is bright and full of potential. The future, however, has yet to be written. Only it will tell us where the field has gone.

REFERENCES

American Psychiatric Association. (2000). *Diagnostic and statistical manual of mental disorders* (4th ed., Text Revision). Washington, DC: American Psychiatric Association.

Baron-Cohen, S., Bolton, P., Wheelwright, S., Short, L., Mead, G., Smith, A., & Scahill, V. (1998). Autism occurs more often in families of physicists, engineers, and mathematicians. *Autism, 2,* 296–301.

Baron-Cohen, S., Wheelwright, S., Skinner, R., Martin, J., & Clubley, E. (2001). The Autism-Spectrum Quotient (AQ): Evidence from Asperger syndrome/high-functioning autism, males and females, scientists and mathematicians. *Journal of Autism & Developmental Disorders, 31,* 5–17.

Baron-Cohen, S., Wheelwright, S., Stone, V., & Rutherford, M. (1999). A mathematician, a physicist, and a computer scientist with Asperger syndrome: Performance on folk psychology and folk physics tests. *Neurocase, 5,* 475–483.

Brewer, W. F., & Samarapungavan, A. (1991). Child theories vs. scientific theories: Differences in reasoning or differences in knowledge? In R. R. Hoffman & D. S. Paler (Eds.), *Cognition and the symbolic processes: Applied and ecological perspectives* (pp. 209–232). Hillsdale, NJ: Erlbaum.

Bull, R., & Scerif, G. (2001). Executive functioning as a predictor of children's mathematics ability: Inhibition, switching, and working memory. *Developmental Neuropsychology, 19,* 273–293.
Cattell, R. B., & Drevdahl, J. E. (1955). A comparison of the personality profile (16PF) of eminent researchers with that of eminent teachers and administrators, and the general population. *British Journal of Psychology, 46,* 248–261.
Choynowski, M. (1948a). Life of science. *Synthese, 6,* 248–251.
Choynowski, M. (1948b, May 31). *Report to the United Nations educational scientific and cultural organization: Natural sciences section,* Occasional Paper No. 11. Institute for the Science of Science. Retrieved May 15, 2010, from http://unesdoc.unesco.org/images/0015/001540/154044eb.pdf
Cole, J. R., & Cole, S. (1973). *Social stratification in science.* Chicago: University of Chicago Press.
Cole, S. (1996). Voodoo sociology: Recent developments in the sociology of science. In P. R. Gross, M. Levitt, & M. W. Lewis (Eds.), *The flight from science and reason* (pp. 274–287). New York: New York Academy of Sciences.
Dunbar, K., Fugelsang, J., & Stein, C. (2007). Do naïve theories ever go away? Using brain and behavior to understand changes in concepts. In M. C. Lovett & P. Shah (Eds.), *Thinking with data* (pp. 193–205). Mahwah, NJ: Lawrence Erlbaum.
Eiduson, B. T. (1962). *Scientists: Their psychological world.* New York: Basic Books.
Fancher, R. E. (1983). Alphonse de Candolle, Francis Galton, and the early history of the nature-nurture controversy. *Journal of the History of the Behavioral Sciences, 19,* 341–352.
Fay, A. L., Klahr, D., & Dunbar, K. (1990). Are there developmental milestones in scientific reasoning? In *Proceedings of the 12th Annual Conference of Cognitive Science Society.* Hillsdale, NJ: Lawrence Erlbaum.
Feist, G. J. (1993). A structural model of scientific eminence. *Psychological Science, 4,* 366–371.
Feist, G. J. (1994). Personality and working style predictors of integrative complexity: A study of scientists' thinking about research and teaching. *Journal of Personality and Social Psychology, 67,* 474–484.
Feist, G. J. (1997). Quantity, impact, and depth of research as influences on scientific eminence: Is quantity most important? *Creativity Research Journal, 10,* 325–335.
Feist, G. J. (1998). A meta-analysis of the impact of personality on scientific and artistic creativity. *Personality and Social Psychological Review, 2,* 290–309.
Feist, G. J. (2006a). The development of scientific talent in Westinghouse finalists and members of the National Academy of Sciences. *Journal of Adult Development, 13,* 23–35.
Feist, G. J. (2006b). The past and future of the psychology of science. *Review of General Psychology, 10,* 92–97.
Feist, G. J. (2006c). *The psychology of science and origins of the scientific mind.* New Haven, CT: Yale University Press.
Feist, G. J. (2006d). The psychology of science: Development and personality in scientific thought, interest and achievement. *Review of General Psychology, 10,* 163–182.
Feist, G. J. (2008). The psychology of science has arrived. *Journal of Psychology of Science and Technology, 1,* 2–5.
Feist, G. J., & Gorman, M. E. (1998). Psychology of science: Review and integration of a nascent discipline. *Review of General Psychology, 2,* 3–47.

Feist, G. J., & Gorman, M. E. (2009). Ensuring a next generation of psychologists of science. *Journal of Psychology of Science and Technology, 2,* 2–4.

Feist, G. J., & Rosenberg, E. (2012). *Psychology: Perspectives and connections.*(2nd ed.). New York: McGraw-Hill.

Fugelsang, J. A., & Dunbar, K. N. (2004). Brain-based mechanisms underlying complex causal thinking. *Neuropsychologia, 43,* 1204–1213.

Fugelsang, J. A., Roser, M., Corballis, P., Gazzaniga, M., & Dunbar, K. (2005). Brain mechanisms underlying perceptual causality. *Cognitive Brain Research, 24*(1), 41–47. doi:10.1016/j.cogbrainres.2004.12.001

Fugelsang, J. A., Stein, C., Green, A., & Dunbar, K. (2004). Theory and data interactions of the scientific mind: Evidence from the molecular and the cognitive laboratory. *Canadian Journal of Experimental Psychology, 58,* 86–95.

Fuster, J. M. (2002). Frontal lobe and cognitive development. *Journal of Neurocytology, 31,* 373–385.

Galton, F. (1874/1895). *English men of science: Their nature and nurture.* New York: Appleton.

Gaston Bachelard. (n.d.). Retrieved June 2, 2010, from http://en.wikipedia.org/wiki/Gaston_Bachelard

Gholson, B., Shadish, W. R., Neimeyer, R. A., & Houts, A. C. (Eds.). (1989). *The psychology of science: Contributions to metascience.* Cambridge: Cambridge University Press.

Gopnik, A., Meltzoff, A. N., & Kuhl, P. K. (1999). *The scientist in the crib: Minds, brains, and how children learn.* New York: William Morrow.

Gorman, M. E. (1992). *Simulating science: Heuristics, mental models and technoscientific thinking.* Bloomington: Indiana University Press.

Gorman, M. E. (2009). Why the science of science policy needs psychology of science and technology: Notes on the science of science policy workshop, Washington, DC, December 2, 2008. *Journal of Psychology of Science and Technology, 2,* 38–39.

Gough, H. G. (1987). *California psychological inventory manual.* Palo Alto: CA: Consulting Psychologists Press.

Gruber, H. E. (1981). *Darwin on man: A psychological study of scientific creativity* (2nd ed.). Chicago: University of Chicago Press.

Helson, R. (1971). Women mathematicians and the creative personality. *Journal of Consulting and Clinical Psychology, 36,* 210–220.

Hemlin, S. (2009). Peer review agreement or peer review disagreement: Which is better? *Journal of Psychology of Science and Technology, 2,* 5–12.

Howell, M., & Prevenier, W. (2001). *From reliable sources: An introduction to historical methods.* Ithaca, NY: Cornell University Press.

Hull, D. L. (2001). *Science and selection: Essays on biological evolution and the philosophy of science.* Cambridge: Cambridge University Press.

Inhelder, B., & Piaget, J. (1958). *The growth of logical thinking from childhood to adolescence* (A. Parsons & S. Milgram, Trans.). New York: Basic Books.

Karmiloff-Smith, A. (1992). *Beyond modularity: A developmental perspective on cognitive science.* Cambridge, MA: MIT Press.

Klaczynski, P. A., & Narasimham, G. (1998). Development of scientific reasoning biases: Cognitive versus ego-protective explanations. *Developmental Psychology, 34,* 175–187.

Klaczynski, P. A., & Robinson, B. (2000). Personal theories, intellectual ability, and epistemological beliefs: Adult age differences in everyday reasoning biases. *Psychology and Aging, 15,* 400–416.

Klahr, D. (2000). *Exploring science: The cognition and development of discovery processes.* Cambridge, MA: MIT Press.

Koerber, S., & Sodian, B. (2009). Reasoning from graphs in young children: Preschoolers' ability to interpret covariation data from graphs. *Journal of Psychology of Science and Technology, 2,* 73–86.

Koslowski, B. (1996). *Theory and evidence: The development of scientific reasoning.* Cambridge, MA: MIT Press.

Kruglanski, A. W. (1994). The social-cognitive bases of scientific knowledge. In W. R. Shadish & S. Fuller (Eds.), *The social psychology of science* (pp. 197–213). New York: Guilford Press.

Kuhn, D. (1989). Children and adults as intuitive scientists. *Psychological Review, 96,* 674–689.

Kuhn, D. (1993). Connecting scientific and informal reasoning. *Merrill-Palmer Quarterly, 39,* 74–103.

Kuhn, D., Amsel, E., & O'Loughlin, M. (1988). *The development of scientific thinking skills.* Orlando, FL: Academic Press.

Kuhn, D., & Pearsall, S. (2000). Developmental origins of scientific thinking. *Journal of Cognition and Development, 1,* 113–129.

Kwon, Y., & Lawson, A. E. (2000). Linking brain growth with the development of scientific reasoning ability and conceptual change during adolescence. *Journal of Research in Science Teaching, 37,* 44–62.

Liberman, S., & Wolf, K. B. (1998). Bonding number in scientific disciplines. *Social Networks, 3,* 239–246.

Ludwig, A. (1995). *The price of greatness: Resolving the creativity and madness controversy.* New York: Guilford Press.

Matarazzo, J. D. (1987). Relationships of health psychology to other segments of psychology. In G. C. Stone, S. M. Weiss, J. D. Matarazzo, N. E. Miller, J. Rodin, C. D. Belar, . . . J. E. Singer (Eds.), *Health psychology: A discipline and a profession* (pp. 41–59). Chicago: University of Chicago Press.

McCullagh, C. B. (1984). *Justifying historical descriptions.* New York: Cambridge University Press.

Menon, V., Rivera, S. M., White, C. D., Eliez, S., Glover, G. H., & Reiss, A. L. (2000). Functional optimization of arithmetic processing in perfect performers. *Cognitive Brain Research, 9,* 343–345.

Merton, R. K. (1973). *The sociology of science: Theoretical and empirical investigations.* Chicago: University of Chicago Press.

Miller, A. I. (1996). *Insights of genius: Imagery and creativity in science and art.* New York: Springer Verlag.

Mullins, N. (1973). *Theories and theory groups in contemporary American sociology.* New York: Harper & Row.

Nersessian, N. J. (2002). The cognitive basis of model-based reasoning in science. In P. Carruthers, S. Stich, & M. Siegal (Eds.), *The cognitive basis of science* (pp. 133–153). Cambridge: Cambridge University Press.

Park, G., Lubinski, D., & Benbow, C. (2008). Ability differences among people who have commensurate degrees matter for scientific creativity. *Psychological Science, 19*(10), 957–961. doi:10.1111/j.1467–9280.2008.02182.x

Piaget, J. (1952). *The child's concept of number.* New York: W. W. Norton.

Roe, A. (1952). *The making of a scientist.* New York: Dodd, Mead.

Roser, M., Fugelsang, J., Dunbar, K., Corballis, P., & Gazzaniga, M. (2005). Dissociating processes supporting causal perception and causal inference in the brain. *Neuropsychology, 19*(5), 591–602. doi:10.1037/0894–4105.19.5.591

Rourke, B. P., & Conway, J. A. (1997). Disabilities of arithmetic and mathematical reasoning: Perspectives from neurology and neuropsychology. *Journal of Learning Disabilities, 30*, 34–46.

Sarton, G. (1952). *A guide to the history of science.* Waltham, MA: Chronica Botanica.

Schulze, A. (1990). On the rise of scientific innovations and their acceptance in research groups: A socio-psychological study. *Social Studies of Science, 20*, 35–64.

Schulze, A., & Meischner, K. (2008, July 18). *Conflict emergence and conflict management in developing innovations.* Paper presented at the Second Biennial Conference for the International Psychology of Science and Technology (ISPST), Berlin, Germany.

Science, Technology, and Society (STS) (n.d.). Retrieved May 11, 2010, from http://www.nsf.gov/funding/pgm_summ.jsp?pims_id=5324&org=NSF&sel_org=NSF&from=fund

Shadish, W. R. (1989). The perception and evaluation of quality in science. In B. Gholson, W. R. Shadish, R. A. Neimeyer, & A. C. Houts (Eds.), *Psychology of science: Contributions to metascience* (pp. 383–426). Cambridge: Cambridge University Press.

Shadish, W. R., & Fuller, S. (1994). *The social psychology of science.* New York: Guilford Press.

Shadish, W. R., Fuller, S., & Gorman, M. E. (with Amabile, T., Kruglanski, A., Rosenthal, R., & Rosenwein, R. E.) (1994). Social psychology of science: A conceptual and empirical research program. In W. R. Shadish & S. Fuller (Eds.), *Social psychology of science* (pp. 3–123). New York: Guilford Press.

Shrager, J., & Langley, P. (Eds.). (1990). *Computational models of scientific discovery and theory formation.* San Mateo, CA: Morgan Kaufmann.

Simonton, D. K. (1988). *Scientific genius: A psychology of science.* Cambridge: Cambridge University Press.

Simonton, D. K. (1989). Chance-configuration theory of scientific creativity. In B. Gholson, W. R. Shadish, R. A. Neimeyer, & A. C. Houts (Eds.), *Psychology of science: Contributions to metascience* (pp. 170–213). Cambridge: Cambridge University Press.

Simonton, D. K (1991). Career landmarks in science: Individual differences and interdisciplinary contrasts. *Developmental Psychology, 27*, 119–130.

Simonton, D. K. (1992). The social context of career success and course for 2,026 scientists and inventors. *Personality and Social Psychology Bulletin, 18*, 452–463.

Simonton, D. K. (1995). Behavioral laws in histories of psychology: Psychological science, metascience, and the psychology of science. *Psychological Inquiry, 6*, 89–114.

Simonton, D. K. (1999). *Origins of genius.* New York: Oxford University Press.

Simonton, D. K. (2000). Methodological and theoretical orientation and the long-term disciplinary impact of 54 eminent psychologists. *Review of General Psychology, 4*, 13–21.

Simonton, D. K. (2004). *Creativity in science.* Cambridge: Cambridge University Press.

Simonton, D. K. (2008). Scientific talent, training, and performance: Intellect, personality, and genetic endowment. *Review of General Psychology, 12*(1), 28–46. doi:10.1037/1089-2680.12.1.28

Simonton, D. K. (2009). Applying the psychology of science to the science of psychology: Can psychologists use psychological science to enhance psychology as a science? *Perspectives on Psychological Science, 4*, 2–4.

Staats, A. (1999). Unifying psychology requires new infrastructure, theory, method, and a research agenda. *Review of General Psychology, 3*(1), 3–13. doi:10.1037/1089–2680.3.1.3

Stevens, S. S. (1939). Psychology and the science of science. *Psychological Bulletin, 36,* 221–263.

Sulloway, F. J. (1996). *Born to rebel: Birth order, family dynamics, and creative lives.* New York: Pantheon.

Sulloway, F. J. Introduction: The science of science. In F. J. Sulloway (Ed.), *Testing theories of scientific change.* Cambridge, MA: Harvard University Press. Manuscript in preparation.

Taylor, C. W., & Barron, F. (1963). *Scientific creativity: Its recognition and development.* New York: John Wiley.

Terman, L. M. (1954). Scientists and nonscientists in a group of 800 men. *Psychological Monographs, 68* (Whole No. 378).

Tweney, R. D., Doherty, M. E., & Mynatt, C. R. (Eds.). (1981). *On scientific thinking.* New York: Columbia University Press.

Volodarskaya, E. (2005). Russia: Psychology of science. In S. P. Restivo (Ed.), *Science, technology, and society: An encyclopedia* (pp. 478–480). New York: Oxford University Press.

Wai, J., Lubinski, D., & Benbow, C. (2005). Creativity and occupational accomplishments among intellectually precocious youths: An age 13 to age 33 longitudinal study. *Journal of Educational Psychology, 97*(3), 484–492. doi:10.1037/0022-0663.97.3.484

Walentynowicz, B. (1975). The science of science in Poland. Present state and prospects of development. *Social Studies of Science, 5,* 213–222.

Westrum, R. (1989). The psychology of scientific dialogues. In B. Gholson, W. R. Shadish, R. A. Neimeyer, & A. C. Houts (Eds.), *Psychology of science: Contributions to a metascience* (pp. 370–382). New York: Cambridge University Press.

Zuckerman, H. (1977). *Scientific elite.* New York: Free Press.

2 Psychology of Science

Influence on the Philosophy of Science

E. J. Capaldi and Robert W. Proctor

Although philosophy dates back to antiquity, recognition of the philosophy of science as a separate discipline is more recent, beginning perhaps with the Vienna Circle in the first half of the 20th century. It is recognized that philosophy has had a major influence on psychology. Our concern in this chapter goes in the opposite direction, with the influence of psychology on the philosophy of science. This endeavor is prompted by the new branch of psychology called the psychology of science (Feist, 2006; Gholson, Shadish, Neimeyer, & Houts, 1989). As an approach to understanding science, the psychology of science joins several other disciplines, including the history of science, the philosophy of science, and the sociology of science (Feist, 2006; see Chapter 1, this volume), adding an emphasis on the psychological processes of the individual scientist.

Although it is true that, as an approach, the psychology of science is relatively new, a variety of prior developments link psychology either to the various fields of science or to the philosophy of science. These developments, as concern the philosophy of science, have been overlooked and little appreciated both by psychologists and by other interested parties, a problem that we seek to correct in this chapter. This treatment has two major benefits. On the one hand, it can supply information as to how useful psychology has been in the past and is at present, and thus may be in the future, in the effort to understand science. On the other hand, since it can be demonstrated that psychology has had a highly significant influence on the philosophy of science, that demonstration indicates the potential value of psychology in illuminating matters in disciplines other than science. Although we indicate how psychology has influenced various philosophic approaches to science, we do not fully explicate any of these approaches in detail in this chapter, an undertaking well beyond its scope.

Psychology and philosophy of science have been intimately linked since the 1800s. Indeed, in Germany, where psychology as a field was initially developed, the distinction between psychology and philosophy hardly existed up through World War II (Cat, 2007), psychologists such as Oswald Külpe being philosophers as well. Given the close association between the two fields, it is hardly surprising that the influence of the philosophy of science on psychology has been profound. In the United States, behaviorism was influenced by operationism (Stevens, 1935) and logical positivism (Spence, 1960), among other fields. More recently, many psychologists have been influenced by the philosophies of science associated with Norwood Hanson and Thomas Kuhn (e.g., Lachman, Lachman, & Butterfield, 1979). Less well known is that the influence has gone in the other direction as well, from psychology to the philosophy of science. Understanding the influence of psychology on the philosophy of science, which is the main concern of this chapter, will enable us to better understand science.

PREEMPIRICAL PSYCHOLOGY: KANT AND THE BRITISH EMPIRICISTS

Prior to becoming an empirical discipline, psychology influenced the philosophy of science. In this phase, we examine Immanuel Kant (1724–1804) and the British empiricists because of their historical importance. The very influential philosopher Kant regarded psychology as the empirical investigation of the laws of mental operations (Immanuel Kant, 2007). His view was that mental operations are not physical, have only a temporal dimension, and cannot be measured, thus denying the possibility of mathematical laws for experience (Berkson & Wettersten, 1984). However, Kant's views of psychology varied considerably from time to time (Makkreel, 2001). At times he was unfavorable to the science of psychology, concluding that it is not possible for psychology to be a quantitative, experimental science. At other times, his views toward psychology were more favorable, as in his influential *Critique of Pure Reason* in which he advocated an approach called transcendental psychology, which is essentially the study of thinking. In that work, Kant attributed cognition to the soul, which lies beyond experience, the Kantian categories being transcendental. Regardless of whether Kant's statements were favorable or unfavorable to psychology, discerning precisely what he meant is difficult and requires considerable interpretation.

Kant appears to have identified psychology with inner events and physics with outer events (Sturm, 2001). It seems rather likely that Kant did not grasp the idea that explanations of psychological phenomena could be investigated by looking at behavior (i.e., outer events). How seriously to take Kant's views is difficult to say because, at the time he philosophized about psychology, it had not yet become empirical psychology. What Kant seems to have been reacting negatively to is the undeveloped view of psychology advocated by his contemporaries, based almost exclusively on philosophic analyses. Psychologists such as Gustav Fechner and Wilhelm Wundt had not yet appeared on the scene. It is possible that had Kant examined psychology during the period in which empirical and laboratory psychology was being developed, his view of the possibility of a scientific psychology may have been more favorable.

With the British empiricists, including Locke, Berkeley, and Hume, among others, who held sway from roughly 1650 to 1850, philosophy and psychology were highly intertwined. Generally speaking, the British empiricists suggested that thinking is accomplished via images that become associated through experience. Frequently mentioned by the British empiricists as determining the formation of an association is contiguity, a view similar to that suggested by Aristotle, who heavily influenced them. Through the ages, contiguity was paramount, although depending on the particular philosopher, other causes were suggested, such as frequency and similarity. The British empiricists' theory of mind, which suggested that thinking consisted of a series of associations between images, was closely related to their views of science. According to them, science was largely an inductive enterprise, the inductions consisting of associations formed on the basis of experience.

Among the British empiricists, Alexander Bain (1818–1903), one of the last of that school, deserves to be singled out for special consideration. Far more than any other British empiricist, he devoted himself to psychology. According to Murphy (1949), "He was genuinely and consistently a psychologist; he might fairly be described as the first to write a comprehensive treatise having psychology as its sole purpose" (p. 107). More to the point of this chapter, his influence on philosophy was immense. On the one hand, he influenced the understanding of British empiricism, emphasizing the role of physiology on psychology, unlike Locke, and the role of causation in the formation of associations, unlike Hume (Young, 2005). On the other hand, Bain may be regarded as one of the original contributors to American pragmatism. It is to be noted that Charles Sanders Peirce (1839–1914), generally regarded as the founder of pragmatism, said of Bain, based on Bain's view of truth, "I am disposed to think of him as the grandfather of pragmatism" (Peirce, 1998, p. 399). Among his many contributions, Bain appears to be the first to suggest that thinking employs a trial and error process: "All creative, productive processes were relegated (for example by Bain) to the operation of trial and error plus a bit of luck" (Mandler & Mandler, 1964, pp. 130–131). This trial and error view was later adopted by a number of individuals, both in psychology and in the philosophy of science.

Bain, despite being so early in the history of psychology, had a modern conception of the relationship between psychology and philosophy of science, more specifically, epistemology. According to Bain (1902, p. viii), "What is now termed epistemology has psychological relationships, but is pursued into issues of a specific character, lying outside of pure psychology."

INFLUENCE OF EMPIRICAL PSYCHOLOGY ON THE PHILOSOPHY OF SCIENCE

We now leave more philosophically oriented psychology to examine the influence of empirical psychology on the philosophy of science, beginning with the earliest investigators. That influence has come from psychophysics, behaviorism, perception (including Gestalt psychology), and cognitive psychology.

PSYCHOPHYSICS

Gustav Fechner (1801–1887), along with Wilhelm Wundt (1832–1920) and Hermann von Helmholtz (1821–1894), is given credit as one of the founders of modern empirical and laboratory psychology. One of Fechner's greatest contributions was to demonstrate in his book *Elements of Psychophysics* (1860/1966) that psychology could be quantified. Psychophysics may be defined as the psychological study of the relation between physical stimuli and their sensations. To appreciate Fechner's contribution, recognize that a variety of theorists, most notably, Kant, had suggested that quantification of psychological phenomena was not possible. Fechner built upon Weber's law, which suggested that an arithmetical increase in a sensation required a geometrical increase in the stimulus. Fechner developed from this law the general idea that sensations could be quantified, and thus a psychology of science was possible. Among the giants of philosophy that Fechner influenced were Ernst Mach (1838–1916), Charles Sanders Peirce, and William James (1842–1910).

In physics, Mach criticized Newton's idea of absolute space and time, and his treatment exercised considerable influence on Albert Einstein, who indicated that Mach provided the philosophic instigation to relativity theory. In psychology, Mach examined the relation between sensations and external stimuli, including space, time, color, and sound. These stimuli, once restricted to physics, now became the province of psychology as well.

In the philosophy of science, Mach was a naturalist who believed that there is nothing beyond nature, and this would include the organic, the mental, and evolution as components of the natural. Mach believed that our minds and cultures have an evolutionary basis, and he considered that simple reflexes and responses gradually evolved into more complex capacities for responding. One process that evolved in this manner was memory, and then memories were communicated culturally. Mach considered that the mind evolved, and that you could quantify sensations, and that this could lead to science. Mach desired to unify physics and psychology, which should not be too surprising, given that he was an active researcher in both fields. As Szaz (1959) noted in his introduction to a reprint of Mach's (1959) book *The Analysis of Sensations*, "Although Mach was a physicist, he wrote the present work as a contribution, in part at least, to psychology. The scope and method of this field of knowledge he conceived on the model of Fechner's *Psychophysics*" (p. v). Mach was heavily influenced by Fechner and Darwin. According to Pojman (2009), "The single greatest influence upon Mach was Fechner; this cannot be overstated. Fechner's *The Elements of Psychophysics* was published in 1860, but Mach was aware of the ideas in it before" (p. 9).

As noted, Fechner's psychophysical work also influenced two prominent American philosophers, the philosopher C. S. Peirce and the psychologist/philosopher William James. Peirce, perhaps America's foremost philosopher and, as previously mentioned, the founder of pragmatism, was an early advocate of psychophysics, and can literally be regarded as the first American experimental psychologist (Cadwaller, 1974). In 1877, he published a psychophysical experiment on color vision in the *American Journal of Science and Arts*, and in an 1885 article with the soon to be famous psychologist Joseph Jastrow, introduced use of a 4-point confidence scale, instead of a single yes or no judgment, to provide more sensitive measurement of sensations (Peirce & Jastrow, 1887). Peirce's interest in psychology extended well beyond psychophysics to consciousness, memory, and cognition. In fact, Peirce (1887) authored an article in the first issue of the *American Journal of Psychology*, called "Logical Machines," that presaged modern computer simulation of cognition.

James, who accepted Peirce's pragmatism, was influenced by Fechner's psychophysics and also his general philosophy. According to Marshall (1974), "James had read Fechner's *Elements of Psychophysics* and his work *On Questions of the Soul* as early as 1870" (p. 304). Although in his earlier writings James was critical of some aspects of Fechner's work, in his later writings he came to embrace the powerful contribution of Fechner's psychophysical and philosophical work (Adler, 1992). James also rejected introspectionism and associationism in psychology. James's knowledge of psychology in general was extensive, as evidenced by his authoring the classic textbook *Principles of Psychology* in 1890, which is still in print. Murphy (1949) notes, "The distinction between psychology and philosophy was never sharp for James, and even when he was working on *The Principles of Psychology* he was making philosophical history" (p. 206). In philosophy of science, James's radical empiricism was influenced by a variety of fields, among them psychology. However, all of these different approaches are combined in a way that the identity of each is lost. According to McDermott (1968), "The philosophy of William James calls for a never-ending series of descriptions and analyses, each from a specific vantage point but no one of them burdened with having to account for everything" (p. xliv).

BEHAVIORISM

Behaviorism is the view that the study of psychology should be based on behavior and not on subjective experience. It was introduced into psychology in a 1913 article in *Psychological Review* by John B. Watson (1878–1958), entitled, "Psychology as the Behaviorist Views It." In its radical form, as practiced for example by B. F. Skinner (1953), behaviorism eschews all forms of mentalism.

Bertrand Russell (1872–1970), a philosopher and mathematician who is regarded as one of the most important logicians of the 20th century (Irvine, 2010), also read extensively in psychology and was strongly influenced by behaviorism. He explicitly acknowledged this influence in his 1921 book *The Analysis of Mind*, indicating at the end of the preface, "My thanks are due to Professor John B. Watson... for reading my MSS. at an early stage and helping me with many valuable suggestions" (p. 6). Russell's knowledge of psychology extended well beyond Watson's behaviorism. For example, he was well acquainted with Thorndike's (1911) work on learning, accepted two of his laws, the law of effect and the law of exercise, saying of them, "There seems every reason to accept these two laws" (p. 53). Although Thorndike established these laws on the basis of animal research, Russell states, "What is true of animals, as regards instinct and habit, is equally true of men" (p. 53). Russell was also influenced by William James's (1890) views on human movement. He accepted James's distinction between three types of movement: voluntary, reflex, and mechanical.

The particular aspect of behaviorism that influenced Russell the most was its emphasis on the measurement of behavior, rejecting introspection. In dismissing introspectionism, he said, "The criticism of introspection has been in the main the work of American psychologists" (p. 110) and singled out for special treatment an article in *Psychological Review* by Knight Dunlap (1912). With regard to behaviorism, Russell says, "Where you might expect a chapter on "thought processes" you come instead upon a chapter on "The Language Habit." It is humiliating to find how terribly adequate this hypothesis turns out to be" (p. 27). However, Russell departed from the behaviorists in one major respect. He put considerable emphasis on images, which followed their own laws, different from those of other psychological processes. Earlier in the preface, Russell writes:

> This book has grown out of an attempt to harmonize two different tendencies, one in psychology, the other in physics, with both of which I find myself in sympathy, although at first sight they might seem inconsistent. On the one hand, many psychologists, especially those of the behaviourist school, tend to adopt what is essentially a materialistic position, as a matter of method if not of metaphysics. They make psychology increasingly dependent on physiology and external observation, and tend to think of matter as something much more solid and indubitable than mind. Meanwhile the physicists, especially Einstein and other exponents of the theory of relativity, have been making "matter" less and less material. Their world consists of "events," from which "matter" is derived by a logical construction. (pp. 4–5)

Watson (1922) wrote a review of Russell's (1921) book, in which he said, "Up to this point the behaviourist and Mr. Russell have travelled a common road chatting amicably together. Mr. Russell suddenly begins to feel that the behaviourist is no longer an adequate companion and leaves him abruptly" (p. 98). What Watson was referring to is the emphasis that Russell puts on images, for which Watson comments, "The behaviourist feels no need of images either for memory or for thought" (p. 98). Russell's departure here may have been due to the influence of another psychologist/philosopher whom Russell praises, William James, saying "The view that seems to me to reconcile the materialistic tendency of psychology with the anti-materialistic tendency of physics is the view of William James" (pp. 5–6). The view to which Russell refers is the radical empiricism of James.

Russell's behaviorism and philosophy, especially in the area of logic, had a large influence on the logical positivists, philosophers of science who attempted to combine empirical evidence with modern logical analysis. As a group, the logical positivists are referred to as the Vienna Circle. In the 1930s, a number of logical positivists, most notably, Rudolph Carnap, Carl Hempel, and Otto Neurath, attempted to unite psychology and philosophy (see Tamminga, 2005). The major concern of this approach was to translate the language of psychology into the language of physics. This approach was called logical behaviorism. According to Carnap and Neurath, for every meaningful cognitive sentence there exists a translation into the language of physics. One of the positive effects that physicalism would have on psychology was that it would enable psychologists from various areas, who employed a variety of different vocabularies, to employ an identical one, the physical language, and so enable them to communicate better with each other. But their major aim was to unify the sciences such that psychology, sociology, physics, and the like would all be members of the same scientific family. Also, it was thought that psychology could be better united with sciences such as physics by translating the language of psychology into the language of physics.

Another philosopher of science who was influenced by behaviorism was Gustav Bergmann (1906–1987), who received his PhD from the University of Vienna and was a member of the Vienna Circle. Bergmann moved to the University of Iowa in the late 1930s, where he collaborated with the behaviorist Kenneth Spence in an attempt to clarify and solve problems associated with operationism (Bergmann & Spence, 1941) and psychological measurement (Bergmann & Spence, 1944). The strong influence of behaviorism on Bergmann is evident in his 1956 article evaluating the contribution of John B. Watson: "John B. Watson is, in my judgment, the most important figure in the history of psychological thought during the first half of the century. Nor is his impact limited to the science of psychology" (p. 265). Elsewhere, with regard to scientific theory in general and in psychology in particular, Bergmann (1953) explicitly links his views and those of the logical positivists to behaviorism, stating,

> Among psychologists, the most articulate proponents of this general view are the "behavior theorists," Hull, Spence, and their fellow students, who have, in a rather limited area, developed what is at present the most elaborate and articulate psychological theory. Historically, these authors stand in the succession of Watson. Among philosophers, the position is most vigorously embraced by logical positivists. It is, therefore, often spoken of as behaviorism, or neobehaviorism, or logical behaviorism. (p. 436)

Perception

The influence of perception on the philosophy of science, particularly that of Gestalt psychology, has been and continues to be substantial. Gestalt psychology is an approach that emphasizes that we perceive figures and objects as wholes, rather than as additive combinations of discrete parts (Hartmann, 1935). Perceptual organization follows a number of principles such as proximity (elements in spatial proximity tend to be grouped together), similarity (elements that are physically similar tend to be grouped together), and good continuation (elements that suggest a continuous line or contour will tend to be grouped together). Gestalt psychology was founded by Christian von Ehrenfels (1859–1932), and some of its more famous members include Max Wertheimer, Wolfgang Köhler, and Kurt Koffka, all of whom did their major work in the first half of the 20th century.

Among those influenced by Gestalt psychology were members of the Vienna Circle, who developed the epistemological view of logical positivism. Rudolph Carnap (1891–1970) was the leading and most famous member of the Vienna Circle, and was among its most important founders. His famous and highly debated *Der logische Aufbau der Welt* (*The Logical Structure of the World*; Carnap, 1928/1961) was "seen as perhaps the crucial document in the formation of the project of logical positivism" (Richardson, 1998, p. 5). The *Aufbau* was not only influential in its own right, but it exercised considerable influence over a number of prominent philosophers, including Putnam, Lewis, Goodman, and Quine (see Richardson, 1998).

In the *Aufbau*, Carnap wished to create a science of experience. This science, according to Carnap, had to identify structures. Carnap chose individual subjective experience as the basis of his system. Experience, of course, was also central to Gestalt psychology. In his view of elementary experiences, in which the whole is perceived, Carnap drew heavily on the Gestalt approach to experience, which was just emerging as a major force in scientific psychology and perception. The influence of Gestalt psychology on Carnap is clear on the basis of his statement that "contemporary psychological research has confirmed more and more that in the various sense modalities, the total impression is epistemically primary, and that the so-called individual impressions, from which one is subsequently accustomed to saying the perception is 'composed', are first achieved through abstraction therefrom" (p. 67).

According to Feest (2007), "This psychological school [Gestalt psychology] was offering a scientific conception of subjective experience, according to which phenomenal experience is structured in accordance with laws, which describe the functional relationship between types of stimulus configurations and types of Gestalt experiences" (p. 4). In accepting the Gestalt view, Carnap disagreed with Mach that the fundamental nature of perception involved individual sensations. That Gestalt psychology influenced Carnap is indisputable, but the precise nature of that influence is debatable (see Feest, 2007). Either Carnap was using Gestalt research findings to bolster an already existing epistemological theory or he was constructing that theory on the basis of the Gestaltists' findings. However, on either assumption, the Gestalt influence on Carnap's view of experience, and thus on logical positivism, was deep and profound.

Norwood Russell Hanson (1924–1967), another well-known and prominent philosopher of science, used several figures from Gestalt psychologists in his major work, *Patterns of Discovery* (Hanson, 1958), to make his point that perception is theory laden. In a footnote, he stated, "Writings by Gestalt psychologists… are many. Yet they are overlooked by most philosophers" (p. 180). He then cited 20 specific articles, and said, "Philosophy has no concern with fact, only with conceptual matters…; but discussions of perception could not but be improved by the reading of these twenty papers" (p. 181). In reviewing Hanson's (1969) posthumous book, *Perception and Discovery,* Achinstein (1972) said,

> There is a unifying Hansonian theme here: theory-dependence. Scientists confront nature with theories when they see, when they report the facts, and when they make causal judgments. Seeing is not simply opening one's eyes and looking but

a rich, theory-infused activity involving seeing that objects would behave in certain ways under appropriate conditions. Facts are not entities which exist completely independently of us, they have human ties, they are affected by our theory laden language. (p. 246)

It would be a mistake to think that Hanson was influenced only by the Gestalt psychologists. It is clear from his 1969 book, that his view of the theory ladenness of perception was influenced by a variety of psychologists, including Bartlett, Külpe, Gibson, and others. Hanson cites extensively from M. D. Vernon's (1952) book on perception, with the following quote indicating the extent to which Hanson used psychological data and theory:

But familiarity is something more than a mere acquaintance with a particular situation or with the material of a laboratory experiment. It denotes the existence of some cognitive scheme of experienced phenomena in which the present percept fits. It is the nature of the relationships to such a scheme which supplies the meaning of the percept. (Hanson, 1969, p. 164; Vernon, 1952, p. 224)

Thomas S. Kuhn (1922–1996), whose characterization of scientific paradigms contributed to the demise of logical positivism and ushered in new postpositivistic approaches to the philosophy of science, was heavily influenced by Hanson, particularly in his view of perception. In his highly influential book, *The Structure of Scientific Revolutions,* Kuhn (1962) also relies on illustrations from Gestalt psychologists to illustrate his views about the way in which paradigms shape perception, determining what one sees. He acknowledges his debt to Hanson, by saying, "N. R. Hanson, in particular, has used Gestalt demonstrations to elaborate some of the same consequences of scientific belief that concern me here" (p. 112). According to Kuhn, in agreement with Hanson,

What a man sees depends both upon what he looks at and also upon what his previous visual-conceptual experience has taught him to see. In the absence of such training there can only be, in William James's phrase, "a bloomin 'buzzin' confusion." (p. 112)

Kuhn (1962) emphasizes psychological factors not only in observation but in theory evaluation as well. He says, "Individual scientists embrace a new paradigm for all sorts of reasons and usually for several at once" (p. 151). He goes on to say that some reasons for accepting a paradigm may "depend upon idiosyncrasies of autobiography and personality. Even the nationality or the prior reputation of the innovator and his teachers can sometimes play a significant role" (p. 152). Another place in which Kuhn stresses psychology in the acceptance or rejection of scientific theories is when he discusses various factors that go into such acceptance or rejection. Examples of such factors are the fertility of a theory, the extent to which it is quantified, its parsimony, and so on. Two individuals may agree completely on the set of factors involved in theory evaluation, but yet they may not agree on their evaluation of a particular theory. The source of such disagreement, according to Kuhn, is the relative weight that each individual ascribes to particular factors. For example, one individual may value fruitfulness more than parsimony, whereas the other individual may have a reverse preference. Clearly, psychological rather than logical factors enter into such judgments. The difference between the two individuals is not the sort of problem that can be solved by logical argument alone.

COGNITIVE PSYCHOLOGY

The Würzburg School

The Würzburg School of German psychologists can be considered one of the important predecessors of contemporary cognitive psychology. Their major concern was with the psychology of thinking. They rejected the notion of the mind as passive that came from British empiricism. The British empiricists also saw thinking as a series of associations between sensations (images). The Würzburgers rejected both associationism and sensations, considering instead that thinking consisted in part of imageless thoughts that were directed by task, or *Aufgabe*. An important view associated with Otto Selz (1881–1943), who had strong ties to the Würzburgers, is that productive thinking consists of trial and error, or "trying-out behavior" (ter Hark, 2004), resulting in learning, a view similar to that earlier suggested by Bain and later accepted by Popper, as considered below.

Karl Popper (1902–1994) studied with Karl Bühler, who himself studied with Oswald Külpe (1862–1915), Külpe being considered the founder of the Würzburg School. Although Popper is known as a philosopher of science, he received his PhD in psychology in 1928, with his dissertation titled *Zur Methodenfrage der Denkpsychologie* (*On the Methodological Problem of Cognitive Psychology*). As we will show below, the Würzburg School, and in particular the views of Selz, intimately influenced Popper's philosophy of science.

Popper is well known for his rejection of the role of induction in science, which was widely accepted at that time, and his acceptance of falsificationism via deduction (Popper, 1959). Popper's falsificationism is a view widely accepted among scientists of various persuasions, including psychologists. He explicitly held a negative opinion of the role of psychology in the philosophy of science, despite his PhD in psychology (Kurz, 1996). Popper's graduate adviser, Bühler, along with other German psychologists, suggested that epistemology and psychology were closely related because both dealt with thought. Popper's early training with Bühler, with whom he disagreed on several matters, probably put Popper on the track to become a philosopher of science (Popper, 1974). Much of Popper's opposition to psychology as a science was based on his rejection of introspectionism and associationism, views with which he became acquainted due of course to his training in the Würzburg school of thought.

While working on his PhD, Popper came not only to embrace philosophy but to explicitly reject psychology. Popper gave up psychology in preference for logic. Popper said that his PhD work in psychology "showed me the priority of the study of logic over the study of subjective thought processes" (1974, p. 61). Popper also stated that on the completion of his dissertation, "I finally turned away from psychology" (p. 61).

Popper put major stock in the distinction between the context of justification, in which he was interested and which was amenable to logic, and the context of discovery, in which he was not interested because it was not in his view amenable to logic. The quote from Popper (1970), which follows, indicates the depth of his expressed opposition to psychology as a vehicle for furthering philosophy. Popper's remarks in this connection were prompted to a considerable extent by Kuhn's (1970) embrace of psychology as one means for better understanding science:

> In fact, compared with physics, sociology and psychology are riddled with fashions, and with uncontrolled dogmas. The suggestion that we can find anything here like "objective, pure description" is clearly mistaken. Besides, how can the regress to these often spurious sciences help us in this particular difficulty? (pp. 57–58)

Popper goes on to complain about Kuhn, "This is why I regard the idea of turning to sociology or psychology as surprising. I regard it as disappointing because it shows that all I have said before about sociologistic and psychologistic tendencies and ways... was in vain" (p. 58). Popper ends his chapter with the statement, clearly directed at Kuhn, "While the Logic of Discovery has little to learn from the Psychology of Research, the latter has much to learn from the former" (p. 58).

Popper's explicit estrangement from psychology was fairly complete, rejecting introspection, behaviorism, and psychoanalysis. Thus, if one reads Popper, it seems to be the case that psychology's influence on Popper was negative in the sense that having studied it and having become a psychologist himself, he decided later that psychology had little usefulness for his system of philosophy. Popper, in common with the Würzburgers, rejected the then common assumption that the mind is relatively passive, taking in sensations and on that basis forming inductions (Humphrey, 1951). Popper suggested that the mind actively seeks information, and his later emphasis on conjectures and refutations (Popper, 1963) can be seen as an active "trying-out behavior" compared to a more passive trial and error learning. Essentially, Popper suggested that a tension arises between generalizations and specific instances that may be incompatible with those generalizations; when this occurs, the scientist should put forth bold conjectures, which then he or she seeks to refute.

As mentioned, although in his writings Karl Popper explicitly rejected psychology as applicable to the philosophy of science, Berkson and Wettersten (1984) present a strong case that Popper's philosophy of science was greatly influenced by the cognitive psychology of the time. If they are correct, Popper's philosophy of science was greatly and intimately influenced by psychology as follows. Through his training with Bühler, he essentially adopted Külpe's idea of an active rather than a passive mind. These ideas that were passed on from Bühler and Külpe to Popper influenced his philosophy of science. This is clearly stated by Berkson and Wettersten, who say:

> Bühler probably was most influential as a transmitter of Külpe's basic view of the mind. The idea that the mind actively creates general theories which grow, change,

and influence further thought and perception at every level is the basic presupposition from which Popper developed all his psychological and methodological theories. (p. 116)

Another major psychological influence on Popper was Selz, whose emphasis on trial and error was, according to ter Hark (2004), influential on Popper's development of conjectures and refutations. Essentially, the ideas of an active mind and trial and error learning were formalized by Popper in his philosophy of science in the form of conjectures and refutations. It is ironic that Popper, who rejected psychology so thoroughly, was so heavily influenced by it. Indeed, according to Berkson and Wettersten, Popper's philosophy of science is better understood as growing out of psychology and his view of learning than out of logic and philosophy. This assessment seems quite reasonable to the present authors.

Popper is considered to be the founder of evolutionary epistemology (see Bradie & Harms, 2008, for an extensive description of evolutionary epistemology). Popper's evolutionary epistemology, like his philosophy of science itself, may be seen as an outgrowth of Selz's view of trial and error. In any event, as developed by Popper, evolutionary epistemology had two major themes. One was that the acquisition of important behaviors such as language depended on innate dispositions. The second was that such behaviors were acquired by trying out a number of behaviors or ideas, eliminating those that did not work and retaining those that led to success. In this formulation, we see the three prongs of natural selection, as suggested by Darwin: variability, selection, and preservation of the solution that works. The view of evolutionary epistemology was developed and elaborated further by Donald Campbell (1960; 1974), a social-personality psychologist, who in fact coined the term evolutionary epistemology. Campbell influenced many philosophers of science, as illustrated by the essays in a book dedicated to him which confirm "that his central epistemological claims were heard, and that they have effected change and excited controversy that will continue to influence debate in the philosophy and sociology of science for generations" (Heyes, 2001, pp. 1–2). It is interesting to see the similarity between some of the ideas of Popper and Campbell and those of current evolutionary psychologists (e.g., Tooby & Cosmides, 2005), who did not come along until much later.

Contemporary Cognitive Psychology and Cognitive Science

In recent times, Ronald Giere, a prominent philosopher of science, has advocated a cognitive approach to that field (see Chapter 4, this volume). His approach emphasizes representations and judgment. In the preface to his 1988 book, *Explaining Science*, Giere explains how he grappled with the best way to proceed to develop a naturalistic approach to science, stating:

The resolution once again came by a fortuitous encounter with other disciplines, this time the cognitive sciences. There one finds models of cognitive agents who develop representations of the world and make judgments about both the world and their representations of it. It took no great leap of imagination to think of scientists as cognitive agents and of scientific models as a special type of representation. Likewise, scientists' decision making about models can easily be seen as an exercise of ordinary human judgment. (pp. xvi–xvii)

More recently, Giere (2009) has said,

Unfortunately, the cognitive study of science is not taken as seriously as it deserves to be, not only in the philosophy of science but in science studies generally.... For those who do take it seriously, however, the cognitive study of science does provide resources for responding to several issues.... I have just noted support for an active, constructive understanding of human visual perception. One can also find support for an explicitly non-computational account of language and language acquisition that emphasizes the communicative intentions of human agents. (p. 223)

Another philosopher of science who employs cognitive science extensively in his analyses of science is Paul Thagard (see Chapter 16, this volume). His approach, called computational philosophy of science, employs computational modeling of human cognition to illuminate questions in the philosophy of science. He says of this approach,

Science is a human enterprise, and understanding the development of scientific knowledge depends on an account of the thought processes of humans. Hence computational philosophy of science overlaps as much with cognitive psychology as it does with scruffy AI. Even its normative prescriptions about how science ought to be done should take human cognitive limitations as starting points. (Thagard, 1988, p. 4)

The influence of cognitive psychology on the philosophy of science does not end with Giere and Thagard. For example, the book *Scientific Understanding*, edited by the philosophers of science de Regt, Leonelli, and Eigner (2009), suggests that understanding science cannot be achieved in the absence of cognitive factors, particularly the intuitive understanding of the individual scientist, a view similar to Giere's. The research of cognitive psychologist Alison Gopnik (e.g., Gopnik et al., 2004), who has provided evidence that even young children construct causal maps that are used to understand the world, is extensively cited in that book. Also cited is the work of psychologist Arthur Reber (1993), whose theorizing about implicit learning has been employed to explain philosopher Polanyi's (1958) influential postulation of tacit knowledge. Other cognitive psychologists contributing to the philosophy of science include Kevin Dunbar (2001) and David Klahr (2002).

CONCLUSION

Psychology has had a continuing influence on the philosophy of science, particularly since it evolved into empirical psychology, and that influence continues to the present, developments that have been largely ignored and are little known. In the earliest days, the influence came from psychophysics. Later, the philosophy of science was influenced by behaviorism and perception, mainly in the form of Gestalt psychology. Cognitive psychology, in the form of the Würzburg School, influenced the philosophy of science in the 1900s and, in its contributions to cognitive science, is continuing to do so at present. Other approaches have had some influence on the philosophy of science as well, most notably Campbell's (1974) contribution to evolutionary epistemology.

It appears that the numerous and varied contributions of psychology to the philosophy of science have been unappreciated both among psychologists and the public at large. These contributions suggest in turn that the fledgling field of the psychology of science has the potential to be a major influence in the future for better understanding both science and the philosophy of science. Indeed, it is difficult to imagine our understanding of the philosophy of science progressing to its full potential in the absence of a psychology of science. Essentially, a better understanding of thinking will lead to a better understanding of the philosophy of science, a view accepted in the earliest days of psychology, by the British empiricists and the Würzburgers, and in the current day by advocates of the psychology of science.

REFERENCES

Achinstein, P. (1972). Reviewed work(s): Perception and discovery by Norwood Russell Hanson. *Synthese, 25*(1/2), 241–247.

Adler, H. E. (1992). William James and Gustav Fechner: From rejection to elective affinity. In M. E. Donnelly (Ed.), *Reinterpreting the legacy of William James* (pp. 253–261). Washington, DC: American Psychological Association.

Bain, A. (1902). *The senses and the intellect* (4th ed.). New York: Appleton.

Bergmann, G. (1953). Theoretical psychology. *Annual Review of Psychology, 4,* 435–458.

Bergmann, G. (1956). The contribution of John B. Watson. *Psychological Review, 63,* 265–276.

Bergmann, G., & Spence, K. W. (1941). Operationism and theory construction. *Psychological Review, 48,* 1–14.

Bergmann, G., & Spence, K. W. (1944). The logic of psychological measurement. *Psychological Review, 51,* 1–24.

Berkson, W., & Wettersten, J. (1984). *Learning from error: Karl Popper's psychology of learning.* La Salle, IL: Open Court Press.

Bradie, M., & Harms, W. (2008). Evolutionary epistemology. *The Stanford encyclopedia of philosophy.* Downloaded from http://plato.stanford.edu/entries/epistemology-evolutionary/

Cadwaller, T. C. (1974). Charles S. Peirce (1839–1914): The first American experimental psychologist. *Journal of the History of the Behavioral Sciences, 10,* 291–298.

Campbell, D. T. (1960). Blind variation and selective retention in creative thought as in other thought processes. *Psychological Review, 67,* 380–400.

Campbell, D. T. (1974). Evolutionary epistemology. In P. A. Schilpp (Ed.), *The philosophy of Karl R. Popper* (pp. 412–463). La Salle, IL: Open Court Press.

Carnap, R. (1928/1961). *Der logische Aufbau der Welt.* Hamburg, Germany: F. Meiner.

Cat, J. (2007). Switching Gestalts on Gestalt psychology: On the relation between science and philosophy. *Perspectives on Science, 15,* 131–177.

de Regt, H., Leonelli, S., & Eigner, K. (Eds.). (2009). *Scientific understanding: Philosophical perspectives*. Pittsburgh, PA: University of Pittsburgh Press.

Dunbar, K. (2001). The analogical paradox: Why analogy is so easy in naturalistic settings, yet so difficult in the psychological laboratory. In D. Gentner, K. J. Holyoak, & B. N. Kokinov (Eds.), *The analogical mind: Perspectives from cognitive science* (pp. 313–334). Cambridge, MA: MIT Press.

Dunlap, K. (1912). The case against introspection. *Psychological Review, 19,* 404–413.

Fechner, G. T. (1860/1966). *Elements of psychophysics* (H. E. Adler, Trans., & D. H. Howes & E. G. Boring, Eds.). New York: Holt, Rinehart & Winston.

Feest, U. (2007). Science and experience/Science of experience: Gestalt psychology and the anti-metaphysical project of the *Aufbau*. *Perspectives on Science, 15,* 1–25.

Feist, G. J. (2006). *The psychology of science and the origins of the scientific mind*. New Haven, CT: Yale University Press.

Gholson, B., Shadish, W. R., Neimeyer, R. A., & Houts, A. C. (Eds.). (1989). *Psychology of science: Contributions to metascience*. New York: Cambridge University Press.

Giere, R. N. (1988). *Explaining science: A cognitive approach*. Chicago: University of Chicago Press.

Giere, R. N. (2009). Scientific perspectivism: Behind the stage door. *Studies in History and Philosophy of Science, 40,* 221–223.

Gopnik, A., Glymour, C., Sobel, D., Schulz, L. E., Kushnir, T., & Danks, D. (2004). A theory of causal learning in children: Causal maps and Bayes nets. *Psychological Review, 111,* 3–32.

Hanson, N. R. (1958). *Patterns of discovery*. New York: Cambridge University Press.

Hanson, N. R. (1969). *Perception and discovery: An introduction to scientific inquiry*. San Francisco, CA: Freeman, Cooper.

Hartmann, G. W. (1935). *Gestalt psychology: A survey of facts and principles*. New York: Ronald Press.

Heyes, C. (2001). Introduction. In C. Heyes & D. L. Hull (Eds.), *Selection theory and social construction* (pp. 11–14). Albany: State University of New York Press.

Humphrey, G. (1951). *Thinking: An introduction to its experimental psychology*. New York: John Wiley.

Immanuel Kant. (2007). Downloaded from Human Intelligence, http://www.indiana.edu/~intell/kant.shtml

Irvine, A. D. (2010). Bertrand Russell. In E. N. Zalta (Ed.), *The Stanford encyclopedia of philosophy* (first published December 7, 1995; substantive revision March 29, 2010), http://plato.stanford.edu/entries/russell/

James, W. (1890). *Principles of psychology*. New York: Holt.

Klahr, D. (2002). *Exploring science: The cognition and development of discovery processes*. Cambridge, MA: MIT Press.

Kuhn, T. S. (1962). *The structure of scientific revolutions*. Chicago: University of Chicago Press. (Rev ed. published in 1970 and 3rd ed. in 1996).

Kuhn, T. S. (1970). Logic of discovery or psychology of research? In I. Lakatos & A. Musgrave (Eds.), *Criticism and the growth of knowledge* (pp. 1–23). New York: Cambridge University Press.

Kurz, E. M. (1996). Marginalizing discovery: Karl Popper's intellectual roots in psychology; Or, how the study of discovery was banned from science studies. *Creativity Research Journal, 9,* 173–187.

Lachman, R., Lachman, J. L., & Butterfield, E. C. (1979). *Cognitive psychology and information processing: An introduction*. Hillsdale, NJ: Lawrence Erlbaum Associates.

Mach, E. (1959). *The analysis of sensations, and the relation of the physical to the psychical* (C. M. Williams & S. Waterlow, Trans.), with foreword by T. S. Szaz. New York: Dover.

Makkreel, R. A. (2001). Kant on the scientific status of psychology, anthropology, and history. In E. Watkins (Ed.), *Kant and the sciences* (pp. 185–201). New York: Oxford University Press.

Mandler, J. M., & Mandler, G. (Eds.). (1964). *Thinking: From association to Gestalt.* New York: John Wiley.

Marshall, M. E. (1974). William James, Gustav Fechner, and the question of dogs and cats in the library. *Journal of the History of the Behavioral Sciences, 10,* 304–312.

McDermott, J. J. (1968). Introduction. In J. J. McDermott (Ed.), *The writings of William James: A comprehensive edition* (pp. xiii–xlvi). New York: Modern Library.

Murphy, G. (1949). *Historical introduction to modern psychology* (Rev. ed.). New York: Harcourt, Brace.

Peirce, C. S. (1877). Note on the sensation of color. *American Journal of Science and Arts, 13,* 247–251.

Peirce, C. S. (1887). Logical machines. *American Journal of Psychology, 1,* 165–170.

Peirce, C. S. (1998). Pragmatism (Ms 318). In Peirce Edition Project (Eds.), *The essential Peirce* (Vol. 2, 1899–1913, pp. 398–434). Bloomington: Indiana University Press.

Peirce, C. S., & Jastrow, J. (1885). On small differences in sensation. *Memoirs of the National Academy of Sciences, 3,* 73–83.

Pojman, P. (2009). Ernst Mach. In E. N. Zalta (Ed.), *The Stanford encyclopedia of philosophy* (Winter), http://plato.stanford.edu/archives/win2009/entries/ernst-mach/

Polanyi, M. (1958). *Personal knowledge: Towards a post-critical philosophy.* London: Routledge.

Popper, K. R. (1959). *The logic of scientific discovery.* New York: Basic Books.

Popper, K. R. (1963). *Conjectures and refutations.* London: Routledge & Kegan Paul.

Popper, K. R. (1970). Normal science and its dangers. In I. Lakatos & A. Musgrave (Eds.), *Criticism and the growth of knowledge* (pp. 51–58). New York: Cambridge University Press.

Popper, K. R. (1974). Autobiography of Karl Popper. In P. A. Schilpp (Ed.), *The philosophy of Karl Popper* (Book 1, pp. 3–181). La Salle, IL: Open Court Press.

Reber, A. S. (1993). *Implicit learning and tacit knowledge: An essay on the cognitive unconscious.* Oxford: Oxford University Press.

Richardson, A. W. (1998). *Carnap's construction of the world.* New York: Cambridge University Press.

Russell, B. (1921). *The analysis of mind.* London: George Allen & Unwin.

Skinner, B. F. (1953). *Science and human behavior.* New York: Macmillan.

Spence, K. W. (1960). *Behavior theory and learning.* Englewood Cliffs, NJ: Prentice-Hall.

Stevens, S. S. (1935). The operational basis of psychology. *American Journal of Psychology, 47,* 323–330.

Sturm, T. S. (2001). Kant on empirical psychology: How not to investigate the human mind. In E. Watkins (Ed.), *Kant and the sciences* (pp. 163–184). New York: Oxford University Press.

Szaz, T. S. (1959). Introduction to Dover edition. In E. Mach, *The analysis of sensations and the relation of the physical to the psychical* (C. M. Williams & S. Waterlow, Trans.) (pp. v–xxxi). New York: Dover.

Tamminga, A. (2005). Introspection and change in Carnap's logical behaviourism. *Studies in History and Philosophy of Science Part A, 36,* 650–667.

ter Hark, M. (2004). *Popper, Otto Selz, and the rise of evolutionary epistemology.* Cambridge: Cambridge University Press.

Thagard, P. (1988). *Computational philosophy of science*. Cambridge, MA: MIT Press.
Thorndike, E. L. (1911). *Animal intelligence*. New York: Macmillan.
Tooby, J., & Cosmides, L. (2005). Evolutionary psychology: Conceptual foundations. In D. M. Buss (Ed.), *Handbook of evolutionary psychology* (pp. 5–67). Hoboken, NJ: John Wiley.
Vernon, M. D. (1952). *A further study of visual perception*. Cambridge: Cambridge University Press.
Watson, J. B. (1913). Psychology as the behaviorist views it. *Psychological Review, 20*, 158–177.
Watson, J. B. (1922). The analysis of mind. *Dial, 72*, 97–102.
Young, R. M. (1990). Alexander Bain: Transition from introspective psychology to experimental psychophysiology. In R. M. Young, *Mind, brain, and adaptation in the nineteenth century: Cerebral localization and its biological context from Gall to Ferrier* pp. 101–133). New York: Oxford University Press. Downloaded from http://human-nature.com/mba/chap3.html

3 Methodological Approaches to Scientific and Technological Thinking

Michael E. Gorman

INTRODUCTION

The psychology of science is a new/old field—new because it has a society and a journal that began a few years ago, old because work in psychology of science has been carried out by individuals and even one or two research groups over an extended period, including a few specialized meetings and edited volumes—but no identity as a research area within psychology.

Historically, both science and technology studies (STS) scholars and psychologists avoided the psychology of scientific and technological expertise. STS scholars generally thought of expertise as an attribution made by society. Psychologists were focused on becoming more scientific and had philosophical arguments about the scientific method.

For example, in graduate school I lived through the so-called crisis in social psychology. At that time, the primary method preferred by social psychologists was experimental studies, typically involving college sophomores. Kenneth Gergen asked whether results from such experiments could be generalized beyond the historical time period in which they were obtained (Gergen & Gergen, 1984). Consider, for example, the famous obedience experiments by Stanley Milgram, which showed that ordinary Americans from New Haven would obey an experimenter's command to shock a helpless victim (Milgram, 1974). Milgram obtained his results in the 1950s. Would the same results have been obtained in the rebellious 1960s?

The answer is probably yes, because Milgram's results have been replicated many times (Blass, 2000). But Gergen was acting like a psychologist of science when he explored the assumptions of the dominant methodological approach in social psychology. He was engaged in what STS scholars call reflexive thinking: studying and reflecting on the practices in his own field, practices in which he had engaged.

ORGANIZATION AND PURPOSE

In this chapter, I:

1. Provide a taxonomy of the methodological approaches psychologists have taken to studying scientific and technological thinking, based on an analogy to biological methods (Dunbar, 2005). I will add suggestions about the roles of implicit and explicit knowledge.
2. Consider a new kind of expertise discovered by Harry Collins and Robert Evans in the course of their work in social studies of science (Collins & Evans, 2002).
3. Consider what it would take to make psychology of science into a field, and whether that is desirable.

A TAXONOMY OF APPROACHES TO PSYCHOLOGY OF SCIENCE

In Vitro

The in vitro approach is the kind of experiments preferred by social psychologists and behaviorists when I was in graduate school. In psychology of science, this approach involves testing participants who are not themselves scientists on laboratory tasks that bear resemblance to aspects of scientific thinking.

Consider Wason's 2, 4, 6 task, in which a participant tries to guess a rule known to the experimenter by proposing number triples. The combination 2, 4, 6 is provided as an example of a triplet that corresponds to the rule. Wason noted that participants would often try several triplets like 6, 8, 10 and 12, 14, 16 and then guess the rule was something like even numbers going up by twos (Wason, 1960). Wason and others proposed this was an instance of confirmation bias because participants did not try to test their rules by proposing instances they thought should be wrong. I found both individual participants and those working in groups could be taught a strategy in which they focused on negative instances, and this worked best when they were urged to test their hypotheses before announcing a guess (Gorman & Gorman, 1984).

The 2, 4, 6 task does resemble scientific reasoning in that participants have to come up with hypotheses and design little experiments to test them, but this task and method have very low ecological validity. None of these psychology of science experiments looked at the role of tacit knowledge in scientific reasoning. In the tacit knowledge area, Reber also did a low ecological validity task involving an artificial grammar, and found that participants could learn aspects of the grammar without being explicitly aware of what they were doing (Reber, 1993).

It is possible to increase the ecological validity in a variety of ways. Tweney and his colleagues gave participants a problem more like a simple video game where they fired particles at shapes on a computer screen, and found when the rules underlying this artificial universe became very complex, it made sense for participants to confirm early in the inference process and disconfirm only later, when they have a hypothesis worth testing (Mynatt, Doherty, & Tweney, 1978). Similarly, when I added the possibility of error to the 2, 4, 6 task by telling participants that 0–20% of the results from their triplet experiments would be erroneous: If a triplet were said to correspond to the rule it actually would not and vice versa. Confirmation is a powerful heuristic for detecting error; one can repeat a pattern to search for anomalous results, and replicate them to see if they are due to error. This process is more time-consuming, of course, and interferes with finding the rule (Gorman, 1989).

Berry and Broadbent (1984) used a simulated sugar production plant to increase ecological validity in a tacit knowledge study. Participants learned to operate the plant more effectively without being aware of how they were doing it. Broadbent demonstrates that the operation of a device or system involves tacit knowledge. Future research could be done on tasks that involve scientific discovery, like the artificial universe created by Tweney and his colleagues. Is there a point where participants are able to fire particles at shapes with some success, but cannot articulate a rule—cannot explain how they are achieving success?

In Vivo Research with Scientific Groups

Experiments with tasks permit the researcher to manipulate features like the optimal strategy for finding the rule and the amount of error. However, experimental studies need to be complemented by in vivo observations of actual scientists and engineers. Some of the best work has been done with scientific groups, because groups provide a record of their thinking as they discuss. Kevin Dunbar studied molecular biology labs by attending their laboratory meetings and analyzing their presentations and discussions (Dunbar, 1997). One of his most provocative findings was that the laboratory directors tended to exhibit a falsification bias, in which they questioned results that appeared to support their current theory and searched for alternate explanations, including errors in the design.

Nancy Nersessian studied biomedical engineering laboratories (Nersessian, 2009). She paid particular attention to the way in which engineering scientists constructed systems that were hybrids of computational models and physical devices to simulate complex physical processes that go on in the organism. These hybrid systems are a combination of in vivo and in vitro methods—they link laboratory devices to organisms in a way that allows both control of a model system and generalization to the in vivo biological system.

Dunbar and Nersessian do not discuss what happens when members of one laboratory move to another, but research on team processes suggests that this is one of the best situations for studying tacit knowledge. High-functioning teams share transactive memory, or knowledge of what other team members know and can do. Taking a member away disrupts this transactive memory, part of which may be tacit in teams that have worked together for a long time (Moreland, 1999). Everyone knows that Susie (say) is the person to call on to do X; when Susie leaves, the team discovers that it has lost the ability to do X as well as before, because aspects of the skill involved in X were tacit. This kind of loss frequently occurs when experienced members of an organization or team retire, and others discover that not all aspects of their skill were documentable. These veterans often have to be called back to do their old job and provide apprentice training to their replacements.

High-functioning teams that work together over long periods also develop shared mental models of the problem domains. Here a shared mental model is akin to Kuhn's notion of a paradigm (Kuhn, 1962), which combines ways of representing the problems appropriate for study in a domain and also techniques for solving them. The term "mental model" is preferred by cognitive psychologists because there is empirical research on the nature and character of mental models (Johnson-Laird, Oakhill, & Garnham, 1996).

One way to discover a shared mental model is to have a new member added to a team. Veteran members of the team may discover aspects of their own shared mental model as they try to explain to newcomers how they know what problems to work on, and what to do next (Levine & Moreland, 1999). What had become tacit knowledge now has to be made explicit as a newcomer is introduced to a new way of thinking.

There will also be laboratory procedures that are at least partly tacit and will be revealed by the new member. An example is Collins's study of the transversely excited atmospheric-pressure (TEA) laser, which some laboratories had built and run but others could not replicate from published documents (Collins, 1985). The only way to replicate was to visit one of the successful laboratories and study its apparatus. It turned out that the key difference was the placement and length of a wire above the laser—even the successful laboratories were not aware that this was an important feature; they had just set up their lasers that way. Here is a case of tacit knowledge being made explicit because of an outsider trying to figure out what a group is doing. More research should be directed at these sorts of encounters.

Ex Vivo

Another methodological approach involves taking scientists out of their laboratories and testing them on complex word problems. This kind of research often includes comparison with novices: The goal is to see what experts do that novices cannot. Key differences include:

- Expert scientist and mathematicians rapidly classify a problem in a way that suggests the best solution path. Novices use a variety of weak or general heuristics (Chi, 2006). Features of a problem or situation that novices do not notice are salient for experts.
- Expert scientists also have skills and heuristics specific to a domain (Kulkarni & Simon, 1988). Expert judgment consists in knowing when specific information and a procedure should be combined to solve a problem (Gorman, 2002).
- Experts are capable of generating sophisticated thought experiments that resemble those of historical scientists like Galileo (Clement, 2009).
- Innovators in science and technology come up with new ways of representing a domain—typically a new mental model that takes advantage of what they

already know but transforms the problem space (Gorman, 1998). Examples are provided in the next section.

Subspecies Historiae

This method involves looking at historical case studies in fine-grained detail—a kind of in vivo study based on documents like detailed notebooks, drafts of publications, letters, and sometimes supplemented by interviews conducted at the time of a discovery. Retrospective interviews are not reliable sources of data because of the reconstructive nature of memory (Ericsson & Simon, 1984).

Michael Faraday kept the most detailed notebooks of any scientist and is therefore the most studied. The best cognitive psychological work has been done by David Gooding and Ryan Tweney. Both tracked Faraday's thinking and experiments in great detail. Tweney constructed problem-behavior graphs of Faraday's processes (Tweney, 1989). Maxwell called Faraday a great intuitive mathematician and Tweney was able to figure out why: He came up with detailed representations of changes in electrical and magnetic fields that served as a kind of geometry on which Maxwell could base his equations (Tweney, 2009).

Gooding took the reconstruction of Faraday's processes a step farther by actually replicating some of his experiments to get a procedural understanding of how he worked, that is, how exactly he build the first generator as a demonstration experiment that he sent to other scientists (Gooding, 1990). Gooding's method holds the possibility of reconstructing aspects of tacit as well as explicit knowledge because there may be aspects of embodied knowledge that cannot be put in the notebooks but become apparent when the experiment is redone.

I adapted the problem-behavior analysis technique to the study of an inventor, Alexander Graham Bell (Gorman, 1995). He left notebooks, letters, and other documents that were not as detailed as Faraday's, but detailed enough to permit a good reconstruction of his path to the invention of the telephone. A particularly important entry is one he made just before submitting a patent that included speaking telegraphy (which we would now call telephony; see Figure 3.1).

This diagram shows the bones of the ear attached to a diaphragm and a speaking tube, vibrating close to two different arrangements of electromagnets. Above, Bell

Figure 3.1 Bell's drawing of the bones of the middle ear ("a") between a speaking tube and two arrangements of electromagnets, from his "Experimental Notebook." The text under "Fig 5" reads "(Helix & core, iron cylinder vibrated in helix)" and at the bottom right Mabel Gardiner, Bell's future wife, notes that she copied the figure on February 21.

announced his approach: that he would follow the analogy of nature by using the ear as a model for the telephone, which means that he would need an armature that would function like the bones of the ear, translating the vibrations of the diaphragm into a current that precisely mimicked the form of the sound wave.

I discovered this diagram in the Library of Congress—I say discovered because no other historian had seen its significance, not even the great Robert Bruce whose biography of Bell is a classic (Bruce, 1973). Bell is here sketching his mental model for what we now call a telephone. He is going to build an electromechanical ear, and his biggest problem will be finding a substitute for the bones of the middle ear. Bell knew exactly how these bones worked because he had built a device to visualize sound waves that incorporated them.

Mental models are partly or even wholly tacit forms of representation. In this case, Bell made his explicit by sketching it and explaining its significance. Mental models can also be embodied in devices like the hybrid systems built by Nersessian's engineering scientists to represent and test their evolving mental models. The weakness of Bell's design is already apparent in his mental model: The ear is a great receiver, but does not produce sound. Other inventors like Edison gave Bell credit for a receiver, but built far superior transmitters.

Explicit and Implicit Knowledge of Processes

One aspect of expert knowledge is metacognitive awareness of problem-solving processes. Bell was very explicit about his thinking and experiments, and generally followed a VOTAT (Vary One Thing at a Time) strategy, varying only one thing at a time.

Edison, in contrast, just dumped sketches when asked in court what he had done to invent a superior telephone transmitter. In his telephone research, he led the first R & D lab at Menlo Park, and could conduct multiple experiments at the same time—and sketch even more possibilities (Carlson, 2000). If he explicitly monitored his thinking at any point, we have no record of it. It is likely he relied heavily on partly tacit judgments to decide on design alterations, and he certainly wanted to portray himself that way—as the wizard of Menlo Park, whose mysterious genius could not be replicated.

Science calls for very explicit description of procedures in journal articles, but these articles are a logical reconstruction of the way science is done, not a veridical account of the actual processes involved (Holmes, 1987). One interesting question for future research is the extent to which expert scientists and inventors have and ought to have metacognitive awareness of their methods. Typically, metacognitive awareness is most necessary when confronting an anomaly or a new kind of problem; on familiar problems it is more efficient just to proceed without reflection, almost automatically.

In Silico

Cognitive psychologists can develop computational models of reasoning processes. Two of the most famous use subspecies historia cases. Krebs's discovery of the ornithine cycle was developed in sufficient detail by the historian of science Larry Holmes to be modeled by Kekada, a program that simulated the way in which Krebs generated and took advantage of surprising results (Kulkarni & Simon, 1988). David Gooding worked with a computer scientist, Tom Addis, to model Michael Faraday's processes in a way that would allow researchers to experiment with the model, proposing new concepts and links (Gooding & Addis, 1999).

In Magneto

Future studies will include techniques like MRI to image the brain of working scientists during problem solving (Dunbar, 1999). One of the most promising avenues for that future work is determining the role of tacit knowledge in scientific and technological thinking. Studies of patients with amnesia show that they can improve their ability to solve certain tasks without remembering having worked on them before—which means they have no explicit memory of any rules or procedures they learned (Knowlton & Greenberg, 2008). For example, Glisky and Schachter (1989) studied a severely amnesic person who had to learn a complex computer task in order to maintain her job. She eventually "acquired over 250 discrete items of new information, many of which were individually complex, and used this knowledge appropriately to perform the data entry task both in the laboratory and the workplace" (p. 117). The fact that she "was able to apply the rules and procedures acquired during training to entirely novel documents, albeit at a slowed rate, indicates that *some* transfer of training occurred" (p. 118). Schachter (1998) concluded, "Amnesic patients, who have selective damage to the inner (medial) regions of the temporal lobes (including the hippocampus and related structures) perform poorly on tests for explicit memory that require them to recall or recognize recently presented information. But the same patients often show normal performance on implicit memory tests, in which they are simply asked to carry out a task and are not required to recollect any past experiences" (p. 59).

Patients with brain damage are a long way from working scientists, making generalization difficult. Brain imaging techniques like FMRI might identify focal areas for aspects of tacit knowledge like implicit memory and implicit learning; then ex vivo studies of scientists and inventors could show when these particular areas were most active, suggesting when and where tacit knowledge was playing a critical role.

It is best to use the approaches above in combination. Consider the use of in magneto techniques to make inferences about various kinds of tacit knowledge in scientific problem solving. Brain-imaging techniques can be used in vitro on problems designed to model tacit components of scientific reasoning, ex vivo by having scientists work on the artificial problems in vivo with scientists working on their research (if appropriate portable technology is developed that does not restrict the scientist)—but not with subspecies historiae, though inferences made from historical cases could be tested experimentally. The point is to use multiple methods to triangulate, which both increases validity and generalizability.

INTERACTIONAL EXPERTISE

Recently, Collins and Evans have identified a new category of expertise, partly to answer one of the fundamental challenges posed to STS researchers: How can one study a science without having expertise in the science? This argument led to heated debates among scientists, STS researchers, and others that came to be called the "science wars" (Gross & Levitt, 1994).

Psychologists of science did not have this concern because it is well known in our literature that experts cannot reliably describe how they work (see Explicit and Implicit Knowledge of Processes section above); methods like protocol analysis and problem-behavior graphs are needed. However, when interpreting the results of these methods, it helps to be able to understand the domain. It is not reasonable to expect every psychologist of science to have both psychological expertise and expertise in all the domains of science she or he wishes to study.

Harry Collins became so immersed in the community of gravitational wave physicists that he could talk in their language, ask intelligent, provocative questions about their research, even understand and tell the sorts of jokes that were shared in this community—all this, without being able to do the research or the mathematics involved (Collins, 2004). It enabled him to do a detailed study of this community in a way that practitioners found credible and interesting. Collins labeled this capability interactional expertise (Collins, Evans, & Gorman, 2007) as distinct from contributory, his label for the expert who can actually do the research. A psychologist or sociologist or anthropologist can gain interactional expertise in one or more domains by interacting with the practitioners.

Collins created an ex vivo experimental procedure called the imitation game—a kind of Turing test for expertise—to prove that he had gained interactional expertise. A gravitational wave physicist submitted questions to Collins and another gravitational wave physicist, and based on the answers, had to decide who was the real expert. The questions were designed so that the two contenders could give quick answers without doing computations or performing experiments, and questions about educational background were off limits. Collins was selected as the one most likely to be the actual physicist, by a small margin.

Immersion in a community is not sufficient for acquiring interactional expertise: While Collins succeeded in gravitational wave physics, his attempts to gain interactional expertise in in amorphous semiconductors failed, despite help from willing members of the community—he gained no useful tacit knowledge.

Collins and his student Rob Evans went on to do additional imitation game experiments. One series tested the hypothesis that those who are color-blind are able to blend into the color-sighted community by gaining interactional expertise. This hypothesis was tested using imitation games that included one participant who was color-blind and another who was not. When the judge could perceive colors, she or he was unable to reliably discriminate between the color-blind and color-perceiving participants. When the judge was color-blind, she or he was able to identify the participant who was truly color-blind. The hypothesis that the color-blind learn how to blend in was confirmed. This led Evans and Collins to propose a more general hypothesis that those in a community that is either a minority or seen as lower status are more likely to gain enough interactional expertise to sound like the dominant community. So in a male-dominated society, one would predict the women would be better at pretending to be men on an imitation game. Evans is testing this hypothesis with a variety of imitation games.

Unfortunately, the new Evans and Collins imitation game studies do not involve scientists and engineers—creating an opportunity for psychologists of science. Interactional expertise may be especially important in multi- and interdisciplinary collaborations, where the person who has interactional expertise across two or more fields could facilitate communication (Gorman, 2010). Interactional expertise is also important for managers of large scientific projects, like the 30-meter telescope, whose director, Gary Sanders, has discussed with Harry Collins his acquisition and use of interactional expertise (Collins, 2007). Sanders had been second-in-command on the Laser Interferometer Gravitational- Wave Observatory (LIGO). Now he had to decide which of three very sophisticated designs should be pursued for the 30-meter telescope. His physics background and prior management experience helped, but he admitted he knew very little about adaptive optics, a critical expertise involved in the design. After about six months of immersion in the project, he could "sit down in a group of adaptive optics experts who will come to me and say

> 'Gary you're wrong, multi-object adaptive optics will be ready at first light and it will give the following advantages'—and others will say, 'no, it's multi-conjugative adaptive optics' and I can give them four reasons why we should go with multi-conjugative adaptive optics based on the kind of science we want to do, the readiness of the technical components, when we need them, and so on. And I will see as I am talking about it that the room is looking back at me and they're saying 'he does have a line, he's thought it through, and yes.'

But if someone said to me, 'OK Sanders, we agree with you, now go and design a multi-conjugative adaptive optics system,' I couldn't do it" (Collins & Sanders, 2007). Here is applied interactional expertise: Sanders knew enough to make and defend a decision without being able to do the design himself.

The Sanders observations were based on interviews with Collins. What is needed to really tease out the development of interactional expertise is in vivo studies of its development. In the Sanders case, this could have been done with a combination of interviews and observations throughout his learning process—difficult to do with a busy expert, but possible if a researcher gains the trust of the participant and acquires interactional expertise. There are also situations where interdisciplinary interactions are recorded, as in the interdisciplinary Mars Rover teams (Paletz & Schunn, 2010; see Chapter 18, this volume). A psychologist of science armed with the right conceptual framework could use this data to investigate whether interactional expertise played any role in the development of the Mars Rovers.

A recent NSF-funded workshop combined psychology of science and this new community of those who study expertise from an STS perspective. One goal was to look at new ways of investigating interactional expertise. Most of the approaches used to study the psychology of scientific thinking could be extended to interactional expertise. For example, as in the Sanders and Mars Rover examples, one could do in vivo studies of the acquisition of interactional expertise that would complement the ex vivo imitation game studies. The latter constitute proof of the concept that interactional expertise exists, but give us little hint as to how it develops and how it is deployed in specific problem-solving situations.

The workshop also improved the theoretical framework by producing new methods for diagramming trajectories of expertise acquisition that permit comparison of interactional with contributory expertise (see Figure 3.2). This kind of diagram allows researchers to articulate and communicate their assumptions across disciplinary boundaries, serving as a kind of creole, or reduced common language. The diagrams played this role in the workshop that connected studies of experience and expertise (SEE) to psychology of science.

The diagram also raises questions, like whether tacit and explicit knowledge can be separated on three-dimensional axes. Tacit knowledge is partly a symbol of our collective ignorance—it is hard to study because by definition it cannot be clearly articulated by those who have it and are using it. But it can be studied in various forms, including shared mental models, implicit learning and memory, and hands-on procedures that involve embodied knowledge. This chapter has provided suggestions for how to do this sort of study using each of the methodological approaches outlined above. The best approach is to triangulate by combining methods to increase validity. Collins has been studying tacit knowledge for years and has come out with a new analysis that highlights three types: collective tacit knowledge, which is akin to a set of implicit, shared mental models; somatic tacit knowledge, which is embodied knowledge of how to perform procedures;

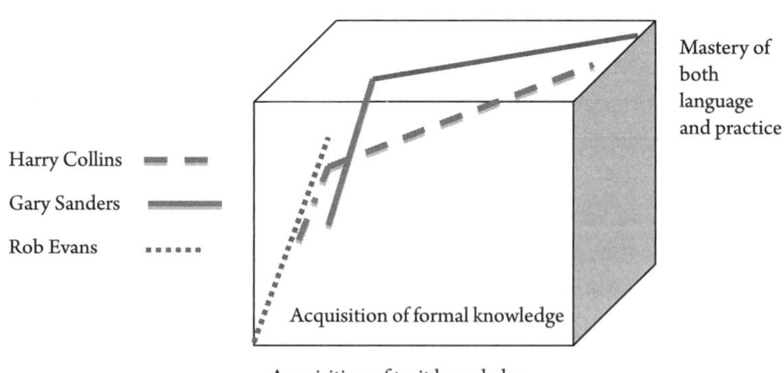

Figure 3.2 Three trajectories of expertise acquisition in gravitational wave physics: Harry Collins's interactional, Gary Sander's contributory, and Rob Evans's explicit knowledge gained from studying primary sources. These trajectories are hypothetical because the data do not exist to plot them, but the diagram suggests directions for future theoretical and empirical work. This figure is based on exchanges between Gorman and Harry Collins.

and relational tacit knowledge, which could be put into explicit form if properly analyzed and described (Collins, 2010). The psychology literature distinguishes between implicit learning and implicit memory; the former involves the process by which the latter is acquired (Stadler, 1998). How do these categories relate to those proposed by Collins? Here is an opportunity for the psychology of science and SEE communities to work together.

I hope this chapter has demonstrated the value of psychology of science. Should it become its own specialty, or remain a kind of meeting ground where different research communities can find ways of communicating as they investigate problems like the acquisition of scientific and engineering expertise, or the role personality plays in scientific success?

MAKING PSYCHOLOGY OF SCIENCE INTO A FIELD

There are at least six critical factors in establishing a new field:[1]

1. A new field must differentiate itself from existing ones. To do psychology of science, one need not be a psychologist—indeed, the work by Collins and Evans promises to contribute to psychology of science, as does the work of historians like Larry Holmes and philosophers like David Gooding. So the differentiation is not based on discipline, but by focus on psychological processes—including groups as well as individuals.
2. To be successful, the new field must not threaten existing ones—it is best if the new field positions itself at the intersection of existing ones. Psychology of science is finding a space between psychology and STS that most practitioners of these communities are not interested in occupying.
3. The new field must be able to demonstrate that it uses existing methods which are recognized as legitimate. This article demonstrates that psychology of science relies on a range of established methods. The new field must develop new approaches as well. The combination of psychology of science and interactional expertise is one promising avenue.
4. Mobilization is also critical, which involves activities like creating a society, founding a journal, and finding sources of funding for research. The International Society for Psychology of Science and Technology is a good start, with over a hundred members and three conferences under its belt. The *Journal of Psychology of Science and Technology* lacks a publisher, but PsycINFO is willing to put it in its database if the journal can be continued. The NSF funded the workshop that linked STS and the group studying interactional expertise, showing that it is possible to seek NSF funding for new efforts in psychology of science—though there is no guarantee of future success.
5. A new field must also have graduate programs in order to ensure that new scholars will be produced. There are currently no graduate degrees in psychology of science.

[1] The first four are adapted from (Hambrick & Chen, 2008).

6. A new field must show it can contribute to the solution of problems that existing fields cannot sufficiently address. Psychology is in a position not only to study science and technology, but also to suggest ways of improving education, policy, and practice. This chapter suggests that a particularly promising area for improvement is the empirical study of tacit knowledge in scientific and technological thinking.

There is an urgent need for study of the ways science and technology are transforming our global civilization and our cognitive, social, and biological capabilities. Those interested in psychology of science and technology from any discipline need intellectual and physical spaces where they can meet and interact. The journal would provide an intellectual space, and biannual meetings create a physical one. If there were no society, there would have been no way to have the joint SEE/psychology of science workshop. There is also room for a virtual space, perhaps an island in Second Life or just a Web forum.

Students interested in psychology of science need places to go to pursue their interest and homes for their continued intellectual endeavors afterward. These student needs could be addressed by forming a network of experts willing to take students for a period of time, and a couple of universities that can grant degrees based on this distributed learning. The risk in this kind of endeavor is small—if the students do not come, then there will be no need to write off major investments in graduate programs at institutions.

REFERENCES

Berry, D. C., & Broadbent, D. E. (1984). On the relationship between task performance and associated verbalizable knowledge. *Quarterly Journal of Experimental Psychology Section A: Human Experimental Psychology, 36*(2), 209.

Blass, T. (2000). The Milgram paradigm after 35 years: Some things we now know about obedience to authority. In T. Blass (Ed.), *Obedience to authority: Current perspectives on the Milgram paradigm* (pp. 35–59). Mahwah, NJ: Lawrence Erlbaum.

Bruce, R. V. (1973). *Bell: Alexander Graham Bell and the conquest of solitude.* Boston: Little, Brown.

Carlson, W. B. (2000). Invention and evolution: The case of Edison's sketches of the telephone. In J. Ziman (Ed.), *Technological innovation as an evolutionary process* (pp. 137–158). Cambridge: Cambridge University Press.

Chi, M. T. H. (2006). Laboratory methods for assessing experts' and novices' knowledge. In K. A. Ericsson, N. Charness, P. J. Feltovich, & R. R. Hoffman (Eds.), *The Cambridge handbook of expertise and expert performance* (pp. 167–184). Cambridge: Cambridge University Press.

Clement, J. J. (2009). The role of imagistic simulation in scientific thought experiments. *Topics in Cognitive Science, 1*(4), 686–710.

Collins, H. (2004). *Gravity's shadow: The search for gravitational waves.* Chicago: University of Chicago Press.

Collins, H., & Evans, R. (2002). The third wave of science studies. *Social Studies of Science, 32*(2), 235–296.

Collins, H., Evans, R., & Gorman, M. (2007). Trading zones and interactional expertise. *Studies in History and Philosophy of Science, 39*(1), 657–666.

Collins, H., & Sanders, G. (2007). "They give you the keys and say drive it!" Managers, referred expertise, and other expertises. *Studies in History and Philosophy of Science, 39*(1), 621–641.

Collins, H. M. (1985). *Changing order: Replication and induction in scientific practice.* London: Sage.

Collins, H. M. (2010). *Tacit and explicit knowledge.* Chicago: University of Chicago Press.

Dunbar, K. (1997). How scientists think: On-line creativity and conceptual change in science. In T. B. Ward, S. M. Smith, & J. Vaid (Eds.), *Creative thought: An investigation of conceptual structures and processes* (pp. 461–493). Washington, DC: American Psychological Association.

Dunbar, K. (1999). How scientists build models: In vivo science as a window on the scientific mind. In L. Magnani, N. Nersessian, & P. Thagard (Eds.), *Model-based reasoning in scientific discovery* (pp. 85–100). New York: Kluwer Academic.

Dunbar, K. N., & Fugelsang, J. A. (2005). Causal thinking in science: How science and students interpret the unexpected. In M. E. Gorman, R. D. Tweney, D. C. Gooding, & A. Kincannon (Eds.), *Scientific and technological thinking* (pp. 57–80). Mahwah, NJ: Lawrence Erlbaum Associates.

Ericsson, K. A., & Simon, H. A. (1984). *Protocol analysis: Verbal reports as data.* Cambridge, MA: MIT Press.

Gergen, K. J., & Gergen, M. M. (1984). *Historical social psychology.* Hillsdale, NJ: Lawrence Erlbaum.

Glisky, E. L., & Schachter, D. L. (1989). Extending the limits of complex learning organic amnesia: Computer training in a vocational domain. *Neuropsychologia, 27*(1), 107–120.

Gooding, D. (1990). Mapping experiment as a learning process: How the first electromagnetic motor was invented. *Science, Technology and Human Values, 15*(2), 165–201.

Gooding, D. C., & Addis, T. (1999). A simulation of model-based reasoning about disparate phenomena. In L. Magnani, N. Nersessian, & P. Thagard (Eds.), *Model-based reasoning in scientific discovery* (pp. 103–124). New York: Kluwer Academic.

Gorman, M. E. (1989). Error, falsification and scientific inference: An experimental investigation. *Quarterly Journal of Experimental Psychology, 41A*, 385–412.

Gorman, M. E. (1995). Confirmation, disconfirmation and invention: The case of Alexander Graham Bell and the telephone. *Thinking and Reasoning, 1*(1), 31–53.

Gorman, M. E. (1998). *Transforming nature: Ethics, invention and design.* Boston: Kluwer Academic.

Gorman, M. E. (2002). Turning students into professionals: Types of knowledge and ABET engineering criteria. *Journal of Engineering Education, 91*(3), 339–344.

Gorman, M. E. (2010). Trading zones, interactional expertise, and future research in cognitive psychology of science. *Topics in Cognitive Science, 2*(1), 96–100.

Gorman, M. E., & Gorman, M. E. (1984). A comparison of disconfirmatory, confirmatory and a control strategy on Wason's 2-4-6 task. *Quarterly Journal of Experimental Psychology, 36A*, 629–648.

Gross, P. R., & Levitt, N. (1994). *Higher superstition: The academic left and its quarrels with science.* Baltimore, MD: Johns Hopkins University Press.

Hambrick, D. C., & Chen, M. (2008). New academic fields as admittance-seeking social movements: The case of strategic management. *Academy of Management Review, 30*(1), 32–54.

Holmes, F. (1987). Scientific writing and scientific discovery. *Isis, 78*, 220–235.

Johnson-Laird, P. N. (1983). *Mental models.* Cambridge: Harvard University Press.

Knowlton, B. J., & Greenberg, D. L. (2008). Implicit learning and memory. In M. J. Aminoff, F. Boller, D. F. Swaab, G. Goldenberg, & B. L. Miller (Eds.), *Handbook of clinical neurology* (pp. 225–236). London: Elsevier.

Kuhn, T. S. (1962). *The structure of scientific revolutions.* Chicago: University of Chicago Press.

Kulkarni, D., & Simon, H. A. (1988). The processes of scientific discovery: The strategies of experimentation. *Cognitive Science, 12,* 139–175.

Levine, J. M., & Moreland, R. L. (1999). Knowledge transmission in work groups: Helping newcomers to succeed. In L. Thompson, D. Messick, & J. Levine (Eds.), *Shared knowledge in organizations* (pp. 267–296). Hillsdale, NJ: Lawrence Erlbaum.

Milgram, S. (1974). Obedience to authority. New York: Harper & Row.

Moreland, R. L. (1999). Transactive memory: Learning who knows what in work groups and organizations. In L. Thompson, D. Messick, & J. Levine (Eds.), *Shared knowledge in organizations* (pp. 3–31). Hillsdale, NJ: Lawrence Erlbaum.

Mynatt, C. R., Doherty, M. E., & Tweney, R. D. (1978). Consequences of confirmation and disconfirmation in a simulated research environment. *Quarterly Journal of Experimental Psychology, 30,* 395–406.

Nersessian, N. J. (2009). How do engineering scientists think? Model-based simulation in biomedical engineering research laboratories. *Topics in Cognitive Science, 1*(4), 730–757.

Paletz, S. B. F., & Schunn, C. D. (2010). A social-cognitive framework of multidisciplinary team innovation. *Topics in Cognitive Science, 2*(1), 73–95.

Reber, A. S. (1993). *Implicit learning and tacit knowledge: An essay on the cognitive unconscious.* New York: Oxford University Press.

Schachter, D. L. (1998). Memory and awareness. *Science, 280*(59), 59–60.

Stadler, M. A., & Frensch, P. A. (Eds.). (1998). *Handbook of implicit learning.* Thousand Oaks, CA: Sage.

Tweney, R. D. (1989). A framework for the cognitive psychology of science. In B. Gholson, W. R. Shadish Jr., R. A. Neimeyer, & A. C. Houts (Eds.), *Psychology of science: Contributions to metascience* (pp. 342–366). Cambridge: Cambridge University Press.

Tweney, R. D. (2009). Mathematical representations in science: A cognitive-historical case history. *Topics in Cognitive Science, 1*(4), 758–776.

Wason, P. C. (1960). On the failure to eliminate hypotheses in a conceptual task. *Quarterly Journal of Experimental Psychology, 12,* 129–140.

PART TWO

AGENCY AND REASONING IN THE PSYCHOLOGY OF SCIENCE

4 The Role of Psychology in an Agent-Centered Theory of Science

Ronald N. Giere

APPROACHES TO THE STUDY OF SCIENCE

Since those whose work might be classified as belonging to a psychology of science have diverging relationships with psychology of science as a discipline, I should begin by circumscribing my own relationship. This can best be done, I think, by stating a fundamental problem that lies behind most of my work.

How are humans, given only their relatively modest evolved capacities (physical, cognitive, etc.), able to do science; in particular, to learn about such things as the movements of the continents, the mechanisms of inheritance, or the age of the universe?

This is implicitly a question about scientific *knowledge,* which reflects my disciplinary background in the philosophy of science. But the question is about learning, the process of acquiring knowledge, not about the ultimate validity of particular knowledge claims. I do not question that we do know the things mentioned, in some ordinary sense of "know." The question also presumes an evolutionary perspective that takes seriously the idea that humans evolved from earlier forms of life. From this perspective, it really is quite remarkable that such evolved creatures as ourselves could come to know what we do in fact know. This is something that demands an explanation. I assume, however, that the explanation must itself be a *scientific* (naturalistic) explanation. This places me in the company of psychologists and sociologists of science who also seek a scientific explanation of the acquisition of scientific knowledge.

My question has two parts. One concerns the *origin* of science. How did humans originally come to do science? The second concerns *contemporary practice.* How do humans now do science? The two questions are, of course, related. In particular, answers to the second question are constrained in that contemporary practice must be something that could arise naturally from our evolutionary past. Except for a few comments about the 17th-century scientific revolution, I will focus on contemporary scientific practice.

Answers to my question are implicit in standard approaches to understanding scientific practice. Here is a brief survey.

A LOGICAL APPROACH

In this approach the focus is on the logical structure of theories and on an inductive logic that provides a logical (epistemological) relationship between observation and theory. In logical empiricism, for example, the logical structure of scientific theories and a logic of justification provide the basis for objective claims to knowledge. Recent historical scholarship reveals this movement to have been a continuation of Kantian and neo-Kantian projects to show how objective (indeed, universal) knowledge is possible (Friedman, 1999; Giere & Richardson, 1996). In this project, the actual psychology of real scientists (as opposed to ideally rational agents) is relevant only to the creation of empirical concepts, not to the validity of their application. This is the well-known distinction between "discovery" and "justification."

A METHODOLOGICAL APPROACH

On this approach, scientists make discoveries by applying a scientific method. Of course, there are lots of candidates for just what constitutes the correct scientific method. One candidate familiar to many psychologists has its roots in pragmatism (Dewey, 1938, esp. Chapter 21). The basic pragmatist position is that one begins with a comfortably held set of beliefs which then conflicts with a new observation. The resulting epistemic discomfort initiates a process of inquiry which begins with the suggestion of hypotheses which, if true, would resolve the conflict (abductive step). From these hypotheses one deduces predictions about possible new observations (deductive step). These predictions are then tested against new observations/experiments. The most successful hypothesis is adopted (induction). Epistemic comfort is thus restored with a revised set of beliefs. This is, of course, just a schema. We need more details about the key parts of the process: abduction, experimentation, and induction. The part of this schema to which psychology seems most relevant is the abductive step. Research on analogical reasoning, for example, finds a place here. But this leaves the role of psychology in the study of science just where the logical empiricists put it: in the context of discovery.

A HISTORICAL APPROACH

There are actually two strains of thought that can be labeled "historical." One is that initiated by Thomas Kuhn (1962) in *The Structure of Scientific Revolutions*. Here scientists are treated as natural agents to whom empirical, psychological categories apply. Thus, invoking the psychology of his time, Kuhn described a scientist adopting a new paradigm as experiencing a "gestalt switch." The second strain consists of philosophical responses to Kuhn by philosophers such as Imre Lakatos (1970), Stephen Toulmin (1972) and Larry Laudan (1977). In these accounts, the logical empiricist idea of "rational justification" is replaced by a semihistorical notion of "rational progress" measured in terms of such things as "solved problems." These are characterized in terms of objective, linguistic entities such as theories, hypotheses, and data. The psychology of real agents is not part of these accounts, so there is again little place for a psychology of real scientists.

A SOCIAL APPROACH

In recent sociology of science, a scientific consensus (which by itself is taken to constitute "scientific knowledge") is a function of social interactions, social networks, and the like (see Biagioli, 1999). Here one is dealing with real human agents, but their psychological characteristics are conceived of in everyday "folk psychological" terms. There is no need for a scientific psychology of science. Indeed, sociologists of science have been explicitly critical of attempted psychological explanations of any aspects of scientific practice. This seems an instance of the traditional border dispute between psychology and sociology. One of the most notorious invocations of this boundary occurred in Latour and Woolgar's "Postscript" to the second edition of *Laboratory Life* (1986, 280) where they proposed "a ten-year moratorium on cognitive explanations of science" and promised that "if anything remains to be explained at the end of this period, we too will turn to the mind!"

A COGNITIVE APPROACH

By the mid-1980s, I personally had become disenchanted with logical approaches to understanding science, but also had reservations about the other then prominent approaches noted above. Inspired by books such as Howard Gardner's (1985) *The Mind's New Science*, I discovered the then newly emerging cognitive sciences. These provided a framework for my work in the philosophy of science. I replaced the philosophy of science notion of the structure of theories with the more general cognitive science notion of representation. Likewise, instead of inductive logic I talked about judgments regarding the fit of models to the world. The overall goal was not (or not directly) to justify science, but to understand it. The result was the book *Explaining Science: A Cognitive Approach* (Giere, 1988). I must admit, however, that the use of actual research in the cognitive sciences in this book is highly selective. The main themes of representation and judgment owe as much to work in the philosophy of science and decision theory as they do to work in the cognitive sciences. Only gradually did it become clear to me that there was an emerging subfield that could be called "The Cognitive Study of Science." At one point, in response to criticism from a constructivist sociologist of science, I went so far as to propose (partly tongue in cheek) a program for "The Cognitive Construction of Scientific Knowledge" (Giere, 1992).

COGNITIVE STUDIES OF SCIENCE

The cognitive study of science developed in the 1970s and 1980s as a multidisciplinary mixture involving the history and philosophy of science and the newly emerging cognitive sciences. Early on there was some European influence due to the work of Piaget (1929) and later Howard Gruber (1981), but most of the early work was American. Beginning in the mid-1960s, Herbert Simon (1966) suggested applying the techniques of artificial intelligence to study the process of scientific discovery, culminating in the book *Scientific Discovery: Computational Explorations of the Creative Processes* (Langley, Simon, Bradshaw, & Zytkow, 1987). In the late 1970s, a group of psychologists at Bowling Green State University in Ohio, led by Ryan Tweney, began a program of systematically studying scientific thinking, including simulated experiments in which students were to try to discover the "laws" governing an "artificial universe." Their work (Tweney, Doherty, & Mynatt, 1981) highlighted such phenomena as "confirmation bias," the tendency of subjects to pursue hypotheses that agreed with some data even in the face of clearly negative data. Tweney himself (Tweney, 1985; Tweney, Mears, & Spitzmuller, 2005) turned his attention to Michael Faraday. Michael Gorman (1992), a psychologist at the University of Virginia, also conducted experiments on scientific reasoning in simulated situations. Around the same time, the psychologist Dedre Gentner and associates began developing theories of analogical reasoning and mental modeling which they applied to historical cases such as the discovery of electricity and the distinction between heat and temperature (Gentner, Holyoak, & Kokinov, 2001; Gentner & Stevens, 1983). A developmental psychologist, Susan Carey (1985), began applying Kuhn's account of revolutionary change in science to the cognitive development of children. Michelene T. H. Chi (1992), a Pittsburgh psychologist, also investigated the phenomenon of conceptual change in the context of differences in problem-solving strategies among novices and experts.

In the wake of Kuhn's *Structure of Scientific Revolutions*, a number of philosophers of science began applying more recent notions from the cognitive sciences to understand conceptual change in science. Nancy Nersessian (1984) applied notions of mental models and analogical reasoning to the development of field theories in physics in the 19th and early 20th centuries. Lindley Darden (1991) applied techniques from artificial intelligence in an attempt to program theoretical and experimental strategies followed in the development of Mendelian genetics. Paul Thagard (1988, 1991, 2000) advocated a full-blown "computational philosophy of science" and went on to develop an account of conceptual change based on a notion of "explanatory coherence" which he implemented in a computer program. Paul Churchland (1989) applied his own "neurocomputational perspective" to philosophy of science topics such as the nature of scientific theories. It will be noted that all this work is quite heterogeneous, involving people in different fields appealing to different aspects of the cognitive sciences and focusing on different topics and different historical periods or figures. The cognitive study of science has been decidedly *multidisciplinary* (See Carruthers, Stitch, & Siegal, 2002; Giere, 1992, 2008; Gorman, Tweney, Gooding, & Kincannon, 2005).

As an example of more recent work in the cognitive study of science I would cite Nersessian's (2008) *Creating Scientific Concepts*. Nersessian has been the most prominent promoter of the idea of the cognitive study of science, and this work represents 25 years of sustained research on the topic of conceptual change in science. Nersessian thinks that the most important part of conceptual change is the creation of the new concepts that will become part of the later conceptual structure. Moreover, she thinks that this process of concept creation has a micro-structure that can fruitfully be studied in actual historical and contemporary settings by selectively applying notions from contemporary cognitive science. For Nersessian, new concepts are generated from old concepts in specific problem situations by a process of "model-based reasoning." Individual agents begin with a "mental model," in her words "a structural, behavioral, or functional analog representation of a real-world or imaginary situation, event, or process" (2008, p. 93). She emphasizes three components of model-based reasoning: analogy, visualization, and thought-experimenting. In the case of dynamic systems, for example, thought-experimenting is a matter of mentally simulating the operation of an imaginary system. Individual mental models may be "coupled" with public visual models in the form of drawings or graphs, as in the case of Newton's famous drawing of a cannon ball shot around the world from off the top of a mountain. She also emphasizes the importance of intermediate "hybrid" models incorporating features of both a source and target domain.

Rich content for Nersessian's book is provided by two "exemplars" of model-based reasoning: Maxwell's development of electromagnetic theory following Faraday's researches on magnetic induction, and a contemporary think-aloud problem-solving protocol experiment designed by the cognitive scientist and education researcher John Clement. In the latter exemplar, a person with a PhD involving both mathematics and physics was asked to explain whether, keeping all other variables constant, a spring with twice the diameter would stretch more, less, or the same as the original spring, both loaded with the same small weight. In the course of this long experiment, the subject developed a new (for him) concept of torsion in springs. Both exemplars are explained in chapter-long presentations and referred to throughout the book.

THE COGNITIVE STUDY OF SCIENCE AND THE PSYCHOLOGY OF SCIENCE

The cognitive study of science is obviously related to the psychology of science. Figure 4.1 shows how I think they are related both to each other and to neighboring disciplines. Both the cognitive study of science and the psychology of science are represented as components of science studies. That they strongly overlap is indicated by the fact that cognitive psychology appears twice in the diagram, once as part of the cognitive sciences and once as part of psychology. I suspect this duplication may

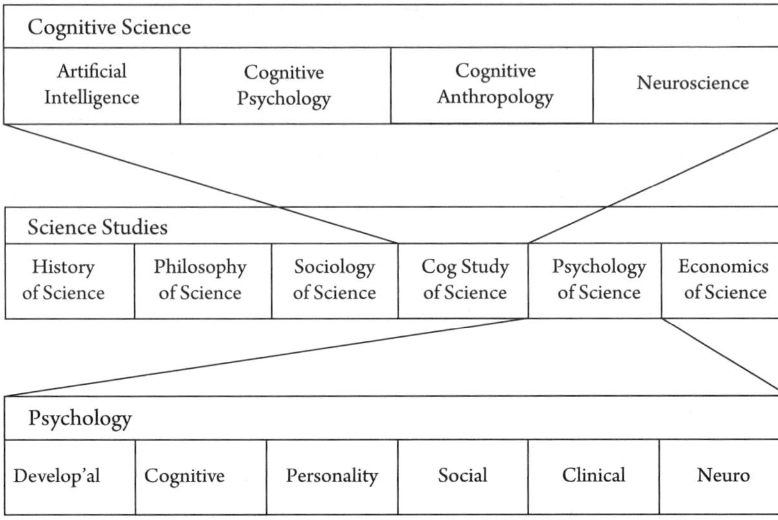

Figure 4.1

reflect a real disciplinary tension among those whose primary identification is with cognitive psychology.

The main difference shown here is that the cognitive study of science looks to cognitive science for inspiration while the psychology of science looks to psychology as a whole (Feist, 2006). The cognitive study of science has strong positive ties with both the history of science and the philosophy of science while being in tension with the sociology of science. The psychology of science, being less well established, has fewer ties with the other science studies disciplines.

Scientific psychology has typically been regarded as the study of fundamental psychological mechanisms that manifest themselves in different contexts. Yet it is clear that the fundamental psychological mechanisms possessed by Europeans did not change appreciably between 1400 and 1900, during which period Western culture went from being theologically based to being to a significant extent scientifically based. In 1400, many of the best minds were focused on theology. In 1900, many were focused on science. So one cannot explain the Scientific Revolution, the Enlightenment, or later developments, in terms of fundamental psychological mechanisms themselves. We have to look beyond the mere existence of cognitive mechanisms to make psychology relevant to any answer to my fundamental question. This means that the psychology of science cannot be a component of psychology in the way that developmental psychology and cognitive psychology are components. Rather, it is more like educational psychology or organizational psychology. Here there are well-established cultural practices, and the psychology of participants is relevant to understanding the practice. Which parts of fundamental psychology are most relevant depends on the practice being studied.

The same holds for the cognitive study of science. It cannot be part of the cognitive sciences in the way that artificial intelligence and neuroscience are parts. It is applied cognitive science. Nersessian's work well illustrates the idea that the cognitive study of science involves studying the operations of fundamental cognitive capacities in the special context of doing science. Thought experiments, for example, are not unique to science, but utilize the same basic cognitive functions as deliberating regarding future actions and anticipating the movements of other bodies. In these studies, however, one does also learn *something* new about the fundamental cognitive capacities; namely, that they *can* be deployed in these new ways, and *how* they function in these new contexts.

DISTRIBUTED COGNITION

Thus far I have discussed one way in which basic human cognitive capacities enter into the practice of science; namely, when basic capacities are deployed in new ways in a scientific context, as when scientists use the basic human resource of reasoning by analogy. There is also a second general way in which basic human cognitive capacities enter into the practice of science; namely, when scientists use these basic capacities to create artifacts that make possible wholly new capacities for learning about the world. There are two sorts of such artifacts, material and symbolic. Both types played major roles in the Scientific Revolution. Microscopes, telescopes, and the air pump were among the most influential material artifacts. Analytic geometry and the calculus are prime examples of symbolic artifacts. The use of such artifacts has recently been incorporated into the cognitive sciences under the umbrella of "Distributed Cognition." As will shortly become evident, I think this incorporation may create difficulties for the still reigning paradigm according to which cognition is nothing but computation.

A standard source for the concept of distributed cognition within the cognitive sciences is Ed Hutchins's (1995) study of navigation. This is an ethnographic study of "pilotage," that is, traditional navigation near land as when coming into port. Hutchins argues that individual humans may be merely components in a complex cognitive system. No one human could physically do all the things that must be done to fulfill the cognitive task, in this case repeatedly determining the relative location of a traditional navy ship as it nears port. For example, there are sailors on each side of the ship who telescopically record angular locations of landmarks relative to the ship's gyrocompass. These readings are then passed on, for example, by the ship's telephone, to the pilothouse where they are combined by the navigator on a specially designed chart to plot the location of the ship. In this system, no one person could possibly perform all these tasks in the required time interval. And only the navigator, and perhaps his assistant, knows the outcome of the task until it is communicated to others in the pilothouse.

One might wish to treat Hutchins's case merely as an example *collective cognition*. The cognitive task—determining the location of the ship—is performed by a collective, an organized group, and, moreover, in the circumstances, could not physically be carried out by a single individual. Hutchins's conception of distributed cognition, however, goes beyond collective cognition. He includes not only persons but also instruments and other artifacts as parts of the cognitive system. Thus, among the components of the cognitive system determining the ship's position are the telescopic devices used to observe the bearings of landmarks and the navigational chart on which bearings are drawn with a protractor-like device. The ship's position is determined by the intersection of two lines drawn on the chart using bearings from the two sightings on opposite sides of the ship. So parts of the cognitive process take place not in anyone's head, but in an instrument or on a chart. The cognitive process is distributed among humans and material artifacts. This incorporation of material artifacts makes the notion of distributed cognition conceptually powerful but also troubling for a paradigm that equates cognition with computation.

Given the notion of a distributed cognitive system, we can see the history of science since the Scientific Revolution as a progression of every more powerful such system for producing scientific knowledge: from Galileo and his telescope, Torricelli and his barometer, to E. O. Lawrence and his cyclotron. In fact, I think we can scale up Hutchins's example to consider archetypical cases of contemporary science such as the Hubble telescope. Hubble has produced genuinely revolutionary observations. For example, in January 2003, the Space Telescope Science Institute released a remarkable image produced by the Advanced Camera for Surveys aboard the Hubble. The process that produced this image involved electronic detectors sensitive to light in the infrared part of the electromagnetic spectrum. The output of the detectors was fed into an onboard computer and put into a form in which it could be transmitted to a Tracking and Data Relay Satellite from which it was retransmitted to the White Sands Complex near Las Cruces, New Mexico, from which it was again retransmitted by domestic satellite to the Data Operations Control Center at the Goddard Space Flight Center in Greenbelt, Maryland. From there it was routed to the Data Capture Facility and finally on to the Space Telescope Science Institute in Baltimore where it was studied by astronomers and other space scientists. Each step in this process in some way modifies the initial input and contributes to the construction of the final image.

A remarkable feature of this particular image is that it involved gravitational lensing. During the exposure, the Hubble telescope was pointed directly at a massive cluster of galaxies estimated to be 2.2 billion light-years away. In accordance with the general theory of relativity, this mass acts like a lens by warping space around it and thus effectively bending light passing by. Scientists who have studied the data claim that the image records light emitted from galaxies roughly 13 billion years ago, when the universe was only one billion years old.

In designing, building, and operating a distributed cognitive system, humans of course employ the basic cognitive capacities studied in cognitive psychology. But there is much more going on in a distributed cognitive system. I would argue that a distributed cognitive system is a *hybrid* system that contains three distinct types of elements: physical, computational (symbolic), and human. In a broad sense, of course, the whole system is a physical system, but computers and humans are importantly different from the systems studied in the physical sciences such as physics and chemistry. That is enough for my distinctions here.

More serious, from my point of view, is the presumption, part of the computational paradigm in both psychology and the cognitive sciences, that the whole system is a *computational* system. In spite of its paradigmatic status, this seems to me just plain mistaken (Giere, 2003, 2006). When we look at a crucial physical component, the interaction between the detectors and incoming light is just a physical interaction. There are no symbols in this system, and thus no rules by which any symbols are manipulated, so, on a strict understanding of computation, there is no computation in this system. Of course, this interaction can be described by quantum theory, but that does not make the interaction itself computational even though the theory may be expressed in the form of a computer program. The output from the detectors, however, does become computational as soon as it is fed into the onboard computers. And the signal, supported by various physical systems, remains computational all the way to the final computer generated images studied by astronomers and astrophysicists.

Here again there is controversy. It is also part of the computational paradigm in cognitive science that human cognition is computational, so the scientists who study the images produced by the telescope are themselves computational systems. This paradigm has been challenged in recent years, for example, by advocates of "dynamic systems theory" (Thelen & Smith, 1994). And even though there is a subject called "computational neuroscience," I think neuroscience supports the dynamic systems approach. Again if we invoke a strict understanding of computation as the manipulation of symbols according to explicit rules, there is no computing going on in a human brain simply because there are no symbols in the brain, only neurons. Describing the brain in computational terms does not automatically make it a computational system. One has to take the additional step of taking the description literally rather than merely metaphorically. Most cognitive scientists seem more than willing to take this step. Even Hutchins, who recognizes that the instruments aboard his navy ships are analog rather than digital devices, nevertheless insists on describing the whole navigational system as computational. I do not think we have to make this move. I realize, of course, that this is currently a controversial stance in both psychology and cognitive science. It may even go against some understandings of a commitment to scientific realism (Giere, 2006).

There exists another way of denying the genuinely hybrid character of distributed cognitive systems, this time from the human side. Some enthusiastic supporters of the idea of distributed cognitive systems advocate treating such systems as having an "extended mind" (Clark, 1997, 2008). So the smart phones of people who keep lots of information in their phones are regarded as literally part of the memory of these people. The mind, which remembers, extends to and includes the smart phone. One can make this idea seem somewhat plausible in the local context of a single person with a phone. It makes little sense when applied to a scientific distributed cognitive system such as the Hubble telescope. Here one would have to say that the mind of the Hubble system extends 2.2 billion light-years out into space to include the cluster of galaxies incorporated into the system as a gravitational lens. And one would have to imbue the system as a whole with the desire to determine the age of the universe, the belief that this age is around 14 billion years, and the epistemic responsibility to justify this belief. These are the kinds of things that creatures with minds are expected to do. But none of this makes much sense applied to the Hubble Telescope System as a whole (Giere, 2004, 2006).

It is important to be clear that these claims in favor of extended minds are not matters of scientific discovery. It is not as if someone could empirically investigate the Hubble system to determine whether or not it has a mind. Rather, this is a suggested revision in how we think about such things. It is a recommendation to extend our commonsense notion of mind to more inclusive systems. It is also a pronouncement on how cognitive science should develop. I am not alone in questioning the wisdom of taking this path (Menary, 2010). Most of the criticism, however, remains at the level of ordinary individuals. Here cognitive studies of science and the psychology of science can contribute by introducing consideration of the larger scale systems that are now common in the sciences.

AN AGENT-CENTERED APPROACH

I think that the concept of a distributed cognitive system is very useful for understanding how humans produce scientific knowledge. Yet, as argued above, this concept rests somewhat uneasily within the cognitive study of science because of the commitment to the paradigm of cognition as computation within cognitive science. I have therefore come to embrace a somewhat broader view for understanding how doing science produces the remarkable knowledge we now have, an agent-centered approach. This approach takes as fundamental what is common to the approaches of Kuhn and his followers and also to both historians and sociologists of science, namely, that the fundamental unit of interest is a human scientist. Theories are conceptual structures created by scientists. Performing experiments is an activity carried out by scientists. What counts as scientific knowledge reflects the judgments of a scientific community. Questions of method and social interaction will, of course, come up in such an approach, but in the context of the activities of individual agents. It is taken for granted that the cognitive makeup of humans must be part of this account, though by no means the whole story.

What is most different about an agent-centered approach is that it privileges the human component of a distributed cognitive system. It treats humans not merely as a locus of more computation, but as genuine agents with such attributes as beliefs, intentions, responsibility, and, yes, consciousness. This means granting humans a degree of autonomy, though without insisting that humans somehow operate outside the causal structure of the world. Importantly, in an agent-centered approach, it is humans, and only humans, who serve as a locus of agency in a distributed cognitive system. It is human agents who design such systems for various purposes and determine how and when they will be used. And it is human agents who determine what has been discovered and the significance of discoveries both within the total body of scientific knowledge and beyond.

A consequence of making humans the locus of agency in distributed cognitive systems is that the humans, and only the humans, bear epistemic responsibility for the claims that emerge from the deployment of such systems. I think it is important for human society that there be some definite place to put responsibility for making claims to knowledge of the world. The best place to put that responsibility, I think, is with individual scientists. Since the practice of science is now largely a collective enterprise, perhaps this can be expanded somewhat to include a scientific community. But it must be clear that the community has no moral standing that does not derive from the individual members. Extending agency to a distributed cognitive system as a whole would excessively dilute responsibility. It would be hard to fix blame for claims that emerged without adequate empirical support. Blaming the whole system has few consequences. No one loses his or her reputation or employment. But the threat of such losses to individuals is conducive to the proper functioning of the scientific enterprise.

There are two challenges to an agent-centered understanding of scientific knowledge production that are grounded in the contemporary practice of science itself. One, as noted above, is that science is now very much a collective enterprise. The existence of scientific papers with a hundred "authors" is no joke. Thus, even if individual scientific agents are taken to be primary, some account must be given regarding the interactions among individuals in scientific groups. There has been some work along these lines within the cognitive study of science. Dunbar (2002) has examined scientific reasoning taking place in weekly lab meetings in molecular biology and immunology labs. Nersessian (Nersessian, Kurz-Milcke, Newstetter, & Davies, 2003) has investigated reasoning and representational practices employed in problem solving in biomedical engineering laboratories. But, by and large, the study of scientific collectives has been left to sociologists of science (Knorr-Cetina, 1999). Here there is an opportunity for a social psychology of science drawing on principles of social psychology rather than cognitive psychology. Thus far, however, a social psychology of science remains more promise than reality (Feist, 2006, Chapter 6).

As is clear from my description of the Hubble telescope, computers are a ubiquitous feature of modern science. Indeed, computers are the most significant scientific artifact created perhaps since the Scientific Revolution. In a half to a quarter of a century, they have massively changed the way science is done. It is now obvious in ways it was not quite so obvious before World War II that to do science is to become part of a distributed cognitive system in which computers play a central role. Yet it seems that neither the cognitive study of science nor the psychology of science has had much to say about the cognitive or psychological aspects of computer use in the sciences. There is much room here for an increased role for psychology in an agent-centered approach to science.

ACKNOWLEDGMENTS

I would like to thank the editors for inviting me to participate in the conference that led to this volume and particularly for comments on an earlier draft of this chapter that resulted in considerable improvements.

REFERENCES

Biagioli, M. (1999). *The science studies reader.* New York: Routledge.
Carey, S. (1985). *Conceptual change in childhood.* Cambridge, MA: MIT Press.

Carruthers, P., Stitch, S., & Siegal, M. (Eds.) (2002). *The cognitive basis of science*. Cambridge: Cambridge University Press.

Chi, M. T. H. (1992). Conceptual change within and across ontological categories: Examples from learning and discovery in science." In R. N. Giere (Ed.), *Cognitive models of science*, Minnesota Studies in the Philosophy of Science (Vol. 15, pp. 129–186). Minneapolis: University of Minnesota Press.

Churchland, P. M. (1989). *A neurocomputational perspective: The nature of mind and the structure of science*. Cambridge, MA: MIT Press.

Clark, A. (1997). *Being there: Putting brain, body, and world together again*. Cambridge, MA: MIT Press.

Clark, A. (2008). *Supersizing the mind: Embodiment, action, and cognitive extension*. Oxford: Oxford University Press.

Darden, L. (1991). *Theory change in science: Strategies from Mendelian genetics*. New York: Oxford University Press.

Dewey, J. (1938). *Logic: The theory of inquiry*. New York: Holt.

Dunbar, K. (2002). Understanding the role of cognition in science: The science as category framework. In P. Carruthers, S. Stitch, & M. Siegal (Eds.), *The cognitive basis of science* (pp. 154–170). Cambridge: Cambridge University Press.

Feist, G. J. (2006). *The psychology of science and the origins of the scientific mind*. New Haven, CT: Yale University Press.

Friedman, M. (1999). *Reconsidering logical positivism*. Cambridge: Cambridge University Press.

Gardner, H. (1985). *The mind's new science*. New York: Basic Books.

Gentner, D., Holyoak, K., & Kokinov, B. (Eds.). (2001). *The analogical mind: Perspectives from cognitive science*. Cambridge, MA: MIT Press.

Gentner, D., & Stevens, A. L. (Eds.) (1983). *Mental models*. Hillsdale, NJ: Lawrence Erlbaum.

Giere, R. N. (1988). *Explaining science: A cognitive approach*. Chicago: University of Chicago Press.

Giere, R. N. (Ed.). (1992). *Cognitive models of science*. Minnesota Studies in the Philosophy of Science (Vol. 15). Minneapolis: University of Minnesota Press.

Giere, R. N. (2003). The role of computation in scientific cognition." *Journal of Experimental & Theoretical Artificial Intelligence, 15*, 195–202.

Giere, R. N. (2004). The problem of agency in scientific distributed cognitive systems." *Journal of Cognition and Culture, 4*(3–4), 759–774.

Giere, R. N. (2006). *Scientific perspectivism*. Chicago: University of Chicago Press.

Giere, R. N. (2008). Cognitive studies of science and technology." In E. J. Hackett et al. (Eds.), *The handbook of science and technology studies* (pp. 259–278). Cambridge, MA: MIT Press.

Giere, R. N., & Richardson, A. (Eds.) (1996). *Origins of logical empiricism*. Minnesota Studies in the Philosophy of Science (Vol. 16). Minneapolis: University of Minnesota Press.

Gorman, M. E. (1992). *Simulating science: Heuristics, mental models and technoscientific thinking*. Bloomington: Indiana University Press.

Gorman, M. E., Tweney, R., Gooding, D., & Kincannon, A. (Eds.). (2005). *Scientific and technological thinking*. Mahwah, NJ: Lawrence Erlbaum.

Gruber, H. E. (1981). *Darwin on man: A psychological study of scientific creativity*. Chicago: University of Chicago Press.

Hutchins, E. (1995). *Cognition in the wild*. Cambridge, MA: MIT Press.

Knorr-Cetina, K. (1999). *Epistemic cultures: How the sciences make knowledge.* Cambridge, MA: Harvard University Press.

Kuhn, T. S. (1962). *The structure of scientific revolutions.* Chicago: University of Chicago Press.

Lakatos, I. (1970). Falsification and the methodology of scientific research programmes. In I. Lakatos & A. Musgrave (Eds.), *Criticism and the growth of knowledge* (pp. 91–195). Cambridge: Cambridge University Press.

Langley, P., Simon, H. A., Bradshaw, G. L., & Zytkow, J. M. (1987). *Scientific discovery: Computational explorations of the creative processes.* Cambridge, MA: MIT Press.

Latour, B., & Woolgar, S. (1986). *Laboratory life* (2nd ed.). Princeton, NJ: Princeton University Press.

Laudan, L. (1977). *Progress and its problems.* Berkeley: University of California Press.

Menary, R. (Ed.) (2010). *The extended mind.* Cambridge, MA: MIT Press.

Nersessian, N. J. (1984). *Faraday to Einstein: Constructing meaning in scientific theories.* Dordrecht, Netherlands: Nijhoff.

Nersessian, N. J. (2008). *Creating scientific concepts.* Cambridge, MA: MIT Press.

Nersessian, N. J., Kurz-Milcke, E., Newstetter, W. C., & Davies, J. (2003). Research laboratories as evolving distributed cognitive systems. In R. Alterman & D. Kirsch (Eds.), *Proceedings of the Cognitive Science Society, 25.* Hillsdale, NJ: Lawrence Erlbaum. The Cognitive Science Society, pp. 857–862.

Piaget, J. (1929). *The child's conception of the world.* London: Routledge & Kegan Paul.

Simon, H. A. (1966). Scientific discovery and the psychology of problem solving. In R. Colodny (Ed.), *Mind and cosmos* (pp. 22–40). Pittsburgh, PA: University of Pittsburgh Press.

Thagard, P. (1988). *Computational philosophy of science.* Cambridge, MA: MIT Press.

Thagard, P. (1991). *Conceptual revolutions.* Princeton, NJ: Princeton University Press.

Thagard, P. (2000). *How scientists explain disease.* Princeton, NJ: Princeton University Press.

Thelen, E., & Smith, L. B. (1994). A dynamic systems approach to the development of cognition and action. Cambridge, MA: MIT Press.

Toulmin, S. (1972). *Human knowledge.* Princeton, NJ: Princeton University Press.

Tweney, R. D. (1985). Faraday's discovery of induction: A cognitive approach. In D. Gooding & F. A. J. L. James (Eds.), *Faraday rediscovered: Essays on the life and work of Michael Faraday, 1791–1867* (pp. 189–210). New York: Stockton.

Tweney, R. D., Doherty, M. E., & Mynatt, C. R. (Eds.) (1981). *On scientific thinking.* New York: Columbia University Press.

Tweney, R. D., Mears, P., & Spitzmuller, C. (2005). Replicating the practices of discovery: Michael Faraday and the interaction of gold and light. In M. E. Gorman, R. Tweney, D. Gooding, & A. Kincannon (Eds.), *Scientific and technological thinking* (pp. 137–158). Mahwah, NJ: Lawrence Erlbaum.

5 The Acting Person in Science Practice

Lisa M. Osbeck and Nancy J. Nersessian

THE ACTING PERSON AS UNIT OF ANALYSIS

In any effort to establish the parameters of a psychology of science, we face preliminary questions that are not empirical in a strict sense, beginning with how to demarcate the core concepts in play (science, psychology). A related question is one we take up in this chapter: What is the appropriate *unit of analysis* for a psychology of science? This question relates to but can be distinguished from that which concerns the appropriate unit of measurement (Meehl, 1978) or even the framework of investigation (Milton, 2010). The unit of analysis concerns the nature of the object under investigation in psychology of science; it is a question about where to turn ones gaze and how widely to extend it, and how best to organize analysis in relation to the level of complexity of the subject matter. In this chapter we propose a unit of analysis we find promising for the psychology of science, namely, the *acting person*. We situate the conceptual advantages of this unit of analysis against what we identify as *the integration problem* in both psychology and science studies. We also examine what is required methodologically. We then demonstrate how we have employed the acting person as a unit of analysis to interpret the psychological practices of biomedical engineering research scientists in two innovation-focused laboratory communities. We focus on two dimensions of activity that have relevance to the integration problem we identify: emotional expression and identity formation through social positioning. We then return to a discussion of the conceptual limitations and advantages offered by adoption of the acting person as an analytic unit for the psychology of science.

The Integration Problem in Psychology and Science Studies

The importance of the question of the unit of analysis, but also difficulties faced in addressing it, is amplified by the state of the two academic contexts in which the psychology of science is situated. Both psychology and science studies have emerged historically as internally divided disciplines. The problematically fragmented state of the discipline of psychology has attracted much theoretical attention and prompted calls for various strategies of unification (Henriques, 2003; Staats, 1983, 1991,1999, 2004; Sternberg & Grigorenko, 2001). Concerns about the "string of raw facts" produced by experimental psychology were expressed early on by William James (1892), the "disturbing absence of that *cumulative* character that is so impressive in disciplines like astronomy, molecular biology, and genetics" by Paul Meehl (1978), and the lack of a coherent paradigm by Stephen Toulmin (1972).[1] Staats goes so far as to label the situation a "crisis" "because the disunification feeds on itself and, left unchanged, will continue to grow" (Staats, 1991, p. 899). Others have been more skeptical about the possibility or desirability of unification in psychology (Koch, 1981; Kukla, 1992; McNally, 1992).

Although internal fragmentation in psychology complicates the issue of the unit of analysis for a psychology of science, our own interest is not in the unification of psychology per se, but rather with an adequate conceptualization of science. Academic study of science polarizes around differing emphases. At one end are accounts reifying an essential inner logic common to all forms of science, a foundation of enduring rational structures to which the norms of particular sciences ultimately appeal (e.g. Carnap, 1935; Hempel, 1952). More recent efforts ground these structures in the sound operations of mechanisms detailed by cognitive science and neuroscience research. At the other end are views of science as a fundamentally social, economic, or political enterprise, structured by institutional and economic forces that infiltrate representational practices and ultimately determine questions asked, explanations offered, and forms of analysis and justification that sanction some activities as "science" (Feyerabend, 1975; Latour & Woolgar, 1986; Rorty, 1979). Thus both Helen Longino (2002) and Nancy Nersessian (2005) point to the science studies community's implicit acceptance of a rational-social divide, such that science tends to be interpreted as *either* a cognitively *or* a socially powered phenomenon, and accounts of science typically are rational/cognitive *or* social/cultural accounts. Of concern is that any such divide between social and rational accounts of science is at odds with the complexities of science revealed by historians, biographers, and ethnographers of science, the intricate interdependency of cognitive, social, affective, material, and agentive dimensions in a "mangle of practice," as Pickering (1995) has called it. We are thus concerned with the *integration problem* in the psychological study of science. By integration problem we refer to an ongoing challenge to adequately characterize the functional interrelationships of social and cognitive domains in everyday moments of actual scientific practice. It is in view of this challenge that we offer an explicit statement of the unit of analysis we find most promising: *the acting person*. We examine below what is entailed in each aspect of this conception and examine historical precedents that inform it, then discuss our application of this unit of analysis in reference to our study of biomedical engineering research laboratories.

[1] Sternberg and Grigorenko (2001) provide brief review of unification efforts.

THE PERSON

Mark Bickhard (2008) and William Smythe (1998) are examples of contemporary psychologists wrestling with the concept of person and its implications. As Smythe points out, "the concept of person has never received much sustained systematic treatment by psychologists, and is becoming increasingly problematic in contemporary psychology, with its steady advance toward depersonalized views of its subject matter" (1998, p. xi). He views the depersonalization as coming from both an emphasis on mechanisms and extreme versions of cultural/constructionist psychology that view the person as a "faceless node in a network of social and cultural relations" (p. xii). But what is it to "personalize" rather than depersonalize our subject? The nature of "person," what is implied in ascribing this label to some objects and not others, is a long-standing philosophical problem of particular difficulty, overlapping with questions about the nature or essence of human beings, which notoriously leads to forms of muddle. Yet that "talk about persons" in comparison to other things—"person-talk"—has unique features is illustrated by Toulmin in what he calls a "classic protest":

> Don't treat me as a type-cast student (black/foreigner/cipher/anonymous statistic): for God's sake treat me as a *person*. (1974, p. 212)

The distinctive nature of person-talk leads to the claim espoused by personalistic theorists that a "person" is a metaphysical primary (Brightman, 1943) involving a differential "field of projection" (Stern, 1906 p. 11), a unified set of capacities and phenomenal point of view. Philosophical "person-talk" typically includes reference to one or more of the following: intentionality, rationality, language-use, rule-following, or individuality/particularity, depending on the context in and purpose for which "person" is invoked. The meaning and function of each of these can be debated endlessly, complicating the project of "personalizing" psychology.

AGENCY AND DISCOVERY

Further complications are introduced by the nature of science practice. Consider the familiar suggestion that "person" implies intentionality, such that consciousness is always "about" or directed toward something, and the related idea of agency, such that "a person" deliberately directs at least some of what she does and says toward goals or ends subjectively experienced as chosen. In all domains of human activity, but most obviously in science, constraints on agency are called into question by an inflexible and recalcitrant material reality: "To hold a natural law to be true is to believe that its presence will manifest itself in an indeterminate range of yet unknown and perhaps yet unthinkable consequences. It is to regard the law as a real feature of nature which, as such, exists beyond our control" (Polanyi, 1946/1964, p. 10). Similarly, scientific discovery is seldom a matter to be willed into occurrence:

> The conditions in which discovery usually occurs and the general way of its happening certainly show it in fact to be a *process of emergence rather than a feat of operative*

action. . . . In this light it may appear perhaps more appropriate to regard discovery in natural sciences as guided not so much by the potentiality of a scientific proposition as by *an aspect of nature seeking realization in our minds.* (Polanyi, 1946/1964, pp. 33–35)

Polanyi is an especially important authority here precisely because he is best known for his effort to ground scientific knowledge in the particular capacities of persons; hence his title *Personal Knowledge* aims at a "fusion" of the personal and the objective. Polanyi's analysis makes clear that "personhood" and "personal" are not equated with a notion of agency limited to explicit, deliberate, predetermined goals, to conscious experience, or a subjective reality independent of the natural world: "Comprehension is neither an arbitrary act nor a passive experience, but a responsible act claiming universal validity. Such knowing is indeed *objective* in the sense of establishing contact with a hidden reality; a contact that is defined as the condition for anticipating an indeterminate range of yet unknown (and perhaps yet inconceivable) true implications" (Polanyi, 1958/1973, p. viii).

The earlier and less well-known "Scientists are Human" by D. L. Watson (1938) similarly affirms "The Scientist" as the principal problem of science, the nature and value, that is, of what is produced as scientific knowledge. Like Polanyi, Watson consults Gestalt psychology to assert that the "unverbalized discerning of intrinsic resemblances" or similarities of form or idea are "the chief kind of knowledge we have." For Watson, what the mind adds to nature "to organize what it sees" includes contributions of the psyche of which the scientist is not aware, *unplanned* operations such that new configurations seem to jump out toward the solution of problems. Watson calls these "unconscious operations" a "rich storehouse of potential insights into the impersonal behavior of the *non-human world*, and therefore an aid for the problems of science" (1938, p. 189, emphasis original). These views of implicit processes are consistent with classical accounts of discovery (Poincaré, 1914; Wallas, 1926) and thus we must say that implicit or tacit or unconscious operations are part of the "person-talk" applicable to the psychology of science, despite the fact that the "automatic" nature of these processes seems to call into question the very agency traditionally regarded as definitive of persons.

ACTING, ACTIONS, ACTIVITIES

Yet spontaneous reorganization and emergence rely on activities that *are* effortful: "The intuitive phase of natural discovery and extra-sensory perception have it in common that they rely on an effort of mental concentration to evoke the knowledge of a real thing never seen before" (Polanyi, 1946/1964, p. 35). Additionally, once evoked, the scientist must recognize "knowledge of a real thing" as having some value, must decide on its importance in relation to an overarching goal or problem of interest, and, ultimately, its relation to truth, even before it is presented to the scientific community for validation:

Viewed from the outside as we described him the scientist may appear as a mere truth-finding machine steered by intuitive sensitivity. But this view takes no account of the curious fact that he is himself the ultimate judge of what he accepts as true.... It is like a game of patience in which the player has discretion to apply the rules to each run as he thinks fit. Or, to vary the simile, the scientist appears *acting* here as detective, policeman, judge, and jury all rolled into one. (Polanyi, 1946/1964, p. 39)

The active nature of all cognition is an important feature of the pragmatist tradition and activity theory, for which reason Jean Lave (1988) establishes "persons-acting" as central to her analysis of cognition in "just plain folks" engaged in everyday reasoning practices.

In relation to science, concentrating in a way that enables discovery and judging the value of what is discovered are activities of persons. Thinking and imagining are forms of activity for Dewey (Dewey, 1910; Dewey & Bentley, 1949), as for James, who notes: "Cognition, in short, is incomplete until discharged in act... the active nature asserts its rights to the end" (James, 1896, p. 85). The meaning of active for both Dewey and James is akin to a stance of engagement in relation to the natural world. This engagement positions the actor as a *participant with* the active natural—and technological—world toward the construction of scientific knowledge. We have elsewhere called this active participation "cognitive partnering" (Osbeck & Nersessian, 2006). This idea of participation is similarly tied to the practice framework in science studies, as articulated clearly in McGuire and Tuchańska's *Science Unfettered*:

Being a subject of scientific activity is to participate in a particular subpractice that belongs to the entire practice of certain historical societies. This participation is a form of being-together-in-the-world: through participating in research an individual takes part in social relations that form the structure of being-together and with other scientists and with the life-world of science. (2001, p. 150)

For Dewey, knowledge involves an active transformation of a situation rather than a passive reception of the world, as is expressed most clearly in his scathing critique of the "spectator" view of knowledge (Dewey, 1930), but also in his concept of experience:

An experience always is what it is because of a transaction taking place between an individual and what, at that time, constitutes his environment, whether... persons with whom he is talking... the subject talked about... the toys with which he is playing, the books he is reading or... *an experiment he is performing*. The environment, in other words, is whatever conditions interact with personal needs, desires, purposes, and capacities to create the experience which is had. (Dewey, 1938, pp. 43–44, emphasis added)

Thus one reason it is important to include "acting" with "person" as a unit of analysis is that it underscores the dynamic, directed participation of the scientist. Such participation, in addition to tacit apprehension of patterns, inevitably includes an affective or motivational component, as Polanyi explains that the scientist

apprehends certain clues as suspect; formulates the charge and examines the evidence both for and against it, admitting or rejecting such parts of it as he thinks fit, and finally pronounces judgment. While all the time, far from being neutral at heart, he is himself passionately interested in the outcome of the procedure. He must be, for otherwise he will never discover a problem at all and certainly not advance toward its solution. (p. 39, emphasis added)

Polanyi's appeal to the emotional involvement, the passionate commitment of the scientist in problem solving is important to the conception of "acting" as "participating." Examining theorists for whom activity is central to the conception of knowledge reveals an additional feature of "act" that is important to the specification of "acting person" as a unit of analysis: the coordination of operations evident in the direction of efforts toward problem solving. Of course, affective processes are engaged in problem solving, not as tacked on to reasoning but as integral to it. This coordination of processes in real practice is essential to pragmatism and activity theory. Even in Dewey's critique of the reflex arc concept we see the act portrayed as a coordinated event, in his charge that psychology treats the reflex arc not as "a comprehensive, or organic unity, but a patchwork of disjointed parts, a mechanical conjunction of unallied processes" (p. 358). The "defect" in psychology's theoretical treatment of the reflex is the assumption that "sensory stimulus and motor response [can be seen] as distinct psychical existences, while in reality they are always inside a coördination and have their significance purely from the part played in maintaining or reconstituting the coördination" (Dewey, 1896, p. 360). Subtext to the emphasis on coordination is an assumption that through a turn to action/activity we are able to overcome dichotomized conceptions of subject and object, mind and world, body and mind. This is summarized effectively in an excellent overview of activity theory in Russian psychology:

To be able to analyze such complex interactions and relationships . . . there is a demand for a new unit of analysis. Activity theory has a strong candidate for such a unit of analysis in the concept of object-oriented, collective, and culturally mediated human activity, or activity system. Minimum elements of this system include the object, subject, mediating artifacts (signs and tools), rules community, and division of labor." (Engeström & Miettinen, 1999, p. 9)

Acting Implicates Sociality

The fuzzy conceptual relation between "acts," "actions," "and "activity/activities" can cause some confusion, despite the implication of coordination expressed by each of these. A rough classification of act and actions as solo, goal-directed performances and activities as collective stems from the roots of "act" and action in Aristotle (Anscombe, 1957) and "activity" in Marx (Engeström & Miettinen, 1999). Yet counterexamples complicate this classificatory scheme, and it is important to recognize the social dimension inherent in the concept of acting.[2] Tolman (1998) traces

[2] Counterexamples include the "whole social act" Mead discusses in the *Philosophy of the Act*, Dewey's analysis of acts in relation to technology, and the more recent incorporation of speech acts into activity theory, drawing especially from Wittgenstein's *Philosophical Investigations* (1953); See Engeström and Miettinen, (1999).

in Aristotle and Hegel the idea that "acting," even though attributed to a person, has an irreducibly social connotation by its connection to the moral order. More recently, actions are viewed "behaviors that are intended performances" and acts are the social meaning of the performances (Louis, 2008, p. 25). In other contexts, "activity" is owned as belonging to a single person engaged in a goal-directed "doing"; for example, Bridgman's claim that "science is activity. Science does not begin until my activities begin" (Bridgman, 1950, p. 50).

The claim that the acting person is the unit of analysis appropriate for science studies includes the qualification that acting always takes place within normative contexts established by the scientific community and its historical claim to epistemic authority. Norms also include those relevant to a particular field of specialization, and the local community of each laboratory, wherein peculiar patterns of language-use and social interaction are established.

Relation to Practice Theory

The centrality of normative structures is important to the concept of practice, which has important similarities to the *act* and *activity* in psychology, and is indeed overlaid on the concept of practical activity (e.g. Garfinkel & Sacks, 1970). Although the term "practice" has been criticized as vague and insufficiently explanatory (Turner, 1994), Rouse emphasizes its philosophical status as normatively constrained performance (Heidegger, 1962; Wittgenstein, 1953), distinguishing it in this way from the looser notions of social regularity, commonality, or custom: "Not all practitioners perform the same actions or have the same presuppositions, but practitioners and other constituents of a practice are accountable for performances or presuppositions that are inappropriate or otherwise incorrect" (p. 168).

Like "act," "practice" implies a coordination of various dimensions of functioning, and also coordination of dimensions or realms (personal, social, material):

> In the case of scientific inquiry, of course, the relevant practices include ways of encountering, responding to, and being resisted by what scientists are dealing with. The intertwining of language, social practice, and reality cannot be neatly bounded at the points where they run up against the natural world, for encounters with the world, and indeed the very boundaries between self and world, belong to interpretive human practices. (Rouse, 2002, p. 85)

Emphasis on the act*ing* person encompasses both the intentional quality of action and the social meaning or force of acts accomplished through the actions, for the intentional performances of persons (actions) always take place within socially negotiated or inherited contexts of social meaning, including those established by the community of scientists. Similarly, practices are instantiated through the activities of intentional persons. The epistemic standards upheld through the practices of scientific communities are internalized in each scientist according to Polanyi, and provide a standard against which the scientist evaluates the merits of a particular discovery. Thus when "actively engaging in the pursuit of discovery and passionately absorbed in solving a problem, he must strive against self-deception and for a true feeling of reality.... Before claiming discovery he must listen to his scientific conscience" (p. 55).

ADDITIONAL FORMS OF NORMATIVITY

The internalization of scientific norms described by Polanyi provides one way of understanding the historical tie of rule following to the distinctive properties of human rationality, as in Descartes's "Rules for the Direction of the Mind" (1628/1985; see also Toulmin, 1967). But the origin of these demands remains in question, whether as innate requirements, for example, for order and coherence, or as values adopted through participation in a community in which they are held in common. Our view is that innate demands and constraints should be considered a kind of cognitive normativity. As Polanyi says of the scientist: "His brain labors to satisfy its own demands according to criteria applied by its own judgment" (1946/1964, p. 38). The specific nature of these demands is not a question to be settled through a priori accounts but by empirical analysis of practicing scientists, historical and contemporary. For example, Nersessian's (2008, 2009) analysis of model-based reasoning practices provides room to suggest that these practices reflect cognitive demands for organized representations of the scientist's domain of problem solving, the affordances of which include both novel conceptions of the material domain and more accurate predictions from observations. Yet that the demand has the label of "cognitive" does not preclude the involvement of the social domain, or an intricate interaction of social and cognitive forms of demand, as has been argued effectively by distributed cognition theorists and cognitive anthropologists (Hutchins, 1995; Lave, 1988).

Finally, any effort to understand the scientist must also face up to forms of demand and constraint imposed by the objects and technology central to that scientist's problem solving. We suggest that the notion of "material normativity" is a more useful concept for psychology of science than is "natural law" to describe these forms of demand, because "material normativity" encompasses the behavior of manufactured or engineered artifacts and devices as well as natural objects. Similarly, the habit of drawing hard lines around "norms" and "laws" artificially divides the social and material domains of scientific inquiry, when scientific accountability is both natural and social in origin. As Rouse explains:

> Kuhn's concern was not whether there "really" are objects, natural kinds, or causal powers independent of scientific practices. Rather, his point was that what scientific practices are normatively accountable to are objects and phenomena manifest through scientists' engaged, meaningfully situated practices of experimentation and theorizing. Muon-detecting is in this respect no less "intersubjectively meaningful" than voting or bargaining, and grasping the significance of an experimental result need be no less affective than recognizing one's own humiliation. (Rouse, 2002, p. 84)

We have also found in our study of biomedical engineering laboratories that anthropomorphizing talk, in particular the projection of emotion onto cells, bears important relations to complex problem-solving processes, and imposes demands upon researchers of varying levels of expertise. This is a kind of "person-talk" extended to the entities and artifacts of scientific practice, and analysis of its epistemic functions, as we have tried to provide, underscores the complicated nature of demarcating the social domain from the material domain from the cognitive domains of practice.

These fuzzy boundary lines in scientific practice—around cognitive, affective, social, and material dimensions—reflect the level of complexity with which we are dealing when imagining a psychology of science. Our intent in forwarding the "acting person" as the unit of analysis for a psychology of science stems from concern with doing justice to the complexity. As we have tried to explain, the "acting person" should not be conflated with "isolated individual." We view "acting person" as an inherently integrated category and reflective of the participatory nature of the scientist with other persons and with the world in producing scientific knowledge.

IMPLICATIONS FOR METHOD

Despite our interest in promoting an integrated unit of analysis for the psychology of science, we share with Howard Kendler the worry that too much openness on the part of psychologists, instead of doing justice to the subject matter, serves only to obscure it with an "amorphous conception" (2002, p. 1125), for which reason he is critical of unification strategies based on the claim that psychologists "need adhere to no particular set of methods, to no particular field, and to no particular paradigm" (Sternberg & Grigorenko, 2001, p. 1078). So here we detail what is required by establishing the acting person as unit of analysis.

It is first important to note that establishing the acting person as unit of analysis is consistent with several influential trends in cognitive science and learning science. As summarized by Anderson (2003), there has been a "rethinking" of the nature of cognition taking place within cognitive science, one that

> focuses attention on the fact that most real-world thinking occurs in very particular (and often very complex) environments, is employed for very practical ends, and exploits the possibility of interaction with and manipulation of external props. It thereby foregrounds the fact that cognition is a highly *embodied* or *situated activity*... and suggests that thinking beings ought therefore be considered first and foremost as acting beings. (Anderson, 2003, p. 91)

The situated approach to cognition construes intelligent behavior as arising among interactions within particular settings. The position holds that "problem solving is carried out *in conjunction with the environment*" (Brown, Collins, & Duguid, 1989, p. 36, emphasis added). It relates closely to frameworks assuming cognition to be *distributed*, meaning that brain and environment are construed as co-constituting a single complex system, in as much as the forms of problem solving that occur would not be possible in isolation from that environment, including the social environment. In so framing our study of the laboratories as situated and distributed we connect it to other investigations of real-world problem solving that implicate the environment in cognition in important ways (e.g. Greeno, 1998; Hutchins, 1995; Resnick, Pontecorvo, & Säljö, 1997).

Consistent with new developments in cognitive science, then, a first methodological implication of the acting person framework is that it requires (1) analysis of real-world thinking, in particular, complex environments in which science is practiced. The "acting person" focus shifts the psychologist from studying reasoning under artificial constraints to investigating persons engaged in actual scientific problem solving in settings required for their work (e.g., laboratories). The second implication is that (2) interpretive activities on the part of students of science is required to clarify the nature of the interactions between social, personal, cognitive, and material domains in natural habitats of practice. The interpretation should of course undergo various social checkpoints to insure plausibility and rigor. A third implication is that (3) "the acting person" requires some attention to the particular style, affective expression, learning history, and motivational flux of scientists, aspects that have traditionally been left out of academic accounts of science and that are not fully appropriated by either cognitive or social explanations. Indeed, we believe that even the "new" cognitive science could do more to integrate personal aspects of scientists into the accounts of science offered—for example, how emotion influences styles of interaction and ways of working in the laboratory as a whole— and to hold as an open question, at least, how the idiosyncrasies and unique experiences of scientists may play important roles in the discovery process and possibly beyond it.

We now turn to a description of our own efforts to address these implications through an in vivo investigation of laboratory practices in two biomedical engineering labs. Before providing the details of our study, we first compare it briefly with the in vivo studies of Kevin Dunbar (1995; Dunbar & Blanchette, 2001), which employ somewhat similar methods. Dunbar, who conducted in vivo investigations of molecular biology and immunology laboratories, includes interviews and analysis of "online creativity" in research meetings. Thus we share Dunbar's conviction that a valid means by which psychologists might study and contribute theoretically to science studies is by entering laboratories and harvesting ecologically grounded data. The qualitative analysis of scientists' practices, including their accounts of what they are doing and why, is vital to a comprehensive psychology of science. Our project differs from that of Dunbar, however, in that our analysis is not limited to cognitive practices as traditionally conceived. We are expanding the range of topics covered to include categories traditionally excluded from analysis of cognition, namely agency, identity, motivation, and sociality, including "relationships" with the objects and artifacts of practice (the material composition of the laboratory). Because the focus of our work was on the day-to-day activities of researchers, the research meetings assumed less importance than they had for Dunbar's analysis. We needed more to elicit from researchers their understanding and perceived relations to others and to artifacts, and to see how these functioned within the life of the labs—aspects that are better addressed through interviewing and extensive field observations. Thus, we differ from Dunbar in theoretical framework, in that we treat cognition as a phenomenon distributed across a cultural space, in this case a research laboratory. We approach data collection and interpretation with the assumption that cognition, and hence science practice, comprises a complex system of persons, devices, artifacts, instruments, texts, and traditions. The analytic framework for the study also draws from previous work of Nersessian's, including the results of prior research centered on interpreting model-based reasoning practices of scientists and situated reasoning practices in science education (e.g., Nersessian, 1995, 2002, 2005).

THE RESEARCH CONTEXT: BIOMEDICAL ENGINEERING LABORATORIES A AND D

Nancy Nersessian and Wendy Newstetter directed a five-year ethnographic investigation of cognitive and learning practices in two research laboratories in the interdisciplinary field of biomedical engineering. These are laboratories populated by "engineering scientists": a breed of researcher aiming to make fundamental contributions to basic science as well as to create novel artifacts and technologies. The labs are cutting-edge research environments located at a Research I institution.

Lab A is a tissue engineering laboratory, with the overarching research goal to understand mechanical dimensions of cell biology such as gene expression in endothelial cells, and to engineer living substitute blood vessels for implantation in the human cardiovascular system. Examples of intermediate problems that contributed to the daily work during our investigation included designing and building living tissue—"constructs"—that mimic properties of natural vessels; creating endothelial cells (highly immune-sensitive) from adult stem cells and progenitor cells; designing and building environments for mechanically conditioning constructs; and designing means for testing their mechanical strength. During our study, the main participants included a director, one laboratory manager, one postdoctoral researcher, seven PhD graduate students, two MS graduate students, and four long-term undergraduates (two semesters or more). Additional undergraduates from around the country participated in summer internships and international graduate students and postdocs visited for short periods. All of the researchers came from engineering backgrounds, mainly mechanical or chemical engineering, and some were currently students in a biomedical engineering program. The tissue engineering laboratory, as an institution, had been in existence nearly 20 years when we entered. The laboratory director was an established, renowned pioneer in the field of biomedical engineering.

Lab D is a neuroengineering laboratory, with the overarching research problem of understanding the mechanisms through which neurons form connections in the brain and, potentially, using this knowledge to develop aids for neurological deficits and "to make people smarter" (PI). Examples of intermediate problems that contributed to the daily work included developing ways to culture, stimulate, control, record, and image neuron arrays; designing and constructing feedback environments (robotic and simulated) in which the "dish" of cultured neurons could "learn"; and using electrophysiology and optical imaging to study "plasticity."

During our study the main participants of Lab D included a director, a laboratory manager, a postdoctoral researcher, four PhD graduate students in residence (one left after two years; three graduated after we concluded formal collection), a PhD student at another institution who periodically visited and was available via video link, one MS student, six long-term undergraduates, and one volunteer for nearly two years, not pursing a degree (with a BS), who helped out with breeding mice. When we began, the laboratory director was a new tenure-track assistant professor, fresh from a postdoc during which he had developed techniques and technologies for studying cultures of neurons. He already had achieved some recognition as a pioneer. The backgrounds of the researchers in Lab D were more diverse than Lab A and included mechanical engineering, electrical engineering, physics, life sciences, chemistry, and microbiology, and some were currently students in a biomedical engineering program. As an institution, the neural engineering laboratory was in existence for a few months and still very much in the process of forming when we entered.

DATA COLLECTION

Several members of our interdisciplinary research group became participant observers of the day-to-day practices in each lab, spending what we estimate as over 800 hours in the labs. Researchers took *field notes* on observations, audio taped interviews, and video and audio taped research meetings. *Interviews* were conducted by members of the research team involved in collecting data with researchers in both laboratories. For some participants we had interviews occurring at regularly scheduled intervals, usually two weeks, and in other cases the schedule was more sporadic. Interviews were of different types and had different foci. All interviews were unstructured, but most had specific aims. Some were aimed at obtaining the lab members' accounts of their learning and social experiences since entering the laboratory; some were focused on personal history prior to entering the laboratory; some had the aim of understanding the development and current use of technology and equipment specific to the laboratory; some (most) were focused on the current research problems researchers are working on and how these relate to the larger laboratory goals. Many interviews took place within the laboratories, invoking the space with reference to the objects and representations (e.g., recordings of network activity) as projected on a computer screen) relevant to the problem or issue under discussion.

DATA ANALYSIS

We do not assume that individual interviews provide us with a telescope into the inner world of participants. Neither do we assume that what researchers say they do when describing their practices necessarily corresponds to what they actually do—the "say-do problem" (Jordan & Henderson, 1996). However, the narrative provided by an interview (in some cases a "situated interview" in that it takes place in the environment—the laboratory—in which the cognitive activities of interest occur) is among the best available data available for analyzing these personal contributions of the scientist to the research process, both in discovery and in justifying and "marketing" discoveries. In his effort toward a "better understanding of the nature and limit of our intellectual tools" in *The Way Things Are*, Percy Bridgman adopts a first person singular to give "immediacy" to his account of science and to be a logical implication of his insight that analysis of science should entail analysis of activities, of "doings and happenings" (1959, p. 3). Similarly, our interviews represent first person accounts of doings and happenings in cutting-edge biomedical engineering research.

Our interest is in both the (1) ways in which personal dimensions of activity impact problem solving in the laboratory as a whole, for example, how emotional expression and identity negotiations intertwine with cognitive practices, but also (2) in the specific forms of integration evident in the activities of a single researcher over time. At this point in the analysis the task is to descriptively illustrate the coordination rather than to demonstrate or explain it causally. To date we have coded approximately 16% of the interviews.

(1) As an overarching strategy for understanding general effects of "personal dimensions," our process was informed by principles of grounded theory (Corbin & Strauss, 2008; Strauss & Corbin, 1998) in that we endeavored to use constant comparison and analytic induction to develop coding schemes and conceptual categories to characterize accounts of practice recorded in the interviews. Selected interviews were analyzed progressively, line by line, with the aim of providing an initial description for most if not all passages in an interview, in an effort to "take the data apart" and begin to characterize it in terms of conceptual or descriptive dimensions (open coding). For each coded passage, the aim was to formulate a succinct and plausible description given its particular context within the interview and to some extent the larger context of the laboratory's workings and recent history. These descriptions were arrived at through detailed discussion about the possible significance and any alternative interpretations of the passage. A code was only assigned when both researchers were in full agreement about its fit and relevance to the given passage. Theoretical memos were also kept. Codes and memos were then compared across interviews, and were revisited in keeping with new thinking on and discussion of text, as well as new observations from the fieldwork. Many codes were refined, involving further distinctions; some were added, and some eliminated as thinking evolved. Codes were then analyzed for conceptual similarity, overlap, and distinction. Those that did not fit easily into one of the main headings were analyzed further for possible overlooked meanings or fit with other categories. This process was repeated until no further reductions could comfortably be made. A written description of main code categories was also developed and revised, with examples of text passages assigned to each category. Main categories, descriptions, and examples were brought to the main research team for feedback, and were revisited and in some cases revised after the feedback was received.

(2) In addition to sampling across texts, another analytic strategy used was to focus coding and analysis on interviews with one particular lab member over time. The interviews, conducted at regular two week intervals, enabled a form of case study analysis of a few researchers in particular whereby the trajectory is analyzed

chronologically from a point very soon after the researcher first enters the laboratory. For case study analysis a coding system similar to grounded theory was also used in that the codes emerged from careful line by line reading and interpretation of interview text, and some of the content was assigned to categories guided by questions that specifically concerned learning.

For both forms of analysis, rigor was sought in three phases of accountability: (1) *Collaborative Coding:* Coding was done by at least two members of the interdisciplinary team to ensure that any codes assigned seemed reasonable and important to at least two persons familiar with the laboratories and the aims of the overall project. Some of these coding sessions were tape-recorded so the process could be evaluated during external audit;(2) *Group Feedback*: Updates on coding were presented at regular research meetings of the interdisciplinary research team. Codes were retained, adjusted, or abandoned in line with feedback from these sometimes difficult group meetings; (3) *External Audit*: An outside assessment specialist familiar with qualitative methods but not involved with the project itself evaluated the coding process and the codes assigned. We received a very favorable report on both the process followed and the codes produced through our collaborative efforts.

THE ACTING PERSON IN BIOMEDICAL ENGINEERING LABORATORIES

Overview

In determining our unit of analysis to be *the acting person in normative contexts of practice*, we intend to emphasize the coordination rather than isolation of processes: cognitive, social, cultural, emotional, agentive, and material. Nevertheless, as Anscombe's (1957) philosophical analysis clarifies, people act under different descriptions of the action in question. Anscombe was referring to descriptions people give of their own actions, yet we benefit from the insight that for the sake of analysis it is frequently helpful to describe actions with different points of emphasis. We take for granted that the different aspects analyzed coordinate and, indeed, are mobilized in action itself.

Some of the higher order categories that emerged from our grounded coding appeared to us to represent different categories of human action that, intriguingly, bear a relation to some of the traditional categories of a general psychology text. Thus our book (Osbeck et al., 2011) analyzes the activity of scientists in the biomedical engineering laboratories under the following descriptive categories as chapter titles. We concentrate in the text on two broad categories of acting: sense-making and identity negotiating. We use the term "sense-making" as shorthand for the continual efforts of any person to sort, understand, plan, and evaluate experiences of every kind, thereby giving them meaning. "Sense-making" is a recognized category in organizational studies, wherein Weick (1995) highlights the irreducibly social *and* cognitive aspects of sense-making. Weick also emphasizes that no sense-making occurs without a sense-maker, thereby tying sense-making in organizations to embodied and intentional actors, and explicitly linking sense-making with identity formation. Under sense-making we examine the *problem-solving* person and the *feeling* person in laboratory practice. Under identity we examine the *positioning* person and the person negotiating culture by means of *enacting race and gender relations* in the laboratory. In the context of discussing positioning we also consider attributions of agency to and cognitive partnering with objects and artifacts that contribute meaningfully to the problem-solving act at hand. We then examine how all of these aspects of sense-making and identity formation come together in relation to a single acting person by examining the *learning* person through a case study of one researcher's experience over time, as analyzed in the context of interviews conducted at regular two-week intervals after she first joined Lab A. Across chapters there is overlap and cross-reference as is consistent with our theme of integration.

Given current space limitations we provide only brief examples of our analysis of the acting person in the laboratories investigated, one under the general category of emotion and one under identity.

The Feeling Person

When we began to analyze interview data, we did not approach coding with the explicit goal of identifying "acting person" expressions within the interviews conducted, but rather to "describe cognitive practices" such as modeling. However, as we were seeking to describe and code cognitive practices as evidenced in the interviews, we found striking examples of interview text that did not fit traditional cognitive categories, but seemed rather to have an affective or emotional coloring. Other passages seemed expressive of desires, goals, and aspirations. Although not clearly "cognitive," passages of these kinds seemed intimately related to and interwoven with accounts of problem formulation and solving, leading us eventually to code "affect/motivation" as a higher order category of sense-making activity.

Because we have conducted very fine-grained coding on only a small subset of our large collection of interviews, our emphasis is not on counting the presence of emotion words or expressions but on characterizing the variety of functions such expressions play in the context of the interviews analyzed and illustrating with examples from different researchers and across laboratories. Our analysis of emotion is still in comparatively early stages, but we have categorized different forms of what we consider expressions of emotion or motivation: I. overt expressions of emotion (positive and negative); II. figurative or metaphorical expressions; III. anthropomorphizing expressions.

The class of emotional expression we found especially interesting in interview transcripts is the attribution of emotional states to the objects and artifacts central to the sense-making practices in the laboratories. The most striking and theoretically important example of this practice is the attribution of happiness to cells in Lab D:

D4 Cell density is important, because for one cells survive more if they, if they're connected to each other. A lone cell by itself is not very happy. (2003-03-20-D-D4)

D6 *So we also have microscopes in here that as you can see, they're always closed up in these boxes that we can keep warm in there, and we can modulate the atmosphere so the cells are staying happy* . (2007-02-14-D-D6)

D24 *This is problem solving on a whole new level because it's like how do we build a device that you put the microscope in that's gonna keep the humidity and the temperature in and we can deliver this to it and keep our cells happy.* (2003-09-16-D-D24)

D24 *There's neurons in a dish and just count them and keep them happy and hope they don't get sick and die* (2003-09-16-D-i-D24)

D28 *Imaging with cells is also non-trivial because there is, you know they're not happy being zapped with laser beams.* (2005-03-15-D-D28)

Happiness is occasionally extended to the dish as a collective:

D4 So yeah there're, there are a couple of noise sources in there.... So yeah, that's the deal, it's pretty good, works very well; **today the dish is happy.** (2003-09-15-D-D4)

And to the cell culture:

D15 Because the glia in the culture... maybe the neurons too... they're producing factors that **keep the culture happy.** (2004-10-28-D-D15)

Researchers across levels of expertise exhibit this practice of attributing happiness to cells, from the PI to D24, an undergraduate who admits little previous experience with or knowledge of MEAs. Therefore the attribution of happiness seems to indicate a local cultural practice (local to Lab D, the field of neuroengineering, or possibly biomedical engineering).

Particularly intriguing is a segment of an interview with D4 that reveals a *normative* component to the concern with happy cells (neurons). This passage reflects *an expectation* for researchers to keep the cells happy, and to *care* about keeping them happy.

I So tell me about these neurons, because you guys are always talking about neurons in the lab.

D4 *What about neurons?*

I Well, yesterday for example you killed a whole batch of neurons, you know when you were trying to get the MEA to work? And uh, remember Grad A came in and you guys, and um, what was the girl's name? She was so upset when the neurons were killed.

D4 *Right, Grad B.*

I Yeah, and you guys talk about the neurons with the undergrads all the time because they're culturing the neurons, so, why do you guys care so much about neurons?

D4 *If you don't essentially care about the nodes in the network,...* ***you gotta care about them!***

I But there's so many of them!

D4 *Right there's so many of them, but that doesn't mean that you want to. . . . So they make up the network,* ***each of them has a part to play****, in the network property, so you want to keep as many as you can. You know, because they make up essentially, as I said, they make some basic rules for the way the network works,* ***so you want to keep them happy.*** (2003-04-09-D-D4)

The implication here is that it is not only the happiness of the cells as a collective that is important but even the happiness of individual cells, given that each of them contributes in a particular way to the functioning of the whole. In a later interview with another interviewer, D4 provides more clues as to what is meant by happy cells:

D4 *So what we see is actually summation of the activity of all the cells around it and-uh so, and this flow of activity is basically because of the connectivity of the network. . . . But then there are some cells which are always active like this guy here, and that guy there***. They are just happy firing all over on their own.**

I Which one, on the grid 70?

D4 *Yeah, on the grid 70. This* ***seems to be happy all the time,*** *you know,* ***it's happy, it's happy.*** *I can't really say. May be they're two cells connected to each other, they go bing-bing-bing all the time, I don't know. So there are some cells which are very active, most of the time. So, this is a* ***pretty active network*** *because I get activity almost like on 80% of recording, recordable channels, which is, which is very good***.** (2003-09-15-D-D4)

D4's comments in the context of this interview help to clarify that "happy" cells are not simply alive but that they are actively forming connections, which is essential for the research.

This passage in particular invites analysis in terms of the social dimension of emotion, not merely in terms of the contribution of cultural and linguistic practices to emotional expression, but more strongly to the idea that emotions themselves constitute interactions or *transactions*. Transactional models of emotion emphasize the dynamic interrelations of person and environment or person and situation in the production of an emotional response (Fridlund & Russell, 2006; Lazarus & Folkman, 1987; Parkinson, Fischer, & Manstead, 2005). As Lazarus (2006) puts it, "Each emotion has a different story to tell about an ongoing relationship with the environment" (2006, p. 34). Philosophers Griffiths and Scarantino (2008) draw from the psychological literature on emotion as transactional in offering what they call a situated theory of emotion, wherein "behaviors which have traditionally been viewed as involuntary expressions of the organism's psychological state are instead viewed as signals designed to influence the behavior of other organisms, or as strategic 'moves' in an ongoing transaction between organisms" (pp. 439–440).

Contemporary transactional theories of emotion find precedent in Dewey's treatment of emotion in the later period of his work. Within this context, "the situation" is always the focus of analysis; the dynamically integrated nature of the situation is made clear by the participation of the actor's body (including dispositional tendencies) and personal/cultural history with the environment (including the social environment—other people) in constituting and transforming each situation: "Emotion in its ordinary sense is something called out by objects, physical and personal; it is response to an objective situation.... Emotion is an indication of *intimate participation*, in a more or less excited way in some scene of nature or life" (Dewey, 1925/2008, p. 390).

How the transactional perspective on emotion might inform the practice of *attributing* emotion is the task we face in trying to understand the practice of describing neurons as happy exhibited in Lab D. One aspect emphasized in some transactional theories of emotion is that an emotional expression constitutes *a demand* for a response. This is one way in which it can be understood as shaping or affecting the future of the relationship, by which it might be understood as a transaction. In attributing the ability to feel happiness to cells, the researcher attaches *a demand* to those cells, a demand to tend them and care for them in order to keep them alive and active. This implicates the normative dimension noted in relation to the interview with D4: There is a laboratory mandate to care about and for the cells precisely because they will die if not cared for. Moreover, as D4 notes, happy neurons are active in forming connections with other neurons. The study of these activity patterns is the business of Lab D. In Lab A, we found similar attributions of happiness to cells, but the meaning of "happy" differed in relation to Lab A endothelial cells. In addition to helping cells to proliferate as in Lab D, in Lab A keeping them happy can also be associated with limiting their number, though this control is conducted in the interests of keeping a subset of cells alive and healthy. For both labs, "happy cells" are vitally important to the problem-solving goals of individual researchers, to the work of the laboratory as a whole, and thus to the construction and dissemination of knowledge within the wider field. Of interest here is a statement made by Freud in a lecture to the Vienna Psychoanalytic Society, referenced by Mischel in an essay for a collection he edited entitled "Understanding Other Persons": "Our understanding reaches as far as our anthropomorphism" (Freud, 1907, quoted in Mischel, 1974).

If we assume that the attribution of emotional states to cells bears some relation to the emotional capacity and thus the embodiment of the researcher, we can say that emotion is implicated in day-to-day cognitive practices within these laboratories, and further that there is a social, specifically a locally normative aspect to this pattern. The study of emotion, then, is one way of organizing analysis around the acting person of science.

The Positioning Person

Similarly, the topic of identity overlaps loosely with philosophical and psychological categories such as self, subject, ego, consciousness, individuality, and agency. The central relation of identity to the very conception of "person" is also emphasized in contemporary discussions of self and subjectivity (Atkins, 2005). Moreover, the topic is critical to our interest in avoiding both social and cognitive reductionism in analyzing the practices of the biomedical engineering laboratory communities.

According to Wenger, "The concept of identity serves as a pivot between the social and the individual, so that each can be talked about in terms of the other. It avoids a simplistic individual-social dichotomy without doing away with the distinction.... It is the social, the cultural, the historical with a human face" (1998, p. 145). However, even within the general framework that acknowledges the close relation of identity and community or identity and practice, multiple ways of articulating their reciprocity have emerged in recent scholarship. Social identity theory, for example, recognizes that an aspect of self-definition is rooted in group membership, and that emotions and valuations relating to the self accompany self-categorization (Tajfel, 1982; Tajfel & Turner, 1979). In contemporary social theory, identity frequently emerges as a form of action, accomplishment, display, or performance situated within networks of meanings and practices; people *identify* as or with various options. One prominent approach with several offshoots and varieties is positioning theory.

Positioning is "the discursive process whereby people are *located in conversations* as observably and subjectively coherent participants in jointly produced storylines" (Davies & Harré, 1999, p. 37). Positioning analysis seeks to describe changes in relations (between persons or persons and groups) occurring through discursive strategies. It concerns social effects of speech, a form of activity (Austin, 1962. Our speech affects others and we are affected by what they say. Positioning represents the action though which these effects are accomplished and a position can be understood as an effect. Although the main applications of positioning theory fall within the framework of "social psychology," that positioning theory concerns "cognitive" or "epistemic" dimensions is also emphasized: "Positioning theory is a contribution to the cognitive psychology of social action. It is concerned with revealing the explicit and implicit patterns of reasoning that are realized in the ways that people act toward others" (Harré, Moghaddam, Cairnie, Rothbart, & Sabat, 2009, p. 5).

Positioning theory builds on the dramaturgical theory of Goffman (1981) and the analysis of rule-following behavior by persons in social groups (Harré, 1974). However, the concept of position is more fluid than that of the social *role*. Role implies a static location in a social order ("a scientist"); positions, in contrast, are always dynamically shifting and renegotiated, both tacitly and explicitly. Also, different positions can be identified for a single speaker in the context of a single record of interaction, such as an interview passage.

Two features of positions make positioning theory a powerful analytic tool for understanding interactions in science practice. First, positions serve to establish the possibilities of action (i.e., analysis focuses on what actions are "socially possible for any social actor at any moment in the flux of social life") (Harré & Moghaddam, 2003, pp. 4–5). A second, related point is that a position "can be looked at as a loose set of *rights and duties* that limit the possibilities of action" (p. 5). It "may also include prohibitions or denials of access to some of the local repertoire of meaningful acts" (p. 6). Hence different positions are understood to *make possible* different forms of practice and *require* some forms of practice: Positions serve to establish the possibilities for action, broadly defined. Positioning is thus closely tied to identity formation *and* emotion in that shifts in position effect changes in the sets of practices in which one can and should participate and feels competent or obligated to do so.

In Osbeck and Nersesssian (2010), we illustrate the usefulness of positioning theory for understanding practices in our biomedical engineering laboratories by analyzing identified positions in terms of rights and duties, cognitive and social—what we call the "epistemic effects" of various positioning strategies. We organize selected examples into three categories to emphasize the qualitatively different ways in which positioning can function in our settings. The categories include (1) statements in which researchers explicitly refer to professional or disciplinary affiliations (e.g., biologist, engineer), which we analyze in terms of their implications for epistemic division of labor; (2) statements that relate to ways in which ideas or methods are justified, extended, or modified in relation to the work of other researchers or scholars; and (3) statements that involve positioning in the realm of researchers' relationships with the artifacts and entities central to their practice, in this case living cells. This broad division into three categories of positioning is intended to distinguish three forms, levels, or dimensions of identity negotiation in laboratory practice, for which there are implications concerning what gets done, by whom, and how it gets done. We have space below for only one example, relating to the category of explicit disciplinary identifications:

In this interview passage A10, a graduate student researcher characterizes engineering in terms of a high degree of experimental control, in part through having well-defined variables:

I So is there a force in vivo that you're trying to simulate with this?

A10 *Right. So the blood vessel is a complex biaxial strain....* ***But, it's important as an engineer to really define your variables or your parameters****, or whatever.... If I just stretched the cells... I would be able to say, you know, my cells did x, in my device, but you know, I don't want to make it device dependent. You know, I want to say [to address/inform] the strain pattern.*

I So you think that your engineering background may be motivating your...

A10 Oh yeah!

I... desire to control that?

A10 *Oh yeah! Oh yeah—a lot of people will just want to stretch them, and see what happens when they're stretched... a lot of people will extract that kind of stuff, prematurely*. (2002-07-05-A-i-A10)

In this example, the higher degree of control that A10 associates with engineering also enables more robust inferences and scientifically relevant generalizations, that is, those that relate to the strain pattern of the blood vessel rather than the specific device used to test (stretch) the cells.

In a later interview, A10 echoes the idea that engineers are concerned with high levels of experimental control as the overarching goal of approximating a blood vessel's environment:

I So why don't you briefly tell me what the flow loop is, and what it's for, in your own words.

A10 *So we use the flow loop as um, a first order approximation of a blood vessel environment, is like, in that, um, as the blood flows over the lumen, the endothelial cells experience a shear stress.* ***Well, um, as engineers, we try to emulate that environment. But we also try to eliminate as many extraneous variables as possible, so we can focus on the effect of one. Or perhaps two, such that our conclusions can be drawn from change in only one variable****. So, um, we've come up with this flow loop as a way to impose a* ***very well-defined shear stress*** *across a very large population of cells such that their aggregate response will be due to that well-defined shear stress*. (2002-07-09-A-i-A10)

A10 reiterates this emphasis on the engineer's superior control by way of contrast with "the biologist's perspective"

I You just have more control with the flow loop.

A10 *Exactly.* ***It's well defined, you can change one thing and therefore whatever happens it's because of that one thing****. No matter what… Whereas like a biologist's perspective, would be, you know, let's see what's happening, but the problem is the conclusion…* ***a lot of times biologists will just try to draw conclusions****.*

I1 Like a causal model, or

A10 *Right*

I1 And that's problematic…

A10 ***Very*** *problematic*

I2 Why?

A10 *Because you* ***do not have a firm control over all your variables****. And you have so many variables, they all could be changing all at once, there's no way to know, all you have is a snapshot here, a snapshot here*

I And so are—I mean, the biologists are, are content with the kind of systemic cause and effect that they describe?

A10 *They* ***can get away with it in their circles****. Engineers, you know, don't really like that too much.*

I1 So do the biologists make progress with that method?

A10 *Well, I mean, it's kind of like paleontology versus microbiology. Paleontology you see a bigger skull and a little skull and you say this came from that. Microbiology [you say] these genes are totally different….* ***But a biologist would say this is this****. You know, the really,* ***engineering mindset*** *would be let's look at genes, and let's see if can we actually draw this kind of relationship.* (2002-07-09-A-i-A10)

In these passages A10's efforts to align with engineering practices and "mind-set" and to distance from those of biology position him as a more rigorous scientist; they establish epistemic rights and duties consistent with that position. Although he does not explicitly criticize biologists, he claims a more rigorous method and solid logical foundation for his own work by identifying with engineers and claiming that engineers are more systematic in drawing experimental conclusions on the basis of eliminative induction.

Disciplinary identifications and their associated rights and duties for laboratory responsibilities take on more importance in the increasingly interdisciplinary landscape of contemporary science. They are assumed by each researcher in ways that reflect historical understandings associated with categories like engineer, as in the A10 example, but also ways of working that are taken up within a given practice context, for example, a laboratory. In turn, these configurations affect self-understandings, which in turn impact the responsibilities a researcher assumes. For example, A11 characterizes the differences between biology and engineering as entailing a division in both focus and task in Lab A and positions herself as "between these":

A11 *I guess you could say it's split because we have* ***some people that do mechanical testing of pumps*** *or just looking at micro C-T or figuring out better ways to use micro C-T. That's about half of our lab. The* ***other half does lots of cell culture. So I consider myself in the middle of these two.*** (2002-03-22-A-A11).

In interdisciplinary science there is no single set of norms to negotiate but rather multiple, sometimes conflicting norms, or conflicting ideas about what constitutes good science. We cannot understand these norms without examining how they are understood and enacted by actual scientists working within these new settings.

CONCLUSIONS

We have been able to provide only a sampling of our efforts to understand biomedical engineering through the analytic unit of the "acting person." But the examples provided illustrate not merely the close relation of emotion and identity to the work of our laboratories, but to problem solving and dedication to the work of the laboratory, as they implicate social and cultural dimensions of emotion and identity expression.

As we have also attempted to make clear, our emphasis on the acting person is not meant to convey the impression that science is a "subjective" enterprise. Neither is the framework "individualistic" in the traditional sense, for at every level the activity of the person is situated in structuring, normative social contexts and inherited conceptual and social categories: "participation within an ontic circle of individuality and conformity" as is demanded by the very nature of scientific creativity (McGuire & Tuchańska, 2001, p. 150).

Full understanding of this participation and the complementary participation of the scientist with the material and technological domains require first person accounts from scientists practicing in their natural settings as an essential complement to controlled investigations of scientific reasoning. We by no means suggest that we have achieved "full understanding"; our claim is that the analytic unit of the acting person points us in the right direction. "Attention to activities and the first person" notes Bridgman, "emphasizes the insight that we never get away from ourselves," though the problem of how to deal with this insight is "perhaps the most important problem before us," linked to, but "infinitely more complicated than, the problem of the role of the observer to which quantum theory has devoted so much attention and regards as so fundamental" (1959, pp. 5–6). The complications Bridgman references implicate complications in data collection and analysis relevant to the psychology of science focused on the acting person. But we do not consider their burden sufficient reason to turn away from the pursuit of adequately integrated accounts of science practice, by which we treat the scientist as a person.

ACKNOWLEDGMENTS

We gratefully acknowledge the significant support we have received from the National Science Foundation, ROLE Program of the Division of Research on Learning: REC0106733 and DRL0411825 (Nersessian, PI; Newstetter, co-PI; Osbeck, Research Scientist); the National Endowment for the Humanities (Nersessian); and Georgia Institute of Technology, Faculty Development Grant (Osbeck). The opinions expressed in this chapter are our own and not those of the funding agencies.

We thank our participants as well as members of the extended research team who helped to make this manuscript possible, especially Elke Kurz-Milcke.

REFERENCES

Anderson, M. (2003). Embodied cognition: A field guide. *Artificial Intelligence, 149,* 91–130.

Anscombe, G. E. M. (1957). *Intention*. Cambridge, MA: Harvard University Press.

Atkins, K. (2005). Introduction. In K. Atkins (Ed.), *Self and subjectivity* (pp. 1–4). Oxford, UK: Blackwell.

Austin, J. L. (1962). *How to do things with words*. Cambridge, MA: Harvard University Press.

Bickhard, M. H. (2008). Are you social? The ontological and developmental emergence of the person. In U. Mueller, J. I. M. Carpendale, N. Budwig, & B. Sokol (Eds.), *Social life and social knowledge* (pp. 17–42). New York: Taylor & Francis.

Bridgman, P. (1950). *Reflections of a physicist*. New York: Philosophical Library.

Bridgman, P. (1959). *The way things are*. Cambridge, MA: Harvard University Press.

Brightman, E. S. (1943). Personality as a metaphysical principle. In E. S. Brightman (Ed.), *Personalism in theology* (pp. 40–63). Boston: Boston University Press.

Brown, J. S., Collins, A., & Duguid, P. (1989). Situated cognition and the culture of learning. *Educational Researcher, 18,* 32–42.

Carnap, R. (1935). *Philosophy and logical syntax*. London: Kegan Paul.

Corbin, J., & Strauss, A. (2008). *Basics of qualitative research: Techniques and procedures for developing grounded theory* (3rd ed.). Thousand Oaks, CA: Sage.

Davies, B., & Harré, R. (1999). Positioning and personhood. In R. Harré & L. van Langenhove (Eds.), *Positioning theory* (pp. 32–52). Oxford, UK: Blackwell.

Descartes, R. (1985). Rules for the direction of the mind. *The philosophical writings of Descartes* (Vol. 1). Cambridge: Cambridge University Press. (Original work assumed written 1628).

Dewey, J. (1910). *How we think*. Boston: D.C. Heath & Co.

Dewey, J. (1896). The reflex arc concept in psychology, *Psychological Review, 3,* 357–370.

Dewey, J. (1930). *The quest for certainty: A study of the relation of knowledge and action*. London: George Allen & Unwin.

Dewey, J. (1938). *Experience and education*. New York: Collier Books.

Dewey, J. (2008). *Experience and nature.* Carbondale: Southern Indiana University Press. (Original work published 1925).

Dewey, J., & Bentley, A. (1949). *Knowing and the known.* Boston: Beacon Press.

Dunbar, K. (1995). How scientists really reason: Scientific reasoning in real-world laboratories. In R. J. Sternberg & J. E. Davidson (Eds.), *The nature of insight* (pp. 365–395). Cambridge, MA: MIT Press.

Dunbar, K., & Blanchette, I. (2001). The in vivo/in vitro approach to cognition: The case of analogy. *Trends in Cognitive Science, 5,* 334–339.

Engeström, Y., & Miettinen, R. (1999). Introduction. In Y. Engeström & R. Miettinen (Eds.) *Perspectives on activity theory* (pp. 19–28). Cambridge: Cambridge University Press.

Feyerabend, P. (1975). *Against method.* London: Verso.

Fridlund, A. J., & Russell, J. A. (2006). The functions of facial expressions: What's in a face? In V. Manusov & M. L. Patterson (Eds.), *Sage handbook of nonverbal communication* (pp. 299–320). Newbury Park, CA: Sage.

Garfinkel, H., & Sacks, H. (1970). On formal structures of practical action. In J. C. McKinney & E. A. Tiryakian (Eds.), *Theoretical sociology: Perspectives and developments* (pp. 338–366). New York: Appleton-Century-Crofts.

Goffman, E. (1981). *Forms of talk.* Philadelphia: University of Pennsylvania Press.

Greeno, J. G. (1998). The situativity of knowing, learning, and research. *American Psychologist, 53,* 5–26.

Griffiths, P., & Scarantino, A. (2008). Emotions in the wild: The situated perspective on emotion. In P. Robbins & M. Aydede (Eds.), *The Cambridge handbook of situated cognition* (pp. 437–453). Cambridge: Cambridge University Press.

Harré, R. (1974). Some remarks on 'rule' as a scientific concept. In T. Mischel (Ed.), *Understanding other persons* (pp. 143–184). Oxford: Basil Blackwell.

Harré, R., & Moghaddam, F. (2003). Introduction: The self and others in traditional psychology and in positioning theory. In R. Harré & F. Moghaddam (Eds.), *The self and others: Positioning individuals and groups in personal, political, and cultural contexts* (pp. 1–11). Westport, CT: Praeger.

Harré, R., Moghaddam, F., Cairnie, T., Rothbart, D., & Sabat, S. (2009). Recent advances in positioning theory. *Theory and Psychology, 19*(1), 5–31.

Heidegger, M. (1962). *Being and time* (J. Macquarrie & E. Robinson, Trans.). New York: Harper & Row.

Hempel, C. (1952). *Fundamentals of concept formation in empirical science.* Chicago: University of Chicago Press.

Henriques, G. (2003). The tree of knowledge system and the theoretical unification of psychology. *Review of General Psychology, 7,* 150–182.

Hutchins, E. (1995). *Cognition in the wild.* Cambridge, MA: MIT Press.

James, W. (1892). *Psychology: Briefer course.* New York: Holt.

James, W. (1896). The sentiment of rationality. Reprinted from *The will to believe and other essays.* Cambridge, MA: University Press, 63–110.

Jordan, B., & Henderson, A. (1995). Interaction analysis: Foundations and practice. *Journal of the Learning Sciences, 4*(1), 39–103.

Kendler, H. H. (2002). Romantic versus realistic views of psychology. *American Psychologist, 57*(12), 1125–1126.

Koch, S. (1981). The nature and limits of psychological knowledge: Lessons of a century qua "science." *American Psychologist, 36,* 257–269.

Kukla, A. (1992). Unification as a goal for psychology. *American Psychologist, 47,* 1054–1055.

Latour, B., & Woolgar, S. (1986). *Laboratory life: The construction of scientific facts.* Princeton, NJ: Princeton University Press. (Original work published 1979).

Lave, J. (1988). *Cognition in practice.* Cambridge: Cambridge University Press.

Lazarus, R. S. (2006). *Stress and emotion: A new synthesis.* New York: Springer.

Lazarus, R. S., & Folkman, S. (1987). Transactional theory and research on emotions and coping. *European Journal of Personality, 1,* 141–169.

Longino, H. (2002). *The fate of knowledge.* Princeton, NJ: Princeton University Press.

Louis, W. R. (2008). Intergroup positioning and power. In F. M. Moghaddam, R. Harré, & N. Lee (Eds.), *Global conflict resolution through positioning analysis* (pp. 21–39). New York: Springer.

McGuire, T., & Tuchańska, B. (2001). *Science unfettered: Philosophical study in sociohistorical ontology.* Athens: Ohio University Press.

McNally, R. J. (1992). Disunity in psychology: Chaos or speciation? *American Psychologist, 47,* 1054.

Meehl, P. (1978). Theoretical risks and tabular asterisks: Sir Karl, Sir Ronald, and the soft progress of slow psychology. *Journal of Consulting and Clinical Psychology, 46*(4), 806–834.

Milton, N. R. (2010). A new framework for psychology. *Review of General Psychology, 14,* 1–15.

Mischel, T. (1974). Understanding neurotic behavior: From "mechanism" to intentionality. In T. Mischel (Ed.), *Understanding other persons* (pp. 216–259) Oxford: Basil Blackwell.

Nersessian, N. J. (1995). Opening the black box: Cognitive science and the history of science. *Osiris, 10,* 194–211.

Nersessian, N. J. (2002). The cognitive basis of model-based reasoning in science. In P. Carruthers, S. Stich, & M. Siegal (Eds.), *The cognitive basis of science* (pp. 133–153). Cambridge: Cambridge University Press.

Nersessian, N. J. (2005). Interpreting scientific and engineering practices: Integrating the cognitive, social, and cultural dimensions. In M. E. Gorman, R. D. Tweney, D. C. Gooding, & A. P. Kincannon (Eds.), *Scientific and technological thinking* (pp. 17–56). Hillsdale, NJ: Lawrence Erlbaum.

Nersessian, N. J. (2008). *Creating scientific concepts.* Cambridge, MA: MIT Press.

Nersessian, N. J. (2009). How do engineering scientists think? Model-based simulation in biomedical engineering research laboratories. *Topics in Cognitive Science, 1,* 730–757.

Osbeck, L., & Nersessian, N. J. (2006). The distribution of representation. *Journal for the Theory of Social Behaviour, 36*(2), 141–160.

Osbeck, L., & Nersessian, N. J. (2010). Forms of positioning in interdisciplinary science practice and their epistemic effects. *Journal for the Theory of Social Behaviour, 40,* 136–161.

Osbeck, L., Nersessian, N., Malone, K.,& Newstetter, W. (2011). *Science as psychology: Sense-making and identity in science practice.* Cambridge: Cambridge University Press.

Parkinson, B., Fischer, A. H., & Manstead, A. S. R. (2005). *Emotions in social relations: Cultural, group, and interpersonal processes.* New York: Psychology Press.

Pickering, A. (1995). *The mangle of practice.* Chicago: University of Chicago Press.

Poincaré, H. (1914). *Science and method* (F. Maitland, Trans.). New York: Dover.

Polanyi, M. (1964). *Science, faith, and society.* Chicago: University of Chicago Press. (Original work published 1946).

Polanyi, M. (1973). *Personal knowledge: Towards a post-critical philosophy.* Chicago: University of Chicago Press. (Original work published 1958).

Resnick, L. B., Pontecorvo, C. & Säljö, R. (1997). Discourse, tools, and reasoning. In L. B. Resnick, R. Säljö, C. Pontecorvo, & B. Burge (Eds.), *Discourse, tools, and reasoning: Essays on situated cognition* (pp. 1–20). Berlin: Springer-Verlag.

Rorty, R. (1979). *Philosophy and the mirror of nature.* Princeton, NJ: Princeton University Press.

Rouse, J. (2002). *How scientific practices matter: Reclaiming philosophical naturalism.* Chicago: University of Chicago Press.

Smythe, W. (1998). Introduction. *Toward a psychology of persons* (pp. xi–xxii). Mahwah, NJ: Lawrence Erlbaum.

Staats, A. W. (1983). *Psychology's crisis of disunity: Philosophy and method for a unified science.* New York: Praeger.

Staats, A. W. (1991). Unified positivism and unification psychology: Fad or new field? *American Psychologist, 46,* 899–912.

Staats, A. W. (1999). Unifying psychology requires new infrastructure, theory, method, and a research agenda. *Review of General Psychology, 3,* 3–13.

Staats, A. W. (2004). The disunity-unity dimension. *American Psychologist, 59*(4), 273.

Stern, W. (1906). *Person und Sache: System der philosophischen Weltanschauung: Band l. Ableitung und Grundlehre* [Person and world: A system of a philosophical worldview: Vol. 1. Introduction and basic principles]. Leipzig, Germany: Barth.

Sternberg, R. J., & Grigorenko, E. L. (2001). Unified psychology. *American Psychologist, 56*(12), 1069–1079.

Strauss, A., & Corbin, J. (1998). *Basics of qualitative research techniques and procedures for developing grounded theory* (2nd ed.). London: Sage.

Tajfel, H. (1982). *Social identity and intergroup relations.* Cambridge: Cambridge University Press.

Tajfel, H., & Turner, J. C. (1979). An integrative theory of intergroup behavior. In W. G. Austin & S. Worchel (Eds.), *Psychology of intergroup relations* (pp. 33–47). Monterey, CA: BrooksCole.

Tolman, C. W. (1998). Sumus Ergo Sum: The ontology of self and how Descartes got it wrong. In W. E. Smythe (Ed.), *Toward a psychology of persons* (pp. 3–24). Mahwah, NJ: Erlbaum.

Toulmin, S. (1967). Neuroscience and human understanding. In G. Quarton, T. Melnechuk, & F. O. Schmitt (Eds.), *Neurosciences* (pp. 822–832). New York: Rockefeller University Press.

Toulmin, S. (1972). *Human understanding.* Princeton, NJ: Princeton University Press.

Toulmin, S. (1974). Rules for understanding human behavior. In T. Mischel (Ed.), *Understanding other persons* (pp. 185–215). Oxford: Basil Blackwell.

Turner, S. (1994). *The social theory of practices.* Chicago: University of Chicago Press.

Wallas, G. (1926). *The art of thought.* New York: Harcourt, Brace.

Watson, D. L. (1938). *Scientists are human.* London: Watts.

Weick, K. E. (1995). *Sensemaking in organizations.* Thousand Oaks, CA: Sage.

Wenger, E. (1998). *Communities of practice: Language, meaning, and identity.* Cambridge: Cambridge University Press.

Wittgenstein, L. (1953). *Philosophical investigations* (G. E. M. Anscome, Trans.). Oxford, UK: Blackwell.

6 Inference to the Best Explanation (IBE) and the Causal and Scientific Reasoning of Nonscientists

Barbara Koslowski

INTRODUCTION

The question of what constitutes either explicit or implicit reasoning (either in general or in the psychology of science) can be a thorny conceptual issue, and one that is made even thornier by the fact that, while some instances of the implicit-explicit distinction are clear-cut, others are not. In the present chapter, I take the pragmatist's (or the coward's) way out and focus on operational definitions. I address the question of implicit and explicit reasoning in the psychology of science by focusing specifically on how people reason about causal explanations. The particular stance in the present chapter is that reasoning about explanations is an important aspect of scientific or causal reasoning and, indeed, of reasoning in general. Furthermore, the approach I will be summarizing argues that sound reasoning is based on two things: a set of *heuristics* that make a good outcome increasingly likely (although not certain) and relevant *background information that is approximately accurate and that includes information about explanation*, which makes it increasingly likely that the heuristics will be applied successfully.

The importance of background information, including information about mechanism or explanation, plays into the explicit-implicit distinction in several ways. One is that information that might otherwise remain in the background can come to be seen as explicitly relevant if there is a theory that can incorporate it into a broader causal framework. Put differently, information that we might have already acquired but might not be actively thinking about will be brought to bear on an issue if the relevant theory is present. Conversely, a focus on certain information can prompt us to consider explanations (including alternative explanations) that we might not otherwise have thought about. Finally, in terms of sound reasoning (whether by scientists or nonscientists), background information (including information about theories or explanations) constrains how people apply various seemingly formal rules in actual practice; we may have access to formal rules (either implicitly or explicitly), but how we apply them in practice depends on considerations of background information.

IDEALIZED MODELS OF SCIENCE VERSUS ACTUAL SCIENTIFIC PRACTICE

One's view of causal explanation and of how it ought to be studied depends, in part, on one's view of science. In the psychological literature, typically (but not necessarily), different views of scientific inquiry map onto different approaches to research on causal reasoning in general and to the respective roles of seemingly formal rules and background information in particular (Gopnik & Schulz, 2007; Klahr, 2000; Koslowski, 1996; Kuhn, Amsel, & O'Loughlin, 1988; Zimmerman, 2000, 2007).

Idealized models of science: relying on formal rules and identifying causal events. The different views of causal explanation and of formal rules are also bound up (though not perfectly) with whether the psychologists's conception of scientific inquiry is intended to be an idealized model of science or a description of actual scientific practice. The best known idealized view of scientific inquiry is the hypothetico- or nomothetic-deductive (H-D or N-D) system from philosophy. On this view, scientific inquiry can be construed as a set of steps that can be followed in a linear and precise fashion and that involve the application of a set of rules treated as formal. Once a hypothesis has been formulated, one proceeds algorithmically to collect data with which to test (or preferably, to try to disconfirm) the hypothesis, to either reject or not reject the hypothesis, depending on what the data look like (Popper, 1959).

In the psychological literature, research that is motivated by the idealized view of science usually has two characteristics. One is that the type of explanation studied consists of identifying the causal event in a situation. The other is that the typical research question asks whether, when people identify causal events, they do so simply by relying on the Humean indices of temporal priority, spatial and temporal contiguity, and covariation. For example, researchers who study how causal events are identified might ask why it is that Jones periodically has a rash. In trying to identify the causal event of Jones's periodic rash, one might consider the hypothesis that Jones sometimes gardens and touches poison ivy when he does so. To test this hypothesis, one can then ask about the Humean indices. Is touching poison ivy temporal prior to the appearance of the rash; is it temporally and spatially contiguous with the rash; and, is it the case that touching poison ivy is followed by a rash while avoiding poison ivy is not? Researchers who study scientific reasoning by asking whether people rely on the Humean indices of priority, contiguity, and covariation typically, at least in terms of how the causation is operationalized, assume that relying on Humean indices is (and ought to be) sufficient for identifying causal events (Kuhn, Amsel, & O'Loughlin, 1988). Put differently, researchers operationally define causation as the explicit application of Humean indices and measure causal reasoning in terms of whether the indices are applied either implicitly or explicitly (Gopnik & Schulz, 2007; Haith & Benson, 1998; Schulz & Gopnik, 2004).

Problems with Idealized Models

Idealized models focus on causal events rather than causal mechanisms. Identifying causal events can certainly be important and often the first step in explaining an effect, but it is only the first step. In most cases, scientists also want to know about the mechanism or process by which the causal event brings about the effect. *How* does touching poison ivy cause a rash? What is the *mechanism* by which touching poison ivy operates to bring about a rash?

Idealized models (mistakenly) treat seemingly formal rules as algorithms that guarantee success. A more important problem is that, in idealized approaches to science, various rules are often treated as formal. By "formal," I mean rules that can be applied without taking background information into account. A clear example of a formal rule is that 2 + 2 = 4. It is formal, because it yields the correct answer regardless of whether one is adding coffee cups or monkeys. In that it guarantees the correct answer, it is an algorithm, rather than a heuristic. In the psychological literature, many rules, including the Humean indices, have been treated as though they were algorithmic formal rules when, in fact, they are merely heuristics.

Furthermore, treating Humean indices as though they were formal rules affects not only how research is designed, but also how the results are often interpreted. For example, sometimes laboratory experiments are designed so that certain events covary with effects (even though they do not cause the effects in the world outside the laboratory), and "correct" performance is stipulated to consist of treating the covarying events as causing the effect. Thus, an experiment might be designed so that the presence of a small toy car on the experimental apparatus either facilitates or inhibits a causal event in the situation (Shultz & Mendelson, 1975), while in another, eating bread in the form of rolls rather than slices might be presented to the participant as having causally affected susceptibility to colds, because type of bread covaried with colds (Kuhn et al., 1988).

Thus, in such experiments, the participants are often required to do two things to be counted as having performed "correctly." One is that they need to take account of the Humean indices they are presented with and to treat them as though they guaranteed a good decision. To do this, participants also often need, in turn, explicitly to ignore background information acquired outside of the laboratory. For example, they need to ignore the background knowledge that proximity of toy cars would likely have no effect on whether pulling down a lever is likely to make a light go on and that the form in which bread is eaten does not change the "essential" (Gelman, 2003) composition of bread and is thus unlikely to affect illness. That is, to be counted as having performed the task correctly, participants need to treat Humean information as the only information that should be taken into account—even if it conflicts with other, equally important background information that has been acquired outside of the laboratory. Furthermore, they need to treat the Humean indices as though they were algorithms that would guarantee the correct decision, rather than heuristics or rules of thumb that will make sound reasoning increasingly likely.

Actual Scientific Practice: Heuristics and Background Information

In actual scientific practice, as opposed to idealized formal models, strategies such as relying on Humean indices are heuristics, not algorithms; they make a sound decision increasingly likely, but they do not guarantee it. The reason is that even when rules can be framed as though they were formal (as the Humean rules often are), they cannot be applied *successfully* unless background information (including information about mechanism or explanation) is taken into account (Koslowski, 1996; Koslowski & Thompson, 2002). Philosophers of science of all persuasions (Humeans as well as non-Humeans) agree that in practice scientists choose explanations and theories from among a handful of relevant alternatives suggested, not just by Humean indices, but also by their theoretical conceptions of the relevant phenomena, including mechanism information (Boyd, 1989; Fine, 1984; Lipton, 1991, 1993; Psillos, 1999; Salmon, 1990; Van Fraassen, 1980). For an especially thorough discussion of the role of mechanism considerations see Darden (2006).

For example, in identifying the causal event that brings about neural tube defects, epidemiologists do consider insufficient folic acid, but do not consider whether the pregnant woman prefers Bach or the Beatles. The reason is background information in the form of possible mediating mechanisms. There is a possible mechanism (or explanation) by which insufficient folic acid can cause neural tube defects; there is no known mechanism by which musical preference can do so.

Furthermore, there are pragmatic as well as conceptual reasons why the Humean indices (or any seemingly formal rule) must be applied in conjunction with background information. Pragmatically, there is no way, during one researcher's lifetime, to compute all the *possible* correlations that exist between possible causal events and the effect; the number of possible correlations is too large to compute (even with a sophisticated computer) and the typical researcher's lifetime is too short to be able to digest the resulting printout. This is also true of other seemingly formal rules, as well. For example, the notion of ruling out alternative hypotheses must be constrained by background information to help one decide which alternatives are not merely possible, but also *plausible* and which are not. For example, one would likely not control for the possible alternative hypothesis that cloud cover or the experimenter's hair color affected a chemical reaction; though possible, such alternative hypotheses are not plausible in this context. (One can imagine that cloud cover might affect a biochemical process that is sensitive to sunlight, but that simply illustrates the point about the role of background information.)

In psychology, rules that have been *treated* as formal include parsimony (Bonawitz & Lombrozo, 2011), internal consistency (Samarapungavan, 1992), probability (Lombrozo, 2007), and (as already noted) Humean indices such as covariation, priority, and contiguity (Bullock, 1991; Bullock, Gelman, & Baillargeon, 1982;

Bullock & Ziegler, 1999). When rules are treated as formal, the idea is that, for example, an event that covaries with an effect should be treated as having caused the effect regardless of whether one is talking about neural tube defects or susceptibility to colds. It is worth noting that when Vosniadou and Brewer (1992) studied internal consistency, they asked children about unfamiliar events and were thus able to control for background information. Similarly, when Lombrozo (2007) studied parsimony, she too controlled for background information.

Inference to the best explanation (IBE) models actual scientific practice. In commenting on idealized descriptions of science, such as the H-D account, the philosopher Kaplan notes that, in such models, the description is such that "most of the important incidents in the drama of science are enacted behind the scenes" (Kaplan, 1964, p. 10). In the present chapter, I try to focus on research that is framed in terms of recent analyses of actual scientific *practice*, in particular, the description of scientific practice termed "inference to the best explanation" (IBE), sometimes called "abduction." IBE attempts to identify some of the important incidents that lie behind the scenes in actual (rather than idealized) scientific practice.

IBE argues that an explanation is chosen over its competitors because it provides a better causal account of the data (Lipton, 1991,1993; Magnani, 2001; Thagard, 1989; Thagard & Verbreugt, 1998). IBE has identified some of what happens "behind the scenes." For example, in actual scientific practice (rather than idealized descriptions of it), an important step consists of deciding what sorts of data or observations should count as relevant. In idealized descriptions, there is a clear-cut distinction between theory (or hypothesis or explanation) and choice of relevant data. In actual scientific practice, however, the distinction is less clear-cut; theory and data gathering are interdependent. Similarly, in actual scientific practice, not only do alternative explanations play a central role, but they also affect whether certain information is treated as evidence, because good evidence distinguishes between competing explanations, and thus whether information is evidential depends, in part, on what the competing explanations are (Boyd, 1989; Salmon, 1990). Again, theory and data evaluation are interdependent.

The classic example of IBE is evolutionary theory. The argument that different species evolved has been judged, not in isolation, but with respect to the alternative that the individual species arose by divine design. In addition, IBE also notes that whether a piece of information is treated as evidentially relevant may well depend on whether an explanation or theory makes it so. Evolutionary theory (in contrast to the alternative) makes the existence of intermediate fossil forms and niche-specific adaptations evidentially relevant, because it incorporates them into a broader causal framework (in a way that the spontaneous rise of species does not) and illustrates the interdependence of theory and evidence. To be sure, the observations can be distorted to be consistent with each of the explanations. One could, for example, attribute niche-specific adaptations to coincidence. However, the evidence is more *plausibly* (and causally) consistent with evolution. That is, IBE suggests that we evaluate competing explanations by assessing the *relative weight* of the evidence in favor of each and judge which explanation the evidence is more plausibly consistent with. Furthermore, evolution (but not the alternative) is causally consistent with the network or web of related background information, that is, with well-established background beliefs about, for example, plate tectonics, population genetics, and animal breeding (Quine & Ullian, 1970). Finally, IBE draws attention not only to the fact that, in actual scientific practice, theory and data are interdependent, but also to the fact that related background information can involve theory (or explanation) as well as data. Clearly, the idealized model of scientific inquiry, criticized by Kaplan, would fail to address many of the issues reflected in the controversy about how different species arose.

The emphasis, in IBE, on the role of background knowledge also draws attention to the point mentioned earlier, namely, the limitations of formal principles or rules for doing science. Consistent with IBE, to apply seemingly formal rules *successfully*, background information (including information about mechanism or explanation) has to be taken into account. A full understanding of scientific inquiry requires understanding the interdependence of seemingly formal rules and background information.

The approach to scientific reasoning taken in this chapter is not the idealized one.

Constraints on treating rules as formal also affect explanations outside the laboratory. The limitation of treating formal rules as though they guaranteed good science has implications beyond the academy. For example, consider the seemingly formal rule that one should prefer explanations that are internally consistent over those that are not. Until the flurry of research (e.g., Block, 1995; Block & Dworkin, 1974; Kamin, 1974; Steele, 1997; Steele & Aronson1995 carried out as a response to Herrnstein's (1973) and later Herrnstein and Murray's (1994) claims, it was easy to generate an internally consistent argument that blacks were genetically inferior to whites in terms of intelligence. The argument noted, roughly, that blacks in contrast to whites had, on average, less education; earned less when one controlled for years of schooling; scored lower on standardized tests; had lower-paying jobs; and so forth. That is, there was an internally consistent argument congruent with the claim of genetic inferiority. However, though internally consistent, this argument failed to take account of several alternative hypotheses, the most obvious being pervasive and institutionalized racism (Fraser, 1995). One might argue that, at the time, scientists should have applied two formal rules, one about internal consistency and the other about alternative hypotheses. However, even adding the rule, consider alternative hypotheses, would not have guaranteed success, because awareness of institutionalized racism was, at the time, so much outside the prevailing cultural zeitgeist—even for people who did not agree with Herrnstein's conclusions—that it would likely not have occurred to many people. Plausible alternative hypotheses are background information whose availability depends on what the culture makes available. This example also illustrates the point that the background information that is available also changes with historical time.

When possible, I will try to use examples from outside the laboratory, as well as from academic research.

There is a caveat about distinctions among background information, explanation, and data. In some of the research described below, I talk about three aspects of reasoning: theory (or mechanism or explanation); data (or observation or evidence); and background information. In fact, however, the distinctions among the three are difficult to pin down because of the way scientific inquiry works. Evidence is not independent of theory; a theory or explanation can affect whether we treat a particular observation or data point as evidence (see below). Furthermore, background information can include theory as well as data or observations. Therefore, I use different terms, not because the terms refer to clearly delineated constructs, but because to make various arguments below, the terms are operationalized in different ways. Finally, I do not try to distinguish among theory, mechanism, and explanation, as all require comparable evaluation strategies.

RESEARCH BASED ON SCIENTIFIC PRACTICE, RATHER THAN IDEALIZED MODELS OF SCIENCE

Identifying a Causal Event Requires That Background Information Be Taken into Account

Consider again explanations that consist of identifying a causal event. For many situations, our background knowledge includes a kind of catalogue of causal events that we can rely on regardless of whether we understand the mechanism by which an event brings about an effect. For example, if we are trying to explain a car accident, information that the driver had been drinking would certainly be relevant, even if we do not know the specifics of how alcohol affects driving (is it reaction time, attention, etc.) and even if we have not kept a running record of covariation, that is, how many times alcohol consumption was followed by a car accident and car accidents occurred without alcohol being involved.

However, when the relevant catalogue is absent, we often rely on background information about mechanism to help us decide which correlations to treat seriously, as likely to be causal. For example, if color of car correlates with gas mileage, one is nevertheless unlikely to view color as causal. However, if provided with a possible mechanism by which color might operate (for example, that car color affects alertness, which makes people drive in a gas-conserving way) then one is more likely to treat car color as causally related to gas mileage (Koslowski, 1996, Chapter 8). I am not arguing that explicit consideration of mechanism underwrites every causal judgment. The catalogue of likely causes that people have are often based on distributed cognition, or a division of cognitive labor; we believe that penicillin kills germs, not necessarily because we understand the mechanism involved, but because experts have told us that it does (Giere, 2002; Koenig & Harris, 2005; Lutz & Keil, 2002). In addition, people (including children) may well not even think about (at least early in learning about an apparatus) the mechanism by which some causes operate (Koslowski, Spilton, & Snipper, 1981); we turn on the TV without necessarily wondering *why* clicking the remote control makes a picture appear.

The point is that, in addition to a catalogue of likely causes—whether or not based on other people's expertise—background information about mechanisms plays a role in causal judgments, especially when the possible causes in question are, at first glance, implausible, as is car color as a cause of gas mileage, even when the two covary.

Reasoning about Mechanistic Explanations

Explanation can render information evidential. In the psychological literature on causal/explanatory reasoning, tasks that ask participants to identify a causal event have typically been operationalized so that there are a few possible causal events, and the goal is to decide among them, usually by relying on Humean indices. In contrast, explanations that involve mechanisms are often embedded in networks of information. One result is that a first step often consists of deciding which potentially relevant pieces of information are actual evidence, likely to be relevant to the explanation, and which pieces are simply noise that can be ignored.

In addressing this question, it is easy to think of theory and evidence as two distinct entities. It is easy to think that information is evidence in an absolute sense, that information either is evidence or it is not. After all, we certainly talk about collecting evidence to evaluate theories, as though the two constructs were separable. What also fosters this notion is that, in many cases, it is so obvious that some information is evidence that we forget that the reason for the judgment is grounded in theory (or at least in the statements of experts, as noted above). It is, in a sense, implicit knowledge.

However, in actual scientific practice, there often is not a clear-cut distinction between theory and evidence; rather, whether we treat information as evidence depends on whether there is a theory or an explanation available that can incorporate the information into a broader causal framework. That is, in many cases, information becomes evidentially relevant, because theory or explanation renders it relevant. Whether fossils are seen as intermediate forms leading up to an extant species or whether they are seen as a set of independent species depends on whether evolutionary theory is available to account for how they might be intermediate forms.

Using a much less rarified set of examples, we asked people to decide whether information was evidentially relevant to explaining an event. One event was that, in a certain area, people who live in the mountains are on average shorter than those who live on the plains. We told participants that one possible explanation is that the village water supply is polluted, which interferes with the absorption of nutrients. We also told participants about two pieces of information about this event. One was that, according to the oral history of the mountain dwellers, they migrated from the plains to the mountains to escape invaders from across the sea. The other is that the language of the mountain dwellers has grammatical structures and expressions that are different from the language of the plains dwellers. Some participants were then told of a control alternative (that the two groups might have access to different foods); others were given the target alternative (that the two groups might come from different gene pools). Participants given the target alternative rated the pieces of information as more evidentially relevant than those who were given the control alternative. Considering the possibility of different gene pools made causal sense of the two pieces of information. It was consistent with the oral history of flight in the face of invaders from across the sea, and it was consistent with the different grammatical structures and expressions (Koslowski, Marasia, Chelenza, & Dublin, 2008).

This study demonstrated that evidential relevance is not a fixed characteristic of information; it is theory-dependent. Put in the terms of the present volume, information that might otherwise remain in the background can come to be seen as explicitly relevant if there is a theory that can incorporate it into a broader causal framework.

Whether information is evidential also depends on the alternative with which an explanation is being contrasted. There is also another sense in which theory affects whether information is seen as evidentially relevant. Recall that, according to IBE, theories or explanations are not evaluated in isolation; rather, they are evaluated with respect to competing alternative theories. Indeed, this is the basis for control conditions. Sometimes the alternative is simply the complement of the target explanation; it is the claim that the target explanation is wrong—that it is not operating. We have been examining alternative explanations that are substantive, in that they describe a competing mechanism.

If an explanation is judged with respect to competing explanations, then whether information is treated as evidentially relevant should also depend on what the particular alternative is with which the target explanation is being contrasted. And, in fact, this is what we have found. For example, we asked people to reason about why the average age of women at first marriage has been increasing, and we also told all participants that the baseline explanation was that improved fertility technologies enable women to have children at older ages. We then presented people with two pieces of information: (1) that more women are achieving high positions in their chosen fields and (2) that attendance at religious services has been declining. In addition, some participants were presented with alternative (a) that more women are pursuing advanced degrees. These participants found information (1) about women achieving high positions in their chosen fields as especially likely to be evidential, presumably, because high positions are consistent with pursuing advanced degrees. Other participants were presented with alternative (b) that there is decreased negative stigma attached to living together before marriage. These participants found information (2) about declining attendance at religious services especially likely to be evidential, since it is consistent with decreased negative (and presumably religious) stigma attached to unmarried couples living together. That is, whether a piece of information that might otherwise be in the background is seen as evidentially relevant to explaining an event depends on the particular alternative to which the target explanation (in this case, the baseline explanation) is being contrasted (Koslowski, Hildreth, Fried, & Waldron, in preparation.)

Now consider an example from life outside the laboratory. Japan has very little in the way of natural resources, including oil. This might be seen as an interesting aspect of geography that is not very relevant to why the Japanese government attacked Pearl Harbor. If, however, one learns that the Japanese wanted to end the pre–World War II US blockade that prevented shipments of oil and other natural resources from reaching Japan, then the fact that Japan has very little oil becomes very relevant, indeed (Worth, 1995). I am not arguing that the attack on Pearl Harbor was justified, any more than that the blockade of Japan was justified. What I am arguing is that once one considers an explanation that consists of access to raw materials, then information about the lack of minerals in Japan becomes relevant in a way that it might otherwise not be. In addition, whether one has access to this information often depends on culture, in particular, whether one has read about this particular aspect of global history or whether one was educated in Japan. In short, whether information is seen as explicit evidentially relevant depends, not only on the explanation being considered, but also on the alternative against which the explanation is evaluated. Different alternative explanations can make information seem explicitly relevant that might otherwise be relegated to the background.

Explanation affects what information people search for as relevant. In the previous two studies, people were presented with the possibly relevant information and with the possible explanations. In subsequent studies, we examined situations in which people were asked to generate or to search for the possibly relevant information on their own. There is a sense in which possibly relevant information is a part of what we know implicitly; we cannot after all search for relevant information in particle physics if we know nothing about particle physics. The information is in the implicit background; the question is, how do we bring it into the explicit foreground. We found that here, as well, what people searched for as explicitly relevant to evaluating an explanation was heavily influenced by theory. We told college students and college-bound adolescents that a hospital administrator wanted to find out whether having parents stay overnight with their hospitalized children would make the children recover faster. We asked the participants, "What sorts of information or evidence would you consider?" Given the instructions, of course, all of the participants requested information. What is relevant to the current chapter is that 94% of the participants explicitly verbalized mechanism information as the motivation for the information they wanted. "You could have some parents stay overnight and some not, *because parents might help their kids by telling the doctors if there were problems*," or "I'd look for age, *because young kids might need their parents more than older kids*" (Koslowski, 1996, Chapter 12; Koslowski, Cohen, & Fleury, in preparation; Koslowski, P'Ng, Kim, Libby, & Rush, in preparation). Note that we did not ask the participants to consider mechanism; they did so because we asked them what sort of evidence they would consider, and considerations of mechanism—or theory, broadly construed—motivated their decisions about whether some information might be evidentially relevant and thus worth considering.

Outside of the laboratory, the rise of feminist social theory in archaeological and anthropological research (Gero & Conkey, 1991) led people to look for information that had been largely ignored. One of the most salient had to do with relative contributions of men and women to food intake among hunter-gatherers. It was (and probably still is) easy to think of groups of males providing the bulk of the food for hunter-gatherer societies, relying on primitive tools and traps to bring down mammals large enough to feed an entire village. Indeed, this picture is central to many museum dioramas of early hunter-gatherers. It is also a picture that is internally consistent with other beliefs; it "made sense" to think that women, because they would be pregnant and nursing, would be tied to a home settlement and thus dependent on males, who were not only mobile but also stronger, for the bulk of their food. However, Slocum's (1975) work demonstrated that, in extant hunter-gatherer societies, women are at least as mobile as men and, more to the point, provide the bulk of the group's dietary intake. That is, doing anthropology from a feminist perspective (an explanation in the broad sense of the term) led to an explicit search for data that might otherwise have remained in the shadows.

Background information can help people generate relevant alternative explanations. Consider again the study that asked people how they could find out whether children's hospital stays were shorter when their parents were allowed to stay with them. In that study, people certainly had the background knowledge that enabled them to generate possible explanations that guided their search for relevant information. Even if they had never been hospitalized, they could nonetheless remember how they had felt when they had been ill, and they knew enough about how hospitals work to be able to generate possible mechanisms that helped them search for relevant information. However, in some situations, such background information can be lacking. For example, I suspect that, for several of us, being put in a physics lab and asked to evaluate an explanation would be impossible, if for no other reason than that we would lack the background to be able to generate possibly relevant information. If this is so, then providing people with background knowledge might enable them to generate possibly relevant information, in particular, possible alternative explanations. We examined this in the following study.

Both experimental and control groups were first presented with background information about, for example, oriental carpets. Control information mentioned the history of such carpets, how they are made, how they came to be exported, and so forth. Experimental information mentioned instead that, in sheep, the type of wool produced depends on the sheep's diet, the climate, how much lanolin the sheep produce, and so forth. The same participants were also asked to participate in a separate study that was arranged to seem unrelated. (This was possible because at the time construction problems made space scarce, and we were able to tell participants that we were running experiments back-to-back.) In the second (seemingly unrelated study), both groups were told of reports that, for example, carpets seem to last longer when people remove their shoes before walking on them. All participants generated alternative explanations. However, people in the experimental condition were more likely than in the control condition to generate an alternative that dealt with the characteristics of the wool from which the carpets had been made. That is, being provided with explicit background information can enable someone either to retrieve an alternative that was already in the background and consider it explicitly or can suggest an alternative that would otherwise not have been considered.

Background information helps distinguish strong from weak anomalies. Few (if any) explanations are perfect; there are frequently anomalies to an explanation. However, some anomalies are easy to dismiss, while others are more problematic for an explanation. How do people decide whether an anomaly is weak or strong? When the task involves relying on Humean indices to identify a causal event, an anomaly is an event that violates the Humean indices. For example, it may be a situation in which the covariation of causal event and effect is less than perfect. In such situations, one way of resolving the anomaly can be to point out that the likelihood of event Y given event X is probabilistic and the probability is less than 1.00. Much of the research that examines what might be construed as anomalies is research on Humean indices in which the covariation, for example, is less than perfect or in which one potentially causal event is confounded with another possibly causal event.

In other situations, however, especially those that deal with mechanistic explanations, anomalies are events that would not be expected to occur if the assumed causal mechanism were operating; such anomalies do not merely suggest that a causal relation is probabilistic; rather, the anomaly undermines or calls into question the assumed explanatory mechanism.

Brewer and his colleagues have studied mechanistic explanations in general and anomalies to mechanistic explanations in particular. The main aim of some of their research was to document the various strategies that people rely on to deal with anomalous data. However, their research also documents the extent to which consilience with background information plays a role, not only in recognizing information as anomalous, but also with (in some cases) trying to resolve the anomaly. For example, Chinn and Brewer (1998) asked college students to assess the theory that dinosaurs were cold-blooded in light of data anomalous to the theory, namely, that that the bone density of dinosaurs was comparable to the bone density of extant warm-blooded animals. In reasoning about the anomalies, the students relied on their related knowledge of, for example, the effects of calcium on bones to suggest that bone density could be affected by factors other than warm-bloodedness.

We were interested in the related question of what distinguishes weak anomalies (that undermine an explanatory mechanism but that are easily dismissed) from strong anomalies (that seriously call an explanation into question). Like Brewer and his colleagues, we also presented participants with anomalies to mechanistic explanations. In addition, we asked participants how they would resolve the anomalies. Afterward, we asked them to rate how plausible their resolutions were. For example, we presented college students with an event (e.g., over the past 50 years, the average height of Americans has decreased while Europeans have on average gotten taller) along with a possible explanation for the event (the proliferation of "junk food" in the United States). Participants were then presented with one of two types of anomalies, weak (the height difference occurred even though a larger percentage of Americans rather than Europeans take vitamins to compensate for nutritional deficiencies) or strong (the decrease in US height is found even in individual groups based on income, education, race, or ethnic origin, that is, even in groups expected to eat very little junk food).

Participants generated comparable *numbers* of resolutions for each anomaly type. However, they rated resolutions for strong anomalies as less plausible than those for weak anomalies. In a separate study, participants rated the explanations themselves as significantly less plausible in the face of strong rather than weak anomalies. That is, the difference in plausibility ratings of the resolutions for the anomalies did, indeed, map onto how problematic the anomalies were for the explanations (Koslowski, Libby, O'Connor, Rush, & Golub, in preparation). Roughly, then, it was fairly easy to generate resolutions even for strong anomalies; the problem was that the resolutions generated were not very plausible. Put differently, the resolutions were not very consistent with the network of related information that people had about, for example, adoptive parents. People could explicitly generate resolutions but knew, explicitly, as well, that the resolutions were deficient.

Merely thinking about an issue can also make background information explicitly relevant. One of the points that Langer (1989, 1997) makes is that mindfulness (an explicit awareness of what one is doing) has profound effects on memory, learning, and so forth. We have suggestive evidence that mindfulness, or explicitly thinking about an issue, can also affect how one evaluates an explanation as well as how one evaluates information that is anomalous to it.

Masnick (1999) interviewed college students about their beliefs about affirmative action in general as well as about various related subbeliefs. During the interview, Masnick noted the inconsistencies in each participant's responses. In the experimental group, each inconsistency was pointed out (as something that "some people" have identified as inconsistent) and the participant was asked "to give some thought to that." In the control condition, inconsistencies were not pointed out. Instead, participants were asked whether they might have "anything further to say after they spent a bit more time thinking about some of the initial questions." However, this was not in the context of mulling over inconsistencies.

Approximately two weeks later, participants in both groups were reinterviewed. There was no significant difference in the general attitude about affirmative action between the experimental or "challenged" and the control or "think about" group. However, the absolute change on the *general* question about affirmative action could be predicted by the change on the *subquestions,* and this was true for both groups—those that were challenged about the inconsistencies, and those who were simply told to give some thought to the initial questions.

These results are relevant for several reasons. One is that they demonstrate that small, local changes can accumulate and lead to larger, global changes. Another is that having attention drawn to inconsistencies in one's reasoning can lead to changes in reasoning. The third is that even examining (thinking about) one's reasoning can lead the reasoning to change. We cannot tell for certain why it was that just thinking about the argument lead to belief change. However, it might be that focused thinking led people to be more mindful or explicitly aware of their reasoning and, thus, more explicitly aware of how it might be flawed. Swiderek's (1999) research, below, also speaks to these issues.

BACKGROUND INFORMATION AFFECTS ISSUES RELEVANT TO CONFIRMATION BIAS

The Role of Explicit Motivation and Alternative Explanations in Confirmation Bias

In the previous studies, theory (or explanation or mechanism) as well as background knowledge affected whether people would realize that information was evidentially relevant. However, in the psychological literature, there is also the suggestion that motivation can affect whether we see information as explicit evidence. A frequently cited argument in both psychology and previously in philosophy is that the gold standard for evaluating an explanation is to try explicitly to disconfirm it. Furthermore, the standard conclusion in the literature on confirmation bias is that attempts to disconfirm (rather than to confirm) beliefs are the exception rather than the rule; that is, people are said to suffer from a widespread confirmation bias. However, there are reasons for thinking that the standard conclusion ought to be framed in a more nuanced way (Capaldi & Proctor, 2008; Koslowski et al., 2008).

Much of the psychological literature operationalizes confirmation bias in terms of how people identify the set to which certain instances belong. For example, such tasks might present participants with three numbers (2, 4, 6) and ask them to discover what the set is (increasing numbers, even numbers, etc.) that the numbers represent (Wason, 1960), or with three cities and ask them to discover the set (for example, coastal cities) to which the cities belong. However, conclusions from this literature are often treated as being applicable to thinking in general (such as judicial reasoning, rationalization of public policy, etc.) and to scientific reasoning and to confirmation bias in particular (Baron, 1985; Kuhn et al., 1988; Nickerson, 1998; see Chapters 12 and 13, this volume).

In the psychological literature, confirmation bias, or the motivation to confirm, is said to affect reasoning in several ways. It prevents people from considering alternative explanations, and it blinds them to data that disconfirm or are inconsistent with their explanation. Thus, according to much of the literature, a motivation to disconfirm is seen as the gold standard for evaluating explanations, because it is a sound strategy for enabling people to see appropriate information as explicitly evidential.

However, there are several reasons for thinking that the standard conclusions from the confirmation bias literature ought to be more nuanced than they are. Here, I note only those that are relevant to the present chapter. One reason is that confirming a belief, disconfirming it, and considering an alternative are often treated as distinct, but they are not. For example, one way of disconfirming an explanation is to amass evidence that it was actually another explanation that accounts for the data; one way of confirming an explanation is to demonstrate that alternative explanations have been ruled out or controlled for. That is, both confirmation and disconfirmation require taking account of alternatives (Koslowski & Maqueda, 1993; Mynatt, Doherty, & Tweney, 1977.

Another reason is that behavior that might appear to be confirmation bias might actually be a bookkeeping strategy. For example, in Chapman and Chapman's (1967 a, b) classic studies of illusory correlations, participants were shown fabricated data in which patient diagnoses were paired with sets of symptoms in a way that was in fact random. Despite the randomness, clinicians as well as college students mistakenly reported that there was a correlation, for example, between drawing atypical eyes and feeling suspicious. However, the incidence of reported correlations decreased by almost half when participants were allowed a pencil and paper in case they wanted to keep records (Chapman & Chapman,1967 b). Note, also, that the relation between diagnoses and symptoms was random. Note, further, that if one lacks the cognitive wherewithal to generate an alternative to account for both initial data and anomalous data, then one will be left with no bookkeeping prop if one rejects the only explanation available. The notion of bias and bookkeeping will come up again, below.

Furthermore, it is very difficult to infer motivation from behavior; it is not always possible to tell whether the motivation for doing a particular test is to see whether the result of the test would confirm the hypothesis or would disconfirm it (McKenzie, 2006; Wetherick, 1962).

In addition, as noted earlier, people sometimes lack the background knowledge to be able to generate alternative explanations and likely to generate what disconfirming evidence would look like.

To avoid inferring motivation from behavior, we manipulated motivation experimentally, explicitly telling participants either to confirm ("find information that will help support"), to disconfirm ("find information that will help to undermine"), or to evaluate an explanation for each of several events. To deal with the possibility that our participants would lack the background information to be able to generate alternatives and to generate disconfirming evidence, we presented them with both. Some participants were presented with one possible explanation (target explanation) for each event, and some with two (target and alternative), and participants rated their agreement with each explanation.

We then presented 12 pieces of information for each event: 4 pieces consistent with the target, 4 pieces consistent with the alternative, and 4 pieces that were causally neutral. For example, one event concerned the mass extinction of large mammals (such as saber-toothed tigers) that took place during the human migration to North America. In this example, the target explanation was overhunting by the migrating humans, and the alternative was that the animals lacked immunity to new diseases carried by the humans. Information consistent with (and thus supportive of) the hunting explanation was that the sites included evidence of tools, which would have been useful for hunting. Information consistent with disease transmission was that small animals such as dogs and birds were brought to North America by the migrating humans and preyed upon by the large mammals, which suggested a vector by which diseases might have been transmitted. Causally neutral information included the observation that this was a time when flower-bearing plants were proliferating.

Note also that information that supported the hunting explanation (the presence of tools) simultaneously undermined the disease explanation, because the disease explanation was unable to account for it. Put differently, information consistent with one explanation was less plausibly consistent with the other. That is, such information was diagnostic. The results were not what confirmation bias would have led us to expect.

Specifically, participants who had been told to disconfirm were more likely than those told to confirm to identify, as relevant, information that was in fact causally neutral. In addition, being told to confirm an explanation did not confer an advantage; it did not make people more likely to discern information that was supportive. In particular, people told to confirm were no more likely than people told to evaluate to identify, as supportive, information that was consistent with the explanation. Finally, in both the confirm and the disconfirm conditions, the presence of an alternative explanation reduced the likelihood of choosing inappropriate information (that is, supportive information in the disconfirm condition and undermining information in the confirm condition). In terms of an earlier point, an alternative made it clear that, although information could be distorted to be consistent with one explanation, it was more *plausibly* consistent with the alternative. That is, the presence of an alternative provided a framework for assessing the *relative* weight of the evidence in favor of each of the two explanations. In short, when motivation is explicitly manipulated, a motivation to disconfirm is not necessarily the gold standard of reasoning (Koslowski, Marasia, & Liang, 2008). Adopting an explicit motivation to disconfirm was less successful than explicitly considering an alternative.

Effects of Reduced Information-Processing Load on Whether Information Is Seen as Evidential

One criterion of sound scientific reasoning is the ability to realize that diagnostic information, which distinguishes between two competing explanations, is more useful that information that is compatible with both. In the previous study, people were more likely to identify, as relevant, information that was diagnostically consistent (or inconsistent) with an explanation rather than information that was causally neutral. However, the information presented to participants did not include any information that was nondiagnostic in the sense of being causally compatible with *both* explanations. In the following study, we did include such information.

The literature on confirmation bias typically concludes that people often have difficulty realizing that diagnostic information is more useful than nondiagnostic information. For example, when primed to ask questions to decide whether someone was an introvert (or an extrovert), people often asked questions whose answers would fail to distinguish between the two possibilities (Snyder, 1981; Snyder & Swann, 1978). (Of course, in terms of an earlier suggestion, one possibility was that people were using the primed personality type as a bookkeeping aid, that is, as a framework for keeping track of which questions to ask.)

We presented people with events to be explained along with two possible explanations for each event, a target and an alternative. After people rated how likely they found each explanation, we presented them with three types of information: some was diagnostic and distinguished between the two explanations by being more plausibly consistent with the target explanation; some was nondiagnostic and thus consistent with both explanations; and some was causally neutral. Thus, in terms of the disease versus hunting explanations for the extinction of large prehistoric mammals, diagnostic information was (depending on condition) consistent with either disease transmission or hunting, as in the study described above. Nondiagnostic information was consistent with both (humans built their settlements close to where many of the large animals roamed), while neutral information was causally relevant to neither explanation (the gravesites of the migrating humans contained decorative beads). After they read the three types of information, we then asked people to rate how useful each piece of information was likely to be in helping to decide which of the explanations was the more accurate. (Note that we presented participants with the various types of evidence, because we were interested, not in their ability to generate it, but in whether they could discern that it was useful.)

What was not surprising was that neutral information always received the lowest rating. What was surprising was that whether people rated diagnostic information as more useful than nondiagnostic information depended on what their initial ratings were of the target explanation. Roughly, the higher they rated the target explanation, the more likely they were to find diagnostic information more useful than nondiagnostic information. For example, when the target explanation received a fairly low rating of 3 (on a 7-point scale), people did not distinguish diagnostic from nondiagnostic information. However, when people found the target explanation very likely, that is, when they gave it a fairly high rating of 6, then they did rate the diagnostic information as more useful than the nondiagnostic information.

This is not what confirmation bias would have predicted. Confirmation bias would have suggested that, the more one agreed initially with an explanation, the more one would be tempted to find relevant information that was consistent with it. And, since nondiagnostic information was compatible with both explanations, it was compatible with the target. However, as initial ratings of the target increased, ratings of the nondiagnostic information did not.

As for why ratings of diagnostic information did increase as initial target ratings increased, consider that there was a fair amount of information for the participants to deal with: one event, two possible explanations, and three pieces of information. In addition, to distinguish the three types of information, participants needed to keep in mind which piece of information was causally consistent with which explanation—or with neither. This likely resulted in a fairly heavy information-processing load. Participants who rated the target explanation as highly plausible in the first place might well have done so because they already had a network of background information into which they could situate the explanation. This, in turn, might have served as a framework they could rely on to evaluate the new information that was presented in the experiment, and this in turn might have made it easier for them to discriminate among the three types of information in terms of utility. To test this possibility, we ran a second study in which we asked people to reason about only two pieces of information.

In one condition, participants were presented with nondiagnostic as well as diagnostic information; in the other condition, they received nondiagnostic as well as neutral information. In this study, diagnostic information was rated as more useful than nondiagnostic information and nondiagnostic information was rated as more useful than neutral information. Especially relevant, given the preceding study, was that initial target ratings in this study had no effect on how relevant people rated any of the types of information, including the diagnostic information. That is, when initial target ratings affected ratings of diagnostic information, we suggest it was because using the target explanation as a framework (which one could do if one knew enough about it to give it a high rating) enabled people to better manage the cognitive load of dealing with a fair amount of additional information and a fair number of decisions.

CHANGING EXPLANATIONS BY UNDERMINING SUPPORTING BACKGROUND INFORMATION

Undermining the Mechanism Component of a Covariation

Many of us hold fairly general beliefs about all manner of things. However, beliefs are typically not isolated. Rather, they are embedded in and judged with respect to a network of related background information. Recall the example mentioned earlier about the network of information in which evolution is embedded. With respect to the construct of evolution, because it is a construct in science, it is likely that the network of related information is fairly explicit and that people can probably retrieve much of it if asked. However, this is probably not likely to be true for many general beliefs, because the related information that supports the belief may well be so standard, or implicit, that we rarely think of it. Thus, trying to disconfirm a belief, if it is a very general one, might be ineffective if one focuses only on the general belief. Rather, it might be more effective to disconfirm the individual subbeliefs that the general belief is supported by. That is, undermining the implicit subbeliefs might make it more likely that people will explicitly question the general belief.

By way of illustrating this point, consider that, often, when we have beliefs about what is causal, the belief consists of two components: a belief that two events covary and a belief—however implicit—that there is a mechanism that might make the covariation causal. Thus, we hypothesized that, to disconfirm a belief, one might need to disconfirm both components. To test this hypothesis, we interviewed college students and adolescents who held two widespread beliefs about covariation: that eating high-sugar rather than low-sugar candy before bedtime makes it difficult for children to fall asleep and that drinking low-fat rather than whole milk has no effect on sleeplessness. The mechanism that participants gave for believing that sugar caused sleeplessness was some variant of an increase in energy ("Candy gives you a second wind" or "It revs you up"). To disconfirm only the covariation component of each belief, we noted that eating high-sugar candy in fact does not cause sleeplessness, but that drinking whole milk does. To disconfirm the mechanism component for (causal) candy, we noted, in addition, that candy is typically consumed as chocolate, which has caffeine, and it is actually the caffeine that causes sleeplessness. To disconfirm the (noncausal) milk belief, the mechanism we provided was that consuming high-fat milk causes the brain to produce serotonin, which induces sleep. We also confirmed each of the beliefs, with either covariation information alone or else with covariation and mechanism information (Koslowski, 1996, Chapter 10).

When the evidence was confirming, covariation plus mechanism information (as opposed to covariation information alone) did more to confirm the causal belief than to confirm the noncausal belief. (This was not surprising, as the confirming condition simply reinforced a belief that participants already held.) Similarly, when the evidence was *dis*confirming, covariation plus mechanism information (as opposed to covariation information alone) did more to disconfirm a causal belief (about candy) than to disconfirm a noncausal belief (about milk). What was surprising, in light of the assumed prevalence of confirmation bias, was the fact that it did not matter whether the evidence confirmed the initial belief or disconfirmed it; people were as receptive to mechanism information when one of their beliefs was disconfirmed as when it was confirmed.

Undermining the Network of Information in Which a Belief Is Embedded

Two other studies are also relevant to the question of why people often fail to reject theories in light of anomalous data. The reasoning behind both studies was that, if any single belief is embedded in, and judged with respect to, a network of related beliefs, then to disconfirm the belief, one may need to chip away at the network of related beliefs that support it. To some extent, we examined this in the previous study, by taking account of the mechanism information that supported a belief. In the following two studies, we examined beliefs embedded in much wider networks and of substantially more importance than beliefs about sugar and milk.

One study was already discussed in a different context, namely, Masnick's study of participants whose subbeliefs about affirmative action were challenged. Recall that, in her study, the absolute change on the *general* question about affirmative action could be predicted by the change on the *subquestions*. Furthermore, this was true for both the experimental condition (in which participants were challenged) and in the control condition (in which participants were simply told to think about their subbeliefs).

Swiderek (1999) interviewed college students about their opinions on capital punishment. Participants were also interviewed about several subsidiary beliefs that underlie the general beliefs (for example, whether capital punishment in fact reduces the incidence of capital crimes, is cost-effective, applied in a nonbiased way, etc). Instead of pointing out inconsistencies in the participants' network of beliefs, Swiderek (1999) presented people with evidence that challenged or disconfirmed each subbelief—regardless of whether participants agreed or disagreed with the subbelief. For example, participants who believed that capital punishment did *not* reduce the incidence of capital crimes were shown data that the rate of capital crimes was reduced after a state introduced capital punishment. Those who believed it *did* reduce capital crimes were given the same information but told that the decrease only held for the first three months after capital punishment was introduced. In the control condition, participants were not presented with disconfirming evidence, but were merely asked to think further about the issue.

Participants were brought back for a second interview approximately two weeks after the first interview and were asked again about their general belief as well as about the subbeliefs. There were no differences between experimental and control groups on any of the subsidiary questions about capital punishment. However, there were differences between the two groups in the magnitude of attitude change, from the first to the second interview, on the overall question about capital punishment. Furthermore, the sum of the change on the individual subbeliefs predicted 85% of the variance of the change on the general belief in capital punishment.

> Although this finding may appear counterintuitive, it is not if viewed from the perspective that any single belief is embedded in and judged with respect to a web of related background knowledge. From this perspective, when presented with disconfirming evidence, small local shifts may occur (that is, changes on the sub-beliefs) that are non-significant when considered individually. However, when taken together, the changes may be cumulative and result in a significant overall change in the general belief. And, indeed, when the local changes (on the more specific questions) were taken as a whole, there was a strong relation between the change on the overall question and the local changes. The results echo the point that any explanation is embedded in a network of related information and, thus, to change any particular belief, one must also change the related beliefs in which it is embedded. (Swiderek, 1999)

Put in terms of the explicit/implicit distinction, when people are made explicitly aware of the shortcomings of the individual subbeliefs that underwrite a general belief, they are more likely to question, explicitly, how compelling the general belief is.

SUMMARY AND CONCLUSIONS

Descriptions of actual scientific practice (as opposed to idealized models of science) stress the importance of explanation or mechanism, the effects of possible alternative explanations, and the network of related background information in which any explanation is embedded. In addition, in actual scientific practice, explanation and evidence are not distinct entities, but are interdependent. It would appear that, in broad outline, this description also characterizes the causal reasoning of nonscientists, as well.

Specifically, people are more likely to realize that information is evidentially relevant to accounting for an event when there is a possible explanation that can incorporate the information into a broader causal framework. Additionally, a possible explanation can also affect what sorts of information we search for in trying to account for an event. Conversely, the presence of background information can also help us generate alternative explanations and can help us decide whether anomalies to an explanation are strong or weak.

Furthermore, background information, possible alternative explanations, and the networks of related information are also relevant to questions related to how we come to realize that an explanation is deficient—a question that is also relevant to confirmation bias. In that regard, background information can help us judge how problematic anomalous information is for an explanation. Specifically, the presence of alternative possible explanations helps us assess a target explanation, because alternatives help us avoid distorting information to make it consistent, after the fact, with a target explanation, because alternatives can make it clear that information might be more *plausibly* consistent with the alternative. That is, the presence of an alternative can provide a framework for assessing the *relative* weight of the evidence in favor of each of the two explanations. Furthermore, in general, one way of disconfirming an explanation is to undermine the network of information in which the explanation is embedded.

There are also certain things that can interfere with our ability to treat background information and explanation as interdependent. Information-processing limitations, for example, can interfere with our ability to recognize discriminating evidence as more useful than nondiscriminating evidence. Similarly, a lack of background information can prevent us from considering (or generating) a plausible alternative explanation. This, in turn, can be affected by both the culture and the historical time period in which the causal reasoning is taking place. In short, the principles of sound science are heuristics, not guarantees, for generating good outcomes.

Finally, one might ask what the reason or mechanism is for why background information in necessary for causal attribution? Another way of framing this question is to ask what one might rely on if considering background information were not an option. The classic answer, as noted earlier, is that one might rely on idealized models of science based on various types of formal rules, such as the Humean indices of causation, and that is exactly why background information is so crucial to anyone who must make pragmatically constrained decisions. In an ideal (or a simple) world, every correlation might indicate a causal relation. However, the actual world is neither ideal nor simple; rather, the actual world is rife with correlations and with events that are temporally and physically contiguous with particular effects. The problem is that few of these are actually causal. Because it is simply not possible to consider all of them, we rely on background information to winnow out those likely to be causal from those likely to be artifactual. Analogously, relying on Bayesian approaches is also not a tractable approach without considering background information; how else to decide which of the many possible priors should actually be considered? Without relying on background information, we would be awash in a sea of data. Relying on background information (including information about theory or mechanism as well as alternative explanations) allows us to decide which data are most likely to indicate causality. Of course, this strategy does not guarantee success; the background information may be incomplete (often because it remains to be discovered) or it may be downright wrong. However, given the world we live in (and in which we try to make causal judgments), relying on background information may be the only option we have.

ACKNOWLEDGMENTS

Tremendous thanks to Richard N. Boyd for discussion, comments, and suggestions.

REFERENCES

Baron, J. (1985). *Rationality and intelligence.* Cambridge: Cambridge University Press.
Block, N. J. (1995). How heritability misleads about race. *Cognition, 56,* 99–128.

Block, N. J., & Dworkin, G. (1974). IQ: Heritability and inequality, Part 2. *Philosophy and Public Affairs, 3*(4), 331–409.

Bonawitz, E. B., & Lombrozo, T. (2011). *Occam's rattle: Children's use of simplicity and probability to constrain inference. Developmental Psychology, 44,* 186–196.

Boyd, R. (1989). What realism implies and what it does not. *Dialectica, 43,* 5–29.

Bullock, M. (1991). *Scientific reasoning in elementary school: Developmental and individual differences.* Paper presented at the biennial meetings of the Society for Research in Child Development, Seattle, Washington.

Bullock, M., Gelman, R., & Baillargeon, R. (1982). Development of causal reasoning. In J. W. Friedman (Ed.), *Psychology of Time* (209–253). New York: Academic Press.

Bullock, M., & Ziegler, A. (1999). Scientific reasoning: Developmental and individual differences. In. F. E. Weinert & W. Schneider (Eds.), *Individual development from 3 to 12: Findings from the Munich longitudinal study* (pp. 38–54). Cambridge: Cambridge University Press.

Capaldi, E. J., & Proctor, R. W. (2008). Are theories to be evaluated in isolation or relative to alternatives? An abductive view. *American Journal of Psychology, 121*(4), 617–641.

Chapman, L. J. & Chapman, J. P. (1967 a). Illusory correlation in observational report. *Journal of Verbal Learning and Verbal Behavior, 3,* 151–155.

Chapman, L. J., & Chapman, J. P. (1967 b). Genesis of popular but erroneous psychodiagnostic observations. *Journal of Abnormal Psychology, 3,* 193–204.

Chinn, C. A., & Brewer, W. F. (1998). An empirical test of a taxonomy of responses to anomalous data in science. *Journal of Research in Science Teaching, 35,* 623–654.

Darden, L. (2006). *Reasoning in biological discoveries: Essays on mechanisms, interfield relations and anomaly resolution.* Cambridge: Cambridge University Press.

Fine, A. (1984). The natural ontological attitude. In J. Leplin (Ed.), *Scientific realism* (pp. 83–107). Berkeley: University of California Press.

Fraser, S. (1995). *The bell curve wars: Race, intelligence and the future of America.* New York: Basic Books.

Gelman, S. A. (2003). *The essential child: Origins of essentialism in everyday thought.* New York: Oxford University Press.

Gero, J. M., & Conkey, M. W. (Eds.). (1991). *Engendering archaeology.* Oxford, UK: Basil Blackwell.

Giere, R. (2002). Scientific cognition as distributed cognition. In P. Carruthers, S. Stich, & M. Siegel (Eds.), *The cognitive basis of science: Multidisciplinary approaches* (pp. 285–299). Cambridge: Cambridge University Press.

Gopnik, A., & Schulz, L. (2007). *Causal learning.* Oxford: Oxford University Press.

Haith, M. M., & Benson, J. B. (1998). Infant cognition. In D. Kuhn & R. Siegler (Eds.), *Handbook of child psychology: Vol. 2. Cognition, perception, and language* (pp. 199–254). Hoboken, NJ: John Wiley.

Harris, P. L. (2000). On not falling down to earth: Children's metaphysical questions. In K. S. Rosengren, C. N. Johnson, & P. L. Harris (Eds.), *Imagining the impossible: Magical, scientific, and religious thinking in children* (pp. 157–178). Cambridge: Cambridge University Press.

Herrnstein, R. J. (1973). *I.Q. in the meritocracy.* Boston, Atlantic Monthly Press.

Herrnstein, R. J., & Murray, C. (1994). *The bell curve.* New York: Free Press.

Kamin, L. J. (1974). *The science and politics of I.Q.* Potomac, MD: Lawrence Erlbaum.

Kaplan, A. (1964). *The conduct of inquiry: Methodology for behavioral science.* Scranton, PA: Chandler.

Klahr, D. (2000). *Exploring science: The cognition and development of discovery processes.* Cambridge, MA: MIT Press.

Koenig, M., & Harris, P. L. (2005) The role of social cognition in early trust. *Trends in Cognitive Science, 9,* 457–459.

Koslowski, B. (1996). *Theory and evidence: The development of scientific reasoning.* Cambridge, MA: MIT Press.

Koslowski, B., Beckman, L., Bowers, E., DeVito, J., Wonderly, B., & Vermeylen, F. (2008, July). *The cognitive basis for confirmation bias is more nuanced than one might expect.* Paper presented at the meetings of the International Society of the Psychology of Science and Technology, Berlin, Germany.

Koslowski, B., Cohen, L., & Fleury, J. (2010). *The conceptual understanding of ruling out alternatives does not guarantee the technical ability to do so.* Paper presented at the biennial meetings of the International Society for the Psychology of Science and Technology, Berkeley, California.

Koslowski, B., Hildreth, A., Fried, L., & Waldron, C. Whether information is seen as evidence depends on what the competing alternative explanation is. Manuscript in preparation.

Koslowski, B., Libby, L. A., O'Connor, K., Rush, K., & Golub, N. What makes some anomalies more problematic than others. Manuscript in preparation.

Koslowski, B., & Maqueda, M. (1993). What is confirmation bias and when do people have it? *Merrill-Palmer Quarterly, 39*(1), Special issue on: *The development of rationality and critical thinking,* 104–130.

Koslowski, B., Marasia, J., Chelenza, M., & Dublin, R. (2008) Information becomes evidence when an explanation can incorporate it into a causal framework. *Cognitive Development, 23,* 472–487.

Koslowski, B., Marasia, J., & Liang, V. (2008, July). *Confirming a belief is not necessarily problematic; disconfirming a belief is not necessarily good.* Paper presented at the meetings of the International Congress of Psychology, Berlin, Germany.

Koslowski, B., & Masnick, A. M. (2002). Causal reasoning. In U. Goswami (Ed.), *Handbook of child cognitive development* (pp. 257–281). Oxford, UK: Blackwell.

Koslowski, B., & Pierce, A. (1981, April). *Children's spontaneous explanations and requests for explanations.* Paper presented to the Society for Research in Child Development, Boston.

Koslowski, B., P'Ng, J., Kim, J., Libby, L.A., & Rush, K. When told to evaluate an explanation, people request information about alternatives. Manuscript in preparation.

Koslowski, B., Spilton, D., & Snipper, A. (1981). Children's beliefs about instances of mechanical and electrical causation. *Journal of Applied Developmental Psychology, 2,* 189–210.

Koslowski, B., & Thompson, S. L. (2002). Theorizing is important, and collateral information constrains how well it is done. In P. Carruthers, S. Stitch, & M. Siegal (Eds.), *The cognitive bases of science: Multidisciplinary approaches* (pp. 171–192). Cambridge: Cambridge University Press.

Kuhn, D., Amsel, E., & O'Loughlin, M. (1988). *The development of scientific thinking skills.* Orlando, FL: Academic Press.

Langer, E. J. (1989). *Mindfulness.* Reading, MA: Addison-Wesley.

Langer, E. J. (1997). *The power of mindful learning.* Reading, MA: Addison-Wesley.

Lipton, P. (1991). *Inference to the best explanation.* London: Routledge.

Lipton, P. (1993). Is the best good enough? *Proceedings of the Aristotelian Society, 93*(2), 89–104.

Lombrozo, T. (2007). Simplicity and probability in causal explanation. *Cognitive Psychology, 55*, 232–257.

Lutz, D., & Keil, F.C. (2002). Early understanding of the division of cognitive labor. *Child Development, 73*, 1073–1084.

Magnani, L. (2001). *Abduction, reason, and science: Processes of discovery and explanation.* Dordrecht, Netherlands: Kluwer Academic.

Masnick, A. M. (1999) *Belief patterns and the intersection of cognitive and social factors.* Unpublished doctoral dissertation, Cornell University.

McKenzie, C. R. M. (2006). Increased sensitivity to differentially diagnostic answers using familiar materials: Implications for confirmation bias. *Memory & Cognition, 34*(3), 577–588.

Mynatt, C. R., Doherty, M. E., and Tweeney, R. D. (1977). Confirmation bias in a simulated research environment: An experimental study of scientific inference. *Quarterly Journal of Experimental Psychology, 29*, 85–95. Excerpts reprinted in P. N. Johnson-Laird & P. C. Wason (Eds.), *Thinking: Readings in cognitive science* (pp. 315–325). Cambridge: Cambridge University Press.

Nickerson, R. S. (1998) Confirmation bias: a ubiquitous phenomenon in many guises. *Review of General Psychology, 2*, 175–220.

Popper, K. R. (1959). *The logic of scientific discovery.* London: Hutchinson.

Psillos, S., (1999). *Scientific realism: How science tracks truth.* New York: Routledge.

Quine, W., & Ullian, J. (1970). *The web of belief.* New York: Random House.

Salmon, W. (1990). Rationality and objectivity in science or Tom Kuhn meets Tom Bayes. In C. W. Savage (Ed.), *Scientific theories*, Minnesota Studies in the Philosophy of Science (Vol. 14, pp. 175–204). Minneapolis: University of Minnesota Press.

Samarapungavan, A. (1992). Children's judgments in theory choice tasks: Scientific rationality in childhood. *Cognition, 45*, 1–32.

Schulz, L. E., & Gopnik, A. (2004). Causal learning across domains. *Developmental Psychology, 40*, 162–176.

Shultz, T. R., & Mendelson, R. (1975). The use of covariation as a principle of causal analysis. *Child Development, 46*, 394–399.

Slocum, S. (1975). *Woman the gatherer: Male bias in anthropology.* In R. Reiter (Ed.), Toward an anthropology of women (pp. 36–50). New York: Monthly Review Press.

Snyder, M. (1981). *Seek and ye shall find: Testing hypotheses about other people.* In E. T Higgins, C. P. Heiman, & M. P. Zanna (Eds.), *Social cognition: The Ontario symposium on personality and social psychology* (pp. 277–303). Hillsdale, NJ: Lawrence Erlbaum.

Snyder, M., & Swann, W. B. (1978). Hypothesis testing processes in social interaction. *Journal of Personality and Social Psychology, 36*, 1202–1212.

Steele, C. M. (1997). A threat in the air: How stereotypes shape the intellectual identities and performance of women and African-Americans. *American Psychologist, 52*, 613–629.

Steele, C. M., & Aronson, J. (1995). Stereotype threat and the intellectual test performance of African-Americans. *Journal of Personality and Social Psychology, 69*, 797–811.

Swiderek, M. R. (1999) *Beliefs can change in response to disconfirming evidence and can do so in complicated ways, but only if collateral beliefs are disconfirmed.* Unpublished doctoral dissertation, Cornell University.

Thagard, P. (1989). Explanatory coherence. *Behavioral and Brain Sciences, 12*, 435–502.

Thagard, P., & Verbeurgt, K. (1998). Coherence as constraint satisfaction. *Cognitive Science, 22*, 1–24.

Van Fraassen, B. (1980). *The scientific image.* Oxford: Oxford University Press.

Vosniadou, S., & Brewer, W. F. (1992). Mental models of the earth: A study of conceptual change in childhood. *Cognitive Psychology, 24,* 535–585.

Wason, P. C. (1960). On the failure to eliminate hypotheses in a conceptual task. *Quarterly Journal of Experimental Psychology, 12,* 129–140.

Wetherick, N. E. (1962). Eliminative and enumerative behaviour in a conceptual task. *Quarterly Journal of Experimental Psychology, 14,* 246–249.

Worth, R. H., Jr., (1995). No choice but war: The United States embargo against Japan and the eruption of war in the pacific. Jefferson, North Carolina: McFarland.

Zimmerman, C. (2000). The development of scientific reasoning skills. *Developmental Review, 20,* 99–149.

Zimmerman, C. (2007). The development of scientific thinking skills in elementary and middle school. *Developmental Review, 27,* 172–223.

7 Detecting, Classifying, and Remediating

Children's Explicit and Implicit Misconceptions about Experimental Design

Stephanie A. Siler and David Klahr

It is well established that children do not come to their science classes as "blank slates" but rather, as learners with preconceptions—sometimes deeply entrenched—about the natural world and ways to explore that world in order to learn more about it (Stathopoulou & Vosniadou, 2007). Children's experiences may also lead to biases—both implicit and explicit—that only come to be revealed in the context of instruction. To the extent that preconceptions about both scientific knowledge and scientific processes are *mis*conceptions, they are important to identify and remediate (Vosniadou, Ioannides, Dimitrakopoulou, & Papademetriou, 2001). Once an implicit misconception has been identified by appropriate assessments, it may be possible to craft instruction that not only renders the implicit misconception explicit, but also directly engages, builds upon, and/or remediates that misconception.

Research on early science learning has focused on two relatively distinct classes of misconceptions. One line of work has focused on *misconceptions in specific domains,* such as heat and temperature (e.g., Slotta, Chi, & Joram, 1995; Wiser & Carey, 1983), the solar system (e.g., Vosniadou, 1994; Vosniadou & Brewer, 1992, 1994), mass and density (e.g., Smith, Carey, & Wiser, 1985), or forces and motion (e.g., Clement, 1983; diSessa, 1993; McCloskey, 1983; Minstrell, 1982). A second line of work has examined children's misconceptions in the abstract and *domain-general scientific procedures* that produce and verify that domain knowledge, such as discriminating determinate from indeterminate evidence in order to make causal inferences (Bindra, Clarke, & Schultz, 1980; Braine & Rumain, 1983; Byrnes & Overton, 1986; Fay & Klahr, 1996; Klahr, Fay, & Dunbar, 1993; Masnick & Klahr,

2003; Morris & Sloutsky, 2002; Pieraut-Le Bonniec, 1980), interpreting anomalous data (e.g., Chinn & Brewer, 1993), understanding the uses of scientific models (e.g., Grosslight, Unger, Jay, & Smith, 1991), and constructing coherent and consistent arguments (Kuhn, 1991; Zohar & Nemet, 2001). In this chapter, we focus on these latter types of difficulties: specifically, the errors and misconceptions that students bring to the task of learning how to design and interpret simple experiments.

Our analysis is motivated by, and based on, a corpus of instructional dialogues produced during a series of extensive tutorial sessions with elementary and middle school students engaged in learning about experimental design concepts and procedures. These tutorial dialogues were generated as part of our ongoing project aimed at building an adaptive computer tutor to teach the basic procedural and conceptual knowledge associated with experimental design (e.g., Klahr, Triona, Strand-Cary, & Siler, 2008; Siler, Klahr, Strand-Cary, Magaro, & Willows, 2009). Because diagnosing students' knowledge and beliefs is an essential prerequisite for instruction in an adaptive tutor, a major goal of the project has been to identify children's conceptual and procedural misconceptions about experimentation. In pursuing this goal, we have drawn on earlier work on this topic by others (e.g., Kuhn, Garcia-Mila, Zohar, & Andersen, 1995; Schauble, Klopfer, & Raghavan, 1991), as well as past work done in our laboratory (e.g., Chen & Klahr, 1999; Klahr & Nigam, 2004; Strand-Cary & Klahr, 2008; Toth, Klahr, & Chen, 2000). We start by outlining an expert model of basic experimental design. Then we describe the phases of our computer development project from which we draw our data, and finally, we describe the types of misconceptions about experimental design expressed by the fifth- through seventh-grade students who participated in the various project phases. We comment on the nature of these misconceptions, including their level of explicitness and their relationship to students' domain-specific beliefs.

Expert Model of Experimental Design

Schauble et al. (1991) were the first to observe the important distinction between two classes of goals that children bring to their initial experimental activities—"science" goals, and "engineering" goals. In order to design an experiment that will support valid causal inferences about a particular variable, one needs to understand the "science" goal underlying experimentation, which is to *find out about* the causal status of that variable. For example, when a child justifies an experimental design by saying: "I wanted *to see if* the different engines would affect the speed of the rockets," she is explicitly expressing the science goal of trying to find out about the causal relation between engine type and rocket speed. A science goal may be considered abstract in two senses. First, it may be applied in any specific context, in or out of school. Secondly, because a science goal involves finding out about one or more variables, it is abstract in the sense that students must be willing to suspend any beliefs about the effect(s) of the variable(s) they are testing in order to design and perform that experiment.

This contrasts a common alternative goal—termed an "engineering goal" by Schauble et al. (1991)—Students with engineering goals attempt to "engineer" a particular desirable outcome (e.g., make the fastest rocket) rather than attempt to find out about those effects via experimentation, as with a science goal. As with science goals, engineering goals are abstract in the sense that students can and do apply them across contexts. However, because applying an engineering goal may involve one's beliefs about some domain-specific variables to produce a domain-specific result, an engineering goal may also be concrete.

In addition to holding a science goal, one needs to know, and correctly execute, the abstract procedural steps involved in designing an unconfounded experiment. We call this procedure the Control of Variables Strategy (CVS) and its abstract rules are shown in Table 7.1.

To apply the basic procedure, one must identify the particular variable whose causal status is being investigated, the "target" variable (Rule 1), contrast the levels of the variable being tested across *at least two* conditions to compare their relative effects on the outcome (Rule 2), and control all other variables (Rule 3)[1]. Only after designing such an experiment can one compare the outcomes of the contrasting conditions (i.e., A1 and A2) to infer whether there was an effect of the target variable.

This procedure is abstract in two ways. First, like engineering and science goals, CVS is domain-general and can in principle be applied in any context. Second, the procedure is not dependent on the specific variables and values in the context in which CVS is applied. In addition to this procedural knowledge, we believe it is important for students to also understand the underlying rationale of CVS—that if one runs the experiment and finds different results, then one can logically attribute that result to the one variable that is contrasted. This logical understanding is also not dependent on the specific variables involved.

Overview of Identified Difficulties

Although the core procedure for CVS can be captured in the few simple rules listed in Table 7.1, their simplicity makes them no easier to master and apply in a wide

Table 7.1 Abstract representation of rules for designing, running, and evaluating an unconfounded experiment

In a multivariable situation, if your goal is to determine whether or not a variable plays a causal role in outcome A, then

Rule 1: Identify that variable (X) and its values:

Rule 2: Create a contrast:

a. In Condition 1, Set X to Value 1.

b. In Condition 2, Set X to Value 2.

Rule 3: Set all other variables (Y, Z, W) to the same values in both conditions.

"Run" the experiment: measure A1 and A2.

If A1≠ A2, then X is causal.

[1] If variable effects are not assumed to be independent, but rather, interact, this procedure must be repeated for all combinations of nontarget variable levels.

range of contexts than does the simplicity of Newton's Second Law make it easy to master, recognize when it is applicable, and apply correctly to physics problems. Kuhn et al.'s (1995) classic study demonstrated that—in a variety of scientific discovery tasks in which participants explored the effects of several variables—even after 20 sessions spread over 10 weeks, fewer than 25% of fourth graders' inferences were valid. For this reason, the procedural—as well as the conceptual or logical—underpinnings of CVS have been the focus of the CVS instruction conducted in our laboratory (e.g., Chen & Klahr, 1999; Strand-Cary & Klahr, 2008).

Similarly, we have found that many children—especially younger children and those with weak science backgrounds—failed to learn from the simple nonadaptive or "straight-line" instruction that we used in our earlier studies with middle-class children (e.g., Klahr & Li, 2005). This instruction focused on the logical underpinnings of the CVS procedural steps.

Over the past several years we have been incrementally developing a computer tutor intended to aid students who had difficulty with the "straight-line" instruction. Our project involves human tutoring of those students who failed to learn from the given instruction at different phases of development. Those instructional moves found to be effective for different knowledge states inform tutor development. Extensive analyses of students' responses during instruction and dialogues from these individualized tutoring sessions revealed that students make characteristic mistakes that interfere with their learning and transfer of CVS.

In addition to these errors, there were three general schema-related misconceptions related to the goal of instruction found in our data:

- Engineering goals, described earlier;
- "Fairness" goals, in which students designed comparisons that were "fair," or could produce the same results. There are several different forms these first two schema-related misconceptions may take, which will be elaborated on in the results section;
- "Domain knowledge" goals, in which students misinterpret the purpose of instruction as a discussion of variable effects within the given domain (e.g., that balls roll faster on a smooth ramp than on a rough ramp), rather than as learning about a domain-general procedure for designing experiments.

Thus, competence in experimental design requires more than mastering the three simple rules in Table 7.1. An inability to apply CVS might be caused by failure to: (a) understand the goal of the task—which may be related to underlying difficulties with abstract concepts and procedures, (b) consider the variable level when testing for causality, (c) understand the need to "compare and contrast" variable values across conditions, and (d) understand that the causal status of only one variable at a time can be investigated in a single experiment.

In the following sections we first describe the sources of our data. Then we provide a detailed taxonomy—organized by goal type—of the approaches students took when they were challenged to design experiments. From a human-cognition perspective, we were interested in the relationship between goal type (e.g., science or engineering) and students' expressions of beliefs about variable effects (i.e., application of domain knowledge goals). That is, when students applied the relatively more abstract science goal, were they also more likely to suspend their beliefs about variables other than the one(s) they were testing? For each approach, we consider whether or not it involves application of domain-specific beliefs. We also consider differences in the level of explicitness of different underlying goals. Of practical interest to us was how those approaches, which reveal failures of one or more of the knowledge components listed above, could inform the instructional and remedial strategies used by the tutor.

METHODS

Data Sources for Errors and Goal Misconceptions

Data were collected during several phases of an ongoing design-based research project aimed at developing a computer tutor (TED, for "Training in Experimental Design") that provides individualized, adaptive instruction on experimental design to elementary and middle school students. The starting point for the project was the instruction procedure—designed by Klahr and colleagues in studies previously reported (e.g., Chen & Klahr, 1999; Klahr & Nigam, 2004; Strand-Cary & Klahr, 2008; Toth, Klahr, & Chen, 2000)—given by humans using physical materials. This instruction involves asking students to evaluate several experiments, and, for each, further prompting them to consider whether the experiment would allow them to know if the target variable caused an outcome and explain why it would or would not. Following a series of such questions, students are then given an explanation for why the experiment could or could not allow for conclusions about the effect of the target variable.

During the course of TED's development, the instructional steps, student prompts and queries, and CVS assessment procedures have been incrementally replaced with computerized components. In the course of the project, we have worked with students in the classroom, in individual or small-group tutoring sessions, and have developed a more complete taxonomy of the types of misconceptions about the purpose and process of experimental design that students bring to the science classroom. Throughout this process, we collected student data in the form of the experimental designs they created or evaluated, their written responses to probes and queries, and from the explanations they gave during one-on-one and small-group tutorial interactions. These data—collected in four separate development phases of the tutor—provide the basis of the analyses presented in this chapter. Because the nature of instruction varied across development phases, it is important to note that the data reported across phases cannot sensibly be compared (e.g., in terms of frequencies). Rather, the purpose of reporting student responses across phases—and within phases but across schools—was to demonstrate the possible range of student responses and their distributions, given the different instructional as well as assessment features.

The primary features of each phase are summarized in Table 7.2, briefly listed here, and described more fully later. In Phase 1, teachers delivered CVS

Table 7.2 Phases of TED project and data sources

	Phase 1	*Phase 2*	*Phase 3*	*Phase 4*
Number of classrooms (and type)	3 (2 low-SES: L1 & L2; 1 middle-SES: M1)	1 (low-SES) (L3)	2 (middle-SES) (M2a1 & M2a2)	4 (2 middle-SES: M2b1 & M2b2; 2 low-SES: L4 & L5)
Grade(s)	6	6	5 & 6	5
Number of students	73 (L1: 23; L2: 17; M1: 23)	21	58 (27 fifth; 31 sixth)	80 (M2b1/2: 50; L4: 16; L5: 14)
Type of instruction	Classroom	Classroom	One-on-one or small group	One-on-one (human & computer)
Remedial tutoring	One-on-one	One-on-one	(n/a)	(n/a)
Data sources	• In-class written responses • Oral and written responses during remedial tutoring	• In-class written responses • Oral and written responses during remedial tutoring • Written/typed explanations on follow-up Story-eval/design test (one year later) [a]	• Oral and written responses during individualized tutoring	• Oral and written responses during tutoring • Written/typed explanation responses on Story-eval/design pre and posttests and follow-up[a]

[a] Response classifications from this assessment point are shown in Table 7.6.

instruction in their classrooms using physical ramps apparatuses; in Phase 2, a teacher led instruction using Flash-based virtual ramps, guided by PowerPoint instructional slides; in Phase 3, human tutors provided computer-supported instruction with virtual materials. Finally, in Phase 4, students completed "straight-line" TED computer-delivered instruction, with no instruction-related human interaction. All student interface actions were recorded in log files, which were saved on a server and later analyzed.

Phase 1 included 73 sixth-grade students and three teachers at three different local K–8 Catholic schools. Two of these schools (classrooms L1 & L2) served primarily low-SES students (95% and 59% of students were eligible for free or reduced lunch, respectively), and one (with classroom M1) served primarily middle-SES students (about 20% were eligible for free or reduced lunch) and incorporated an inquiry-based science curriculum. Teachers were trained on CVS instruction similar to the one-to-one instruction given to students in Chen and Klahr (1999), which focused on the rules and rationales for setting up informative (i.e., unconfounded) experiments.

On the first day, all students completed a "Story-evaluation" pretest that required them to evaluate six experiments in three different domains (selling drinks, flying rockets, and baking cookies; an example item from each domain is shown in Figure 7.1), and to correct any experiments they evaluated as "bad." All variables had two possible values (e.g., for drinks, students could choose either lemonade or iced tea). As with assessments used in past studies (e.g., Chen & Klahr, 1999; Klahr & Nigam, 2004), students were not required to give explanations for their responses. Students completed this test in approximately 15 minutes.

The next day, the teacher introduced a ramps apparatus (Figure 7.2) and demonstrated the four variables, or ways it could be changed (slope—steep or not steep; starting position of the ball—at the top or middle; surface—smooth or rough; and ball type—yellow or pink). Students individually designed an experiment to test each of these four variables by entering values for all variables for each ramp into a table on the paper-pencil ramps pretest on their worksheets. Then the teacher led the lesson on CVS by first presenting a confounded experiment using two ramps and asking students to individually evaluate the experiment as a "fair" or "unfair" way to find out about the target variable on their worksheets. Following this evaluation, the teacher led a class-wide discussion on whether or not the experiment under discussion was informative (i.e., whether it enabled one to make inferences about effects of the target variable) and explained why, in fact, the initial (confounded) setup was not informative. As students identified the confounded variables, the teacher controlled them until eventually the experiment was unconfounded (i.e., informative for the target variable). Then students wrote explanations for why the corrected experiment was a fair way to find out about the target variable. This sequence was repeated two more times with other ramps setups and target variables. On the final day of instruction, students completed the ramps posttest in which they again designed experiments to test each of the four ramps variables.

Those students who created fewer than three out of four unconfounded ramps experiments on the posttest were tutored by a member of the research team.

These two pictures show how they tested whether or not the ***age of the child*** selling the drinks made a difference in how much they sell.
Look carefully at the pictures. Each one shows a time of day (Morning or Afternoon), a child (Older or Younger), and a drink (Iced Tea or Lemonade).
Do you think this is a good way to find out whether the *age of the child* (Older or Younger) makes a difference in how much they sell?

(a) **If you think it is a good way, then circle the word "Good" below. If you think it is a bad way, circle "Bad."**

Good
Bad

(b) ***If you circled "Bad," change the picture(s) above to make it a Good comparison.***
(For example, you might want to change the age of the seller, the type of drink, or the time of day in one or both of the set-ups.)

Figure 7.1a Story-evaluation questions used in pre- and post-test during Phases 1 and 2. Drink sales question. (This is a "bad" experiment because age of child is confounded with type of drink.)

Table 7.3 provides a summary of the questions discussed during the tutoring session in this and subsequent phases.

For classroom L1, only 33% of students designed at least three unconfounded experiments on the ramps posttest—much lower than in prior studies with middle-SES students (e.g., Toth et al., 2000). To rule out the possibility that students had simply not paid close attention in class, the instruction given to the full class was repeated in tutoring. However, this repeated instruction was not successful, suggesting that inattentiveness was not the problem. In fact, even after intensive one-to-one tutoring, none of the 10 tutored students showed mastery on the Story-evaluation posttest. Thus, in this and subsequent tutoring sessions in L2 and M1, as well as in all subsequent phases, we sought to identify preconceptions that might have affected students' ability to learn CVS. In L2 and M1, we used several alternative pedagogical methods, including simplifying instruction

These two pictures show how they tested whether or not the ***engine direction*** made a difference in how high the rockets fly.

Look carefully at the pictures. Each rocket has a certain body shape (Curved or Straight), number of windows (One or Four), and engine direction (Down or Tilted).

Do you think this is a good way to find out whether the *engine direction* (Down or Tilted) makes a difference in how far the rockets fly?

(a) **If you think it is a good way, then circle the word "Good" below. If you think it is a bad way, circle "Bad."**

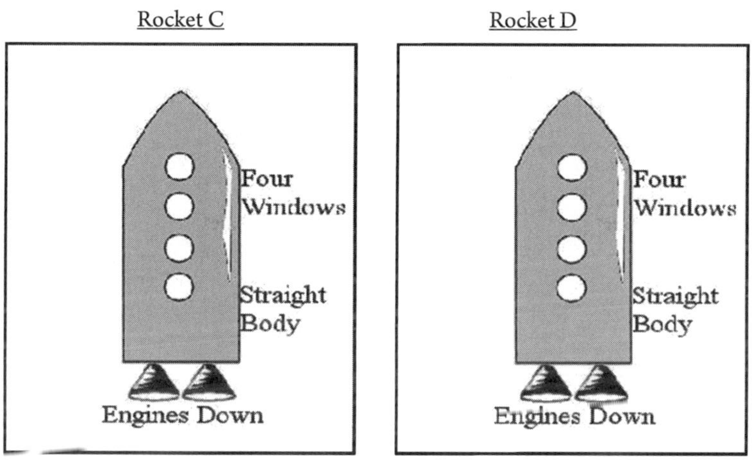

Good

Bad

(b) ***If you circled "Bad," change the picture(s) above to make it a Good comparison.***
(For example, you might want to change the body shape, the number of windows, or engine direction for one or both of the rockets.)

Figure 7.1b Story-evaluation questions used in pre- and post-test during Phases 1 and 2. Rocket design question. (This is a "bad" experiment because the target variable is not varied.)

by breaking it down to focus on one CVS rule (as listed in Table 7.1) at a time. Approximately two weeks after the tutoring phase, all students completed a Story-evaluation posttest, identical to the Story-evaluation pretest. Because of similar mastery rates and response patterns for L1 and L2 (e.g., mastery rates were 33% and 36%, respectively versus 80% for M1), these classes were combined in later analyses.

Phase 2 was conducted with 21 sixth-grade students in classroom L3 at a school where 80% of students received free or reduced lunch. All students first completed the Story-evaluation pretest. One science teacher was trained to administer CVS instruction using a procedure that we had found to be productive with some students in the remedial tutoring of Phase 1. This procedure presented CVS in a more incremental fashion: All CVS rules were discussed sequentially, and prior to discussing the rationale behind CVS. In this four-day teacher-led classroom instruction, the teacher introduced students to the topic of experimental design, which

These two pictures show how they tested whether or not the ***oven temperature*** made a difference in which cookies people like.

Look carefully at the pictures. Each one shows the oven temperature (High or Medium), sweetener (Honey or Sugar), and number of eggs (One or Three) used to make the cookies.

Do you think this is a good way to find out whether the *oven temperature* (High or Medium) makes a difference in which cookies people like better?

(a) **If you think it is a good way, then circle the word "Good" below. If you think it is a bad way, circle "Bad."**

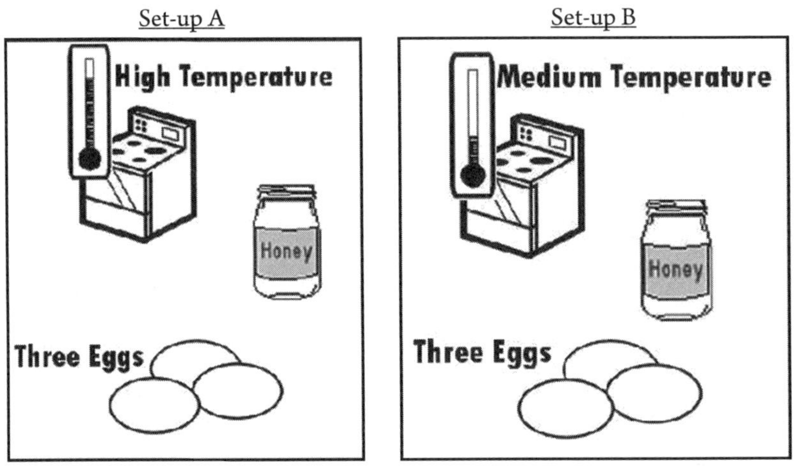

Good

Bad

(b) ***If you circled "Bad," change the picture(s) above to make it a Good comparison.***

(For example, you might want to change the number of eggs, oven temperature, or type of sweetener in one or both of the set-ups.)

Figure 7.1c Story-evaluation questions used in pre- and post-test during Phases 1 and 2. Cookies question. (A "good" experiment because only the target variable is contrasted and all other values are the same in both conditions.)

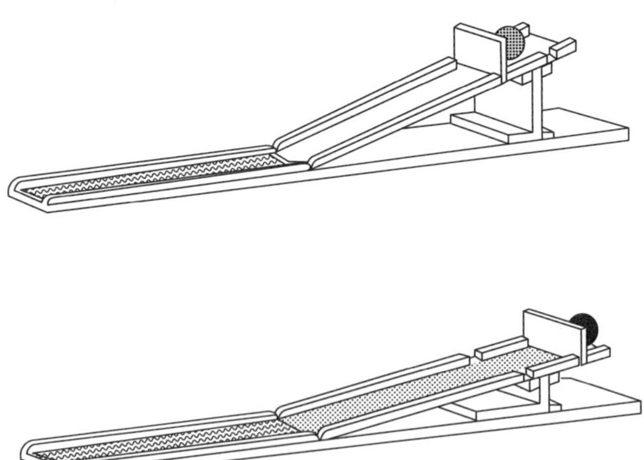

Figure 7.2 Representation of ramps apparatuses used across development phases.

Table 7.3 Contexts of questions serving as data sources by phase

Phase 1	*Phase 2*	*Phase 3*	*Phase 4*
During in-class lesson:	*During in-class lesson*:	*Questions tutors could choose among*:	Story-evaluation & design w/explanations pre-, post-, follow-up tests.
Evaluate ramps setups (written on worksheets)	Evaluate study experiment (i.e., Does X affect grades?)	Design & evaluate study experiment (i.e., Does X affect grades?) Design of abstract experiment.	Ramps design pre- & posttest w/explanations.
Tutoring questions included:	Evaluate ramps experiment.	Design & evaluate running experiment.	
Design & evaluate ramps experiments.	*Immediate Posttest*:	Design & evaluate shooting basketballs experiment.	Explicit instruction: Evaluate ramps setups (typed or spoken).
	Evaluate basketball experiment.	Evaluate restaurant experiment.	
Design study experiment (i.e., "Does X affect grades?")	*Tutoring questions included*:	Design of carts experiment (TIMMS ...).	Standardized item post & follow-up tests (evaluate beetles—NAEP & design carts experiment—TIMMS questions).
	Design & evaluate ramps experiments.	*In the following contexts, tutors could choose among questions addressing R1, R2, and R3 individually, or all three rules (as in the design or evaluate questions)*:	
Design fish experiment. Design running experiment.	Evaluate basketball experiment.	Hens laying eggs experiments. (for R2:) A farmer wants to know whether older or younger hens lay more eggs. She has some hens that are 1 year old and some hens that are 3 years old. Which ones should she study if she wants to answer her question?	
	Evaluate drink experiment.	a. the 1-year-old hens only	

(continued)

Table 7.3 (Continued)

Phase 1	*Phase 2*	*Phase 3*	*Phase 4*
	Delayed posttest: Story-evaluation & design delayed posttest (drink sales, rocket design, and cookie baking experiments)	b. the 3-year-old hens only c. some of the 1-year-old hens and some of the 3-year-old hens Plant growth experiment. (for R1:) James is doing an experiment with plants to find out if the kind of water used with them affects their growth. How should he go about doing that? a. Do something with the soil the plants are living in b. Do something with the type of light the plants receive c. Do something with the type of water given to the plants d. Do something with the location of the plants Dog-calling experiment. (for R3:) James is watching Brittany do her experiment from next door. He can see that for both trials in her experiment, she is standing up, wearing a white shirt, and he can hear her calling for Bandit with the word "Here!" James can't see what she is holding in her hand for each trial, he is too far away. What can James know about the variables he can see Brittany using? a. She is not testing those variables. b. She is testing those variables. c. She's not doing her experiment correctly. d. It doesn't matter what she's holding in her hands.	

included brief discussions of the purpose of science, definitions and examples of key terminology (e.g., "variables," "good experiment"), followed by some CVS rules (e.g., only test one variable at a time). The teacher led the class in designing an experiment (on factors affecting the effectiveness of studying) in a highly scaffolded manner (i.e., by choosing different values for the target variable and the same values for the other variables across conditions).

Students individually designed an experiment for a different target variable on their worksheets and then evaluated an experiment. All worksheet responses were given feedback and returned to students the following day, when students were introduced to virtual ramps apparatus (using the TED computer-based interface) and designed an experiment testing one ramps variable on their worksheets. Students evaluated a confounded experiment and wrote explanations for their evaluations. Then the teacher led a discussion of why the experiment was not informative and how to improve it. Immediately after instruction, students designed four ramps experiments. As in Phase 1, students who designed fewer than three out of four unconfounded experiments were given remedial tutoring. Approximately two weeks after remedial tutoring was completed, all students took the Story-evaluation posttest. One year later, students (now seventh graders) were given a delayed "Story-evaluation/design" posttest that required them to design and then evaluate experiments in three different domains (selling drinks, flying rockets, and baking cookies—see Figure 7.1), and to provide written explanations for each item.

In Phase 3—unlike in previous phases—there was no whole-classroom instruction. First, 27 fifth- and 31 sixth-grade students in classrooms M2a1 & M2a2 of a middle-SES school, where 11% of students received free or reduced lunch, completed the Story-evaluation pretest. Students who corrected fewer than four of the six confounded experiments on the Story-evaluation pretest were tutored—either individually or in small groups—by a member of the research team. The tutor led the student(s) through the computerized instruction that included an introduction to the lesson topic and some introductory definitions (e.g., "variable"), and then required students to evaluate experiments and derive the (CVS) rules for informative experiments. Afterward, tutors selected CVS problems for students from a database of problems, categorized by CVS rule (Rules 1, 2, and 3 from Table 7.1). Tutors selected these problems based on their assessments of students' particular weaknesses. For example, if the tutor believed the student could identify the target variable and vary it across conditions (Rules 1 & 2) but did not understand the need to control nontarget variables (Rule 3), the tutor chose questions targeting Rule 3. All tutoring sessions were recorded and transcribed.

In the final phase described in this chapter, Phase 4, 50 fifth-grade students in classrooms M2b1 and M2b2 from the same middle-SES school as in Phase 3 (but a year later) worked one-to-one with either a member of the research team or used the TED tutor. Students first completed the "Story-evaluation/design" pretest—described in Phase 2—that required them to design and evaluate experiments in three different domains and give explanations for their responses. The instruction began with an introduction to the topic of the lesson, then a ramps pretest in which they designed an experiment for each of the four variables and gave explanations for their designs. Then they evaluated three different experiments as a "good" or "bad" way to find out about the target variable and explained their responses. Students were given feedback for each evaluation and told why the experiment was or was not informative. After this, students completed the ramps posttest, identical to the pretest. The next day, they completed the Story posttest, identical to the Story pretest. About three weeks later, students completed a follow-up Story posttest. This same procedure was repeated in two low-SES classrooms (L4 and L5, with 16 and 14 students, respectively).

In sum, our analysis of preconceptions about experimental design is based on students' explanations collected during these four tutor development phases. There are three primary sources for these explanations: (a) written explanations given during the teacher-led CVS instruction, (b) written and spoken explanations during one-to-one remedial tutoring sessions, and (c) written and typed explanations on the ramps and Story-evaluation/design pre- and posttests. Table 7.3 provides information about the types of questions given to students across phases.

This explanation-based approach extends and enriches an earlier analysis of CVS misconceptions in which Toth et al. (2000) used a "rule assessment" approach (Siegler, 1981). They classified each student as holding one of four distinct reasoning strategies by examining each student's pattern of evaluations for a 10-item battery of experimental setups including the following four types: unconfounded, singly confounded, noncontrastive, and maximally contrastive (cf., Toth et al., 2000, Table 7.4). However, their analysis rested on the assumption that students approached each of the items in the test battery with a stable—albeit incorrect—strategy. As discussed in Siegler (1996) and also found in our data, the assumption of a stable strategy is tenuous, particularly with students struggling to understand CVS, and whose approaches to it are likely to be changing even within a given assessment. Moreover, even if students had applied the same strategy across problems, it is not clear from this analysis what strategy they had applied, since, as will become apparent, different underlying strategies can lead to the same setup designs or evaluation responses.

Results

In this section, we present an analysis of students' misunderstandings, based on the explanation sources described earlier. The presentation is organized in terms of the goals students held—science, engineering, "fairness," or "domain knowledge"—and the approaches they took within those goals. The analysis is based on five contextual aspects of the student-tutor interaction: (a) student explanations (uttered, typed, or written) in conjunction with the referential setup; (b) our inference about goal type; (c) our inference about error within goal type; (d) the knowledge components addressed in the tutor's response to this information, and (e) the impact of the tutor's responses (i.e., the extent to which the response was effective in remediating the inferred student error, thereby supporting our inference about specific

Table 7.4 Percentage of responses by goal type, phase, and assessment

	Phase 2: L3	*Phase 4: M2b*			*Phase 4: L4 & L5*		
	Story one-year follow-up	*Story pretest*	*Story post*	*Story follow-up*	*Story pretest*	*Story post*	*Story follow-up*
Response classification	*Q1*	*Q1*	*Q1*	*Q1*	*Q1*	*Q1*	*Q1*
Science goal	33.3	12	50	55.3	0	20	26.9
CVS procedure[a]	8.3	2	11.4	17.0	0	13.3	15.4
Engineering goal	50	52	22.7	14.9	56.7	33.3	23.1
Domain knowledge goal	8.3	0	0	0	3.3	10	0
Guess/don't know	0	26	9.1	8.5	26.7	16.7	23.1
No response	0	2	2.3	0	0	0	3.8
Unknown (descriptive only)	0	6	4.5	4.3	13.3	6.7	7.7

[a] *Explanation includes CVS procedure with no explicit science goal.*

knowledge component deficits). Later in the chapter, we discuss possible reasons why students hold different goals.

For an overview, Table 7.4 presents the percentages of responses to the first question of the Story pre-, post-, and follow-up tests that indicated a science goal, an engineering goal, an expression of domain beliefs (i.e., "domain knowledge" goal), or another type of response. Initial responses were chosen because students tended to elaborate most on their first response.[2] A second coder categorized a random 10% sample of Story test responses for goal type; intercoder agreement was good (kappa =.90).

CONCEPTUAL AND PROCEDURAL ERRORS WITHIN SCIENCE GOALS

As noted in our task analysis, to understand the point of instruction on experimental design, a student must understand the science goal of the task. That is, the student must realize that the underlying (abstract) goal of designing the comparison is to find out about the causal status of a variable based on the experimental outcome. However, even if a student holds a science goal, errors may occur in the pursuit of that goal due to incomplete or faulty conceptions. Thus, we first discuss the errors students made when designing and evaluating experiments when they held a science goal. As mentioned previously, our inference that a student held a science goal is based on evidence in the explanation indicating that the student was *trying to find out about* the effects of a variable (or variables). For example, we interpreted statements such as "I wanted to see whether [variable X] made a difference" or "I was trying to figure out if downward or straight engines were better" as evidence of science goals. In contrast, statements such as "I wanted to make the balls roll far" or "I wanted to make the balls roll the same," would be explicitly indicative of engineering goals.

Our discussion of these incorrect approaches (the features of which are summarized in Table 7.5) is roughly ordered in terms of decreasing similarity to the expert approach to experimental design. For each approach, we briefly discuss associated underlying knowledge deficits and/or alternative beliefs along with suggestions for remediation. Though the approaches listed do not include every possible feature combination, they include those detected in more than a single instance in our data corpus and/or reported in previous research. Table 7.6 presents a summary of the frequencies of student responses on the first Story test items in Phases 2 and 4 falling into different science goal categories. Table 7.6 also includes responses that indicated CVS application, where students set up unconfounded experiments and gave at least partial CVS responses.

Due to unique response patterns, data from different classes is presented separately in Table 7.6. The first column provides the frequencies of the different approaches for L3 students in Phase 2 on the first design question of the delayed follow-up Story

[2] Additionally, students designed experiments and explained their designs on this initial question, which allow greater opportunity for students to indicate their goals than when evaluating others' experiments.

Table 7.5 Summary of science goal experimental approaches and feature(s) erroneous or absent in that approach

	Aspect of CVS missing or misconceived						
Approach	*Focus on variable level*	*Test one variable*	*Identify correct target variable*[a] *(R1)*	*Compare across conditions/outcomes*	*Vary what testing (R2)*	*Control variables not testing (R3)*	*Understand setup/outcome causal relationship*
(a) CVS for wrong variable			X				
(b) CVS-vary-"noncausal"						X (Noncausal belief)	
(c) Test multiple variables		X	X			X (No CVS logic)	X
(d) Ignore confound(s)						X (No CVS logic)	X
(e) Whole condition comparison	X	X	X			X (No CVS logic)	X
(f) Noncontrastive target variable					X	X	X
(g) Single-condition experiment				X	X	X	X

[a] *Failing to identify the correct target variable may occur in each of the approaches, but is only a necessary criterion in three.*

Table 7.6 Percentage of approach within science goal and CVS responses (and overall percentages) by test and phase

	Phase 2: L3	*[a]Phase 4: M2b*			*Phase 4: L4 & L5*	
Approach	*Story one-year follow-up Q1*	*Story pretest[b] Q1*	*Story post Q1*	*Story follow-up Q1*	*Story post Q1*	*Story follow-up Q1*
Number of science goal & CVS responses (%)	5 (41.6%)	7 (14%)	27 (61.4%)	34 (72%)	10 (33%)	12 (42%)
CVS proc + logic	0	14.3 (2.0)	11.1 (6.8)	17.7 (12.7)	0	16.7 (7.0)
CVS proc; science goal	0	14.3 (2.0)	55.6 (34.1)	50 (36.0)	20 (6.6)	16.7 (7.0)
CVS proc (no goal stated)	20 (8.3)	0	14.8 (9.1)	11.8 (8.5)	40 (13.2)	25 (10.5)
CVS setup (science goal but incomplete)	0	0	0	2.9 (2.1)	0	8.33 (3.5)
Total CVS	**20 (8.3)**	**28.6 (4.0)**	**81.5 (50.0)**	**82.4 (59.3)**	**60 (19.8)**	**66.7 (28.0)**
(a) CVS but test wrong variable	20 (8.3)	0	3.7 (2.3)	0	10 (3.3)	0
(b) CVS-vary- “noncausal”	0	0	0	0	0	0

(c) Test multiple variables	40 (16.6)	0	11.1 (6.8)	5.9 (4.2)	0	8.3 (3.5)
(d) Ignore confound(s)	0	14.3 (2.0)	0	2.9 (2.1)	0	8.3 (3.5)
(e) Whole condition comparison	20 (8.3)	57.1 (8.0)	0	5.9 (4.2)	20 (6.6)	0
(f) Noncontrastive	0	0	0	0	0	0
(g) Single-condition experiment	0	0	0	0	0	8.3 (3.5)
Vague science goal	0	0	3.7 (2.3)	2.9 (2.1)	10 (3.3)	8.3 (3.5)

[a] Phase 3 data was not included in this table because unlike in the other phases, there were no assessment events that were common to all students.

posttest taken one year after instruction. Next, frequencies are given for the Phase 4 fifth-grade M2b students' responses on the first Story pretest design question and on the first design question of the immediate and follow-up Story posttests—administered three weeks after instruction. The same data is provided for L4 and L5 (because they did not express any CVS or science goal responses on the first Story design pretest item, this column is not included in the table). Data from this table is referenced throughout the discussion of various science goal approaches.

As shown in Table 7.6, among students expressing science goals on the pretest, CVS responses were relatively rare. However, given that students held a science goal, Phase 4 students in both M2b and L4/5 were most likely to express a CVS understanding on the first Story posttest question (82% and 60%, respectively). However, the L3 Phase 2 students were only 20% likely to give a CVS response on the one-year follow-up. Furthermore, M2b students were relatively more likely than L4 and L5 students to explicitly indicate a science goal when they expressed a CVS explanation (e.g., "I only made the age of the child different because I want to find out if the age makes a difference") on the posttest. CVS explanations given by the M2b students generally included explicit indications of science goals whereas those given by L4 and L5 students did not. It is possible that L4 and L5 students were more likely to learn the CVS procedure without understanding its purpose in serving science goals.

Types of Error-Prone Approaches within a Science Goal

Here we describe the set of experimental design errors exhibited by children expressing a science goal at that moment (summarized in Table 7.6), along with corresponding remediation for each approach. The first of these (a) is an unconfounded experiment but for the "wrong" factor. The next four (b–e) are different types of errors that led to confounded designs. In these, students held science goals of designing experiments to find out about a variable (or variables) and understood the need to "compare and contrast" the tested variable(s) across conditions, but failed to control the other variables. The final two (f & g) are other types of errors. A second coder categorized a random sample of science goal responses; intercoder reliability was good (kappa =.86).

(a) CVS for wrong variable (R1 failure): This approach occurred when students were attempting to find out about a single variable and set up an experiment that was unconfounded, but with respect to the wrong variable (i.e., a variable other than the one given in the question).[3] For example, when asked to design an experiment to test time of day in the drink sales story problem (Figure 2a), one student designed an unconfounded experiment to test age instead. His justification was:

I made one variable so I will find out if age sells more.

[3] In our current work, the variable to be tested was always included in the problem statement, and not left to the student to choose (in contrast, the Kuhn et al., 1995 & Schauble et al., 1991, tasks allowed students to determine which variable(s) to find out about).

Students may design unconfounded experiments to test something other than the target variable because they believe they already know the causal role of the (teacher)-intended target variable, and substitute it for one that they would prefer to find out about. This is an action they would likely take when engaged in open-ended science inquiry, and not an unreasonable one. The "error" here is bringing to bear beliefs about likely causal factors in a domain rather than following the "academic" exercise of suspending prior beliefs in the service of a kind of Platonic approach to experimental design. Another possible cause of this error is simply that children did not read the problem statement carefully.

Remediation: In general, when students who made this mistake were prompted to reread the problem statement, they realized that the variable to be investigated was given. They subsequently design unconfounded experiments to test that variable with minimal help from the tutor, as shown in the following exchange between a tutor (T) and student (S):

T: Ok, so is this [CVS but wrong target variable] a good experiment?
S: Ah, yes?
T: Ok... all right, so let's think about that a little more. The first thing you want to do is find out which of the variables they want to find out about. OK? So which variable do they want to find out about?
S: Where they study?
T: Exactly. So which variable would that be?
S: Location?
T: Exactly, so they want to find out about location. OK? So now that you know that, what do you think? Is that a good experiment?
S: No.
T: OK, why not?
S: Because they're both the same?
T: OK, and how do you want to change that?
S: Shayna should study on the couch.
T: OK, all right, so now we made this different. OK? So is this a good experiment now?
S: No.
T: OK, what do you wanna change about it?
S: The time of day?
T: OK, so what do you want to do?
S: Make both of them... late?
T: OK, so it doesn't matter, they could be both late or both early. But the important thing is to make them both...
S: The same. (unconfounded experiment for the target variable now)
T: Exactly. And anything else you want to change?
S: [No]

As shown in Table 7.6, given that students held a science goal (or applied CVS), they were between 0 and 20% (on average across the 6 points of analysis, about 6%)

likely to make this error on the Story tests. In all of these cases, there was no indication from students' written explanations that this error was more than a slip—as opposed to students' desire to test a different variable due to their beliefs about specific variable effects. Thus, converging evidence suggests that this error was *not* generally caused by students applying their beliefs about variables.

(b) CVS but vary "noncausal": Kuhn et al. (1995) described a type of behavior in a task where participants (fourth graders and adults) had to choose evidence from which to infer the effect of a particular variable where participants simply ignored variables they believed were noncausal when making their choices of which evidence to compare. This behavior led to confounded comparisons.

However, we failed to find evidence of students explicitly indicating they did not control a variable because they did not think it mattered. For example, as shown in Table 7.6, identified instances of this approach were nonexistent in the pre- and posttest responses to the first Story question, as well as in responses to the other items. This may indicate that, when students are applying science goals, they tend to suspend their beliefs about variable effects. As discussed later, varying what one believes to be noncausal was more common within engineering goal applications.

Though infrequent, instances of what may be this approach were identified in other student responses, including those given during tutoring sessions (refer to the excerpt below). In this approach, students understand the causal relationship between the experimental setup and outcome (i.e., to determine whether a variable affects an outcome, only that variable can differ; otherwise, the cause cannot be determined) when making inferences about variable effects. However, they fail to realize their beliefs may not be consistent with reality, and that therefore they must control even nontarget variables they believe are not causal.

Remediation: During the tutoring sessions, we used two different methods to remediate this approach when we suspected students of using it. In the first, we told students: "even if you think a variable doesn't matter, you must still control that variable, because you can never be completely certain that it does not affect the outcome." This explanation led to the adoption of the expert approach. When it did not, we used a second method in which we either suggested or asked students for a plausible explanation for why the (confounding) variable *might* make a difference. This method is shown in the example below, where a ramps experiment testing for starting position is confounded by ball type (red or yellow):

T: Now can we tell whether one ball rolls farther than the other because they have different starting positions?

S1: Yes.

T: How about those balls? What if one's a ping-pong ball and the other's a marble? Could we tell whether it's just the starting position of the balls that's affecting how far they go?

S 1 & 2: [shake heads]

T: So how do you want to change that?

S2: I'm not sure.

T: If the red ball's a ping pong ball and the yellow ball's a marble, how would you...
S2: Make them the same ball.

Between these two remedial methods, this particular error was almost always corrected.[4] It is important to note, however, that the same behavior shown in the tutoring excerpt may be due to students ignoring confounding variables (discussed later).

Failures in Understanding CVS Logic

In the next two approaches, students demonstrate lacking an understanding of the indeterminate nature of confounded experiments by varying other variables. Two alternative explanations for this behavior exist. Toth et al. (2000) assumed that students who designated as "good" any experiment that varied the target variable, regardless of the settings for the other variables, were ignoring the confounded variables. However, without the additional evidence provided by participants' explanations, students may have had other reasons for their "good" and "bad" designations of different experimental designs. For example, Toth et al.'s participants may have been trying to find out about more than one variable at a time, as Kuhn and Dean (2005) speculated:

> Students who have developed an understanding of the need to access an available database as a source of information may nonetheless still initially pose ineffective questions, *in particular because they aim to discover the effects of all variables at once. It may be this ineffective intention that leads them to simultaneously manipulate multiple variables (in effect, overattending to them, rather than underattending by failing to control them, as is typically assumed)* [italics added]. (p. 867)

Our explanation database suggests that students both overattend to the uncontrolled variables (by trying to test multiple variables), and underattend to the uncontrolled variables (by simply ignoring them). As shown in Table 7.6, these two response categories were detected at similar rates, though were relatively uncommon overall (e.g., in Phase 4, these responses represented between 0 and 16% of responses—an average of 10% of science goal responses). Additionally, with instruction, the rate of overattending responses increased while the rate of underattending responses decreased, suggesting that variables become more salient over the course of instruction. The corresponding knowledge components comprising these two categories are described in Table 7.5 (c–d), and these categories are further elaborated on below.

(c) Test multiple variables: One reason that students design confounded experiments is that they attempt to find out about *each* of the contrasted variables in

[4] The effectiveness of the second remedial method could be partly due to people's biases toward believing variables are causal (e.g., Kuhn et al., 1995); we found that students have little trouble producing or believing a reason for a variable to be causal (or at least entertaining this possibility).

a single experimental contrast. Students almost always indicated wanting to test all of the variables; thus, they typically did not apply any beliefs about variable effects (such as controlling variables they believed were noncausal), but rather, held a "pure" abstract science goal. This is shown in the following student's explanation for why she evaluated a maximally confounded experiment as good:

> I think this is a good way because testing the differences can probably turn out to be a good experiment and you could see what effect the time, drink, and person [age] have on them.

Kuhn et al. (1995) reported similar conceptual errors in students who attempted to find out about multiple variables simultaneously. This approach reveals two related misconceptions: first, a belief that the effect of the target variable can be determined from a confounded experiment, and second, a belief that the effect of the nontarget variables can be determined from the same experiment. That is, students believe that they can "do it all" in a single experiment. These misconceptions stem from failure to understand the logic of CVS: that the cause of any differences in outcomes cannot be uniquely determined in a confounded experiment.

Kuhn et al. (1995) noted that as their students progressed through a series of experimental design problems, they became less likely to express "an intent to assess effects of multiple features by examining a single instance or pair of instances" (p. 65). Thus, at least some students who hold science goals and are trying to find out about the effects of more than one variable at a time can eventually learn to focus on only one variable when no feedback is provided. However, there was a point in the Kuhn et al. intervention in which the number of features students intended to investigate stopped decreasing, suggesting the need for more instructional support.

Remediation: Because they only lack an understanding of the rationale for controlling the nontarget variables, students applying this approach are ideal candidates for explicit CVS instruction (e.g., as given in Chen & Klahr, 1999), which explicitly focuses on this rationale. In Phase 1, all of the five students (in L1 or L2) who held science goals and understood the need to vary what was being tested but lacked the procedural knowledge of controlling for nontarget variables (as shown by their ramps pretest setups) designed at least three out of four informative experiments on the ramps posttest following explicit CVS instruction. However, for some students, simply designing experiments during the Story or ramps pretest—where no feedback was provided—allowed for their development and application of CVS.

(d) Ignore confound(s): Students may set up maximally confounded experiments or evaluate them as good, but only refer to the target variable, ignoring the other variables. Thus, again, students do not apply their beliefs about the nontarget variables. For example, the following statement was made by a student evaluating a maximally contrastive setup involving drink stands: "I think it's a good way because both of the stands are different and that way you could tell whether the *age of the child* makes a difference." The student did not address the nontarget variables: time of day and

type of drink. Similarly, a student who had designed a maximally contrastive rocket experiment simply explained: "I wanted to see if the *engines* would affect the speed and direction of rockets." Again, the student ignored the nontarget variables (body shape and number of windows).

Such explanations suggest three possibilities: (1) students did not attend to the uncontrolled variables, indicating that they did not recognize the importance of controlling variables either; (2) students did attend to the uncontrolled variables, but did not understand that the (potentially causal) confounded variables prevent valid inferences about the effect of the target variable; (3) students attended to the uncontrolled variables and understood the rationale for controlling variables, but believed they were noncausal and not necessary to control (i.e., they actually held a "CVS but vary noncausal" approach).

Remediation: In the first two possibilities, students are prime candidates for explicit CVS instruction (e.g., as given in Chen & Klahr, 1999, and Strand-Cary & Klahr, 2008), which explicitly focuses on the nontarget variables and their impact on inferences that can be made. For the third possibility, the remediation described for CVS vary noncausal is appropriate.

(e) Whole condition comparison: Another approach within a science goal is comparing two conditions rather than an individual variable or variables. This approach—typically associated with maximally contrastive designs—once again does not involve application of beliefs about variable effects but is an "abstract" science goal application. This approach is exemplified in the following explanation: "I set it up the way I did so everything would be different to see which one [cookie] people like better."

A second example is from a tutoring excerpt in which the tutor and student discuss an experiment about plant growth (in Phase 3):

S: Umm... no [changes experiment from CVS to maximally contrastive]
T: So why did you set it up like that?
S: To be different to see which one would grow better.

The frequency of this approach (shown in Table 7.6) ranged from 0 to 57% of Story test responses within science goals, and was greatest on the pretest responses of Phase 4 M2b students. In addition to not considering the variable level, these students probably also lack an understanding of controlling variables and the rationale for doing so.

Remediation: Perhaps because it is natural for students to think at the concrete level of variable values, simply prompting them to consider the variable level by identifying the given target variable and then prompting them to think through the logic of the relationship between the target variable and outcome resulted in subsequent consistent CVS application. This is shown in the following dialogue (continued from the dialogue above):

T: What is his experimental question?
S: If fertilizer or sand...

T: . . . makes a plant grow better. So if Plant #1 grew better than Plant #2, would he know for sure, the way he has this set up, that it's because of the fertilizer?
S: Yeah.
T: How would he know that?
S: No.
T: How could he set up his experiment so he would know for sure that it's the soil that the plant is in that is making any difference in how the plants grow.
S: These would all have to be the same.
T: So why don't you go back and make those changes.

(f) Noncontrastive target variable(s) (R2 failure): Students sometimes held a science goal but did not contrast the levels of the target variable across conditions. Thus, these students did not seem to understand the need to compare different values of a variable (Rule 2). However, these cases were relatively rare in our data. This may be because, as Kuhn et al. (1995) noted, the idea of comparing things to see if there is a difference is intuitive for students and understood from a young age. In the overwhelming majority of cases in which students designed setups where the target variable was not contrasted, they indicated engineering rather than science goals. Of the six analyzed interventional points shown in Table 7.6, there was *no* evidence of students failing to contrast what they indicated they were testing. In all of the Story evaluation/design pre-, post-, and delayed posttest responses to design questions (from Phases 2 and 4), there were only three instances out of 622 responses (less than .5% of responses) where (two different) students expressed a science goal but did not contrast the variable they indicated they were testing. For example, one student designed a noncontrastive comparison for the cookies item: "so I can see what they like the best." It is possible that such responses were slips, where students mistakenly failed to contrast variable(s) in their setups. Regardless, it is safe to say that cases of failing to contrast the variable(s) of interest when the student held a science goal were rare in our data. Thus, science goals and the idea of contrasting variables appear to be strongly associated.

Remediation: Our remediation for failure to contrast the target variable focused on helping students understand the rationale for comparing outcomes across conditions. Students generally understood that comparing the same value of a variable would not provide information about whether that variable had an effect (because it would always produce the same results, so it would not be possible to find out if the different values have different effects on the outcome). When asked "If the [values of the target variable] were the same, would you be able to tell if [the target variable] made a difference in the result?," students generally came to realize the need to contrast values of the target variable across conditions, as in the following:

T: OK, so now we have two ramps that are identical. If you did this experiment, can you tell if the balls rolled farther just because of the starting position, if there's a difference between them?
S1: No, because they'll go the same [distance].

T: How could you change this so that you could do an experiment to figure out how starting position affects [the result].

S2: You could put one in the middle and leave the other one there [at the top].

(g) Single-condition experiment: The least sophisticated science goal approach found in our corpus is one in which students did not compare results or contrasting variables across conditions, but rather viewed each condition as a separate experiment. As shown in Table 7.6, instances of this approach were rare, identified in only one analysis point (where it was 8% of responses of the low-SES students in Phase 4 on the Story follow-up). In this approach, students consider an outcome from a single condition to allow for inferences to be made about the effects of individual variable values, exemplified by the following justification:

> I set up this experiment like the way I did because I wanted to see if tilted [engine] had an effect on how fast the rocket went and I picked straight [engines] for the other one to see if down had an effect on to [*sic*] see how fast the rocket went also.

Here, the student's use of "also" implies that she isn't comparing the outcomes of conditions to each other, but rather viewing the two setups and their outcomes independently. This type of error was noted by Kuhn et al. (1995), who found that when trying to discover the effects of variables given instances with variable settings and an associated outcome, fourth-grade students did not compare instances for about half of the inferences they made.[5]

There are likely two deficits underlying this approach. First, rather than thinking of outcomes as relative (e.g., that one is "better" than another), students view outcomes as absolute (e.g., "good" or "bad"), without realizing that an absolute outcome is subjective (i.e., how does one determine what a "good" outcome is?). Second, students may ignore the nontarget variables. That is, they may attribute an outcome solely to one variable, ignoring any potential impact of the other variables. However, if the student is attending to the other variables, he or she may either (a) fail to understand the causal ambiguity of the experimental contrast they have created, or that it is impossible to know the effects of the individual variables on the outcome, or (b) understand this, but apply their beliefs about the effects of the other variables to estimate the target variable's contribution to the outcome. For example, if the student believes a rocket's windows and engine direction have no effect on how high it flies, he or she may attribute a "good" outcome to the rocket's body shape.

Remediation: Because of the small number of identified cases of this single-condition experiment approach, evidence for the success rates of remedial methods

[5] One possible reason that this type of noncontrastive science approach was more common in Kuhn et al.'s (1995) task than in ours is that Kuhn's task required students to actively choose both pieces of evidence (condition/outcome) and compare them, whereas for our task, students repeatedly saw experimental conditions (though no outcomes) that were presented side by side. This likely increased the salience of comparison.

are not available. However, at minimum, this remediation would involve addressing all CVS rules.

Summary of Erroneous Approaches within Science Goals

As demonstrated in the prior discussion, students tended to explicitly state their science goal intentions when explaining their designs. Additionally, when students applied science goals, they rarely expressed or applied their beliefs about the variables. The most common error within a science goal orientation was failing to control nontarget variables (refer to Table 7.5). Rather than ignoring variables considered noncausal, this error was due to students not realizing that confounds render experimental results uninterpretable. Why might this be the case? According to Kuhn et al. (1995), in the real world, people do not see many unconfounded experiments. Instead, they typically observe correlations among lots of data—even over time—and make inferences based on those. These correlations are embedded in the noise of confounds that cannot be controlled. From a child's point of view, this turns out to be a relatively effective, thus practical, way to discover causal patterns. However, it can lead to conceptual errors. For example, people commonly infer that heavier objects fall faster than lighter ones based on their correlational observations, but fail to account for confounds such as surface area or object shape. But given its apparent utility, it is not surprising that children draw inferences from confounded experiments, without thinking through the logic of the causal relationship between the setup and possible outcomes.

When students begin with a science goal and understand the idea of comparing outcomes to see whether a variable has an effect, prompting them to think through the logic of a confounded experiment helps them to realize the "power" of CVS. We believe that this is one reason why the CVS instruction developed by Klahr and colleagues, which does just that, promotes more rapid and immediate CVS gains than discovery learning (Chen & Klahr, 1999; Klahr, 2009; Klahr & Nigam, 2004; Kuhn & Dean, 2005; Strand-Cary & Klahr, 2008).

Engineering Goals

As stated previously, students may hold or adopt goals other than the science goal of trying to find out about some variable(s) in tasks that—from the instructor's perspective—involve experimenting. Others (e.g., Kuhn, Amsel, & O'Loughlin, 1988; Schauble et al., 1991; Tschirgi, 1980) have noted this as well. Unlike erroneous approaches within a science goal, which can be compared to the expert model of experimental design in terms of missing and faulty knowledge, approaches involving alternative goals cannot be sensibly mapped onto the expert model due to their fundamentally different knowledge structures.

As discussed earlier, Schauble et al. (1991) termed one alternative class of goals "engineering" goals. Students with engineering goals attempt to "engineer" a particular result rather than attempt to *find out about* those effects via experimentation, as with a science goal. Though engineering goals can be applied across domains, they are domain specific in the sense that students intend to produce an effect that is specific to the particular context. Whether students apply their beliefs about variable effects within various types of engineering goals will be explored throughout this section.

In what follows, we discuss the different kinds of engineering goals expressed by students in various phases of our development project. Unlike with science goals, students rarely stated their engineering goal intentions explicitly (as in, e.g., "I am trying to make the balls roll far"); their goals had to be inferred from the rationales they gave for their design selections. This suggests that, while science goals are generally linked to explicit knowledge, engineering goals are not. Frequencies for the different types of engineering goal approaches were assessed at the same intervention points as in the analyses of science goal approaches and are given in Table 7.7. Intercoder reliability for categorization of engineering goal responses was good (kappa = .90). After discussing the alternative goals, we discuss potential methods of eliciting science goals.

Types of Engineering Goal Approaches

(1) Maximize outcome: According to Scauble et al. (1991), "the main objective of engineering practice is to optimize a desired outcome" (p. 860). One type of optimal outcome is the *maximum* outcome, such as the fastest car, the highest flying rockets, or the best-tasting cookies. As shown in Table 7.7, maximize outcome goals were common, particularly on the pretest, where they accounted for 88% and 96% of all engineering responses for Phase 4 low- and middle-SES students, respectively. However, they still accounted for a fairly large percentage of responses on the posttest, between 71% and 90% of engineering responses for Phase 4 students, and all engineering responses of the L3 students in the Phase 2 follow-up. Students applied maximize outcome goals in three ways. In the first two, students designed setups to maximize effects in *both* conditions, assuming independent or dependent variable effects. In the third, students maximized the outcome in just one condition. Maximize outcome responses were roughly evenly divided among their three types.

(a) Maximize outcomes (independent effects): Students applied the goal of producing a maximum effect in each condition by choosing the "best" value for one or more variables for both conditions. This often resulted in students setting up identical conditions, or designing noncontrastive "experiments." In the following, a student expresses this type of maximizing goal for her design, where both rockets had straight bodies, downward engines, and one window:

> I don't know, it just makes sense to have a straight body and down[ward] engines so it goes straight up.

Unlike with science goals, it was not uncommon for students to set variables they claimed "mattered" the same across conditions, but contrast those variables they either explicitly stated "did not matter" or did not claim to matter. Note that this is

Table 7.7 Percentages of approaches within engineering goals (and overall percentages) on Story test Q1 by test and phase

	Phase 2: L3	*Phase 4: M2b*			*Phase 4: L4&L5*		
Approach	*Story follow-up*	*Story pretest*	*Story post*	*Follow-up*	*Story pretest*	*Story post*	*Story follow-up*
Number (percent) engineering response	6 (50)	26 (52)	10 (23)	7 (15)	17 (56.7)	10 (33)	6 (23)
1a. Maximize both (independent effects)	33 (16.5)	26.9 (14.0)	30 (6.9)	57.1 (8.6)	35.3 (20.0)	30 (9.9)	50 (11.5)
1b. Maximize both via interactions	33 (16.5)	34.6 (18.0)	40 (9.2)	0	11.8 (6.7)	10 (3.3)	0
1c. Maximize one	33 (16.5)	34.6 (18.0)	20 (4.6)	14.3 (2.1)	41.2 (23.4)	40 (13.2)	16.7 (3.8)
Maximize outcome (unclear for one/both)	0	0	0	0	0	0	16.7 (3.8)
Total max outcome	**100 (50)**	**96.2 (50.0)**	**90 (20.7)**	**71.4 (10.7)**	**88.3 (50.1)**	**80 (26.4)**	**83.4 (19.2)**
2. Different outcomes	0	3.8 (2.0)	10 (2.3)	28.6 (4.3)	5.9 (3.3)	10 (3.3)	0
3. Same outcomes	0	0	0	0	0	0	0
Other engineering[a]	0	0	0	0	5.9 (3.3)	10 (3.3)	16.7 (3.8)

[a] For example, "normative" goals, such as "That's how I would do it."

opposite the behavior noted by Schauble et al. (1991), where students *contrasted* those variables they believed mattered. In this form of maximizing goal, students are not *comparing* the two conditions, but rather considering the conditions independently. (It was not uncommon for students with engineering goals to think of one of the conditions *as* an "experiment.") This behavior is demonstrated in the following explanation for what appears to be an unconfounded experiment testing for number of eggs (Cookie A: 350 degrees, sugar, three eggs; Cookie B: 350 degrees, sugar, and one egg)[6]:

I did 350 because it's sort of in the middle. Not a lot of people like honey in cookies. I guessed about the eggs.

(b) Maximize outcomes via interaction: There were also cases where students intended to maximize the outcomes of both conditions, but, because their domain theories involved interactions—where different combinations of variable values produced good outcomes—their resulting setups were contrastive, and often maximally contrastive. For example, one student designed a maximally contrastive drink stand "experiment" (Setup A: noon, older child, and iced tea; Setup B: 3:00 p.m., younger child, lemonade), and explained:

Most younger kids are outside around 3 pm and older kids like to get up in the noon [*sic*]. Also more younger kids like lemonade than iced tea.

In this example, the student's belief that the better values for time and drink depended on the age of the child resulted in two two-way interactions.

(c) Maximize outcome in one condition: The final form of the maximize goal involved maximizing the outcome (either referencing dependencies among variables or not) in only *one* condition. Students who expressed this form of maximizing goal typically did not provide a rationale for setting up the other condition, which they tended to set up as maximally contrastive from the maximized-outcome condition (e.g., 12 of the 14, or 86%, of identified maximize outcome one condition explanations in Phase 4 middle-SES classroom Story pretest responses were associated with a maximally contrastive setup). Less commonly, students set up only one condition, completely ignoring the other.

In the following, a student who set up a maximally contrastive drinks "experiment" (Stand A: noon, younger child, lemonade; Stand B: 3:00 p.m., older child, iced tea) with dependent variable effects, expresses this type of engineering goal:

I thought that noon would be a better time for a younger child to be out and younger children drink more lemonade than an older child would.

[6] This is also an example of how a given setup type (i.e., what appears to be an unconfounded experiment) can result from different underlying goals, and thus why more information is necessary to diagnose the student's specific underlying goal.

This student, typical of those using this approach, did not mention Stand B at all. Perhaps in these instances, students did not address the other condition because they did not consider it essential to achieving their maximize outcome goal. However, it is unclear why students tended to set up the second stand opposite to the first. Perhaps they were setting variables to their opposite values for aesthetic reasons (e.g., for "variety"). Alternatively, perhaps students were contrasting those variables they believed to be causal, the behavior noted by Schauble et al. (1991), implicitly attempting to produce different outcomes (discussed next).

(2) Different outcomes: Student responses sometimes indicated intents of producing different outcomes across conditions. Unlike the maximizing goals just discussed, where students considered the conditions independently, this form of engineering goal is comparative. This type of engineering goal often requires students to apply their beliefs about variables' effects. As shown in Table 7.7, these responses were less common than maximize outcome responses at every analysis point. As with maximizing goals, it was quite rare for students to explicitly state different outcome goals, as in: "because they will both fly a different way." More often, students' different outcomes goals had to be inferred from their explanations, as with the following:

> I set it up the way I did because I think it is more likely for stand A to sell more because she's young and cute, because more people are around at noon and because lemonade sometimes sells more. Then I set up stand B to be the exact opposite.

The behavior of making variables "that matter" different, noted by Schauble et al. (1991), too, may result from implicit goals of producing different outcomes. In our data, students occasionally expressed a desire to [only] vary those variables they believed to be causal. For example, for the drink sales question in which students were asked to design an experiment to test time of day, a student designed an experiment in which only the type of drink differed and explained: "I know that it does not matter whether how old or what time all that matters is the type of drink!" These responses—though again not explicit—may be indicative of different outcome goals.

What is the underlying motivation for different outcome goals? One possibility is that children try to produce results that are consistent with their expectations about variable effects. For example, if the student is asked to set up an experiment to show whether the slope of a ramp affects how far a ball will roll down it—and if the student already believes slope to be causal—then the student may set up a comparison between a steep, smooth, long ramp and a low, rough, short ramp. Not only is the target variable contrasted, but the other "best variable settings" are paired with the "best target variable setting" to ensure the result. For this particular goal, the effort of using one's beliefs about variable effects to "prove" the effect of the target variable makes this a type of engineering goal. However, it differs from other engineering goals in which variable effect beliefs are applied to produce some optimal result. We have not found evidence supporting this possibility; however, this may be related to the nature of the tasks in the current project, where students do not run their experiments.

A more likely explanation for a different outcome goal—supported by our data—is a (perhaps implicit) belief that a "good" experiment is one that produces different outcomes. In other words, some students may not conceive of an "experiment" in terms of science goals, but rather within an engineering interpretation. For example, students often evaluated the unconfounded experiment testing the number of windows on a rocket as "bad" because they believed that the number of windows did not have an effect, as expressed in the following:

Even though they have different amount of windows doesn't mean that if one has 4 windows it might fly more faster and farther than another.

This student changed the setup to maximally contrastive, which (presumably) would be more likely to produce a different outcome. Implicit beliefs that a good experiment is one that produces different outcomes might have been triggered during instruction when students were asked: "Is this a good way to find out whether the balls go different distances just because of the ball?" Consistent with this possibility, the proportion of different outcome goals increased from the Story pretest to immediate Story posttest (Table 7.7).

(3) *Same outcomes*: Some student responses indicated that their specific engineering goal in designing an experiment was to produce the same outcomes in both conditions. For example, students set up two ramps with the goal of making the ball roll the same distance on each ramp. Or they evaluated an experiment as good because they believed that both setups would (or *should*) produce the same outcome (e.g., equal sales). To our knowledge, this specific type of engineering goal has not been previously reported, likely because it is relatively rare. We found no evidence of this in Story pre- or posttest responses (Table 7.7). However, students in Phase 1 expressed this misconception in about 15% of their explanations given during the classroom instruction. Students in Phase 1 may have expressed this type of approach more than in most other phases due to the instructional wording—which asked students if the experiments were "fair tests" of the target variable. Some students may have interpreted "fair" to mean "the same," thus eliciting this particular type of engineering goal. In subsequent versions of instruction, the word "fair" was replaced with "good." No evidence of this type of misconception was uncovered again until the evaluation phase of computerized instruction in Phase 4. When L4 and L5 students worked with the TED ("Training in Experimental Design") tutor, 7.4% (2 out of 27) of responses to "Tell me why this [unconfounded experiment] is a good way to find out whether the balls go different distances just because of the ball" indicated a same outcome goal. Immediately before answering this question, students heard the following explanation:

To fix this experiment [to test for ball type], we would need to make everything about the ramps the same, except for the balls. That means both ramps should have the same surface, should have the same slope, and should have the same starting position.

It is possible that students—especially those students who did not hold science goals and thus did not understand why controlling nontarget variables is necessary—interpreted these explanations as "everything must be the same," eliciting (or elicited by) "same outcome" goals.

(a) Same outcomes via noncontrastive setups: Students can apply "same outcome" goals in several ways. The simplest is by using identical conditions. For example, students might compare a high ramp with a smooth surface to a high ramp with a smooth surface. Note that here it is *not* necessary for students to apply their beliefs about the effects of the variables. This goal is expressed by a student in her written explanation for why an unconfounded experiment testing ramp height "is a fair comparison for height" in the following:

> I think we should have made them both high because I think it would have been really close or the same. I think that's what we should have done.

Interestingly, though, this student mentioned that the ramps should be "high" rather than "the same," indicating that she may have also been applying an implicit maximizing goal.

(b) Same outcomes vary "noncausal": Another manifestation of the "same outcome" engineering goal *is* influenced by students' beliefs about variable effects. In this case, students design setups they believe will produce the same outcome by making the variables they believe "matter" the same across conditions, but, as in previous approaches described, vary the variables they believe "don't matter." Students may express this approach by creating a setup in which the slope and surface of the ramps (which they typically believe to be causal) are the same, but the color of the ball (which students generally say does not matter) is varied. The following student explanation for why an experiment in which only ramp height differs is "a fair way to find out about the height" exemplifies this approach:

> [It's] fair b/c high/low probably doesn't matter because it would probably travel the same length.

It appears that this student's goal of producing the same outcomes actually *overrode* his belief that a high ramp would make the ball roll farther, which he had expressed earlier.

(c) Same outcomes via "balanced" setups: Alternatively, students tried to engineer same outcomes by setting variables to "balance" each other across conditions. This approach requires applying one's beliefs about the effects of context variables. For example, students made one ramp high (better outcome) but with a rough surface (worse outcome) and the second ramp low (worse outcome) but with a smooth surface (better outcome). Because the two variables (surface and slope) presumably balanced each other out across the ramps, students would expect the outcomes (e.g., how far a ball will roll down the ramp) to be the same. Below is an explanation from a student who took this approach for

why he varied the length of the ramp (the target variable), in relation to the surface:

> Because the rough one [inaudible] probably mess the ball up so I made it longer so it could move faster.

This approach may arise when students who hold the engineering goal of producing the same outcome learn to vary the target variable and assimilate this new (correct) information onto their prior beliefs. Such assimilation was identified in Vosniadou and Brewer (1992), where some students developed hybrid models of the earth (e.g., as round, but with a flat interior, representing the earth's surface), by assimilating new information (e.g., that the earth is round) with their intuitive preconceptions (i.e., that the earth is flat).

Goal Hybrids and Goal Consistency

In a few rare instances, students applied *both* engineering and science goals within a given setup. For example, on the first question of the ramps pretest, only 3 out of 70 (4%) responses indicated both types of goal. This "hybrid" goal is shown in the following explanation, where the student applied engineering goals for her choices of slope and expressed a goal of finding out about the starting position:

> I thought that you should use steep because if it was flat it wouldn't be able to roll that well. I don't know why I used [different surfaces], I just thought it would work. I picked the middle because it's a shorter distance and it might roll farther. And I picked the top to see if it might roll farther than the middle.

It was less rare for students (especially those in the Phase 4 human-tutor condition) to express both science and engineering goals on different questions during the ramps pretest, where 16 out of 70 students (23%) expressed both types of responses. Figure 7.3 shows the number of L1&L2 and M2b students expressing only science goals throughout the ramps pretest, only engineering goals, at least one science goal and at least one engineering goal, and neither of these goals (e.g., they gave descriptions of their setups but not the underlying rationale). However, most students (48 out of 70, or 69%) expressed only one goal type throughout the ramps pretest. Thus, in a feedback-free task in a single domain, goals were stable for the majority of students.

Overextended "Fairness" Schema

Closely related to "same outcomes" engineering goals, students occasionally expressed the goal of making the setups "fair" or giving the conditions—rather than target variable—the "same chance." This overextended "fairness" schema differs from the "same outcomes" engineering goal in that its focus is on the setup rather than result. That is, students seem to consider differing outcomes acceptable, provided

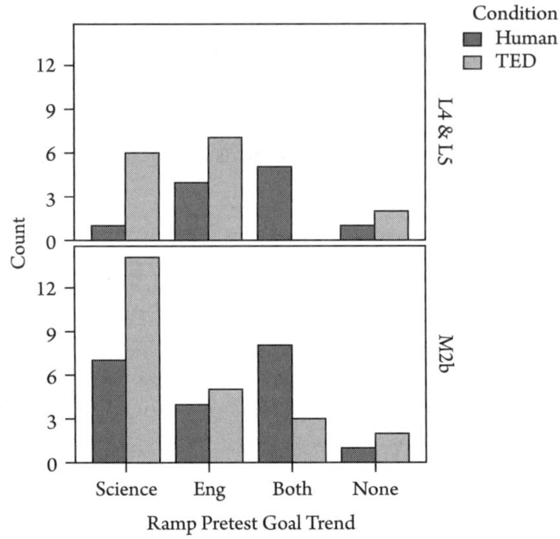

Figure 7.3 Frequency of goal response trends during ramps pretest, by school and condition.

that the setup at least allows for a reasonable possibility of the same outcomes. Additionally, this is not considered a science goal because rather than attempting to discover causal relations in variables, the student is attempting to make the comparison "fair." For this reason, this goal differs from the noncontrastive target variable(s), or Rule 2 failure, science goal approach. This type of response was not detected in students' responses to the Story test questions. However, it was detected in explicit instruction in 10.5% (4 out of 38) of Phase 1 low-SES students' responses to "Why is this a fair comparison for height?" and in 11% (3 out of 27) of responses to "Tell me why this is a good way to find out whether the balls go different distances just because of the ball" during the instruction in Phase 4. This type of response was not found at the same instructional points for the middle-SES students.

This approach is expressed in a student's explanation for why his noncontrastive (NC) experiment is a "fair test of [the target variable]," even though the balls might roll different distances:

> Because the surface . . . one might go farther, one might go [less far] and if the one ball goes less and one ball goes more I think it's fair because the surface is both the same way, the height (slope) is the same, they're both long.

Another example comes from a student responding to why an unconfounded experiment "is a fair comparison for height." This student's overextended fairness schema seems to have overridden his perception of the experimental setup, in which the heights were actually different:

> It is fair because they are the same height [*sic*] and the same track and the same ball and . . . the same surface.

Like the same outcome goal, this misconception may arise when students interpret instructional explanations about controlling variables within the context of a "fairness" schema. Students' explanations such as these indicate an explicit understanding of "fairness," which they apply to their evaluations of ramps setups when they justify the "fairness" of the comparison in terms of whether the variable settings are the same.

"Domain Knowledge" Misinterpretations and Relationship to Engineering Goals

Perhaps the "deepest"—and most likely implicit—misinterpretations that occurred during instruction were when students interpreted questions about experimental design as either asking them to apply their beliefs about the effects of variables to *predict* the experimental outcome or as asking about the effects of specific variables on an outcome (without necessarily predicting the experimental outcome). This misunderstanding is analogous to the behavior of judging the validity of a logical syllogism based on whether the conclusion corresponds to one's beliefs rather than by applying abstract deductive reasoning. As shown in Table 7.4, these responses were relatively rare on the Story tests. This type of response occurred more often during instruction when students were asked to evaluate experiments for whether they would provide evidence of the causal status of a target variable (e.g., "If we ran this experiment, could you tell whether the [target variable] made a difference in how far the balls roll?"). The following is an example of a student misinterpreting the tutor's question as about predicting the outcome:

T: Now they're [the balls] both at the top. If we ran this, could you tell whether there's a difference caused by the starting position?
S: It doesn't gain that much speed.

Students sometimes responded to questions about the determinism of experiments by simply stating their beliefs about the effect of a variable on an outcome, without necessarily using them to predict the specific experimental results. For example, in the following instance, a student simply stated her belief about the effect of the age of the seller on the experimental outcome when asked whether a maximally confounded experimental setup was a good way to find out whether age makes a difference:

"It really should not matter how old the child is. The children are both selling drinks and it should not matter if the older person was 14 and the younger child was 10."

This student was not evaluating the validity of the experimental setup, as asked, but rather was answering the more familiar question: "Does age make a difference?" In a second example (from Phase 3), the student interprets a question about whether the experiment would produce results that could lead to valid claims about the target variable as asking about the effect of that variable on the result:

T: What if I put it this way: can we tell if it's just the slope of the ramp that makes a difference in how far the balls roll? Using this (maximally contrastive) experiment that you have set up here, could you answer that question?
S: Yes. It would affect how far the ball rolls.

Though it is possible that students who expressed these "domain knowledge" misinterpretations actually held science goals and simply "skipped" to the prediction stage of experimentation, this is not supported by our data. Students who gave "domain knowledge" responses on the Story-evaluation/design pre-, post-, and follow-up tests almost always gave engineering explanations but no science explanations in their other responses. Thus, when students gave responses about the effects of variables on outcomes, they most likely held engineering goals.

More evidence of the relationship between domain knowledge and engineering goals comes from Phase 1, where students who expressed engineering or domain knowledge goals in their written evaluations of experiments were as likely to give the same type of response in a subsequent question (i.e., students giving engineering responses to both questions) as they were to give the other type of response (i.e., students giving an engineering response to one question and a domain knowledge response to the other). However, these students generally did not shift from an engineering or domain knowledge goal to a science goal. For example, in Phase 1, of the nine students in L1 who gave engineering or domain knowledge explanations on the first in-class written evaluation, seven (78%) continued to give engineering and domain knowledge responses throughout instruction. Furthermore, three more students in this class expressed engineering or domain knowledge interpretations later in the instruction. Similarly, the one student in L2 who gave an engineering response on the first evaluation question continued giving these explanations throughout the instruction, and one student "regressed" from a science to an engineering goal during the course of the instruction. However, only one of the three students in inquiry-based M1 continued to give engineering/domain knowledge responses during instruction, and no students in this class regressed to engineering goals. This correlation between engineering and domain knowledge goals is not surprising, given that in the vast majority of engineering goal approaches and, by definition, in all "domain knowledge" interpretations, students state or apply their beliefs about the effects of the variables.

These findings also show that, for students not in the inquiry-based science classroom (M1), misconceptions about the nature of the task that surface during CVS instruction persisted throughout instruction and interfered with learning the skills of experimental design.

Remediating Engineering Goal Misinterpretations

These findings of the ways students misinterpreted instruction informed modifications of the TED tutor. For example, instructional wording that students could plausibly misinterpret within an alternate goal were clarified. In addition, the finding of a strong association between science goals and varying the variable(s) one is testing, discussed earlier, informed remediation in the TED tutor. In a recent phase of development, sixth graders from a science and technology magnet school with a diverse student population completed an instructional unit on the TED tutor. After the ramps pretest but prior to the "explicit" instruction portion of the tutor, students identified as not applying science goals or knowledge of the need to contrast the variable(s) one is testing were provided with remedial tutoring, which required them to identify the target variable and select its values across conditions. Students were given immediate feedback on their responses. All students were then asked why it is necessary to vary the target variable then given the explanatory feedback described in the remediation section for R2 failure. This instruction simultaneously enforced the science goal of the activity and its associated contrasting the target variable element. These students showed significantly better transfer performance on an immediate Story posttest than eighth graders from the same school given similar instruction but without the remedial tutoring and wording clarifications.

DISCUSSION

This chapter reports the first in-depth compilation of the errors and misconceptions that arise when students are instructed in experimental design. Our findings of the different errors middle school students make and their misconceptions about experimental design add to those of previous researchers (e.g., Kuhn et al., 1988; Kuhn et al., 1995; Schauble et al., 1991; Tschirgi, 1980) to encompass a wide range of preconception types. On a practical note, the analysis of misconceptions presented here is being used to further inform the development of our adaptive computer-based tutor for simple experimental design (Siler et al., 2009), the TED (Training in Experimental Design) tutor.

Students' motivations underlying their responses can be broadly categorized as employing science goals, engineering goals, or overextended "fairness" goals. Engineering goals—where students attempt to produce desired outcomes—were the most common goal-related misconceptions identified. Most often, students applied the practical engineering goal of maximizing the outcomes of one or both conditions by selecting the variable values they believed would produce the best effect. Somewhat less commonly, students designed setups aimed at producing different outcomes by varying one or more variables they believed to be causal. This goal may be related to (perhaps an implicit) belief that a "good" experiment is one that produces different outcomes, or, relatedly, one that tests a causal variable(s).

Finally, students designed setups intended to produce the same outcomes across conditions. They did this by setting up the conditions exactly the same and by setting only the variables they believed mattered to the same values. They also did this by "counterbalancing" variable settings by using one "good" and one "bad" variable setting in one condition and the opposite values in the other condition, such that the intended net result was for both conditions to produce the same outcomes. "Same outcome" goals—as well as overextended "fairness" schemas—seemed to arise when instructional cues triggered intuitive "fairness" schemas. Such schemas have been identified in students as young as second graders (Wollman, 1977).

Science goals also appear to be tied to the notion of contrasting values of the variable(s) they were testing. However, students often designed experiments that did not allow for valid causal claims, either by intentionally testing multiple variables or by ignoring experimental confounds. Thus, when students adopt science goals, they seem to have most difficulty with controlling variables (and understanding why that is necessary).

Whereas intentions within science goals appear to be largely explicit, as evidenced by students' explicit goal statements such as "I am trying to find out about the [target variable]," engineering goals appear to be largely implicit. That is, when expressing engineering goals, students rarely made analogous explicit statements of intention (e.g., "I am trying to make the balls roll far"). It is plausible that—though students likely have copious experiences applying engineering goals (Schauble et al., 1991), if not accompanied by verbalizations or higher-level thinking, such goals may remain implicit. In contrast, people almost certainly have fewer experiences designing or evaluating experiments (Schauble et al., 1991); however, such experiences may be more likely to be associated with higher-level cognitive processes that result in explicit knowledge. Moreover, students likely hear more explicit statements of science than engineering goals. For example, in their science classes, teachers may say "Today, we are going to do an experiment to find out about X" though may be less likely to say "Today, we are going to try to make the fastest X." The nature of students' "fairness" beliefs are less clear, due to the infrequency with which they were expressed—particularly when explaining their own designs. As with applying engineering goals, students likely have abundant experiences enforcing "fairness" goals, however, these may be more likely to occur in social situations involving verbal negotiations which could contribute to their becoming (or staying) explicit knowledge.

As with the relationship between engineering goals and domain beliefs, engineering goals and domain knowledge "goals," or interpretations of instructional questions, appear to be inter-related. Students who expressed one of these goals during the course of instruction tended to express the other much more often than they expressed science goals. Moreover, that they were more common in the noninquiry science classrooms suggests that domain knowledge and engineering goals may be more likely to be adopted when classroom experiences consist primarily of "learning facts" or "producing results." Because general domain knowledge goals are not likely to be explicitly expressed by the teacher (e.g., by stating "We are going to learn facts," though a teacher might say, "Today, we are going to learn about the layers of the earth"), they may be learned and adopted implicitly.

Furthermore, students generally explicitly expressed their beliefs about variable effects when they applied engineering goals, but rarely did when applying science goals. This suggests that (at least when first learning about experimental design) students' science goals are dissociated from their explicit domain beliefs. Interestingly, though, science goals may be associated with implicit domain beliefs. For example, Tschirgi (1980) found that both children and adults showed biases toward choosing the hypothesized "better" value of a variable when they were supposed to "prove" that that particular variable value was responsible for either a good or bad outcome. That is, when given a good outcome, participants were more likely to choose a second instance with the same value than with a different value of the target variable. However, when given an analogous bad outcome, they were more likely to choose a second setup with the opposite value.

Identifying a range of specific errors and misconceptions—especially those that are resistant to typical classroom activities or instruction on experimental design and thus prevent learning—is an important first step in improving instruction in experimental design because it allows for easier identification during instruction. Knowledge of the ways that students can misinterpret instruction can also alert those involved in instructional design to the kinds of wordings that may elicit engineering goals, fairness schemas, or "domain knowledge" misinterpretations of instruction. Designers of instruction may decide to intentionally elicit such misinterpretations in order to address them or to avoid eliciting them. Our knowledge of how students misinterpreted instruction led to revisions in the on-screen text and audio voice-overs in our intelligent tutor. These "low-tech" revisions were not only time- and cost-effective, but were correlated with significant performance improvements.

Furthermore, identifying the types of misconceptions that arise in this context is relevant in devising means of instruction to *remediate* them. As discussed earlier, remedial tutoring in the TED-tutor that required students to identify and vary the target variable prior to the "explicit" instruction on the rationale for controlling the other variables was associated with gains in transfer performance. We believe this remediation elicited science goals; however, it may not have alerted students of their alternative goal interpretations. Other methods, which do elicit this awareness, may lead to even stronger gains. For example, alternative goals may be directly contradicted when detected in order to both make students' goal assumptions explicit and induce cognitive conflict. And induced cognitive conflict that elicits such cognitive processes as knowledge building may promote conceptual change (Chan, Burtis, & Bereiter, 1997) such as goal shifting. The use of refutational texts designed to directly contradict students' misconceptions has been shown to promote conceptual change in such domains as physics (e.g., Diakidoy, Kendeou, & Ioannides, 2002; Hynd, McWhorter, Phares, & Suttles, 1994), biology (e.g., Mikkilä-Erdmann, 2001), and ecology (e.g., Ozkan, Tekkaya, & Geban, 2004). However, to our knowledge, this method has not targeted goal-related misconceptions.

To apply this remedial method in the TED tutor, when students are asked to design experiments to test a particular variable, they will first be asked to select their goal in setting up the experiment, thereby making explicit any implicit goals. If students select a response indicating an engineering goal, such as "to make the balls roll far," they will receive immediate feedback, such as "Actually, the question is asking you to design an experiment that will let you figure out if the surface affects how far balls roll, not to make the balls roll far." If students persist in selecting responses indicating engineering or other alternative goals, they will receive more in-depth remediation (e.g., they will be asked to categorize questions as either focused on engineering or science goals until they can consistently distinguish between the two).

Finally, identification of these goal misconceptions provides information about possible alternative instructional approaches. For example, students who fail to learn CVS via the "explicit" instruction may fare better with instruction that builds upon their intuitive ideas, such as their understandings of "fairness" of comparisons in combination with their understandings of "comparing and contrasting" a target variable. The relative effectiveness of such instructional methods will be investigated in future studies.

However, we believe that even if these methods are effective in helping students of all backgrounds to develop robust understandings of experimental design, engaging students in scientific investigations on a regular basis is still of vital importance. Such "mindful" experiences may not only reinforce understanding the procedures and logic of CVS, but also the development of other inquiry skills, better enabling students to explore and learn about their world.

ACKNOWLEDGMENTS

This research was supported in part by grants from the Institute of Education Sciences (R305A100404 and R305B090023), and in part by a grant from the National Science Foundation (SBE-0836012). We thank Kevin Willows and Cressida Magaro for their invaluable contributions on many aspects of the project. We also thank the participating schools and teachers who allowed us to conduct research in their classrooms.

REFERENCES

Bindra, D., Clarke, K., & Schultz, T. (1980). Understanding predictive relations of necessity and sufficiency in formally equivalent "causal" and "logical" problems. *Journal of Experimental Psychology: General, 109,* 422–443.

Braine, M. D., & Rumain, B. (1983). Logical reasoning. In J. H. Flavell & E. M. Markman (Eds.), *Cognitive development: Handbook of child psychology* (Vol. 3, pp. 261–340). New York: John Wiley.

Byrnes, J., & Overton, W. (1986). Reasoning about certainty and uncertainty in concrete, causal, and propositional contexts. *Developmental Psychology, 22,* 793–799.

Chan, C., Burtis, P., & Bereiter, C. (1997). Knowledge building as a mediator of conflict in conceptual change, *Cognition and Instruction, 15,* 1–40.

Chen, Z., & Klahr, D. (1999). All other things being equal: Children's acquisition of the control of variables strategy. *Child Development, 70,* 1098–1120.

Chinn, C. A., & Brewer, W. F. (1993). The role of anomalous data in knowledge acquisition: A theoretical framework and implications for science instruction. *Review of Educational Research, 63,* 1–49.

Clement, J. (1983). A conceptual model discussed by Galileo and used intuitively by physics students. In D. Gentner & A. Stevens (Eds.), *Mental models* (pp. 345–362). Hillsdale, NJ: Lawrence Erlbaum.

Diakidoy, I., Kendeou, P., & Ioannides, C. (2002). Reading about energy: The effects of text structure in science learning and conceptual change. *Contemporary Educational Psychology, 28,* 335–356.

diSessa, A. (1993). Towards an epistemology of physics. *Cognition and Instruction, 10*(2–3), 105–225.

Fay, A. L., & Klahr, D. (1996). Knowing about guessing and guessing about knowing: Preschoolers' understanding of indeterminacy. *Child Development, 67,* 689–716.

Grosslight, L., Unger, C., Jay, E., & Smith, C. (1991). Understanding models and their use in science: Conceptions of middle and high school students and experts. *Journal of Research in Science Teaching, 28,* 799–822.

Hynd, C., McWhorter, Y., Phares, V., & Suttles, W. (1994). The role of instructional variables in conceptual change in high school physics. *Journal of Research in Science Teaching, 32,* 933–946.

Klahr, D. (2009). "To every thing there is a season, and a time to every purpose under the heavens": What about direct instruction? In S. Tobias & T. M. Duffy (Eds.), *Constructivist theory applied to instruction: Success or failure?* (pp. 291–310). London: Taylor & Francis.

Klahr, D., Fay, A. L., & Dunbar, K. (1993). Developmental differences in experimental heuristics. *Cognitive Psychology, 25,* 111–146.

Klahr, D., & Li, J. (2005). Cognitive research and elementary science instruction: From laboratory, to the classroom, and back. *Journal of Science Education and Technology, 14,* 217–238.

Klahr, D. & Nigam, M. (2004). The equivalence of learning paths in early science instruction: Effects of direct instruction and discovery learning. *Psychological Science, 15,* 661–667.

Klahr, D., Triona, L., Strand-Cary, M., & Siler, S. (2008). Virtual vs. physical materials in early science instruction: Transitioning to an autonomous tutor for experimental design. In J. Zumbach, N. Schwartz, T. Seufert, & L. Kester (Eds.), *Beyond knowledge: The legacy of competence meaningful computer-based learning environments.* (pp. 163–172). New York: Springer.

Kuhn, D. (1991). *The skills of argument.* New York: Cambridge University Press.

Kuhn, D., Amsel, E., & O'Loughlin, M. (1988). *The development of scientific thinking skills.* New York: Academic Press.

Kuhn, D., & Dean, D. (2005). Is developing scientific thinking all about learning to control variables? *Psychological Science, 16,* 866–870.

Kuhn, D., Garcia-Mila, M., Zohar, A., & Andersen, C. (1995). Strategies of knowledge acquisition. *Monographs of the Society for Research in Child Development, 60*(4), 1–157.

Masnick, A. M., & Klahr, D. (2003). Error matters: An initial exploration of elementary school children's understanding of experimental error. *Journal of Cognition & Development, 4,* 67–98.

McCloskey, M. (1983). Naive theories of motion. In D. Gentner & A. L. Stevens (Eds.), *Mental models* (pp. 299–324). Hillsdale, NJ: Lawrence Erlbaum.

Mikkilä-Erdmann, M. (2001). Improving conceptual change concerning photosynthesis through text design. *Learning and Instruction, 11,* 241–257.

Minstrell, J. (1982). Explaining the "at rest" condition of an object. *Physics Teacher, 20,* 10–20.

Morris, B. J., & Sloutsky, V. M. (2002). Children's solutions of logical versus empirical problems: What's missing and what develops? *Cognitive Development, 16,* 907–928.

Ozkan, O., Tekkaya, C., & Geban, O. (2004). Facilitating conceptual change in students' understanding of ecological concepts. *Journal of Science Education Technology, 13,* 95–105.

Pieraut-Le Bonniec, G. (1980). *The development of modal reasoning: The genesis of necessity and possibility notions*. New York: Academic Press.

Schauble, L., Klopfer, L. E., & Raghavan, K. (1991). Students' transition from an engineering model to a science model of experimentation. *Journal of Research in Science Teaching, 28,* 859–882.

Siegler, R. S. (1981). Developmental sequences within and between concepts. *Society for Research in Child Development Monographs, 46* (Whole No. 189).

Siegler, R. S. (1996). *Emerging minds: The process of change in children's thinking.* New York: Oxford University Press.

Siler, S. A., Klahr, D., Strand-Cary, M. S., Magaro, C., & Willows, K., (2009). *Adapting an effective lesson plan for a computer-based tutor.* Paper presented at the 13th Biennial Conference for the European Association for Research on Learning and Instruction (EARLI), Amsterdam, Netherlands.

Slotta, J. D., Chi, M. T. H., & Joram, E. (1995). Assessing students' misclassifications of physics concepts: An ontological basis for conceptual change. *Cognition and Instruction, 13,* 373–400.

Smith, C., Carey, S., & Wiser, M. (1985). On differentiation: A case study of the development of size, weight, and density. *Cognition, 21,* 177–237.

Stathopoulou, C., & Vosniadou, S. (2007). Conceptual change in physics and physics related epistemological beliefs: A relationship under scrutiny. In S. Vosniadou, A. Baltas, & X. Vamvakoussi (Eds.), *Re-framing the conceptual change approach in learning and instruction,* Advances in Learning and Instruction Series (pp. 145–165). Oxford: Elsevier.

Strand-Cary, M., & Klahr, D. (2008). Developing elementary science skills: Instructional effectiveness and path-independence. *Cognitive Development, 23,* 488–511.

Toth, E. E., Klahr, D., & Chen, Z. (2000). Bridging research and practice: A cognitively-based classroom intervention for teaching experimentation skills to elementary school children. *Cognition & Instruction, 18,* 423–459.

Tschirgi, J. E. (1980). Sensible reasoning: A hypothesis about hypotheses. *Child Development, 51,* 1–10.

Vosniadou, S. (1994). Universal and culture-specific properties of children's mental models of the earth. In L. Hirschfeld & S. Gelman (Eds.), *Mapping the mind* (pp. 412–430). New York: Cambridge University Press.

Vosniadou, S., & Brewer, W. F. (1992). Mental models of the earth: A study of conceptual change in childhood. *Cognitive Psychology, 24,* 535–585.

Vosniadou, S., & Brewer, W. F. (1994). Mental models of the day/night cycle. *Cognitive Science, 18,* 123–184.

Vosniadou, S., Ioannides, C., Dimitrakopoulou, A., & Papademetriou, E. (2001). Designing learning environments to promote conceptual change in science. *Learning and Instruction, 11,* 381–419.

Wiser, M., & Carey, S. (1983). When heat and temperature were one. In D. Genter & A. Stevens (Eds.), *Mental models* (pp. 267–297). Hillsdale, NJ: Erlbaum.

Wollman, W. (1977). Controlling variables: Assessing levels of understanding. *Science Education, 61,* 371–383.

Zohar, A., & Nemet, F. (2001). Fostering students' knowledge and argumentation skills through dilemmas in human genetics. *Journal of Research in Science Teaching, 39,* 35–62.

PART THREE
IMPLICIT AND EXPLICIT PROCESSES IN THE COGNITIVE PSYCHOLOGY OF SCIENCE

8 How to Define and Examine Implicit Processes?

Jan De Houwer and Agnes Moors

Psychology as a scientific discipline is directed at understanding the processes underlying behavior. During the past two decades, more and more psychologists have become interested in the impact that so-called implicit processes have on behavior. This evolution can be seen in the various subdisciplines of psychology, including general psychology (e.g., Schacter, 1987), social psychology (e.g., Greenwald & Banaji, 1995), clinical psychology (e.g., Gemar, Segal, Sagrati, & Kennedy, 2001), and addiction research (e.g., Wiers & Stacy, 2006). In line with this evolution, the current book addresses the topic of implicit and explicit processes in one particular aspect of human behavior, namely scientific behavior.

Despite the huge interest in implicit processes, relatively little attention has been given to the question of what it means to say that a process is implicit. In this chapter, we present an overview of the conceptual work that we have performed in this context over the past years (see De Houwer, 2006, 2011; De Houwer & Moors, 2007, 2010; De Houwer, Teige-Mocigemba, Spruyt, & Moors, 2009a, 2009b; Moors & De Houwer, 2006a, 2006b, 2007; Moors, Spruyt, & De Houwer, 2010). The starting point of our analysis is the postulate that the meaning of the term "implicit" is identical to the meaning of the term "automatic." As we will argue later on, defining implicit as automatic has the advantage that it encompasses many of the earlier, more informal definitions that have been given for the concept implicit. It also allows one to draw on the extensive conceptual work on automaticity that has been undertaken in the past.

In this chapter, we put forward the following ideas. First, automaticity is not an all-or-none property of mental processes but refers to a set of features that do not necessarily co-occur within each automatic process. Hence, it does not make sense to simply say that a process is implicit because there is no agreement about the automaticity features to which this term refers (see Bargh, 1989, 1992; Moors & De Houwer, 2006a, 2007). Second, the heterogenic nature of the concept automaticity calls for a decompositional approach in which researchers need to specify for each process which automaticity features are thought to apply to that process. This requires a precise definition of each automaticity feature. Within a cognitive decompositional approach (Bargh, 1992; Moors & De Houwer, 2006a, 2007), most automaticity features are defined in terms of mental constructs such as goals, consciousness, and processing resources. Hence, when using the term "implicit," one should specify in a precise manner the automaticity features one has in mind (i.e., De Houwer, 2006; De Houwer et al., 2009a).

Third, claims about automaticity features need to be backed up with sound arguments and empirical evidence. Hence, when claiming that a process is implicit in a certain manner, one should be able to justify these claims (De Houwer et al., 2009a, 2009b). Fourth, obtaining evidence for the automaticity of mental processes is complex (see Moors et al., 2010). Hence, studying the implicit nature of mental processes can run into important problems. Finally, we introduce the idea that progress in understanding the nature of implicit processes can be facilitated by a functional decompositional approach to automaticity (De Houwer, 2011). This approach entails that automaticity features are defined strictly in terms of observable elements in the environment rather than unobservable mental constructs. In addition, the primary aim is to examine the automaticity features of effects (i.e., the causal impact of elements in the environment on behavior). Knowledge about the automaticity of effects, however, can be used to guide the construction of models about the mental processes by which elements in the environment influence behavior (see De Houwer, 2011). We end the chapter by illustrating the implications of our analysis for future research on implicit processes underlying scientific behavior.

AUTOMATICITY IS NOT AN ALL-OR-NONE FEATURE

According to the dual-mode or all-or-none view on automaticity, processes are either automatic or nonautomatic. Automatic processes are assumed to have all automaticity features; nonautomatic processes have the opposite features. For instance, all automatic processes are assumed to be unintentional, uncontrolled, unconscious, efficient, and fast whereas all nonautomatic processes are assumed to be intentional, controlled, conscious, inefficient, and slow. According to this view, it is relatively easy to diagnose a process as automatic. It suffices to demonstrate that the process possesses one of the automaticity features. If it has one of the features, it can be assumed to have all other automaticity features and thus to be fully automatic.

It became clear, however, that the different automaticity features do not always co-occur. Evidence from Stroop studies, for instance, suggests that the processing of word meaning is automatic in that it does not depend on the intention to process the meaning of the word. At the same time, word processing is nonautomatic in that it depends on the allocation of attention to the word (see Bargh, 1989, 1992, 1994, and Logan, 1985, 1989, for a discussion of this evidence). Several reasons can be identified for why the all-or-none view on automaticity remained popular despite the evidence against it (see Moors & De Houwer, 2007, for a discussion). Nevertheless, there can be little doubt about the fact that this view is incorrect.

Given the assumption that the concepts automatic and implicit can be used in an interchangeable manner, one can conclude that it makes little sense to simply say that a process is implicit. This means little more than saying that a process possesses one or more features of automaticity, leaving it unspecified which automaticity features apply. Hence, when using the term "implicit," it is best to always make explicit the automaticity features to which one refers. Such an approach can be called decompositional in that the concepts implicit and automatic are actually decomposed into various nonoverlapping features. However, a decompositional approach is meaningful only if one can clearly define the different automaticity features in a nonoverlapping way. In the next section, we discuss the cognitive decompositional approach that does incorporate detailed definitions of these features.

THE COGNITIVE DECOMPOSITIONAL APPROACH TO AUTOMATICITY

Within this section, we first provide an overview of the definitions of automaticity features as put forward by Moors and De Houwer (2006a, 2007). The main purpose of this overview is to show that detailed, nonoverlapping definitions of automaticity features can and have been formulated. Researchers who want to make explicit their ideas about the implicit nature of processes can thus draw upon these definitions. Next we discuss whether the cognitive decompositional approach allows one to encompass the different definitions that have been given in the past for the concept implicit. Finally, we examine the problems that can arise when diagnosing whether a process possesses certain automaticity features.

Defining Automaticity Features

Many automaticity features, such as (un)intentional, goal-directed, goal-(in)dependent, (un)controlled/(un)controllable, and autonomous, are somehow related to goals. Perhaps the most central goal-related automaticity feature is the feature uncontrolled. Uncontrolled is the opposite of controlled. We therefore start with a definition of controlled. To say that a process is controlled implies a proximal goal (i.e., a goal regarding the target process such as the goal to engage in, alter, stop, or avoid the process) that causes the achievement of the end state put forward in the goal (i.e., the actual occurrence, change, interruption, or prevention of the process). To say that a process is uncontrolled can therefore have different meanings. It can refer to the fact that the state of a process changes (i.e., that the process occurs, changes, is interrupted, or prevented) in the absence of a goal to achieve this change. It can also refer to the fact that the goal is present but the desired effect is absent. For instance, a process can be called uncontrolled when the process occurs despite the goal to prevent the occurrence of the process. A process can also be described as uncontrolled when both the goal and the desired effect are present, but the effect was not caused by the goal. Imagine that you have the goal to kneel down in front of your loved one in order to propose marriage. You do kneel down but it happens because someone pushes you. In that case, both the goal and the desired effect are present, but it would be wrong to regard the act of kneeling down as controlled because the goal did not cause the desired effect.

Other goal-related automaticity features can be defined in terms of the feature uncontrolled. To say that a process is intentional means that the goal to engage in a process causes the occurrence of the process. Hence, intentional is identical to controlled in the sense of the goal to engage in (rather than the goal to alter, stop, or avoid the process). Therefore, unintentional processes are a subset of uncontrolled processes. A process can be termed unintentional because the process occurs without the intention to engage in the process or because the goal to engage does not cause the occurrence of the process. The feature autonomous can be defined as uncontrolled in terms of every possible processing goal. That is, an autonomous process is a process that is uncontrolled in every possible sense. Goal-independent processes are processes that operate independently from proximal goals (i.e., goals relating to the target process such as the goal to start, stop, alter, or avoid the process) and distal goals (i.e., goals unrelated to the target process). Therefore, a process might be uncontrolled but still be goal-dependent. Finally, processes can be described as purely stimulus-driven when in addition to being goal-independent (i.e., not dependent on any type of goal), they also do not depend on other factors such as awareness or attention. The occurrence of purely stimulus-driven processes depends only on the presence of a stimulus and certain basic conditions which ensure that the stimulus can be physically registered (e.g., in the case of visual stimuli, that the eyes are not closed).

The feature unconscious is often listed as a core feature of automaticity. Unfortunately, (un)consciousness as a mental state is notoriously difficult to define (see Moors & De Houwer, 2006a, for an analysis). This limits its usefulness as a defining feature of automaticity. It is also important to realize that the term "unconscious" can be used as a predicate of several things. It can refer to (a) the stimulus input that evokes the process, (b) the output of the process, (c) the process itself, or (d) the consequences of the process such as its influence on subsequent processing (e.g., Bargh, 1994). Therefore, when using the term "unconscious," it is important to specify what it is a predicate of. Note that there is no complete overlap between the feature (un)conscious and goal-related features. For instance, our definitions allow for the possibility of unconscious intentional processes, that is, a process that occurs only when there is a goal to engage in the process but that can operate in the absence of awareness of the process, its input, output, or consequences.

A process can also be automatic in the sense of efficient. Efficient processes can be defined as processes that consume few processing resources or attentional capacity. Because efficiency leads to the subjective experience that processing is effortless, the terms "efficient" and "effortless" are often used interchangeably. Again there is no complete overlap between efficiency and other automaticity features. Whereas goals are related to the *direction* of attention (goals may determine the focus of attention), efficiency is related to the *amount* of attention. We also allow for the possibility of a state of consciousness that exists outside of attention and thus for efficient conscious processes (Block, 1995). Likewise, in principle, processes can be unconscious but nonefficient (e.g., Naccache, Blandin, & Dehaene, 2002).

Finally, processes can be automatic in the sense of fast. The feature fast refers to the time that is needed for a process to run to completion. This feature is clearly a gradual feature. There is no objective threshold for calling something fast or slow, so investigators need to rely on commonsense arguments for calling some interval short or long or for deciding whether a process is fast or slow. Processes that are fast also tend to be uncontrolled (because the implementation of goals tends to require time), unaware (because consciousness needs time to develop), and efficient (because efficient processes are typically faster than nonefficient processes). Nevertheless, this overlap is only partial in that some uncontrolled, unconscious, or efficient processes might be slow and some controlled, conscious, or inefficient processes might be fast.

DESCRIBING IMPLICIT PROCESSES IN TERMS OF AUTOMATICITY FEATURES

In the past, researchers have described processes as being implicit because they operate in an unintentional, uncontrolled, efficient, fast, or unconscious manner. Our definition of implicit as automatic, in combination with a decompositional approach of automaticity, allows us to encompass many if not all of the previous definitions of implicit processes. The decompositional approach also clarifies that researchers need to make explicit the automaticity feature that they have in mind when using the concept implicit, and it provides them with the conceptual tools to precisely describe their definition of the crucial automaticity features. Hence, we hope that our analysis of the concepts implicit and automatic will improve communication among researchers when they study implicit processes.

We do not commit ourselves to one particular automaticity feature as being the crucial feature for diagnosing whether a process is implicit. Others have put forward one specific feature such as the feature unintentional (e.g., Fazio & Olson, 2003; Richardson-Klavehn, Lee, Joubran, & Bjork, 1994) or the feature unconscious (e.g., Schacter, 1987) as the one feature that determines whether a process qualifies as implicit. In our opinion, the selection of a particular feature or set of features as central for implicit processes cannot be settled on the basis of a priori arguments but needs to be decided on the basis of empirical data. The distinction between implicit and explicit processes is functional only if it allow us to increase our understanding of (the processes underlying) behavior. If behavior is not differentially affected by whether the underlying process possesses a particular automaticity feature, then there is little added value in distinguishing between implicit and explicit processes on the basis of that automaticity feature. Hence, the concept implicit should be linked to the features that actually matter for behavior. For instance, research on implicit memory has shown that the conscious or unconscious nature of memory retrieval is relatively unrelated to the way in which variables influence memory retrieval (e.g., the impact of level of processing during encoding). What does seem to matter is whether retrieval is intentional (see Richardson-Klavehn et al., 1994).

This suggests to us that it is more functional to define implicit memory in terms of the feature unintentional than in terms of the feature unconscious. It remains to be seen, however, whether these results generalize to processes other than memory retrieval. Until sufficient data are available for selecting one automaticity feature as central for the concept implicit, we prefer a broad definition of implicit as automatic covering several automaticity features. Of course, the selected features need to be made explicit and defined in an unambiguous manner.

One could, however, argue that such a broad definition is problematic (e.g., Nosek & Greenwald, 2009). First, defining implicit as automatic and explicit as nonautomatic might blur the distinction between implicit and explicit processes. We agree that most processes are probably automatic (and thus implicit) in some ways and non-automatic (and thus explicit) in other ways. This does indeed blur the distinction. However, at present we believe that defining implicit as automatic and explicit as nonautomatic is the best available alternative. As indicated above, one could select one automaticity feature as the criterion that provides the dividing line, but it is not clear which criterion should be selected or whether it will ever be possible to reach consensus about the to-be-selected criterion. Hence, for now it is best not to commit to one feature but to adopt a broad definition that allows each researcher to make explicit the specific feature that he or she has in mind.

DIFFICULTIES WITH EXAMINING THE AUTOMATICITY FEATURES OF MENTAL PROCESSES

Although our definition of implicit does allow researchers to specify in great detail the conceptualization that they have in mind, even conceptually precise claims about the implicit nature of processes have little value when these claims cannot be backed up by valid arguments and empirical evidence. Therefore, claiming that a process is implicit requires not only precision, but also justification. Unfortunately, several important challenges need to be overcome in order to obtain empirical evidence for automaticity features of mental processes. In this section, we provide a brief overview of these challenges (for more details, see De Houwer, 2011; De Houwer, Gawronski, & Barnes-Holmes, 2010; Moors et al., 2010).

Each automaticity feature refers to a certain set of conditions. For instance, the feature unintentional refers to a situation in which the goal to initiate a process is absent. A process is said to possess a certain automaticity feature if it operates under the set of conditions specified by the automaticity feature. For instance, the process qualifies as unintentional if it operates in the absence of the goal to initiate that process. Therefore, in order to examine the automaticity features of a mental process, the following steps need to be taken: First, it needs to be established that the set of conditions that is specified by a feature, is actually present (e.g., that a person does not have the goal to initiate the process). Second, one should be able to ascertain that the process operates under those conditions.

When adopting a cognitive decompositional approach, both steps are complicated by the fact that mental constructs cannot be observed directly (see De Houwer, 2011; De Houwer et al., 2010). First, several automaticity features refer to the presence or absence of unobservable mental constructs. For instance, goal-related features refer to goals. Likewise, the feature efficient refers to the construct of mental resources. Goals and mental resources cannot be observed directly. One can at best infer their presence on the basis of observable elements in the environment or observable behavior (e.g., task instructions or performance on a secondary task), but such inferences depend on assumptions that might not always hold (e.g., that participants follow instructions or that a secondary task loads the same resources as the primary task).

Second, many of the mental processes that are studied by cognitive psychologists (e.g., the formation of associations in memory, the activation of nodes and associations in semantic networks, working memory, and reasoning) cannot be observed directly. Consider the process of attitude activation (see Moors et al., 2010, for a more detailed discussion of this example). Attitudes can be conceived of as mental representations that specify a positive or negative disposition toward a certain object (e.g., Eagly & Chaiken, 2007). Unfortunately, attitudes (e.g., toward science) cannot be observed directly. Researchers have therefore developed tasks that provide measures of attitudes. One of these measures is affective priming (Fazio, Sanbonmatsu, Powell, & Kardes, 1986; see De Houwer et al., 2009a, for a review). Affective priming refers to the observation that responses to a target stimulus (e.g., HAPPY) are faster when it is preceded by a prime stimulus with the same affective valence (e.g., SUMMER) than when the prime stimulus has a different valence (e.g., CANCER). Because affective priming can occur only if the attitude toward the prime stimulus has been activated, the observation of affective priming allows one to conclude that attitude activation has taken place. The automaticity of attitude activation can be examined by examining the conditions under which affective priming occurs. For instance, if affective priming occurs even when participants do not have the goal to retrieve their attitudes toward the prime stimuli, one can conclude that attitude activation was unintentional.

There are, however, limitations to the extent to which the automaticity features of mental processes such as attitude activation can be inferred. First, although the presence of an observable effect (e.g., affective priming) allows one to infer the presence of certain mental processes (e.g., attitude activation), the absence or change in the size of the effect does not allow one to infer the absence of or change in the operation of a particular mental process. Affective priming, for instance, depends not only on the activation of the attitude toward the prime, but also on other processes by which the activation of the prime attitude influences responses to the target (e.g., spreading of activation or response activation; see De Houwer et al., 2009a, for a review). Therefore, if affective priming is absent under the conditions specified by a particular automaticity feature (e.g., when participants do not direct attentional resources to the prime; e.g., Hermans, Crombez, & Eelen, 2000; but see Degner, 2009), this does not mean that attitude activation does not possess that automaticity feature (see Gawronski, Deutsch, LeBel, & Peters, 2008, and Moors et al., 2010, for a detailed discussion of this issue). More generally, because observable behavior is often (if not always) determined by multiple mental processes, it is difficult to draw strong conclusions about a single process based on the absence of or change in a particular behavior.

A second limitation is that even the presence of a behavioral effect informs us only about mental processes that are conceptualized at the functional level. Marr (1982) pointed out that mental processes can be described at three levels: the functional, algorithmic, and implementational levels. Functional descriptions of mental processes specify what the processes do, that is, which inputs are transformed into which outputs. For instance, at the functional level, attitude activation can be described as (a necessary part of) the mental process by which stimuli in the environment cause evaluative responses (see De Houwer et al., 2010). The algorithmic level deals with a symbolic description of the way in which the mental process achieves this function. For instance, attitude activation can be attributed to the activation of evaluative nodes in a semantic network (e.g., Fazio, 2007). Finally, the implementation level refers to how the mental process can be physically instantiated, for instance, in a (simulated) human brain.

Importantly, there is a many-to-one relation between the algorithmic descriptions and functional descriptions. For instance, the functional process of attitude activation can be conceived of at the algorithmic level not only as involving the activation of evaluative nodes in a semantic network (e.g., Fazio, 2007), but also in terms of the retrieval of separate episodic memory traces (e.g., Schwarz, 2007). Therefore, although the presence of a behavioral effect (e.g., affective priming) allows for conclusions about the automaticity of functional mental processes (e.g., attitude activation), it does not allow for strong conclusions about the automaticity of algorithmic mental processes (e.g., activation of nodes in a semantic network). For instance, the conclusion that affective priming can occur in the absence of the goal to process the prime does not allow for strong conclusions about whether evaluative nodes can be activated automatically in a semantic network (see Moors et al., 2010, for more details).

SUMMARY

The cognitive decompositional approach has been elaborated in detail at the conceptual level (Moors & De Houwer, 2006a, 2007) and can help researchers to specify their conceptualization of the term "implicit." Unfortunately, problems arise when studying the automaticity features of mental processes empirically. These problems result from the fact that automaticity features are defined in terms of mental constructs and the fact that researchers often aim to determine the automaticity features of (algorithmic) mental processes. In the next section, we therefore explore the possibility of developing a decompositional approach to automaticity that does not refer to mental constructs.

TOWARD A FUNCTIONAL DECOMPOSITIONAL APPROACH TO AUTOMATICITY

De Houwer (2011) recently proposed a functional-cognitive framework for psychological research that allows for a functional decompositional approach to automaticity. Within this approach, automaticity features are defined not in terms of mental constructs but in terms of observable elements of the environment. Although the functional decompositional approach is primarily directed at describing the automaticity of effects, it can also shed light on the nature of the mental processes that underlie behavior. We start this section by briefly explaining the functional-cognitive framework. Afterward, we provide some preliminary ideas about how automaticity features might be defined in a functional, nonmental manner. Next, we explore for the first time how research on implicit processes might benefit from the functional decompositional approach to automaticity.

THE FUNCTIONAL-COGNITIVE FRAMEWORK FOR PSYCHOLOGICAL RESEARCH

De Houwer (2011) argued that psychological research can have two aims. The first aim is to describe which elements of the observable environment influence behavior under which conditions. Within the psychology of science, for instance, the first aim would entail the discovery of those elements in the environment that determine scientific behavior. Examples of such determinants are science education, pictures of famous scientists, and rewards that are contingent upon scientific behavior. This first approach can be described as functional in that it focuses only on discovering effects (i.e., the causal impact of the environment on behavior) and the elements in the environment that moderate these effects. The second aim is to describe the mental processes by which elements in the environment influence behavior. For instance, once it has been verified that science education has an effect on scientific behavior under certain conditions, cognitive theories can be constructed about the mental representations that are formed as the result of science education (e.g., positive attitudes toward science) and how those representations promote scientific behavior. This is the aim of the cognitive approach in psychology.

Importantly, the functional and cognitive approaches are mutually supportive (see De Houwer, 2011). Functional knowledge (i.e., knowledge about the moderators of effects) imposes constraints on ideas about the mental processes by which the environment influences behavior. Therefore, the more we know about the impact of the environment on behavior, the better we are able to formulate ideas about the underlying mental processes. The development of functional knowledge can in turn benefit from the existence of strong cognitive theories because those theories can reveal links between existing pieces of functional knowledge (i.e., the heuristic function of theories) and can lead to predictions about yet unknown functional relations (i.e., the predictive function of theories).

A decompositional approach to automaticity can be developed within the functional approach. The functional approach aims to identify the moderators of effects, that is, the environmental conditions under which certain elements of the environment (e.g., science education) influence behavior (e.g., performance on scientific tasks). When automaticity features are defined in terms of environmental conditions, they can thus be regarded as a subset of all moderators of effects. Hence, a functional decompositional approach of automaticity entails that effects are characterized in terms of whether they occur under the environmental conditions that are specified by the various automaticity features. It differs from a cognitive decompositional approach in that (a) automaticity features are described only in terms of observable elements in the environment (and not mental constructs) and (b) automaticity features are used in the first instance to characterize effects rather than mental processes. Because research on the automaticity of effects increases functional knowledge of those effects, it constrains the development of cognitive theories and thus contributes also to our understanding of the mental processes that underlie behavior. In the following sections, we explore for the first time whether a functional decompositional approach can be developed conceptually (i.e., whether automaticity features can be defined in terms of environmental conditions) and what its possible merits might be.

DEFINING AUTOMATICITY FEATURES IN NONMENTAL TERMS

Within the cognitive decompositional approach, the feature fast was already defined in terms of environmental conditions, more specifically, in terms of the length of time. Effects can be described as fast when they occur even though little time elapses between the presence of the environmental cause (e.g., the prime in an affective priming study) and the observed behavior (e.g., the response to the target).

Other features might be reconceptualized in nonmental terms. The label efficient, for instance, could be used to describe effects that occur even in the presence of other demanding tasks. Let us return to the example of affective priming effects (i.e., the impact of a prime on responses to a target with the same or a different valence as the prime). The fact that affective priming effects can occur while participants memorize complex digits (e.g., Hermans et al., 2000, but see Degner, 2009) can be taken as evidence for the efficient nature of the affective priming effect. Of course, one should also be able to define the nature of demanding tasks without referring to mental constructs. One possibility is to define the difficulty of a task in terms of the quality of task performance (e.g., speed and accuracy) under certain conditions. The task itself can be conceptualized as a set of operant contingencies that is present in the environment (i.e., regularities between responses and outcomes of those responses that occur under certain conditions).

Reconceptualizing goal-related features and (un)consciousness might be more difficult. The names of features such as unintentional, goal-independent, and unconscious intrinsically refer to mental constructs (i.e., intentions, goals, and consciousness). It would thus be difficult to redefine these concepts in nonmental terms (but see Hayes, Barnes-Holmes, & Roche, 2001, for suggestions on how mental concepts could be rephrased in functional, nonmental terms). An alternative is to introduce new labels that do not refer to mental constructs but that capture the environmental conditions associated with cognitive automaticity features. For instance, effects could be described as task-independent (rather than goal-dependent) if they occur regardless of the presence of other tasks in the environment. Uncontrolled effects can be relabeled as proximal task-independent effects, that is, effects that occur regardless of the presence of tasks related to the effect. For instance, affective priming can be described as proximal task-independent when the effect occurs even if participants are instructed not to show the effect. Finally, effects can be described as nonreportable (rather than unaware) when participants cannot recognize the effect of the environmental cause on their behavior. Of course, much more conceptual work is needed before the functional decompositional approach can be considered as a viable alternative for the cognitive decompositional approach. In the next section, we argue that there are enough potential benefits to justify the further development of the functional decompositional approach.

MERITS OF THE FUNCTIONAL DECOMPOSITIONAL APPROACH TO AUTOMATICITY

In what ways can the functional decompositional approach to automaticity help researchers who would like to study the implicit processes that underlie behavior? First, whereas the cognitive decompositional approach focuses on the implicit nature of mental processes, the functional decompositional approach highlights the fact that effects can also qualify as implicit. Within the functional decompositional approach, automaticity features are a subset of all possible environmental moderators of effects. Understanding the (automaticity-related) moderators of effects is important as such because it reveals the nature of the relation between the environment and behavior and thus help us to understand, predict, and control human behavior (see De Houwer, 2011; Hayes & Brownstein, 1986). For instance, assume that future research will show that science education reduces the probability that people will turn to alternative medical practices. Knowing that this effect occurs allows one to reduce the extent to which people call upon alternative medicine by making science education available in schools. If research also shows that the effect of science education on the use of alternative medicine is nonreportable (i.e., the effect occurs regardless of whether people recognize the effect when asked about it), this would suggest that there is no need to draw attention to the link between science education and alternative medicine in order to expect a relation between both. In sum, functional research—including research on the implicitness of effects—has merits as such.

Second, empirical research on the implicit nature of effects is faced with fewer problems than empirical research on the implicitness of mental processes. First of all, functional automaticity features are defined in terms of observable environmental conditions and are thus easier to establish than cognitive automaticity features that are defined in terms of unobservable mental constructs. Moreover, the presence of effects is typically easier to verify than the presence of mental processes. Establishing the presence of an effect entails that a particular change in behavior (e.g., fast responses to a target) is causally attributed to an element in the environment (e.g., the presence of a certain prime). Although causal relations cannot be observed directly, experimental procedures allow one to implement the controls necessary for drawing causal conclusions with great confidence.

Third, because of the mutually supportive nature of the functional and cognitive approaches, improvement in functional knowledge can improve theories about the mental processes that underlie behavior (see De Houwer, 2011, for a detailed justification of this claim). Knowledge about the functional automaticity features that apply to a certain effect imposes important constraints on the nature of the algorithmic mental processes that are assumed to underlie the effect. As such, research on the implicitness of effects can promote the development of cognitive theories about the implicit mental processes that underlie behavior.

Whether the functional decompositional approach will actually turn out to be valuable for research on implicit processes of course remains to be seen. We have introduced only the basics of the approach and much still remains to be done. Nevertheless, we hope to have provided the starting point for these future developments.

IMPLICATIONS FOR THE STUDY OF SCIENTIFIC BEHAVIOR

Based on the analysis presented above, the following general guidelines can be formulated for researchers who aim to study implicit processes in scientific behavior. First, specify the automaticity features that you want to refer to when using the term "implicit." It does not make sense to simply say that a process or effect is implicit because there is no single class of automatic processes that all have the same set of automaticity features. Although an all-or-none view of automaticity is appealing because of its simplicity, empirical evidence shows that different processes can have different features of automaticity. Specifying one's conceptualization of implicit requires not only the selection of certain automaticity features, but also a precise definition of those automaticity features. Our earlier work (Moors & De Houwer, 2006a, 2007) provides the basis for defining these features in a cognitive manner. In the present chapter, we also discussed ways of defining automaticity features in a functional manner.

Second, the claim that a mental process or effect is automatic in a certain sense is actually a claim about the conditions under which the process or effect occurs. Hence, these claims make sense only when they can be backed up by sound arguments and evidence. It is important to realize that claims about cognitive automaticity features of mental processes are not easy to back up. In the future, a functional decompositional approach could provide a more workable alternative. Even in its current state of development, the functional approach highlights the fact that one can describe as implicit or explicit not only mental processes, but also effects. The study of the implicit nature of effects can have merits as such and help develop theories of the implicit nature of mental processes.

To illustrate these guidelines, we will relate them to the chapter of Zimmerman and Pretz (Chapter 10 this volume) about implicit and explicit processing in scientific discovery. They asked one group of participants to solve a real-world physics task by looking for a rule. Other participants were not given this instruction or asked not to think too hard while performing the task. Zimmerman and Pretz referred to the first group as the explicit strategy condition, whereas the second group was referred to as the implicit strategy condition. With regard to our first guideline (i.e., specify the meaning of implicit), they describe dual-process theories that make a distinction between implicit and explicit modes of processing (e.g., Evans, 2008; Sloman, 1996). The explicit mode is characterized as rule-based, deliberate, and analytical and as dependent on awareness, attention, and effort. Implicit processing is described as associative, holistic, unaware, and effortless. However, it is not specified whether these definitions of explicit and implicit were assumed to hold also for the type of processing that was encouraged in their own research, although by referring to the distinctions as different "strategies" and "approaches," the implication is that their main concern is with the distinction between a rule-based/analytical strategy and an associative/holistic one. Moreover, their definitions of implicit and explicit are based on an all-or-none view of automaticity, much like that taken by other authors (e.g., Dijksterhuis & Nordgren, 2006), rather than the decompositional approach that we favor (Moors & De Houwer, 2006b; see also Bargh, 1994, and Keren & Schul, 2009, for a critical discussion of dual-process theories).

With regard to the second guideline (i.e., justify the meaning of implicit), participants were said to have adopted an implicit processing mode when they were not given any specific instructions on how to approach the task (Experiment 1) or when they were instructed not to think too much about the task (Experiments 2 and 3). These instructions could indeed have led to absence of the goal to discover a rule. Processing could thus be described as implicit in the sense of independent of the distal goal to solve the problem by finding a rule. However, processing in the implicit strategy condition was in all likelihood explicit in many other ways. For instance, all participants had the goal to make correct predictions on each trial. Hence, processing was probably deliberative (i.e., intentional in the sense of dependent on the proximal goal to perform well in the task). Also, participants in the so-called implicit conditions were probably aware of the stimuli that were presented and the responses that they gave. They might have been unaware of the reasons for their predictions, but Zimmerman and Pretz do not provide data that speak to that issue. Furthermore, it is possible that (some) participants in the implicit condition did use an intentional strategy to solve the problems (i.e., adopted the goal to solve the task in a certain manner), be it a strategy other than looking for a rule (e.g., responding based on similarity with previous problems), a limitation that Zimmerman and Pretz acknowledge near the end of their chapter. There are also no data about whether performance in the implicit conditions was effortless (e.g., no evidence that performance did not deteriorate as the result of secondary tasks). Participants gave their predictions more quickly in the implicit conditions than in the explicit conditions, but in both conditions, reaction times were longer than two seconds, leaving enough time to allow for slow-acting processes.

In sum, the data of Zimmerman and Pretz support the important conclusion that the quality of scientific problem solving can sometimes benefit from a processing mode that can be described as implicit in the sense of operating in the absence of the goal to find a rule. Although we do not exclude the fact that scientific problem solving can benefit from processes that are implicit in other senses (e.g., unaware or effortless), the evidence provided by Zimmerman and Pretz does not speak to such a conclusion. The fact that Zimmerman and Pretz adopted a broad, all-or-none definition of the concept implicit tends to conceal the specific contribution of their work and could create a general impression that their results generalize to other processes with other automaticity features. We hope that this example illustrates how our analysis can contribute to clarifying research on the role of implicit processes in scientific behavior.

CONCLUSION

The distinction between implicit and explicit processes has become a hot topic in psychological research. Unfortunately, this distinction is more complex than is often assumed. At the conceptual level, it is best not to think of the relation between implicit and explicit processes as an all-or-none distinction between, on the one hand, processes that have all automaticity features and, on the other hand, processes that do not have any automaticity features. Because different processes can possess different features of automaticity, one should specify the automaticity features one has in mind when using the term "implicit". At the empirical level, important problems can arise when trying to back up claims about the automaticity features under study. This is especially the case when automaticity features are defined in terms of mental constructs such as goals and used to describe mental processes. We explored for the first time the possible merits of a functional decompositional approach to automaticity that tries to solve some of these problems. Nevertheless, it should be clear that research on implicit processes can be very complex. One reaction is to simply ignore the complexity by continuing to conceptualize implicitness in an all-or-none manner. Although this strategy certainly simplifies matters, it is bound to create problems in the long run, for instance, when confronted with evidence for automaticity features that do not co-occur. One can also choose to ignore the concept implicit and abandon research on implicit processes all together. Such an approach runs the risk of missing out on important knowledge about the determinants of human behavior. For us the most meaningful strategy is to acknowledge the complexity and to try to deal with it in the best way possible. We hope that despite its complexity, researchers will also adopt this strategy when studying scientific behavior.

REFERENCES

Bargh, J. A. (1989). Conditional automaticity: Varieties of automatic influence in social perception and cognition. In J. S. Uleman & J. A. Bargh (Eds.), *Unintended thought* (pp. 3–51). New York: Guilford Press.

Bargh, J. A. (1992). The ecology of automaticity: Toward establishing the conditions needed to produce automatic processing effects. *American Journal of Psychology, 105,* 181–199.

Bargh, J. A. (1994). The four horsemen of automaticity: Awareness, intention, efficiency, and control in social cognition. In R. S. Wyer & T. K. Srull (Eds.), *Handbook of social cognition* (Vol. 1, pp. 1–40). Hillsdale, NJ: Lawrence Erlbaum.

Block, N. (1995). On a confusion about a function of consciousness. *Behavioral and Brain Sciences, 18,* 227–287.

Degner, J. (2009). On the (un)controllability of affective priming: Strategic manipulation is feasible but can possibly be prevented. *Cognition and Emotion, 23,* 327–354.

De Houwer, J. (2006). What are implicit measures and why are we using them? In R. W. Wiers & A. W. Stacy (Eds.), *The handbook of implicit cognition and addiction* (pp. 11–28). Thousand Oaks, CA: Sage.

De Houwer, J. (2011). Why a cognitive approach in psychology would benefit from a functional approach and vice versa. *Perspectives on Psychological Science, 6,* 202–209.

De Houwer, J., Gawronski, B., & Barnes-Holmes, D. (2010). A functional-cognitive framework for attitude research. Manuscript submitted for publication.

De Houwer, J., & Moors, A. (2007). How to define and examine the implicitness of implicit measures. In B. Wittenbrink & N. Schwarz (Eds.), *Implicit measures of attitudes: Procedures and controversies* (pp. 179–194). New York: Guilford Press.

De Houwer, J., & Moors, A. (2010). Implicit measures: Similarities and differences. In B. Gawronski & B. K. Payne (Eds.), *Handbook of implicit social cognition: Measurement, theory, and applications* (pp. 176–193). New York: Guilford Press.

De Houwer, J., Teige-Mocigemba, S., Spruyt, A., & Moors, A. (2009a). Implicit measures: A normative analysis and review. *Psychological Bulletin, 135,* 347–368.

De Houwer, J., Teige-Mocigemba, S., Spruyt, A., & Moors, A. (2009b). Theoretical claims necessitate basic research. *Psychological Bulletin, 135,* 377–379.

Dijksterhuis, A., & Nordgren, L. F. (2006). A theory of unconscious thought. *Perspectives on Psychological Science, 1,* 95–109.

Eagly, A. H., & Chaiken, S. (2007). The advantages of an inclusive definition of attitude. *Social Cognition, 25,* 582–602.

Evans, J. St. B. T. (2008). Dual-processing accounts of reasoning, judgment, and social cognition. *Annual Review of Psychology, 59,* 255–278.

Fazio, R. H. (2007). Attitudes as object-evaluation associations of varying strength. *Social Cognition, 25,* 603–637.

Fazio, R. H., & Olson, M. A. (2003). Implicit measures in social cognition research: Their meaning and use. *Annual Review of Psychology, 54,* 297–327.

Fazio, R. H., Sanbonmatsu, D. M., Powell, M. C., & Kardes, F. R. (1986). On the automatic activation of attitudes. *Journal of Personality and Social Psychology, 50,* 229–238.

Gawronski, B., Deutsch, R., LeBel, E. P., & Peters, K. R. (2008). Response interference as a mechanism underlying implicit measures: Some traps and gaps in the assessment of mental associations with experimental paradigms. *European Journal of Psychological Assessment, 24,* 218–225.

Gemar, M. C., Segal, Z. V., Sagrati, S., & Kennedy, S. J. (2001). Mood-induced changes on the implicit association test in recovered depressed patients. *Journal of Abnormal Psychology, 110,* 282–289.

Greenwald, A. G., & Banaji, M. R. (1995). Implicit social cognition: Attitudes, self-esteem, and stereotypes. *Psychological Review, 102,* 4–27.

Hayes, S. C., Barnes-Holmes, D., & Roche, B. (Eds.). (2001). *Relational frame theory: A post-Skinnerian account of human language and cognition.* New York: Plenum Press.

Hayes, S. C., & Brownstein, A. J. (1986). Mentalism, behavior-behavior relations, and a behavior-analytic view of the purposes of science. *Behavior Analyst, 9,* 175–190.

Hermans, D., Crombez, G., & Eelen, P. (2000). Automatic attitude activation and efficiency: The fourth horseman of automaticity. *Psychologica Belgica, 40,* 3–22.

Keren, G. B., & Schul, Y. (2009). Two is not always better than one: A critical evaluation of two-systems theories. *Perspectives on Psychological Sciences, 4,* 533–550.

Logan, G. D. (1985). Skill and automaticity: Relations, implications, and future directions. *Canadian Journal of Psychology, 39,* 367–386.

Logan, G. D. (1989). Automaticity and cognitive control. In J. S. Uleman & J. A. Bargh (Eds.), *Unintended thought* (pp. 52–74). New York: Guilford Press.

Marr, D. (1982). *Vision: A computational investigation into the human representation and processing of visual information.* New York: Freeman.

Moors, A., & De Houwer, J. (2006a). Automaticity: A conceptual and theoretical analysis. *Psychological Bulletin, 132,* 297–326.

Moors, A., & De Houwer, J. (2006b). Problems with dividing the realm of cognitive processes. *Psychological Inquiry, 17,* 199–204.

Moors, A., & De Houwer, J. (2007). What is automaticity? An analysis of its component features and their interrelations. In J. A. Bargh (Ed.), *Automatic processes in social thinking and behavior* (pp. 11–50). Hove, UK: Psychology Press.

Moors, A., Spruyt, A., & De Houwer, J. (2010). In search of a measure that qualifies as implicit: Recommendations based on a decompositional view of automaticity. In B. Gawronski, & B. K. Payne (Eds.), *Handbook of implicit social cognition: Measurement, theory, and applications* (pp. 19–37). New York: Guilford Press.

Naccache, L., Blandin, E., & Dehaene, S. (2002). Unconscious masked priming depends on temporal attention. *Psychological Science, 13,* 416–424.

Nosek, B. A., & Greenwald, A. G. (2009). (Part of) the case for a pragmatic approach to validity: Comment on De Houwer, Teige-Mocigemba, Spruyt, and Moors (2009). *Psychological Bulletin, 135,* 373–376.

Richardson-Klavehn, A., Lee, M. G., Joubran, R., & Bjork, R. A. (1994). Intention and awareness in perceptual identification priming. *Memory and Cognition, 22,* 293–312.

Schacter, D. L. (1987). Implicit memory: History and current status. *Journal of Experimental Psychology, Learning, Memory, and Cognition, 13,* 501–518.

Schwarz, N. (2007). Attitude construction: Evaluation in context. *Social Cognition, 25,* 638–656.

Sloman, S. A. (1996). The empirical case for two systems of reasoning. *Psychological Bulletin, 119,* 3–22.

Wiers, R.W., & Stacy A.W. (Eds.). (2006). *Handbook of implicit cognition and addiction.* Thousand Oaks, CA: Sage.

9 How Should We Understand Implicit and Explicit Processes in Scientific Thinking?

Donelson E. Dulany

INTRODUCTION

The question posed in the title leads into several levels of development in the psychology of science. At one level, the standard cognitive theory of the conscious explicit and unconscious implicit is contrasted with a mentalistic theory in which deliberative (explicit) and evocative (implicit) mental episodes carry different forms of conscious contents interrelated by different forms of nonconscious mental operations. Supporting research shows the role of awareness in both forms of learning and the superiority of the deliberative in handling novelty. Other research shows deliberative processing in causal learning. At another level, the question calls for a deliberative logic of competitive support, especially when the issues involve realistically interpreted constructs and strong controversy. At still another level, it is argued that various metatheoretical and ideological positions have led to the standard cognitive view and continue to inspire conceptual and methodological limitations and biases that have provided support for the view.

There is a very common vernacular sense of the "implicit" and "explicit": What is only implied is said to be "implicit" in what is stated, and what is openly stated is said to be "explicit." In fact, something like that is indeed common in scientific thinking. We could say that whatever is "explicitly" laid out in a description of data is deductively implied by—hence "implicit" in—some aggregate of theory, mappings, and auxiliaries. In a related, but somewhat different sense, it can be said that something is "implicit" in arrays of data that may become "explicit" in lawful relations when acceptably inducted—or even "explicit" in theoretical assertions if they are defensibly abducted, in the sense of being a tenable and productive explanation (e.g., Capaldi & Proctor, 2008). At this simple level of description, these usages definitely occur—but that is not where I see the deeper theoretical, and even metatheoretical, issue.

There is a deeply controversial issue that divides two vigorously investigated conceptualizations of the implicit and explicit in ordinary thinking—and by reasonable generalization, in scientific thinking as well: the roles of what is conscious and what is nonconscious, and their relation to the explicit and implicit.

At this stage of the development of the psychology of science (Feist, 2006), there can be great value in observational studies of scientists doing their scientific thinking (e.g., Dunbar, 1995). There is also great value in learning how that kind of thinking develops by observing children confronted with analogous problems, for example, Brewer, Chinn, and Samarapungavan (1998). For investigation of this controversial issue, however, I draw here on the mainstream of theoretical and experimental psychology. Furthermore, in arguing for metatheoretical and ideological influences on conceptions of the explicit and implicit, and in sketching a logic of competitive support, I see this as drawing on forms of descriptive and normative philosophy of this science respectively.

Those influences? How we conceptualize the explicit and implicit involves answers to the fundamental questions that have animated the Grand Sweep of Ideas in the history of psychology. What is the proper place of consciousness and the nonconscious in theoretical explanation? Which can carry symbolic representations—or is it both? And where can the more complex mental activities occur (Dulany, 2009)?

So how *should* we understand those implicit and explicit processes?

STANDARD COGNITIVE VIEW OF THE IMPLICIT AND EXPLICIT PROCESSES

Put simply, the explicit processes are conscious, and the implicit are unconscious—a correlation that would fall into a simple 2 × 2 table. *Explicit* learning, reasoning, memory, and perception are *conscious* processes; and *implicit* learning, reasoning, memory, and perception are *unconscious* processes. This has been a common view from Reber (1967) through Kihlstrom (1987) to Evans (2009) and beyond.

My general suggestion is that the standard view within psychology has been shaped by a succession of powerful metatheoretical and ideological influences from the behaviorist revolution onward—influences that shape the theories and methodologies in controversial ways. With a historical suspicion of consciousness, an independent and unconscious implicit processing has been emphasized.

METATHEORETICAL AND IDEOLOGICAL CONTEXTS FOR THE STANDARD COGNITIVE VIEW

Computational

Let us remember that the "cognitive revolution" was not a return to an earlier science of consciousness. It was based primarily on two things: (a) challenges to stimulus-response theory, provided by various experimental literatures, from reexaminations of learning paradigms to semantic and narrative memory; and (b) provision by Artificial Intelligence (AI) of a more advanced, but still physicalistic, computer metaphor, no longer the telephone switchboard of behaviorism. The mind, like the computer, was viewed as an instantiation of a Universal Turing Machine, and so we were granted Hilary Putnam's (1960) famous analogy: Mind is to brain as software is to hardware. But with suspicion of a causal consciousness, this quickly became *cognition* runs in the brain the way software runs in that hardware. And consciousness? That was only a sometime and noncausal emergent—only an epiphenomenon, as in Jackendoff (1987) and Haugland (1978), something that could be dismissively consigned to "folk psychology," as in Churchland (1993). In fact, we can see it all infused with enthusiasm for Kuhnian paradigm change.

So...processing is "explicit" only within an *epiphenomenal* consciousness, but unconscious and "implicit" otherwise.

Information Processing

With elaboration of that basic metaphor, minds must have a structure of *subsystems* as computers do: A RAM-like working memory connected to a hard-drive-like long-term memory, shuffling information between Input buffers and Output controls. And consciousness? Identify it with a subsystem of WM called "attention," which is limited in "resources" and "scope" (e.g., Posner, 1994), and perhaps only an emergent itself—leaving various complex processes for WM outside consciousness entirely. If we are said to unconsciously "store" symbolic images and propositions and even narratives on a symbolic "shelf," and then "search" and "identify" and "retrieve," it becomes reasonable to think of problem solving, reasoning, and decision making as something that can be accomplished outside attention by "unconscious implicit processes."

So...processing is "explicit" only within conscious *attention*, but unconscious and "implicit" otherwise.

Value on the "Counterintuitive"

Although the counterintuitive can be theoretically significant, its overemphasis can, with imperfect methodology, bring an excess of the counterfactual. This has also been referred to as pursuit of the "cute"—a term we can hear in various presentations of "evidence for the unconscious."

So...processing is "implicit" and unconscious if *counterintuitive*—if "cute," and consciously "explicit" otherwise.

Psychoanalysis

The psychoanalytic influence endures, not only in the culture, but in an APA division and journal, and even in the mainstream psychology of the science: The "cognitive unconscious" (Kihlstrom, 1987) shares the two most fundamental processes with the psychoanalytic unconscious—symbolic representation and complex mental activity. It's just "expurgated." And with Freud's (1923/1935) rank-ordering of the unconscious "*primary* process" and a conscious "*secondary* processes," their relative importance was clear.

So…processing is unconscious and "implicit" within a *primary* process, and conscious and "explicit" only within a *secondary* process.

Behaviorism

What might be the most influential source of all this de-emphasis upon consciousness as opposed to what is nonconscious?

The behavioral revolution was most fundamentally the rejection of consciousness as an acceptable subject matter of the new science of psychology. It was a rejection not only of the unproductive Titchnerian (1898) analysis of conscious states into their elements for a "mental chemistry," or its unacceptable first person data language in a too-demanding method of "classical introspection." It was also rejection of the Wundtian (1896–1907) concept of conscious and causal volition, as well as Fechnerian psychophysics. There was no place for the promising Jamesian (1890) program for analysis of what cognitivists study with behavioristic manipulation-response methods—attention, concepts, memory, reasoning, emotions. James had placed them within an unacceptable "stream of consciousness." It was even the continuing disregard of the functionalist's movement emphasis on the adaptive value of consciousness for the species and the individual (e.g., Carr, 1930). And the Gestalt movement (Koffka, 1935) with its experimentally productive analyses of holistic perception, insight, transposition, and problem solving? For behaviorists there was no place for the conscious states central to those analyses.

Consciousness was rejected within the behavioral revolution for deeply ideological reasons. As Watson (1913/1994) put it in his Behaviorist Manifesto: "This suggested elimination of states of consciousness as proper objects of investigation in themselves will remove the barrier from psychology which exists between it and the other sciences" (p. 253). Why? "Behaviorism claims that "consciousness" is neither a definite nor usable concept; that it is merely a word for the "soul" of more ancient times. The old psychology is thus dominated by a kind of subtle religious philosophy" (Watson, 1924, p. 3).

How then to handle a truly **behavioristic theory**? Postulate "intervening variables," but only as abstract constructs—transforms of manipulations on the stimulus side to be arithmetically combined for learning on the responses side (Hull, 1943). Or fashion a *"quasi-cognitive" behavior theory* if there is compelling evidence for "cognitive expectancies" or "cognitive maps"—as there was, but assure readers that these cognitive states were outside of consciousness (Tolman, 1951). Or hold, as *radical behaviorist* Skinner (1974) did, that it is present, and somehow past, stimuli that control the organism, "operantly" or "respondently." With realistic recognition of conscious states, however, assure the reader that "this does not mean…(and this is the heart of the argument) that what are introspectively observed are the causes of behavior. . . . What are introspectively observed are certain collateral products of those histories" (p. 17). Although epiphenomenalism came to the computational view out of philosophy, for example, Broad (1925), by way of AI, behavioral epiphenomenalism could make it seem more "down home" and natural.

The **focus of research** was then on nonlinguistic organisms—from pigeons to Sprague-Dawley "participants"—unable to report anything troublesome for their metatheoretical-ideological view. And then when there became a need to generalize "the automatic action of reinforcement" to humans, there were attempts to demonstrate "learning without awareness" with insensitive report conditions and neglect of correlated and explanatory conscious contents (Dulany, 1968). The problem even carried over to cognitive studies of implicit learning (Dulany, 1997, 2003).

Continuing Influence on Conceptions of the Explicit and Implicit?

Writing as president of Association for Psychological Science, Roediger (2004) reminded cognitivists that behaviorism is still with us in behaviorist journals, behaviorist societies, and manipulation-response designs. We can also see an intellectual tradition shaping conceptions of the implicit and explicit that was most central to what Watson expressed: a fear of, or at least concern for, (a) ontological nonmaterialism, and (b) "free will" in the sense of indeterminism. For example, we can see the former in Reber (1997) and the latter in the volume edited by Pockett, Banks, and Gallagher (2006). This thinking can, however, be put aside as a mistaken identification of theoretical assertions, which can be empirically evaluated, with metaphysical assertions, that are not subject to empirical evaluation in the science as we now know it.

I believe we can see this array of influences as a type of "social construction of knowledge," not of an empirically unknowable reality (e.g., Gergen, 2001) as critiqued by Capaldi and Proctor (2009), but of an empirically knowable reality represented with metatheoretical and ideological commitments—which *can* methodologically bias.

So . . . a process can be "explicit" if *in some sense* conscious, but "implicit" if in *various senses* unconscious.

THE EXPLICIT AND IMPLICIT IN THE MENTALISTIC METATHEORY

The explicit (deliberative) and the implicit (evocative) are central to the mentalistic metatheory, described with various applications (Dulany, 1991, 1997, 2002, 2004, 2009, 2011). The metatheory has been used by others—"explicitly" so by some, in that vernacular sense, for example, Carlson (2002), Perruchet and Vinter (2000), Tzelgov (1997), and "implicitly" so by many others investigating the roles of conscious states, as well as those critiquing various claims for "unconscious implicit processes." Two recent papers provide useful reviews of a range of studies demonstrating a causal role for conscious states: Baumeister and Masicampo (2010) and Baumeister, Masicampo, and Vohs (2011).

In essence, the "explicit" are *deliberative mental episodes* in the sense that *conscious* propositional contents are interrelated by *nonconscious* deliberative operations. We infer or decide. And the "implicit" are *evocative mental episodes* in the sense that *conscious* nonpropositional contents are interrelated by *nonconscious* associative-activational operations. One idea, or just a sense of something, activates another.

The contrast is clear: On the standard cognitive view, the explicit and implicit are *correlated* with the conscious and nonconscious. On the mentalistic view, the explicit and implicit are *orthogonal* to the conscious and nonconscious, with the conscious contents and nonconscious operations taking different forms in the explicit (deliberative) and implicit (evocative) mental episodes. For a visualization of that contrast, you might see Dulany (1997, p. 191).

Although the mental episodes differ in the forms of contents and operations, within each episode the conscious and nonconscious are indissolubly wedded. As I've said lightly, . . . what nature has joined, let not theorists put asunder. But they have—as we have just seen.

The Aim

The general aim of the metatheory is supportable theories of the causal interactions among conscious states, their antecedents and consequents (Dulany, 2008). Conscious states, including intentions, can comfortably appear within causal theoretical networks—with working assumptions of determinism and material status, assumptions that can be productive. This can amount to assuming that conscious states and brain states are coordinate *in some way*. Contrary to historical and enduring misconceptions, theoretical assertions causally interrelating actions and conscious states, even intentions, in no way entail nonmaterialism or "free will" in the sense of indeterminism.

The specific aim of what follows is an *analysis* of explicit and implicit processes within that mentalistic metatheory.

Symbolic Representation

Most fundamentally, too, the mentalistic metatheory holds that consciousness is the sole carrier of symbolic representations of what lies beyond those mental states: a present in perception, a past in remembrance, and a future in fears through expectations and hopes.

Furthermore, mental symbols can be *identity symbols*, an attentional identification of a thing as such, or *literal symbols*, the contents of the ephemeral awareness that precedes and surrounds that conscious identification.

What is symbolized beyond a current mental state can be something external or internal, real or only imagined. The extraordinary capacity of conscious symbolic representation even permits *higher order representation* of our own mental states and mental episodes.

As a construct, "symbolic representation" is not so simple as to submit to a definition; it is functionally specified. Symbols activate other symbols—as in a thought of "winter" activates a thought of "snow." They may be subjects or predicates in propositional assertions—as in "Amount of snow makes this winter distinctive." And they may participate in that special proposition "This represents that," so long as "this" is a mental content and "that" is anything from a book on my desk, which I can reach and open, through something correctly or falsely remembered, to the real or unreal that is only imagined. For a metatheory, functional specification of symbols is what I believe should derive from Brentano's (1774/1973) complex analysis of "intentionality" and "intentional inexistence."

Thus consciousness is central to and essential for any kind of mental episode—explicit or implicit. And it is all of this unique capacity for symbolic representation that gives consciousness its adaptive significance.

Analysis of the Structure and Interrelations of Mental Episodes

In *deliberative mental episodes (the explicit)*, propositional contents are carried by conscious *modes*, as in the common belief *that* ______, and one propositional content and mode are yielded by a nonconscious deliberative operation from two or more other propositional contents carried by their own conscious modes. In scientific thinking, as in everyday experience, these inferences and decisions are common. Deduction, induction, abduction, as well as ordinary inferences from momentary beliefs, are explicit in this sense.

In *evocative mental episodes (the implicit)*, subpropositional contents are carried by conscious *modes*, as in a sense (perception, idea, feeling) *of* _____, and may be yielded by one or more subpropositional modes and contents by a nonconscious evocative (associative-activational) operation. In scientific thinking, as in everyday experience, one or more ideas or perceptions quite commonly evoke another idea or image. "Automatic" is another vernacular term, with different senses applied as revealed in different experimental paradigms (e.g., Moors & De Houwer, 2006). But this basic form of mental episode is *automaticity of thought*, and this is common in ordinary experience. Furthermore, when the nonconscious operation is transduction within the response system, this evocative episode incorporates this kind of *automaticity of action*.

For both propositional and nonpropositional contents, these may be held with a conscious sense of personal possession, of "agency" in the classical term, but a sense that varies over circumstances, and even over persons—with a loss of possession and attribution to external sources in pathology.

Commonality of Form

For either deliberative or evocative mental episodes, a conscious state i at moment $n + 1$ may be yielded by nonconscious operation on other conscious states, j and k, at prior moments:

$$\text{CsState}_{in+1} \leftarrow \text{NonCs operation } (\text{CsState}_{jn}, \ldots \text{CsState}_{kn-m}).$$

Each conscious state is, however, at the intersection of a number of variables—the mode (e.g., degree of belief *that* ____ or clarity of perception *of* ____), the relation of a propositional subject to what is scored correct, or the predicated degree or likelihood of something, or even that sense of possession at any moment. There is a natural form, then, for quantitative models of any theory with these mental episodes:

$$CS_{in+1} = f(CS_{jn}, \ldots, CS_{kn-m}).$$

The function would then describe a class of mental operations and the equation a group of mental episodes of the same form with mental states taking different values—and with each episode providing an n-tuple within that equation.

The Nonconscious and Nonsymbolic

(a) Mental operations have already been described as nonconscious, and are clearly nonsymbolic. (b) On this view, too, memories inactive (in the sense of out of awareness) are nonsymbolic neural networks that we construct and activate. Despite a too loose vernacular, what we "know," "believe," "intend," or "plan" when in that memory network lacks the functional specification of those states when in consciousness and functioning symbolically. Contrary to the computer metaphor, too—and its ingrained terminology—we don't "store" and "search" and "retrieve" symbolic memories. In fact, these construction and activation processes become the mental *operations* within evocative (implicit) or deliberative (explicit) learning and remembrance episodes. (c) Motor transduction systems convert a conscious state to action, with a possible conscious feedback—which can also function as nonconscious operation for an evocative episode, automaticity of action and influence on RT. (d) Sensory transduction systems convert the nonconscious stimulus into a first conscious representation—in literal awareness.

The Self

It is enough here to say that the "self" is a construct, inferred from a distributed cluster of perceptions, beliefs, and feelings with a subjective "I" as agent, the active self, or directed at a subjective "me," the recipient self—all functioning within deliberative or evocative episodes. The construct would be relevant in any context—certainly including the psychology of science for the scientist.

A Place for Brain States?

Of course there is a place for brain states, on the general assumption that conscious states and nonconscious operations have neural coordinates. This is with recognition of the significant contributions of brain imaging—but also significant problems at this stage in general methodology (Vul, Harris, Winkielman, & Pashler, 2009) and also limitations of current technology, especially for the contents, dimensions, and propositional forms of specific conscious states (Dulany, 2011).

Interrelations between the Explicit (Deliberative) and Implicit (Evocative)

In ordinary and in scientific thinking, these interrelations are important in themselves, but for some thinking we may appropriately speak of what is *predominantly* one or the other.

1. The evocative mental episodes and a proposition that may serve deliberative episodes may each be *transformed* into the other, by common mental processes.
   (a) An evocative mental episode, by *representation in higher order awareness,* may become a proposition within deliberative episodes: (Idea 1 *activates* Idea 2) → (Idea 1 *suggests* Idea 2). (b) Furthermore, *by repetition,* because predication produces activation, a proposition within deliberative episodes may become an evocative episode: (Idea 1 *suggests* Idea 2) → (Idea 1 *activates* Idea 2).
2. Within many significant psychological phenomena the two forms may be *interlaced.* Even an explicit remembrance may be inferred from what is an evoked sense of familiarity. And an implicit remembrance may be revealed only when what is evoked is voluntarily reported.

Since this is the mentalistic metatheory, it would then be useful to consider examples of mentalistic theories it implies. We can then look for the role of conscious states in both the deliberative and evocative mental episodes—in explicit and implicit learning. We can also illustrate an explicit (deliberative theory) and its quantitative model—in causal learning.

EXPERIMENTAL EXAMPLE OF LEARNING ABOUT AN UNOBSERVABLE STRUCTURE

Conscious Representation in Both Implicit and Explicit Learning

Suppose, for example, scientists are presented a problem, as have been experimental subjects, of learning what kind of unobservable structure might generate a finite sequence of events—with nothing at hand analogous to a "magnetic resonance imager." From short sequences of events, they learn at least enough about the structure for some correct categorization of observable event sequences, as generated or not by the structure. The structure? With each transition from State *i* to State *j*, or recursion on State i, the system forms a distinctive event, and these events may appear within observable strings generated by permissible transitions through the system from one of its starts to one of its exits.

In this case, we actually have a finite state grammar (FSG) that generates letter strings to be viewed, under explicit and implicit learning conditions—a task introduced by Reber (1967), with critiques launched by Dulany, Carlson, and Dewey (1984, 1985) The literatures now have many summaries and critiques for the several paradigms—for example, Shanks and St. John (1994), French and Cleeremans (2002), and Perruchet and Vinter (2002), and the problem has been reviewed fairly recently by Shanks (2005) and Pothos (2007). The deeper question throughout has been what is represented in consciousness and what is not—what is the nature of the implicit process.

In the *explicit* learning condition of Dulany et al. (1984, 2009), subjects were informed that the strings were generated by a set of rules, rules that allowed certain letters to follow others, and they should try to learn as much as possible about those rules. In an *implicit* learning condition, subjects were simply instructed to view the items and learn something about the items. At learning, subjects were presented 20 items, three to six letters in length, then at test 50 items with the series repeated, in each case 25 grammatical, with only 5 old items, and 25 nongrammatical items. Subjects categorized the item as whether it followed or violated the rules, but they also reported a consciously represented rule: what it was about the item that implied its categorization. They marked a feature, a set of letters.

Standard Cognitive and Mentalistic Theories of Implicit and Explicit Learning

How might learning about this system proceed and be revealed in this task? On the standard cognitive theory, explicit learning is conscious learning. It would be revealed in conscious judgment that a test item is grammatical or nongrammatical. Implicit learning is unconscious learning—abstraction of the structure generating the strings, the FSG. It would then be revealed by the activation and control of that unconsciously represented grammar.

On the mentalistic theory (Dulany, 1997), *explicit learning* in this paradigm consists of venturing conscious propositional hypotheses, confirmed by observing that the feature in the hypothesis appears in a grammatical item—a deliberative mental episode. It would then be revealed by acting on consciously represented rules, based on variably accurate memory and inference, rules suggesting the item is grammatical or not—another deliberative mental episode. *Implicit learning* would be the strengthening of associative-activation between features presented and a sense of grammaticality, the strengthening of an evocative mental episode. It would then be acted upon when features of a string activate or fail to activate that mental episode—but at test it would require a higher order representation of an evocative episode, a rule implying grammaticality or not, in reflection of the strength of activation.

For a rigorous competitive examination, our program calculated the *validity* of each rule, the probability of a correct response if acted upon. Thus if a subject reports "Feature *i* implies that String *j* is G," *Rule validity* $= P(S_j \in G \mid F_i \in S_j)$. And if the report is that "Feature *i* implies that String *j* is NG," *Rule validity* $= P(S_j \in NG \mid F_i \in S_j)$. In essence, the mentalistic theory predicts that action on these consciously represented rules will produce—within error—an identity function relating mean rule validity to proportion correct judgments over subjects. That was found with a slope of .99, intercept of .01, $r = .83$, with normal distribution of error, and nonsignificant residual for 49 of 50 subjects. Furthermore, proportion correct for experimental subjects was $M = .644$ and reliably exceeded the .555 for controls tested in the same way but without the prior learning. There was no difference in those provided the explicit and implicit learning instructions (Dulany et al., 1984, 2009).

With experimental conditions that should predispose to explicit and to implicit learning, this is evidence that both kinds of mental episodes, with rules learned deliberatively, or evocative episodes that can be represented in higher order awareness, can be acted upon producing the experimental effects.

Transfer to Novel Instances from the Same Structure

Handling of novelty is important both in ordinary experience and in scientific thinking. One kind of transfer is simply generalization across overlapping observable events—already shown in transfer to new strings (as well as in connectionist and classical learning literatures). Although the implicit and explicit are interrelated, by transforms and interlacing, are there tests in which what is *predominantly* explicit and implicit might be discriminated in more demanding transfer? After learning that some strings of events are generated by an unobservable structure and some are not, will learning *transfer to a novel observable strings of events generated by the same unobservable structure*—again the FSG?

The answer from the standard cognitive theory is yes, after either explicit or implicit learning. The answer from the mentalistic theory is yes only when conditions permit deliberative inference.

In Dulany and Pritchard (in preparation), subjects were presented grammatical and nongrammatical items randomly during learning, with explicit learning instructions to "find rules" and categorize; implicit learning instructions were to "say what first comes to mind" and categorize. In each condition there was a "Right" or "Wrong" with only half the time for the implicit trial, 1.5s versus 3s, but twice the number of trials. In essence, these are the findings across two experiments: (a) Subjects learned over trials to the same degree in these explicit and implicit conditions. (b) Again evidence for action on consciously represented rules was shown in prediction of proportion correct from rule validity within error by identity functions—with r's =.93 and .92. (c) Subjects showed transfer to categorization of novel letter strings one letter higher in the alphabet only when informed of that relation. That means transfer occurred only when that rule together with another acquired rule permitted the deliberative inference that this was grammatical or nongrammatical.

Saving Assumptions for the Standard Theory of Implicit Learning?

In either case, the challenged theory is that of unconscious abstraction and control by the FSG, with either of two auxiliary assumptions:

(a) The identity function was a task artifact that would be yielded even by random selection of a rule independent of the FSG. In Dulany et al. (1984, 1985) we reported 100 computer simulations of guessing rule validities for the obtained proportions correct for 50 subjects over 100 trials. This yielded a distribution of functions with a mean slope of 1.65, intercept of –.032—progressive underprediction from rule validity = .5 upward—leaving the obtained identity function one to be expected by guessing about once in 10 billion occasions.

(b) Alternatively, the rules were somehow exuded by the unconscious grammar. It has recently been claimed that consciously represented rules and familiarity that predict constitute only a "judgment knowledge" and there must be an unconscious and controlling "structural knowledge" (e.g., Scott & Dienes, 2008). As explained in Dulany et al. (1984), however, there is no known or plausible set of processes that would permit an unconscious grammar to exude 100 conscious rules for 50 Ss with just the validities needed to yield the obtained identity function. Neither would it explain this kind of transfer to the novel, to new surface forms from the old structure, only with deliberative processing—what should be required for "analogical transfer" in general.

Furthermore, neither alternative assumption would account for the normal distribution of prediction error, the near zero mean of that error, and prediction without significant residual for all but one subject. With the *logic of competitive support* we may draw on the fairly rich array of data over the rule validity values and conditions, as predicted by a fairly rich theory, then call upon the Duhem-Quine thesis and elementary Bayesian inference for relative credibility of the explanations.

CAUSAL LEARNING AND A MENTALISTIC THEORY AND MODEL OF PROPOSITIONAL (EXPLICIT) LEARNING

Learning "what causes what" is certainly not limited to ordinary experience, but is central to this science and many others—and when presented with a *causal problem to be solved* the process should be predominantly explicit (deliberative). On the mentalistic theory, there certainly could be conditions for the associative-activational learning of causal relations to some degree, evocative mental episodes, which could then be propositionally represented in higher order awareness. In science, however, a causal problem clearly calls for deliberative attention.

Thus the theory and model of propositional learning (Carlson & Dulany, 1988; Dulany, 1979) has challenged a dominant view of causal learning as associative learning. These conceptions have been dominated by associative models that vary in the probabilities assigned to cells and marginals of a 2×2 for cause, noncause, effect, noneffect—from Shanks and Dickenson (1987) to Cheng (1997), although not only this approach, but a range of others are more recently examined in Gopnik and Schulz (2007).

Model and Experimental Example

It is common, of course, to use multivariate analysis of variance or regression analysis to examine effects of multiple causes and their interactions. But there is another and significant case—the appearance of an unusual effect, perhaps an illness, an effect that could be caused by any *one* of several causes, perhaps one of four viruses. There can be various analogues in this science and elsewhere. In our experimental work (Carlson & Dulany, 1988) the theory and model of propositional learning (Dulany, 1979) was examined within an analogous task more likely to arouse the interest of undergraduate subjects: The "effect" described on the computer screen was a murder—most foul—at an estate with four weekend guests, each a causal suspect, and with clues on each trial, the evidence, tending to confirm or disconfirm particular causal hypotheses. On every trial a clue was associated with one, two, or three of the suspects, with counterbalancing, and the clue was incriminating for one of the four suspects on 9, 7, 5, or 3 of 12 trials. They solved two 12-trial mysteries.

On every trial, too, subjects reported degree of belief mode value for one of the several beliefs controlling revision of belief in the guilt or innocence of a suspect.

In each case, the scale was certainty, from certain Yes, $\beta = +1$, to certain No, $\beta = -1$ that this was evidence. (This β scale is natural for these beliefs and simplifies these equations.)

One was degree of belief the evidence was associated with a particular suspect, βA—analogous to belief the evidence was associated with the cause represented in a particular scientific hypothesis. Another was the Forward Implication, βF, belief in the relative likelihood that this clue would be true of the guilty or innocent suspect. We can think of that as a belief in the relative likelihood that some piece of scientific evidence would follow from a true cause or some other cause.

Now we could expect the product of these two to predict Subjective Evidence, βE, how strongly they believe the evidence provided by this clue suggests guilt or suggest innocence for that suspect. Put simply,

$$\beta E_{nij} = \beta A_{nij} \times \beta F_{nij},$$

and these predictors strongly interacted in the prediction of Subjective Evidence, βE, $F(1{,}569) = 1379.61$.

Furthermore, we also consider the "Backward Implication" of evidence, βB, the relative likelihood that the murderer (a true cause) would be someone of whom the clue (the evidence) is true or false. What is theoretically, "Convincingness of the Evidence," βV, the parameter controlling percent of belief revision toward certainty, should then come from the interaction of the Subjective Evidence and the Backward Implication of that evidence, both of which were assessed for each clue:

$$\beta V_{nij} = \beta E_{nij} \times |\, \beta F_{nij} \,|.$$

Thus we could say that for revision of Belief in a Causal Hypothesis, βH, each belief reported on a scale ranging from Certain Innocent to Certain Guilty, is yielded in this way:

$$\beta H_{n+1,i} = \beta H_{ni} + /\beta E_{nij} \times \beta B_{nij} / (1 - \beta H_{in}), \quad \text{if } \beta E_{nij} > 0$$

$$\beta H_{n+1,i} = \beta H_{ni}, \quad \text{if } \beta E_{nij} = 0$$

$$\beta H_{n+1,i} = \beta H_{ni} - /\beta E_{nij} \times \beta B_{nij} / (1 + \beta H_{in}), \quad \text{if } \beta E_{nij} < 0.$$

Do they revise their causal beliefs in accord with these equations describing deliberative (explicit) inferences? For all four ratios of confirmation versus disconfirmation, the observed belief values closely tracked the theoretically predicted values (Figure 4 in Carlson & Dulany, 1988). The correlation of predicted and observed, over 12 subjects and 24 trials, was .911 with a slope of .981 and essentially zero intercept. Various other analyses showed consistency of prediction over subjects, modest error of prediction for individuals, no reliable revision of belief when Subjective Evidence was zero, and so forth, with consistency of results over three experiments.

This is an example of *predominantly explicit (deliberative) processing* within a paradigm *analogous to a scientific paradigm.*

Saving Assumptions for an Associative Theory of Causal Learning?

For this case, I see two kinds of auxiliary assumptions—augmenting a theory that association of incriminating and exonerating clues directly modified hypothesis belief values—by some sort of distribution of *p* values in those cells and marginals. (a) The reported propositional belief values were postprocess explanations. This assumption lacks credibility because none of the subjects knew these equations to follow in reporting this fairly rich network of propositional beliefs. (b) The pattern of conscious beliefs was somehow exuded by that associative process. Again we lack a mechanism for exuding rules that would create this fairly rich network of propositional beliefs. As before, the relatively rich mentalistic theory is favored on a *logic of competitive support*—with prediction of relatively rich data, Duhem-Quine thesis, and our own Bayesian thinking.

A LOGIC OF COMPETITIVE SUPPORT: DUHEM-QUINE THESIS, NETWORK LOGIC, BAYESIAN INFERENCE

I sketch this logic in this chapter for two reasons: I used it in describing "competitive support" for the mentalistic over the standard theories of implicit and explicit learning, as well as causal learning. The Bayesian inference component is also an example of quite deliberative (explicit) scientific thinking, despite some interlacing of the evocative. It is something normative that can approach the descriptive in the psychology of this science. As a logic of *competitive* support for theories, it shares that fundamental approach with Capaldi and Proctor's (2008) emphasis on "comparative abduction," Lipton's (2004) "inference to the best explanation," and even Thagard's (2007) identification of "explanatory coherence" (see also Koslowski, Chapter 6, this volume).

With a concern for the interrelations among mental states, this is a logic for gaining confidence in one description, rather than others, of the behavior of *hypothetical unobservables*—because with a physicalistic data language that is what conscious states are. It also focuses on competitive support from single experiments, a logic that can be iterative over experiments. Thus it distinctively combines its three basic characteristics:

1. One is emphasis on a version of the **Duhem-Quine thesis** on which a hypothesis should include not only the theory, but various auxiliaries, as well as mappings of unobservables to observables, Quine (1951). In the famous quotation, "They go to the court of experience as a corporate body." So it is {T & M & A} → D, the data. We know, however, that any data may follow from an indefinite number of hypotheses—the well-known "under determination by data," the "fallacy of affirming the consequent." Furthermore, if some predicted data are *not* found, {T & M & A} → not-D; disconfirmation of a conjunction disconfirms disjunctively. It could be a failure of theory, of mapping validity, of procedural control, or of any combination. There is no absolute disconfirmation either. This says a little formally what we generally know but need to keep in mind because theories according causal roles to conscious states often stand in opposition to theories according those causal roles to unconscious states—and we need a methodology, not simply of confirmation or "falsification" alone, but of competitive support.
2. Competitive support also draws on an elaborated **logic of theoretical networks**. The complexity of the relations among mental states calls for complexity of theory for two reasons: From Hempel (1952) onward—for example, Carruthers (2000); Churchland (1986); Cronbach and Meehl (1955); Dulany (1968)—the meaning of a realistically interpreted theoretical construct, for example, a conscious state term, is conveyed by the set of theoretical and mapping assertions in which it appears. The richer the theory the richer the *meaning*, its functional specification. Intentions *are* what intentions *do.*

Using that rationale and extending it from *meaning to support*, we can also recognize this, as elaborated in Dulany (1968) for the theory of propositional control, and later for other theory: *The richer the theoretical network, the richer the network of data it can predict.* And the *richer* the *data* network, *the narrower the range of credible alternative interpretations with saving auxiliary assumptions.* Put bluntly, the widely celebrated "parsimony" should not apply to our theories themselves—theories of complex mental processes—but to the ratio of unmapped to mapped constructs, and to those implausible saving auxiliaries.

3. We also need a rational model for according relative credibility to theories realistically interpreted, with empirical support from single experiments, which can then be iteratively applied over experiments—**Bayesian inference**. Although various forms of Bayesian analyses are offered in the philosophy of science and psychology (e.g., Shafer, 2001), a relatively simple and common representation will serve our purposes—without more complex "Bayesian nets" and parameter estimations for unmapped parameters.

We can express the relative prior credibilities of two complex hypotheses, their relative predictability of a data set, and the relative credibility of the two hypotheses in the light of that data. We may—with a first step beyond the conventional—elaborate this kind of comparison for Duhem-Quine aggregates in this way:

$$\frac{P[(T,M,A)_1\,|\,D]}{P[(T,M,A)_2|D]} = \frac{P[D\,|\,(T,M,A)]_1\; P(T,M,A)_1}{P[D|(T,M,A)]_2\; P(T,M,A)_2}$$

a posteriori credibilities — likelihood ratio — *a priori* credibilities

These prior credibilities would realistically reflect the usual sources—not only the prior credibility of theory, but also of mappings and procedural control as well. Clearly, too, the *a posteriori* credibility of one aggregate obtains a relative boost in the degree the likelihood ratio benefits that aggregate. Sometimes that is enough.

Still further elaboration can be needed, however, once we recognize the "flexibility" of the Duhem-Quine components. Revised auxiliary assumptions for *simple data* may be highly credible and allow the "disconfirmed" hypothesis to predict as well as the other—to "push the likelihood ratio to one" for now equally credible *a prioris*, with a Bayesian standoff. In fact, ingenuity in doing so seems to reflect the degree to which there can be strong competition between theoretical views within the community—especially between explanations attributing effects to conscious symbolic states or to unconscious symbolic states with strong prior commitments (Dulany, 2003).

To the degree the theory of interest predicts a rich network of data, however, on network logic the credibility of the "saving" auxiliaries diminishes—as we saw in applying this logic to studies of implicit and explicit learning and to causal learning. We may denote an aggregate with that new auxiliary assumption in this way: (T,M,A)′. And in doing so, that can also "push the likelihood ratio to one"—but now there is a Bayesian price to pay.

$$\frac{P[(T,M,A)_1\,|\,D]}{P[(T,M,A)'_2|D]} = \frac{P[D\,|\,(T,M,A)]_1\; P(T,M,A)_1}{P[D|(T,M,A)]'_2\; P(T,M,A)'_2}$$

a posteriori credibilities — likelihood ratio — *a priori* credibilities

If the *a priori* of the hypothesis, with the saving auxiliaries, are now less credible, the *a posteriori* must remain less credible with a likelihood ratio of one.

So $P(D|T,M,A)'_2 < P(D|T,M,A)_1$. That's rational competitive support.

Limitations of the Bayesian Analysis?

There are two commonly mentioned limitations of Bayesian analysis:

Isn't it all too subjective? Credibility of a theory is itself inherently subjective—as is the scientist's evaluation and sense of acceptability that comes out of any comparative evaluation of theories. But it is not all subjective. It is only the case that the inference rules operate on and produce subjective beliefs—with intrinsic responsiveness to empirical data, iteratively over experiments.

How can we use Bayesian analysis if we cannot precisely and accurately specify antecedent or predictive probabilities? There are two kinds of important answers. (a) If we say that whenever Likelihood Ratio ≠ 1 the relative credibility of H1 and H2 in the *a priori* will be modified in comparison to the relative credibility of H1 and H2 in the *a posteriori*, we actually speak "ordinal Bayesian"—which is common and can be useful. In fact, if the inverse of the LR is ordinally greater than the *a priori*, the ordinal relation within the *a posteriori* will be an ordinal reverse of that in the *a priori*. (b) Nevertheless, we may formulate *a prioris* and LR's with as much precision as knowledge of the relevant science permits—and these credibilities are subject to increasing precision, iteratively with empirical evidence.

Bayesian inference rationally and empirically optimizes output for the available input.

Other requirements, such as "exhaustiveness" or "summing to one," are handled simply by this usage being a comparative examination of two competing formulations—what we usually need, or can fashion.

UNCONSCIOUS IMPLICIT THINKING?

Unconscious Implicit Problem Solving in Science?

The belief in unconscious problem solving has been out there in the literature—and in undergraduate math and science classes—at least since Hadamard's (1945) *The Psychology of Invention in the Mathematical Field,* and more recently in Andreasen's (2005) *The Creating Brain: The Neuroscience of Genius,* with special attention to the two most famously cited cases in math and science, Poincaré and Kekulé. On the general topic, even Baars (1988) in *A Cognitive Theory of Consciousness* wrote that "the bulk of spontaneous problem solving is unconscious" (p. 158).

Poincaré?

The famous claim is based on *other's interpretation* of Poincaré's account, written long after the creative moments, as quoted (Poincaré, 2001): "I drank black coffee and could not sleep. Ideas rose in crowds; I felt them collide until pairs interlocked, so to speak, making a stable combination. By the next morning I had established the existence of a class of Fuchsian functions, those which come out from the hypergeometric series" (p. 220). Unconscious problem solving?

The debunking comes from the careful scholarship of Ericsson and Simon (1993) in which this is their conclusion: "A critical reading of Poincaré does not provide any evidence for that interpretation, nor for the belief that Poincaré himself favored such an interpretation" (p. 162). They even quote Poincaré himself saying "In no case do unconscious processes make any calculations" (p. 164).

Kekulé?

The belief in an unconscious creation of the benzene ring comes originally from an account delivered for the first time in an after dinner speech, in March 1890, reporting a "Halbshlaf," a half-sleep reverie before a fireplace 29 years earlier—again not a condition of experimental control. Kekulé said he saw flames licking at flames, making him think of a ring of snakes with tails in mouth, just the image he needed for the benzene ring. His account made the newspaper the next day.

The debunking again comes from careful scholarship, this time by Wotiz and Rudofsky's (1984) *Chemistry in Britain.* They quote Kekulé's own expression of regret for those after-dinner comments, "It is certainly difficult to say something rational at such an occasion,... but it is still more difficult by far to reproduce word for word three weeks later the stupidities one has said.... Whenever I speak in public I am always conscious of having terribly embarrassed myself and then I get a hangover for at least three days [all that good Schnapps].... Today I have the feeling it would be better if one burned the whole rubbish and did not allow anything to be printed" (p. 720). But it was—and has been grist for undergraduate chemistry courses and the popular history of science. As the authors say, "Although we would like to correct the record, we have our doubts that truth will get in the way of a good story" (p. 723).

Unconscious Implicit Problem Solving in Experimental Tasks?

There has been, in fact, an extensive classical literature on unconscious implicit problem solving in experimental tasks, and I can only describe the paradigm here, summarize the results, and quote conclusions of those who examined the literature carefully. Subjects have been presented a variety of problems to be solved, from anagram solution to quite difficult puzzles. On the standard design, a control group is presented the problem, then no time off, followed by a test. An experimental group is presented the problem, given that special time off with various "distractions," followed by the test. Superior performance by the experimental group is called an "incubation effect"—with the interpretation that time off with distraction permitted "unconscious problem solving."

Reviewers agree on a summation of the findings. Sometimes the incubation effect occurs, sometimes not. They also have suggested various artifacts that could explain the incubation effect, for example, "physical refreshment, fruitful forgetting, losing commitment to an ineffective approach, and noticing clues in the environment. Extended unconscious thinking does not occur" (Perkins, 1981, p. 81). Mandler (1994, p. 220) also concludes that "there is no direct evidence that complex unconscious work (new elaboration and creations of mental contents) contributes to the incubation effect."

We then have two possibilities for Bayesian consideration: One or more of these artifacts explains the incubation effect when it occurs. Or evolution has equipped us with an unconscious mechanism for problem solving, but it shows up in only some of these experiments.

Unconscious Implicit Decision Making?

There is a more recent analysis of decision making launched by Dijksterhuis and Nordgren (2006), with the general approach nicely summarized by Dijksterhuis and Aarts (2010) in the *Annual Review of Psychology:* "One way to approach this issue is to propose that, in principle, the operation of higher cognitive processes does not care much about the conscious state of the individual" (p. 475). For example, the subjects were presented descriptions of features of options for purchase or renting decisions—for example, cars or apartments—before a decision was to be made. They identified "unconscious decision" with choice after a period of distraction, and "conscious decision" with choice after a period of nondistraction before the decision. For experts—but not for novices—the decision was better after distraction.

It is enough to say here—where I must be brief—that the work has quickly launched at least four multiexperimental critiques: Cleeremans, Waroquier, David, and Klein (2009); Lassiter, Lindberg, Gonzalez-Vallejo, Bellezza, and Phillips (2009); Rey, Goldstein, and Perruchet (2009); and Waroquier, Marchiori, Klein, and Cleeremans (2009). Among their various findings were failures to replicate and identification of artifacts that can bias the results—for example, best decisions made immediately by the experts, which were revised downward on implied "experimental demand" when they were to decide later. As Lassiter et al. (2009), for example, refer to their findings: "Such judgments are ultimately a product of conscious rather than unconscious thought" (p. 671). And Cleeremans et al. (2009) report that they "find no evidence that one can 'think' without awareness" (p. 82).

Unconscious Implicit Volition—the Wegnerian Leitmotif?

In his final paragraph, Wegner (2002) writes, "We experience willing a walk in the park, winding a clock, or smiling at someone, and the feeling keeps our notion of ourselves as persons intact. Our sense of being a conscious agent who does things comes at a cost of being technically wrong all the time" (p. 342). The evidence ranges from senses of external control in schizophrenia to his cleverly engineered experiment in which his confederate leads subjects to believe the confederate's action with a computer mouse was actually their own (pp. 74–78). In the same way, however, we might conclude that evidence from hallucinations in schizophrenia to cleverly engineered perceptual illusions means that our sense of being a conscious perceiver who sees things is "technically wrong all the time." We can acknowledge the theoretical significance of illusions. But the atypical in no way disconfirms theory of the more typical phenomena.

Unconscious Implicit Memory?

From Kihlstrom (1987) to current textbooks, implicit memory is commonly labeled "unconscious memory"—to distinguish it from explicit memory as conscious memory, a conscious propositional representation of something believed or recollected.

But the classic work reveals a consciously represented prime activating a consciously represented identification, or fragment completion, or semantic association. In fact, there is a natural explanation of each as strengthening the associative-activational relations within an evocative mental episode. I don't believe there is any evidence that a failure to represent the key proposition in consciousness, for example, " 'winter' was on the word list to learn," is represented in an unconscious—thanks to inattention or amnesia.

MAPPING OF CONSCIOUS STATES FOR LAWFULNESS AND FOR "NULL AWARENESS"

The general validity conditions for lawfulness: We can recognize that validity of any mapping, including phenomenal reports, requires that *validity conditions* be met. For the subject reporting, there must be memory for the conscious event; it must be verbalizable; and there must be understanding of what is to be reported and motivation to report accurately. A first person data language would fail to meet a fundamental requirement of data languages, replicability over observers for common experimental events. For common conditions we can vary in conscious experience for numerous reasons. For a physicalistic data language, with mapping to the conscious state, we must meet acceptable validity conditions (Dulany, 2009).

As we have seen with the summarized experiments on implicit and explicit learning and causal learning, a report validity mapping, on the Duhem-Quine thesis, participates in the prediction, and subjects showed quite acceptable lawfulness. In fact, a logic of competitive support provides "construct validation" of reports that is *competing theoretic* (and thus beyond the monotheoretic approach of Cronbach and Meehl (1955).

The General Problem for Determining "Null Awareness"

Validity conditions for an acceptable interpretation of "null awareness" are much more demanding than they are for lawfulness—and establishing that "null awareness" is essential for demonstrating *unconscious implicit processing.* With a little assessment error we have a little error variance in a lawful relation. But with a little assessment error we have a spurious finding of null awareness—and thus a spurious finding of "unconscious" implicit this-that-or-the-other. Although we could also say that this would make it difficult to identify the unconscious implicit if it occurs, we should also recognize that this is not an acceptable condition for a standard and widespread acceptance.

BASES FOR SPURIOUS FINDINGS OF NULL AWARENESS AND UNCONSCIOUS IMPLICIT PROCESSING

There are many bases for spurious findings of null awareness and unconscious implicit processing. With no attempt to "review literatures" I can incompletely list and briefly describe them here, and some have been elaborated in Dulany (1997, 2003, 2004, 2011).

"Operational Definition" of Awareness State

With continued homage to Bridgman, there are still those who would say that an index, usually a report, of a conscious state is to be an *operational definition*—for example, Baars (2003), Destrebecqz and Peigneux (2005), and others. As detailed earlier (Dulany, 1997), the formalization closest to Bridgman's intent, Bergmann's (1951) "explicit definition" (still another meaning of "explicit"), univocally and exhaustively defined the term—so it refers only to that measure in that experiment and not to a conscious state at all. And even longer ago Carnap (1937) formalized the operational definition as a "reduction sentence," recognizing the need for multiple indices of a realistically interpreted construct.

But the inescapable problem is this: The relation of any measure, including a report, to any unobservable state, including a conscious state, is a hypothesis—*not* a definition. It is a mapping hypothesis that can be evaluated, one within the larger Hypothesis examined that included the theory and auxiliaries (on the Duhem-Quine thesis).

So "operational definition" seems sometimes to be offered as merely a *label* of a measure—what is no more than an unscholarly misuse of the term—but it seems also to be offered as a spurious justification for an inadequately valid index of awareness: "But that's the way I operationally defined it."

Failure to Examine Contents of Awareness "Correlated" with the Correct

What the Dulany et al. (1984, 1985) studies demonstrated was the importance of examining those correlated contents—even though subjects had no conscious representation of the formal finite state grammar. In the term we used at the time, the total set of consciously represented rules, by virtue of the set of rule validities, constituted a "correlated grammar."

Dissociation of Higher Order and Lower Order Awareness Treated as Unconscious Implicit

Report of the content of a just prior awareness is in fact a report of higher order awareness of the content in that prior lower order awareness. The failure of that higher order representation, however, can occur when experimental subjects have forgotten, or failed to establish the memory, or failed to understand instructions, or failed to categorize correctly, and so forth.

1. With **post experimental reports**, which are common, this dissociation is especially likely to be a source of report invalidity. With imperfect memory of conscious states on enough trials, and assessment introduced as "postexperimental," subjects

can have not only unacceptably imperfect memory but also a sense that the "real business of science is over"—and lower their own standards for reporting.

2. In the extreme case, the investigator may draw upon one or another species of **higher order thought theory** (Gennaro, 2004), a philosophical view that all mental states are unconscious states unless granted the "property" of consciousness when graced by a higher order representation. Thus whenever a subject fails to report the conscious representation of the prior conscious state—regardless of failure to meet standard conditions of report validity—the conscious state just wasn't there. Analogously, as I said in my critique (Dulany, 2004), if ambient illumination impairs the sensitivity of a microscope, the bacterium just wasn't there.
3. In some cases, the dissociation of higher and lower order awareness may have **special theoretical interest**. With damage to V1 in the striate cortex can, as in the Oxonian subjects, DB and GY, discrimination of stimuli presented to the scotoma may be preserved, despite what Weiskrantz (1997) had characterized as a lack of "any awareness whatsoever of a visual event," p. 65—and we should keep "visual" in mind. We should first note that discrimination is measured with what has been long recognized (Reingold & Merikle, 1988) as the *most sensitive index of awareness,* the "direct objective" index of awareness. The subject reports a conscious representation of the figure as at the right, or at the left, or moved right or left, and so forth, with substantial d's from signal detection theory. When asked to use the "commentary key" to characterize this perception as "seeing" or not—a higher order representation—this can get a much lower even zero d′.

This dissociation of higher order and lower order awareness cannot, however, be properly interpreted, as it generally is, as revealing *"unconscious implicit perception"—of "blindsight,"* on the Oxonian oxymoron. This is, however, a very interesting dissociation, in which awareness of stimulus position, by virtue of damage to V1 and degraded experience, is not given a certain *higher order representation, the categorization of "seeing."* In fact, this was Weiskrantz's (1980) interpretation of EY in a revealingly entitled paper, *Residual Experience*: "When he was asked to report when he *saw* the light coming into the field, he was densely blind by this criterion, but when he was asked to report merely when he was *aware* of something coming into the field, the field was practically filled," p. 378. With the ideological sweep across decades within the discipline, as outlined above, we can understand why a later interpretation like "unconscious perception" has become more common across the discipline.

"Subliminal Processing" Treated as "Unconscious Implicit Processing"

"Subliminal" means "unconscious" is an identification more likely to occur in reviews or citations of other evidence—but can evidently be compelling. It usually refers to "perception"—with claims of unconscious "implicit perception," but can refer to identification of "unconscious" intentions or plans. "Subliminal" does refer to a disparity between two measures, in which an "indirect measure" of some stimulus has a significant score, and what is taken as a "direct measure" of the awareness has a "zero" score of some kind, But "unconscious," of course, is a theoretical interpretation. What follows then is a brief list of common reasons to challenge that identification.

1. **Signal detection theory** has been the gold standard for showing what is a subliminal effect—a reliable d' for the "indirect effect," when the d' is zero for the "direct effect"—interpreted as the "awareness threshold," the "limen." In actuality, however, signal detection theory does not provide a **"threshold."** In the *theory*, d' = 0 for the direct measure reflects normal distributions of signal strength and error with equal means. But clearly if the distributions are random the signal can sometimes exceed the noise. That is, however, only the underlying "theory," and it is more useful and convincing to recognize the list of biases that can lead to erroneous interpretations of "unconscious implicit processes."
2. The measure of awareness can be a **"subjective measure"** subject to well-demonstrated criterion biases, identifiable with a signal detection theory analysis.
3. Even with retreat to the famous **"process dissociation"** paradigm of Debner and Jacoby, (1994), a sophisticated SDT analysis of Snodgrass (2002) showed that "exclusion methods are *actually subjective threshold* paradigms and as such are vulnerable to SDT-based criterion artifact" p. 546.
4. **Masking** of stimuli may be imperfect over trials and subjects, as recognized by many from Holender (1986) to Kouider and Dehaene (2007), biasing the d' for the "direct objective measure."
5. Direct and indirect measures may be taken under somewhat **different validity conditions**, fewer trials for the direct, the direct before or after the indirect, and so forth. A direct measure also may use a larger or less available response set than the indirect measure.
6. A direct response may, by instruction, be characterized as, "I *guess* X (or Y)," and interpreted as random. But a common meaning of "Well, I guess . . . " can be a judgment of **some, but low, certainty for a conscious belief.**
7. Furthermore, there are various studies purporting to show unconscious indirect effects with **brain imaging,** but fMRI and other commonly used brain imaging measures are imperfectly specific and appropriate for a d' measure. It is also now more widely revealed—thanks to Vul et al. (2009) and Vul and Kanwisher (2010)—that some brain imaging can to an unusual degree be "constructed."
8. From Sperling (1960) to its emphasis by Block (2007) and others, it has been clear that a **"literal awareness,"** variously termed, precedes for 200–300 ms and surrounds an attentional state of stimulus identification. As I argued in Dulany (2001), we do not live in attentional tunnels preceded by, and surrounded by, the Grand Scotoma. Baars (1988) put it clearly, "The Sperling experiment serves as a reminder that conscious events may decay in a few hundred milliseconds" (p. 16). Since we know that attentional processing is needed to establish information in memory, for any but the most immediate report of awareness, subjects

could often have effects produced by that literal awareness—effects which are mistakenly interpreted as "unconscious perception."

In the theoretically interesting **prosopagnosia**, there is clear evidence of the **dissociation of identity awareness from literal awareness** caused by damage in the fusiform gyrus area, blocking attentional identification of faces. They nevertheless show (a) greater GSR to familiar than unfamiliar faces (Tranel & Damasio, 1985), and even slowing of a naming response when a familiar face and occupational category disagree (DeHaan, Bauer, & Greve, 1992). But (b) they are as good as normals at matching facial photographs, familiar or unfamiliar (DeHaan, Young, & Newcombe, 1987). With this dissociation, something is preserved, the literal awareness—that can be causal, and if disregarded, lead to an unacceptable interpretation of "unconscious facial identification."

It is an interesting dissociation, of a type that can occur in normals, as a bias or as a problem for further investigation.

HOW THEN SHOULD WE UNDERSTAND THE "IMPLICIT" AND "EXPLICIT" IN OUR OWN SCIENTIFIC THINKING?

Suppose we have a theoretical or experimental problem to solve. It is very common in my own experience, and probably in that of others, that I hear a talk or read a paper and some of the ideas expressed, consciously perceived, together with others that come consciously to mind, can activate other related ideas—evocative mental episodes, some of memory and some activating related ideas. They can then be given the kind of thought that moves us among conscious modes and contents by nonconscious operations, with some transforming and interlacing of the two kinds of mental episodes. For something productive it takes something deliberative.

Even before we are fully satisfied, we can deliberatively decide the ideas are or are not promising. With a loose vernacular at hand, we may characterize some of those preliminary low confidence beliefs as "gut feeling," or "intuition." But an "Intelligence of the Unconscious," as in Gigerenzer (2007)? This is not what I find in our "Tool Box."

With openness of scientific inquiry, however, any of us could try to formulate or evaluate a metatheory, a theory, a model, a methodology, or an experiment with a "blink."

CONCLUSION

So how *should* we understand implicit and explicit processes in the psychology of our science *at large*? I have addressed this question at more than one level. At one level, we can ask what the basic science of psychology offers as competing theories, the standard cognitive theory and a mentalistic theory, and I emphasize what I think is the more acceptable and promising theory —the mentalistic. Two examples of theory and evidence were summarized, a mentalistic theory of explicit (deliberative) and implicit (evocative) learning, as well as of explicit (deliberative) causal learning. Each was an experimental paradigm that makes analogical contact with scientific thinking. But examining this question can raise a question at another level: How have deeper metatheoretical and ideological commitments influenced the formulation and biased the experimental support for that standard formulation in the psychology of this science? There are lists. We can also ask how we might deliberatively and *normatively* evaluate these, and other, strongly competing theories—what constitutes a strong logic of competitive support with Duhem-Quine, the logic of networks, and Bayesian inference. I see all of this as bearing on an understanding of the explicit and implicit in the psychology of this science.

REFERENCES

Andreasen, N. C. (2005). *The creating brain*. New York: The Dana Foundation.

Baars, B. J. (1988). *A cognitive theory of consciousness*. New York: Cambridge University Press.

Baars, B. J. (2003). Introduction: Treating consciousness as a variable: The fading taboo. In B. J. Baars, W. P. Banks, & J. B. Newman (Eds.), *Essential sources in the scientific study of consciousness* (pp. 1–10). Cambridge, MA: MIT Press.

Baumeister, R. F., & Masicampo, E. J. (2010). Conscious thought is for facilitating social and cultural interactions: How mental simulations serve the animal-culture interface. *Psychological Review, 117,* 945–971.

Baumeister, R. F., Masicampo, E. J., & Vohs, K. D. (2011). Do conscious thoughts cause behavior? *Annual Review of Psychology, 62,* 331–261.

Bergmann, G. (1951). The logic of psychological concepts. *Philosophy of Science, 18,* 93–110.

Block, N. (2007). Consciousness, accessibility, and the mesh between psychology and neuroscience. *Behavioral and Brain Science, 30,* 481–499.

Brentano, F. (1774/1973). *Psychology from an empirical standpoint*. London: Routledge & Kegan Paul.

Brewer, W. E., Chinn, C., A., & Samarapungavan, A. (1998). Explanation in scientists and children. *Mind and Machines, 8,* 110–136.

Broad, C. D. (1925). *The mind and its place in nature*. London: Routledge Kegan Paul.

Capaldi, E. J., & Proctor, R. W. (2008). Are theories to be evaluated in isolation or relative to alternatives? An abductive view. *American Journal of Psychology, 121,* 617–642.

Capaldi, E. J., & Proctor, R. W. (2009). Two radically different views of psychology of science. *Journal of Psychology of Science and Technology, 2,* 44–58.

Carlson, R. A. (2002). Conscious intentions in the control of skilled mental activity. *Psychology of Learning and Motivation, 41,* 191–228.

Carlson, .R. A., & Dulany, D. E. (1988). Diagnostic reasoning with circumstantial evidence. *Cognitive Psychology, 20,* 463–492.

Carnap, R. (1937). Testability and meaning. *Philosophy of Science, 4,* 1–40.

Carr, H. A. (1930) Functionalism. In C. Murchison (Ed.) *Psychologies of 1930*. Worcester, MA: Clark University Press.

Carruthers, P. (2000). *Phenomenal consciousness: A naturalistic theory*. Cambridge: Cambridge University Press.

Cheng, P. W. (1997). From covariation to causation: A causal power theory. *Psychological Review, 104,* 367–405.

Churchland, P. M. (1993). Eliminative materialism and the propositional attitudes. In A. I. Goldman (Ed.), *Readings in philosophy and cognitive science* (pp. 255–270). Cambridge, MA: MIT Press.

Churchland, P. S. (1986). *Neurophilosophy*. Cambridge, MA: MIT Press.

Cleeremans, A., Waroquier, L., David, M., & Klein, O. (2009). To think or not to think?: A critique and reappraisal of "unconscious thought theory." *Association for Scientific Study of Consciousness* [Abstracts]. Berlin, Germany, pp. 81–82.

Cronbach, L. J., & Meehl, P. (1955). Construct validation in psychological tests. *Psychological Bulletin, 52,* 281–382.

Debner, J. A., & Jacoby, L. L. (1994). Unconscious perception: Attention, awareness, and control. *Journal of Experimental Psychology: Learning, Memory, and Cognition, 20,* 304–317.

Dehaan, E. H. F., Bauer, R. M., & Greve, K. W. (1992). Behavioural and physiological evidence for covert face recognition in a prosopagnosic patient. *Cortex,* 28, 77–95.

Dehaan, E. H. F., Young, A., & Newcombe, F. (1987). Face recognition without awareness. *Cognitive Neuropsychology, 4,* 385–415.

Destrebecqz, A., & Peigneux, P. (2005). Methods for studying unconscious learning. *Progress in Brain Research, 150,* 69–80.

Dijksterhuis, A., & Aarts, H. (2010). Goals, attention, and (un)consciousness. *Annual Review of Psychology, 61,* 467–490.

Dijksterhuis, A., & Nordgren, L. F. (2006). A theory of unconscious thought. *Perspectives on Psychological Science, 1,* 95–108.

Dulany, D. E. (1968). Awareness, rules, and propositional control: A confrontation with S-R behavior theory. In T. Dixon & D. Horton (Eds.), *Verbal behavior and general behavior theory* (pp. 340–387). New York: Prentice-Hall.

Dulany, D. E. (1979). Theory of propositional learning. Unpublished paper.

Dulany, D. E. (1991). Conscious representation and thought systems. In R. S. Wyer & T. K. Srull (Eds.), *Advances in social cognition* (Vol. 4, pp. 97–120). Hillsdale, NJ: Lawrence Erlbaum.

Dulany, D. E. (1997). Consciousness in the explicit (deliberative) and implicit (evocative). In J. Cohen & J. Schooler (Eds.), *Scientific approaches to consciousness* (pp. 179–212). Mahwah, NJ: Lawrence Erlbaum.

Dulany, D. E. (2001). Inattentional awareness. *Psyche: An Interdisciplinary Journal of Research on Consciousness,* 7(5). Retrieved from http://psyche.cs.monash.edu.au/v7/psyche-7-05-dulany.html. Or Journal PSYCHE ASSC, then Vol. 7

Dulany, D. E. (2002). Mentalistic metatheory and strategies. *Behavioral and Brain Sciences, 24,* 337–338.

Dulany, D. E. (2003). Strategies for putting consciousness in its place. *Journal of Consciousness Studies, 10*(1), 33–43.

Dulany, D. E. (2004). Higher order representation in a mentalistic metatheory. In R. J. Gennaro (Ed.), *Higher order thought theories of consciousness* (pp. 315–338). Amsterdam: John Benjamins.

Dulany, D. E. (2008). How well are we moving toward a most productive science of consciousness? *Journal of Consciousness Studies, 15*(12), 77–100.

Dulany, D. E. (2009). Psychology and the study of consciousness. In T. Bayne, A. Cleeremans, & P. Wilkens (Eds.), *Oxford companion to consciousness* (pp. 540–544). Oxford: Oxford University Press.

Dulany, D. E. (2011). What should be the roles of conscious states and brain states in theories of mental activity? In A. R. Singh & S. A. Singh (Eds.), *Brain, mind, and consciousness: An international and interdisciplinary perspective, Mens Sana Monographs, 9*(1), 93–112.

Dulany, D. E., Carlson, R. A., & Dewey, G. I. (1984). A case of syntactic learning and judgment: How conscious and how abstract? *Journal of Experimental Psychology: General, 113,* 541–555. Reprinted in D. Shanks (Ed.). (2009). *Psychology of learning* (Vol. 2. pp. 131–153). London: Sage.

Dulany, D. E., Carlson, R. A., & Dewey, G. I. (1985). On consciousness in syntactic learning and judgment: A reply to Reber, Allan, and Regan. *Journal of Experimental Psychology: General, 114,* 27–34.

Dulany, D. E., & Pritchard, E. Awareness and novelty in explicit (deliberative) and implicit (evocative) learning and memory of a finite state grammar. Manuscript in preparation.

Dunbar, K. (1995). How scientists really reason: Scientific reasoning in real-world laboratories. In R. J. Sternberg & J. Davidson (Eds.), *The nature of insight* (pp. 115–139). Cambridge, MA: MIT Press.

Ericsson, K. A., & Simon, H. (1993). *Protocol analysis: Verbal reports as data.* Cambridge, MA: MIT Press.

Evans, J., St. (2009). Introspection, confabulation, and dual process theory. *Behavioral and Brain Sciences, 32,* 142–143.

Feist, G. J. (2006). *The psychology of science and the origins of the scientific mind.* New Haven, CT: Yale University Press.

French, R. M., & Cleeremans, A. (Eds.). (2002). *Implicit learning and consciousness: An empirical, philosophical and computational consensus in the making.* New York: Taylor & Francis.

Freud, S. (1923/1935). *The ego and the id.* London: Hogarth Press.

Gennaro, R. J. (Ed.). (2004). *Higher order thought theories of consciousness.* Amsterdam: Benjamins.

Gergen, K. J. (2001). Psychological science in a postmodern context. *American Psychologist, 56,* 803–813.

Gigerenzer, G. (2007). *Gut feelings: The intelligence of the unconscious.* London: Penguin Books.

Gopnik, A., & Schulz, L. (2007). *Causal learning: Psychology, philosophy, and computation.* New York: Oxford University Press.

Hadamard, J. (1945). *The psychology of invention in the mathematical field.* Princeton, NJ: Princeton University Press.

Haugland, H. (1978). The nature and plausibility of cognitivism. *Behavioral and Brain Sciences, 2,* 215–260.

Hempel, C. (1952). *Fundamentals of concept formation in empirical science.* Chicago: University of Chicago Press.

Holender, D. (1986). Semantic activation without conscious identification in dichotic listening, parafoveal vision and visual masking. *Behavioral and Brain Sciences, 9,* 1–23.

Hull, C. (1943). *Principles of behavior.* New York: Appleton-Century-Crofts.

Jackendoff, R. (1987). *Consciousness and the computational mind.* Cambridge, MA: MIT Press.

James, W. (1890). *Principles of psychology* (Vols. 1 & 2). New York: Holt.

Kihlstrom, J. F. (1987), The cognitive unconscious. *Science, 237,* 1445–1452.

Koffka, K. (1935). *Pinciples of gestalt psychology.* New York: Harcourt Brace.

Kouider, S., & Dehaene, S. (2007). Levels of processing during non-conscious perception: A critical review of visual masking. *Philosophical Transactions of the Royal Society, 362,* 857–875.

Lassiter, G. D., Lindberg, M. J., Gonzalez-Vallejo, C., Bellezza, F. S., & Phillips, N. D. (2009). The deliberation-without-attention effect. *Psychological Science, 20,* 671–675.

Lipton, P. (2004). *Inference to the best explanation* (2nd ed.). London: Routledge.

Mandler, G. (1994). Hypermnesia, incubation, and mind popping: On remembering without really trying. In C. Umiltà & M. Moscovitch (Eds.), *Attention and performance, 15* (pp. 4–33). Cambridge, MA: MIT Press.

Moors, R., & De Houwer, J. (2006). Automaticity: A theoretical an conceptual analysis. *Psychological Bulletin, 132,* 297–326.

Perkins, D. N. (1981). *The mind's best work.* Cambridge, MA: Harvard University Press.

Perruchet, P., & Vinter, A. (2002). The self-organizing consciousness. *Behavioral and Brain Science, 25,* 297–329.

Pockett, S., Banks, W. P., & Gallagher, S. (Eds.). (2006). *Does consciousness cause behavior?* Cambridge, MA: MIT Press.

Poincaré, H. (2001). *The value of science: Essential writings of Henri Poincaré.* New York: Random House.

Posner, M. I. (1994). Attention: The mechanisms of consciousness. *Proceedings of the National Academy of Sciences, USA, 91,* 7398–7403.

Pothos, E. (2007). Theories of artificial grammar learning. *Psychological Bulletin, 133,* 227–244.

Putnam, H. (1960). Minds & machines. In S. Hook (Ed.), *Dimensions of mind.* New York: NYU Press.

Quine, W. V. O. (1951). Two dogmas of empiricism. *Philosophical Review, 60,* 20–43.

Reber, A. S. (1967). Implicit learning and artificial grammars. *Journal of Verbal Learning and Verbal Behavior, 5,* 855–863.

Reber, A. S. (1997). How to differentiate implicit and explicit modes of acquisition. In J. Cohen & J. Schooler (Eds.), *Scientific approaches to consciousness* (pp. 137–159). Mahwah, NJ: Lawrence Erlbaum.

Reingold, E. M., & Merikle, P. M. (1988). Using direct and indirect measures to study perception without awareness. *Perception & Psychophysics, 44,* 563–575.

Rey, A., Goldstein, R. M., & Perruchet, P. (2009). Does unconscious thought improve complex decision making? *Psychological Research, 73,* 372–379.

Roediger, H. L. (2004). What happened to behaviorism? *Observer,* 17, 40–42.

Scott, R., & Dienes, Z. (2008). The conscious, the unconscious, and familiarity. *Journal of Experimental Psychology: Learning, Memory, & Cognition, 34,* 1264–1286.

Shafer, M. J. (2001). Bayesian confirmation of theories that incorporate idealizations. *Philosophy of Science, 66*(1), 36–52.

Shanks, D. R. (2005). Implicit learning. In K. Lamberts & R. L. Goldstone (Eds.), *Handbook of cognition* (pp. 202–220). London: Sage.

Shanks, D. R., & Dickenson, A. (1987). Associative accounts of causality judgment. In G. H. Bower (Ed.), *The psychology of learning and motivation. Advances in research and theory* (Vol. 21, pp. 229–261). San Diego, CA: Academic Press.

Shanks, D. R., & St. John, M. F. (1994). Characteristics of dissociable human learning systems. *Behavioral and Brain Sciences, 17,* 367–447.

Skinner, B. F. (1974). *About behaviorism.* New York: Knopf.

Snodgrass, M. (2002). Disambiguating conscious and unconscious influences: Do exclusion paradigms demonstrate unconscious perception? *American Journal of Psychology, 15,* 545–580.

Sperling, G. (1960). The information available in brief visual presentation. *Psychological Monographs: General and Applied, 74,* 1–29.

Thagard, P. (2007). Coherence, truth, and the development of scientific knowledge. *Philosophy of Science, 74,* 28–74.

Titchener, E. B. (1898). The postulates of a structural psychology. *Philosophical Review, 7,* 449–446.

Tolman, E. C. (1951). *Collected papers in psychology.* Berkeley: University of California Press.

Tranel, D., & Damasio, A. R. (1985). Knowledge without awareness: An autonomic index of facial recognition by prosopagnosiacs. *Science, 228,* 1453–1454.

Tzelgov, J. (1997). Automatic but conscious: That is how we act most of the time. In R. S. Wyer Jr. (Ed.), *Advances in social cognition* (Vol. 10, pp. 217–230). Mahwah, NJ: Lawrence Erlbaum.

Vul, E., Harris, C., Winkielman, P., & Pashler, H. (2009). Puzzlingly high correlations in fMRI studies of emotion, personality, and social cognition. *Perspectives in Psychological Science, 4,* 274–290.

Vul, E., & Kanwisher, N. (2010). Begging the question: The non-independence error in fMRI data analysis. In S. Hanson & M. Bunzl (Eds.), *Foundational issues for human brain mapping* (pp. 71–91). Cambridge, MA: MIT Press.

Waroquier, L., Marchiori, D., Klein, O., & Cleeremans, A. (2009). Methodological pitfalls of the unconscious thought paradigm. *Judgment and Decision Making, 4,* 601–610.

Watson, J. B. (1913/1994). Psychology as the behaviorist views it. *Psychological Review, 20,* 158–177. Reprinted in *Psychological Review, 101,* 248–253.

Watson, J. B. (1924). *Behaviorism.* New York: W. W. Norton.

Wegner, D. M. (2002). *The illusion of will.* Cambridge, MA: MIT Press.

Weiskrantz, L. (1980). Varieties of residual experience (Eighth Sir Frederic Bartlett Lecture). *Quarterly Journal of Experimental Psychology, 32,* 365–386.

Weiskrantz, L. (1997). *Consciousness lost and found.* Oxford: Oxford University Press.

Wotiz, J., & Rudofsky, S. (1984, August). Kekulé's dreams: Fact or fiction? *Chemistry in Britain,* 720–723.

Wundt, W. (1896/1907). *Grundriss der Psychologie* [*Outline of psychology*] (C. H. Judd, Trans.). Berlin, Germany: Stechert.

10 The Interaction of Implicit versus Explicit Processing and Problem Difficulty in a Scientific Discovery Task

Corinne Zimmerman and Jean E. Pretz

The goal of the present volume, as stated in the preface, is to explore the "roles played by both implicit and explicit processes in the empirical and theoretical aspects of science." The work we present is situated within the cognitive psychology of science, which focuses largely on the conceptual understanding and problem-solving processes that underlie scientific thinking (Feist, 2006). Within cognitive psychology, there are a number of dual-process theories that focus on the distinction between implicit and explicit processing (e.g., Evans & Over, 1996; Sloman, 1996). The goal of the current chapter is to replicate and extend research on implicit and explicit learning in process-control paradigms to investigations of problem solving in a simulated scientific-discovery paradigm. As previewed in the title, what we intend to show is that accuracy and rule discovery in the balance-scale task depend on both problem difficulty and whether an implicit or explicit approach to problem solving is used. Potential implications for future studies and for real-world scientific practice are then discussed.

DUAL PROCESS MODELS OF COGNITION

Common theoretical models of cognition assume two modes of processing: implicit and explicit. There are a variety of dual-process theories and frameworks that vary with respect to terminology, emphases, and assumptions (e.g., Evans & Over, 1996; Klaczynski, 2000; Sloman, 1996; Stanovich & West, 2000). Some theorists view implicit and explicit modes as distinct cognitive *systems*, whereas others propose that the two simply represent different types of *processes* (for a critical review, see Keren & Schul, 2009). Whether conceived of as different types of processing or completely different systems, the contrast between implicit and explicit has received multiple labels, including automatic versus controlled, System 1 versus System 2, and heuristic versus analytic. Different dual-process theories share similar characterizations of the two modes of processing (Evans, 2008). For example, the implicit mode operates largely without conscious awareness, requires little attention or effort, and processes information associatively and holistically. The explicit mode operates with awareness, attention, and effort. Explicit processing is rule-based, deliberate, and analytical. These two modes of processing have been documented in processes of learning (Reber, 1976), memory (Roediger, 1990), reasoning (Evans & Over, 1996; Sloman, 1996), decision making (Dijksterhuis & Nordgren, 2006; Hogarth, 2005), and problem solving (Pretz, 2008).

APPLYING THE IMPLICIT/EXPLICIT DISTINCTION TO SCIENTIFIC THINKING

In discussing the cognitive psychology of science, Feist (2006) noted that "the process of doing science is inherently a cognitive activity" (p. 84). As such, the thinking processes involved in scientific activity seem far more amenable to the explicit side of the dual-process dichotomy. For example, one of the major frameworks used by cognitive and developmental psychologists who study scientific thinking is Klahr and Dunbar's (1988) *Scientific Discovery as Dual Search* (SDDS) model. This framework is based on classic research in which problem solving is conceived of as involving the search of problem spaces (e.g., Newell & Simon, 1972; Simon & Lea, 1974). Simon and his colleagues discussed the search of "instance space" or "rule space," but when applied to scientific discovery contexts, one searches "hypothesis space" and "experiment space" (Klahr & Dunbar, 1988). Such active problem solving in the service of scientific discovery is clearly an active, conscious, deliberate—and *explicit* process.

Traditionally, cognitive psychology has highlighted the power of the explicit mode and the limitations of the implicit mode. Many models of problem solving assume an analytical style (e.g., Sternberg, 1986), and research on judgment and decision making highlights the errors made by the intuitive, implicit mode of processing (Tversky & Kahneman, 1974). Evans and Over (1996) have shown that explicit processing can lead to more logical conclusions that are less influenced by beliefs, and Pacini and Epstein (1999) have found that individuals with a stronger preference for rational thought are less susceptible to heuristic biases in the ratio-bias paradigm. Some argue that the explicit mode often serves as a necessary check system for the errors made by the intuitive mode (e.g., Epstein, Pacini, Denes-Raj, & Haier, 1996; Evans, 2008).

Although there is a great deal of evidence showing the benefits of analytical processing, there is also research to demonstrate some of the downsides to taking an explicit, analytical approach. For example, strategies such as hypothesis testing are not always optimal in a learning environment. Reber (1976) found that in an artificial grammar learning paradigm, an explicit strategy was more likely to lead to inappropriate problem representations and hypotheses than an implicit approach. Such misleading ideas interfere with the acquisition of additional information about the task structure. Similarly, research on "rational errors" has shown that inductive reasoning can lead to errors that are applied deliberately and systematically (e.g., Ben-Zeev, 1995; VanLehn, 1986). In such cases, participants induce incorrect rules that are based on incomplete knowledge or experience with the problem domain. Ben-Zeev (1995) provided evidence that rational errors can be predicted based on the particular characteristics of the exemplars to which participants were exposed. That is, rational errors are most likely to occur when the learning environment is not representative of the entire problem space.

Research in other domains provides converging evidence for the paradoxical effect of an explicit approach to reasoning and problem solving. Schooler, Ohlsson, and Brooks (1993) found that participants who thought aloud were less successful on an insight problem-solving task than those who did not verbalize. The researchers explained that thinking aloud engaged a more analytical process, which may have led to fixation on inappropriate approaches.

What about the role of implicit processing in higher cognition? Recently, researchers have begun to point out the power of implicit processing of information in learning, decision-making, and problem-solving situations (Berry & Broadbent, 1988; Dijksterhuis & Nordgren, 2006; Klein, 1998; Pozzali, 2007; Wilson & Schooler, 1991). Decades ago, Polanyi (1966) claimed that the tacit system is very powerful, that "we can know more than we can tell" (p. 4). Bargh and Ferguson (2000) have shown that the majority of our behavior is conducted automatically and implicitly, and is fundamentally adaptive. Although there are varying definitions of implicit learning (Frensch, 1998), growing evidence that implicit processing can be uniquely advantageous has been documented in a number of paradigms. For example, participants who articulated reasons for their judgments made objectively less accurate judgments than participants who provided holistic, intuitive responses (McMacklin & Slovic, 2000; Wilson & Schooler, 1991).

There is evidence in the literature that implicitly acquired knowledge is flexible and easily transferred. Berry and Broadbent (1988) found evidence for implicit, but not explicit, transfer of knowledge gained in a process-control task. Schunn and Dunbar (1996) showed convincing evidence for the effect of priming on positive transfer in solving a complex real-world genetics problem. Participants who had solved a problem using the concept of inhibition the day before were more likely to solve the genetics problem, whose solution also relied on the concept of inhibition. However, no participants were aware that their earlier experience had transferred to the genetics task. The evidence from these experiments suggests that implicit knowledge can be applied more flexibly than explicit knowledge in difficult problem-solving tasks.

Dijksterhuis and Nordgren (2006) have proposed an explanation for such findings in their *Unconscious Thought Theory*. They argue that unconscious (implicit) thought has an advantage over conscious (explicit) thought because implicit processing is not limited in capacity, is unaffected by expectancies and schemas, is better at weighting and integrating informational cues, and is more divergent and flexible. For example, the implicit mode should be better when working memory is limited, when schemas are inappropriate, and when a task requires the integration of highly complex information.

PROCESS-CONTROL TASKS: THE INTERACTION OF IMPLICIT AND EXPLICIT PROCESSING

Process-control tasks are commonly used by researchers interested in studying the distinction between implicit and explicit modes of processing (or the interaction of the two) (Sun, Terry, & Slusarz, 2005). Unlike other types of tasks used to study implicit learning (e.g., artificial grammar learning tasks, serial reaction time tasks), process-control tasks have features that are most similar to scientific thinking or simulated scientific discovery tasks.

In a process-control task, the participant interacts with a computer program, offering an input and observing an output. The participant's goal is to achieve a specific target output (e.g., a specific level of sugar produced in a sugar factory) by manipulating inputs (e.g., the number of factory workers). The algorithm used by the computer to calculate the output based on the input given is unknown to the participant. Similar input-output algorithms can be used, but the cover story can be changed to produce isomorphic versions of the task. An isomorphic version of the sugar production factory is the "computer person" task, in which graded adjectives describing the participant's behavior (e.g., "friendly," "polite," "indifferent") are input to maintain a specific target reaction from the "computer person" (e.g., "very friendly").

Such process-control tasks have features in common with the *self-directed experimentation* (SDE) tasks used by researchers who study scientific reasoning (see Zimmerman, 2000a, 2007, for reviews) in which participants try to determine the factors that influence an outcome. SDE tasks may involve either a physical apparatus or computer simulation, but in both cases they are multivariable causal systems in which the goal is to manipulate variables to determine their effects on an outcome. For example, participants may experiment with the size, shape, and weight of a boat to determine their impact on travel time (Schauble, 1996), or on various factors (e.g., bedrock type, S-wave rate) that may or may not predict earthquake risk (Keselman, 2003).

THE INTERACTION OF TASK DIFFICULTY AND PROBLEM-SOLVING APPROACH

An important factor that has been shown to influence the success of implicit and explicit approaches to problem solving is task or problem difficulty. Hogarth (2005) has proposed that as a task becomes more complex or difficult, the accuracy of explicit, analytical processing decreases. Complex tasks may be better approached by relying on the implicit, tacit mode simply because the amount of information overwhelms the capabilities of the explicit, analytical mode. Evidence for this argument comes from Berry and Broadbent's (1984, 1988) experimental work using process-control tasks. One process-control experiment specifically investigated the effect of task approach (implicit or explicit) and problem difficulty (salience of the rule) in the computer person task (Berry & Broadbent, 1988). Participants in the explicit condition were told that they should "try to crack the pattern" of the relationship between input and output. Other participants were instructed to take an implicit approach and simply tried to achieve and maintain a target output by changing the input. Both groups interacted with the task for many trials to achieve the target output.

The salience of the underlying algorithm was also manipulated. One rule was relatively simple; the computer's output value was a direct function of the participant's input value. The other rule was more difficult; the computer's output value was a direct function of the participant's input value *from the previous trial.* Participants who were trying to crack the pattern did better than those who were focused on maintaining the target, but only when the rule was salient. When the rule was more difficult, those with an implicit approach achieved the best performance, whereas those with an explicit approach performed worse.

One explanation for this finding regards the nature of the initial encoding. Berry and Broadbent (1988) explained that explicit participants may be learning selectively, not taking in all the critical information in the task. Implicit participants may learn unselectively, taking in everything, regardless of relevance. Selective learning may encourage explicit participants to use a rule that does not represent all the critical information. However, no post hoc evidence was found for a misrepresentation of rule information among explicit participants.

More recent work in dynamic process-control task paradigms has focused on the effect of *goal specificity* on learning. Researchers have found that participants learn more about the relationship between variables when given a nonspecific rather than a specific goal (e.g., Burns & Vollmeyer, 2002; Geddes & Stevenson, 1997). For example, when participants were given the *nonspecific goal* to "establish the pattern" that governed the output (i.e., similar to the explicit instructions used by Berry & Broadbent), they learned more about these relationships than when given a *specific goal* of simply maintaining a specific target output (i.e., similar to implicit instructions). Burns and Vollmeyer showed that the goal of establishing the pattern (without maintaining a specific output) led to more hypothesis testing, the likely mechanism behind greater learning.

Relating this work to the research contrasting implicit and explicit approaches, we might predict that those with an explicit approach are more similar to those pursuing a "nonspecific goal" in that they are seeking a general pattern among the stimuli rather than trying to maintain a specific target output. In that case, we would expect that those with an explicit approach would be more likely to successfully test hypotheses and induce the rule. However, research in the process-control paradigm found that the effect of the explicit versus implicit approach depended on the salience of the relationship to be discovered, but little goal-specificity research has focused on this potential variable. Notably, in one goal-specificity study, DeShon and Alexander (1996) found that a hypothesis-testing approach to a highly complex task did not improve performance relative to others who were instructed to "do your best," suggesting that explicit processing is not well suited to highly complex or difficult tasks.

This distinction between specific and nonspecific goal instructions has more in common with research by Schauble, Klopfer, and Raghavan (1991), who had children approach tasks as either "scientists" or "engineers." Engineers had a specific goal of changing variables to produce a desired effect (e.g., the longest spring length, the fastest boat). Scientists, in contrast, were given the goal of determining which variables influenced spring length or boat speed without having to produce a specific length or speed (i.e., they were trying to "establish the pattern"). When working as scientists, participants were more systematic, searched the experiment space more broadly, and took a theory-driven approach, but those with an engineering approach used more trial and error testing. As with the goal-specificity work, these instructions to approach the tasks in different ways are potentially consistent with taking different types of explicit or analytical approaches to a task.

Arguably, the work on goal specificity has used well-defined problems in which relationships among stimuli were relatively salient and could be discovered via analytical approaches (e.g., discovering the relationship among a few variables in a water-tank problem [Burns & Vollmeyer, 2002] or kinematics problem [Sweller, Mawer, & Ward, 1983]). In this case, participants pursuing nonspecific goals should have been able to successfully induce the relationship via hypothesis testing. However, if the relationship had been more complex, we would predict that an implicit approach (but not a "nonspecific goal") may have facilitated learning of the structure of the stimuli. In fact, this argument is supported by the findings of DeShon and Alexander (1996), though level of difficulty was not a manipulated variable in that research.

Overall, research using the process-control paradigm suggests that the appropriateness of an implicit or explicit approach to scientific discovery may be influenced by task difficulty. Empirical and theoretical research further suggests that knowledge acquired implicitly may be qualitatively different from that acquired explicitly. Specifically, Berry and Broadbent (1988) found evidence for implicit, but not explicit, transfer of knowledge gained in a process-control task. When participants interacted with two isomorphic process-control systems, the researchers found evidence for transfer when the tasks were perceptually similar.

One exception to this explanation is the study by Geddes and Stevenson (1997). They used the nonsalient version of the computer-person task of Berry and Broadbent (1988), but, contrary to the earlier findings, they found that performance was better among those pursuing a nonspecific goal (i.e., using an explicit approach to establish the algorithm governing the input-output relation). Their data from the initial stages of learning trials were in line with the findings of Berry and Broadbent, showing that participants pursuing a specific goal (i.e., an implicit approach, maintaining a target output) were more successful than those explicitly trying to establish the pattern. However, their results also showed that participants who were seeking the algorithm had acquired explicit rules, as evidenced by successful transfer to new problems in a posttask. It is unclear why the participants in the study by Geddes and Stevenson became successful with a more explicit approach, yet those tested by Berry and Broadbent did not. Perhaps the latter participants would have been successful if given more time (additional trials) to discover the rule.

At the same time, most research on hypothesis testing typically shows a benefit of the explicit, analytical approach with respect to transfer. For example, Burns and Vollmeyer (2002) found greater transfer among those who explicitly sought an abstract representation of the relationship among a set of variables as compared to others who simply sought to reach a specific target or outcome. In fact, Geddes and Stevenson (1997) also found evidence for greater transfer among those using an explicit approach in a process-control paradigm, contradicting the original findings of Berry and Broadbent (1988). Given the lack of consensus in the literature regarding the transferability/flexibility of implicitly versus explicitly acquired knowledge, we also examine this issue in the experiments reported in this chapter.

The current study was motivated in part by the contradictory results shown by various researchers using process-control tasks. We wanted to reexamine the relationship between difficulty and problem-solving strategy (implicit vs. explicit) but in a simulated scientific discovery context. Using a task more similar to those encountered in real-world scientific contexts, we hoped to shed light on the role of implicit versus explicit approaches to scientific problem solving, including one of the boundary conditions (i.e., difficulty) that may influence the success of the two approaches.

THE BALANCE-SCALE TASK PARADIGM

The goal of the present research was to replicate and extend the work of Berry and Broadbent (1988) in a new paradigm. We decided to test the viability of the balance-scale task (e.g., Inhelder & Piaget, 1958; Siegler, 1976) as a paradigm for studying dual-process models of cognition. This task is ideal because the problems can be varied systematically to cover a range of difficulty. The task has been used with children and adults, and a great deal is already known about baseline performance at various stages of development. Compared to process-control tasks based on hypothetical algorithms, however, the balance task has ecological validity and teaches participants real-world physics principles (e.g., torque, mass, center of gravity) over the course of many trials in which feedback is provided to test hypotheses to induce a rule governing the behavior of the apparatus. Most children and adults are aware that balance depends on weight, but most cannot articulate specifically how distance from the fulcrum influences the state of a two-arm balance without receiving feedback (Zimmerman, Raghavan, & Sartoris, 2003). Therefore, the task is sufficiently difficult such that rule discovery is not trivially easy, even for adults. If we are able to successfully replicate the research previously conducted with process-control paradigms, then we can use this new paradigm to test potential mechanisms underlying the counterintuitive interaction of strategy and difficulty in scientific problem solving.

In the balance-scale task, participants view a balance scale on a computer screen and must predict the state of the apparatus (tip left, tip right, or balance) based on the configuration of weights at various distances from the fulcrum. Unlike previous developmental research with this task (e.g., Siegler, 1976), participants can use immediate feedback over the course of a series of trials to become more accurate in making predictions. Some balance-scale problems can be solved perceptually (which side looks like it will tip?) (Klahr & Siegler, 1978), and for many (but not all) balance-scale problems, adding weight and distance units on each side of the scale and comparing the sums will result in a correct prediction (Wilkening & Anderson, 1982). However, the algorithm that can be used to make accurate predictions on every trial is the *product* of weight and distance (i.e., torque; see Figure 10.1). Most college students are not aware of the torque algorithm, whether or not they have learned it previously.

Problem difficulty was manipulated by using problems that could be solved via a visual rule (low difficulty), an additive rule (low or medium difficulty), or the multiplicative rule (low or medium or high difficulty). Because the problem set includes problems that vary in difficulty, we can manipulate the overall difficulty of the learning environment by using proportionally more easy or difficult problem types. All problems were *conflict problems* in which one side has the greater number of weights but the other side has the greater distance (Siegler, 1976). We manipulated implicit or explicit approach using simple instructions. Participants in the explicit condition were instructed to seek a rule that would help them make predictions, whereas participants in the implicit condition were unaware of the existence of a rule. This manipulation is based on that used by Berry and Broadbent (1988).

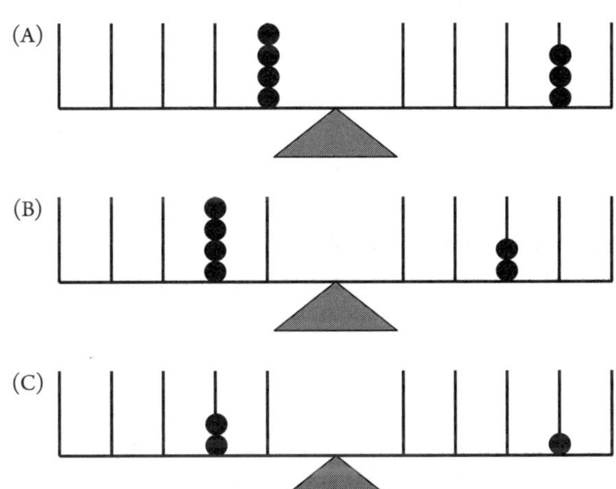

Figure 10.1 Panel A shows a low-difficulty problem (torque difference = 8). This problem can be solved using an additive rule (i.e., 4 weights + 1 peg < 3 weights + 4 pegs) or the multiplicative rule ($4 \times 1 < 3 \times 4$). Panel B depicts a medium-difficulty problem (torque difference = 2) for which either the additive ($4 + 2 > 2 + 3$) or multiplicative ($4 \times 2 > 2 \times 3$) rule can be used. Panel C depicts a high-difficulty problem (torque difference = 0). The additive rule results in an incorrect prediction ($2 + 2 < 1 + 4$); only the multiplicative rule ($2 \times 2 = 1 \times 4$) works.

RATIONALE AND HYPOTHESES

In our first experiment, we attempted to replicate the work of Berry and Broadbent (1988) using the balance-scale task. Our primary goal was to explore whether we could replicate the interaction of implicit versus explicit approaches to problem solving and task difficulty in a paradigm that simulates scientific discovery. By having participants interact with a real-world physics task, we can see the generality of the expected interaction as a first step to exploring the utility of dual-process theories for understanding scientific thinking.

We used the instructions from the original study and manipulated problem difficulty at three levels. With respect to low-difficulty problems that could be solved perceptually via a visual rule, we expected that performance would be similar among participants who approached the task implicitly and those who approached the task explicitly. We predicted that participants in the explicit condition would have higher accuracy than those in the implicit condition for medium-difficulty problems. For highly difficult problems, however, we expected implicit participants to have higher accuracy than those in the explicit condition. In subsequent experiments, we extend this paradigm in order to test hypotheses about potential explanations for the pattern of results.

EXPERIMENT 1

Method

Participants. Participants were 72 undergraduate students (18 male, 52 female) at a Midwestern university who were randomly assigned to either the implicit or the explicit strategy condition. Participants had a mean age of 19.8 ($SD = 1.3$), with a mean of 2.5 ($SD = 1.1$) years of university education. Participants were recruited for cash ($10 for one hour) or credit via the departmental participant pool.

Procedure. Participants were tested individually or in small groups. All participants were instructed to make predictions about the state of a two-arm balance based on configurations of weights at distances from the fulcrum. All problems were presented on a computer screen and participants clicked either the "left," "right," or "balance" button. The 300 balance-scale problems presented varied in difficulty. Our operational definition of difficulty is based both on the rule that can be used to solve the problem (additive or multiplicative) and on torque-difference values (i.e., the absolute value of the difference in torque on the left and right sides). The 100 easy problems could be solved visually. Low-difficulty problems had torque difference values from 5 to 15 (see Figure 10.1A). The 100 medium-difficulty problems could be solved via an additive rule (compare Weight + Distance for each arm) and had torque differences ranging from 0 to 4 (Figure 10.1B). The 100 high-difficulty problems also had torque difference values ranging from 0 to 4 (Figure 10.1C), but these problems could only be solved via a multiplicative (or torque) rule (i.e., compare Weight x Distance for each arm). After each trial, participants were given immediate feedback about whether their prediction had been accurate. The background screen behind the response buttons turned green for accurate predictions and red for inaccurate predictions (the correct response was not presented). Problems were randomly presented in five blocks of 60 trials each (each block was a random set of 20 low-, 20 medium-, and 20 high-difficulty problems).

Participants were randomly assigned to either the explicit or implicit condition. Instructions in the explicit condition were to actively seek a rule that would help make accurate predictions. Instructions in the implicit condition were to make predictions (no rule was mentioned). To ensure that participants in the explicit condition continued to look for a rule, they were asked to report whether they had discovered the rule after each block ("yes," "no," or "not sure"). Both accuracy and response times were recorded.

After the balance task, participants completed a posttask questionnaire designed to probe their knowledge of the rule. Participants were shown a diagram of a scale (high difficulty) and were told that it balances and were asked to explain why it balances. In addition, they were asked to indicate if they were looking for a rule during the task, what the rule was, whether they had known the rule prior to the experiment, and whether they had learned the rule in the past but did not recall it during the experiment. Finally, participants provided demographic information, including major and number of physics and math courses taken in college.

Results

Seven participants in the explicit condition discovered the multiplicative rule during the experiment. Participants were considered rule discoverers if (a) their performance consistently reached 95% accuracy across all problems, (b) they described using the multiplicative rule in the posttask questionnaire, and (c) they used the multiplicative rule to explain the difficult paper-and-pencil problem. All rule discoverers found the correct rule before the end of the third block, and their accuracy for Blocks 4 and 5 was 98.2%.

A preliminary analysis comparing the performance of explicit participants who did and did not discover the rules on prediscovery trials showed no significant differences in accuracy (all p's > .70). Given that the accuracy rates of explicit discoverers and nondiscoverers were effectively identical prior to rule discovery, we concluded that the most appropriate comparison would be between explicit nondiscoverers and implicit nondiscoverers. The main analyses were conducted on nondiscoverers alone because there were fewer discoverers and their postdiscovery performance was at ceiling with no variance, making it difficult to interpret the results of an analysis of variance.

A 2 (Strategy: implicit vs. explicit) × 5 (Block) × 3 (Difficulty: low, medium, high) ANOVA revealed a main effect of Difficulty, Wilk's $\Lambda = .059$, $F(2, 62) = 498.34$, $p < .001$. As difficulty increased, accuracy decreased. As predicted, the Strategy x Difficulty interaction was significant, Wilk's $\Lambda = .882$, $F(2, 62) = 4.14$, $p = .02$ (see Figure 10.2). For low-difficulty problems, there was no difference in accuracy between the implicit and explicit participants, $F(1, 63) = 1.45$, $p = .24$, $d = .31$. For medium-difficulty problems, accuracy was greater for the explicit participants, $F(1, 63) = 4.40$, $p = .04$, $d = .54$. The pattern reversed for the high-difficulty

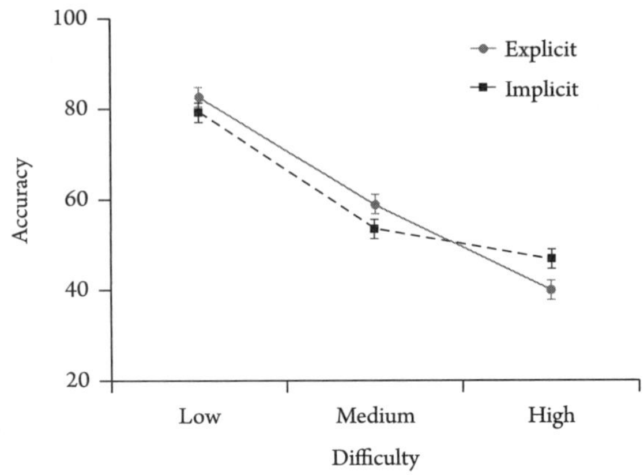

Figure 10.2 Accuracy as a function of problem difficulty and strategy type for Experiment 1.

problems; accuracy was greater for the implicit participants, $F(1, 63) = 4.96, p = .03, d = .52$. All other effects and interactions were nonsignificant (all F's ≈ 1).

Response times were measured as a manipulation check. We expected that if the participants were following the strategy instructions, explicit participants would take longer per problem, regardless of problem difficulty. On average, the implicit participants took 2079.1 ms ($SD = 535.4$), whereas the explicit participants took 2621.9 ms per problem ($SD = 1610.4$). Rule discoverers spent 4322.6 ms ($SD = 526.8$) per problem. Notably, the performance of those who were looking for the rule but did not find it was the most variable. A key factor distinguishing those who discovered the rule was the additional amount of time spent per problem across all blocks and problem types. That is, they spent additional time on the easy problems—those who did not discover the rule answered these quickly, likely with a visual heuristic, whereas the discoverers may have been taking additional time to analyze if a current hypothesized rule under consideration could be used to solve them.

Discussion

Using a real-world physics task that involves the induction of a rule, we replicated and extended previous research by Berry and Broadbent (1988). Although participants who discovered the rule were most accurate in making predictions, the pattern of performance among participants who did not discover the rule depended on both the difficulty of the problem and the strategy used to approach the task. Specifically, we found that accuracy was generally higher when participants sought a rule to govern their predictions, with the exception of problems that were highly difficult. On the most difficult problems, participants who were passively engaged with the task, making predictions without awareness of the existence of an underlying rule, performed better than those explicitly seeking the rule. The response times corroborate these findings. Rule-seeking instructions led to increased response time overall. In the implicit condition, however, response times revealed how little effort participants were expending even as their predictions were significantly more accurate than their counterparts in the explicit condition. Zimmerman (2000b) found that when participants were instructed only to respond as quickly and accurately as possible, there was a clear association between fast responding and increased accuracy for most medium- and high-difficulty problems. This pattern of results does not follow those found by Geddes and Stevenson (1997) despite the fact that participants had hundred of trials in which to induce the rule.

Notably, seven participants discovered the multiplicative rule during the experiment, performing at ceiling on postdiscovery blocks. Previous research using process-control and implicit learning paradigms rarely reports accuracy levels of 100%. Process-control task algorithms typically include a source of random error that prevents perfect performance. In our experiment, we did not create a source of random error because the task is based on real physics principles, so it was possible to discover the actual rule and apply it flawlessly to subsequent trials.

There are two explanations for our findings. First, performance in the explicit condition may have been harmed by a fixation on an incorrect rule, such as an additive or weighted additive rule (Wilkening & Anderson, 1982) or other strategy (see Zimmerman, 2000b for a review) that could be used to solve the less difficult problems. If these participants discovered an additive rule, for example, they would be accurate on low- and medium-difficulty problems, though accuracy on the most difficult problems (which can be solved only via the multiplicative rule) would decrease. This is what was observed. This pattern of performance may reflect a *rational error* on the part of explicit participants (e.g., Ben-Zeev, 1995; VanLehn, 1986). Explicit participants took longer to make predictions for high-difficulty problems. This is most likely because participants were receiving negative feedback on responses based on an incorrect rule, prompting them to spend more time hypothesis testing on difficult problems. This contrasts with the rule discoverers, who spent more time on all problems, regardless of difficulty.

Another way to explain these results is that there is a benefit of processing complex information in a manner that is implicit, passive, and "unselective." Recent empirical and theoretical work on dual-process theories suggests that complex information is better processed implicitly. Dijksterhuis and Nordgren (2006) have argued that when it comes to integrating a great deal of information (e.g., feedback on many problems varying in complexity), passive, unconscious thought is better than active, conscious thought. Hogarth (2005) also predicted that, when feedback is clear and immediate, the tacit system will be more successful than the deliberate system at dealing with complex information.

One concern that exists in any experiment contrasting implicit and explicit processing, however, is the degree to which implicit processing is truly implicit. This is a notoriously difficult issue (Berry, 1997; Holyoak & Spellman, 1993), and our approach cannot guarantee that participants in the implicit condition were processing the problems implicitly. Although our manipulation in Experiment 1 was based on that used by Berry and Broadbent (1988), we decided to strengthen the instruction in Experiment 2. Specifically, we attempted to dissuade participants from "thinking too hard" about their predictions and asked them to trust that their performance would improve over time. By strengthening this manipulation, we hoped to rule out some of the concerns about any unknown strategies that may underlie performance in the implicit condition.

In order to test our two possible explanations for the findings of Experiment 1, two additional experiments were conducted. In Experiment 2, we attempted to induce fixation on the incorrect, additive rule by presenting problems of primarily low and medium difficulty. If fixation is responsible for diminished performance among high-difficulty problems, this pattern should be exaggerated in Experiment 2. This experiment will allow us to examine the degree to which processing mode (implicit or explicit) is affected by fixation. We expect that explicit participants will be more susceptible to fixation than implicit participants due to the overt nature of the processing mode. If implicit participants learn unselectively, they should pick up on all stimuli, including those that are infrequently represented in the stimulus set. In Experiment 2, we also tested the hypothesis that implicit processing leads to more flexible knowledge by including a posttest and transfer items. If our hypothesis is correct, participants in the implicit condition will perform better on these measures than those in the explicit condition.

EXPERIMENT 2

Method

Participants. Participants were 77 undergraduate students (26 male, 51 female) at a Midwestern university. Participants had a mean age of 19.02 ($SD = 1.01$), with a mean of 1.77 ($SD = 1.13$) years of university education. Participants were recruited for credit via the departmental participant pool.

Materials and Procedure. We made three key changes to the procedure. First, we changed the instructions for participants in the implicit condition to enhance our manipulation. Rather than being asked to simply make predictions, implicit participants were told, “We have found that people tend to do better if they don’t think too much during the task. Don’t dwell too much on each problem—just trust that your performance will improve over time as you get a sense of the task.” Explicit instructions remained unchanged. Second, we changed the distribution of problems to reflect primarily low- and medium-difficulty problems. Four blocks of 60 trials were presented. Each block contained a randomly selected order of 25 low-difficulty, 25 medium-difficulty, and 10 high-difficulty problems.

Finally, we added a posttest and near transfer items for all participants. The posttest consisted of 15 difficult problems presented without feedback. The near transfer items were added to the beginning of the posttask questionnaire. These five items presented balance scales with weights on two different pegs on the left side of the fulcrum. Participants were instructed to draw nonsymmetrical configurations of weights on the right side of the fulcrum that would make the scale balance. To complete these items, participants would need to transfer knowledge about torque to these similar but novel problems. Participants also rated their confidence in their response on each transfer item on a 4-point scale from “wild guess,” “unsure,” “fairly confident,” to “certain.” The posttask questionnaire probed participants’ knowledge of the rule.

Results and Discussion

Under fixating conditions, only six participants discovered the torque rule (two implicit, four explicit; one math major) and all did so by the end of Block 2. Accuracy on Blocks 3 and 4 was 97.3% among discoverers. As with Experiment 1, rule discoverers spent more time per problem. The reaction times of the two discoverers in the implicit condition were consistent with explicit, rule-seeking behavior, indicating that they disregarded our instructions. Averaged across all problems, discoverers spent 3893.9 ms (SD = 891.5), compared to 1729.9 ms (SD = 505.5) for implicit and 3244.3 ms (SD = 1480.4) for explicit nondiscoverers. Again, a preliminary analysis on the presdiscovery trials showed no differences in accuracy between those in the explicit condition and those who discovered the rule (all $ps > .35$). Therefore, our main analysis to test for the interaction of difficulty and strategy includes only those participants who did not discover the rule.

A 2 (Strategy: implicit vs. explicit) × 4 (Block) × 3 (Difficulty: low, medium, high) ANOVA revealed a main effect of Difficulty, Wilk's $\Lambda = .08$, $F(2, 68) = 390.6$, $p < .001$, which was qualified by a significant interaction between Strategy and Difficulty, Wilk's $\Lambda = .76$, $F(2, 68) = 390.6$, $p < .001$. As can be seen in Figure 10.3, the accuracy of explicit participants was greater than that of implicit participants for both low-difficulty problems, $F(1, 69) = 3.91$, $p = .052$, $d = .47$, and medium-difficulty problems, $F(1, 69) = 14.59$, $p < .001$, $d = .91$. The pattern reverses for high-difficulty problems, with greater accuracy for implicit participants, $F(1, 69) = 16.04$, $p < .001$, $d = .95$. The accuracy of explicit participants for high-difficulty problems (33.6%) was not significantly different from chance-level responding, $t(34) = .10$, $p = .92$, but the accuracy of implicit participants (46.9%) was, $t(35) = 6.65$, $p < .001$. All other effects and interactions were non-significant (all F's ≈ 1).

With respect to posttest performance, discoverers were at ceiling performance. The accuracy of implicit participants ($M = 7.0$) was significantly better than that of explicit participants ($M = 5.0$), $t(69) = 2.58$, $p = .012$. Moreover, the accuracy of implicit participants was significantly different from chance, $t(35) = 4.19$, $p < .001$, which was not the case for explicit participants, $t(34) = .05$, $p = .96$. There were no differences in the number of correct multipeg transfer problems or average confidence ratings (p's > .05). However, a similar number of participants (approximately 27%) in both conditions got two or more multipeg transfer problems correct. The discoverers solved all five multipeg problems correctly.

In this experiment, we successfully induced fixation among our explicit participants. That is, the rate of rational errors increased. Their performance on low- and medium-difficulty problems was much better than that of implicit participants, but their performance on difficult problems was even worse than the accuracy rate found in Experiment 1. Generally, the effect of strategy condition at each level of difficulty was exaggerated compared with Experiment 1. Small effects became medium effects, and medium effects became large effects. Notably, performance of implicit participants in the second experiment was essentially the same as in the first experiment. The change in the distribution of items did not have much impact on their performance.

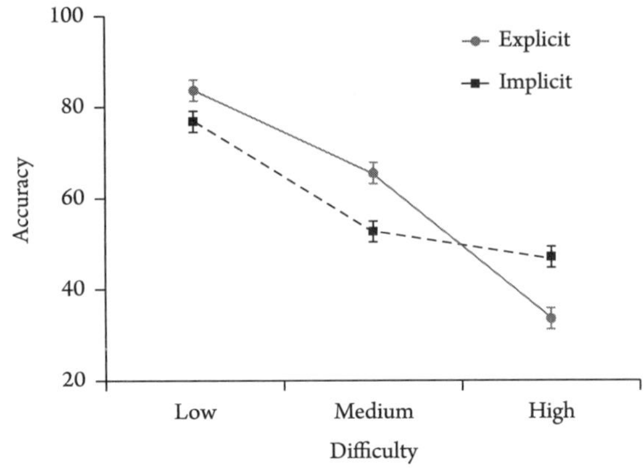

Figure 10.3 Accuracy as a function of problem difficulty and strategy type for Experiment 2. Experiment 2 included proportionally more low- and medium-difficulty problems.

Because of our success in inducing fixation in explicit participants, it is possible that their poor transfer performance is not due to inflexibility of knowledge but rather fixation on inappropriate knowledge. To facilitate acquisition of the correct rule and maximize the chance of successful transfer, a third experiment was conducted. We attempted to *reduce* fixation by presenting more diagnostic exemplars (mostly high- and medium-difficulty problems). We proposed two sets of competing hypotheses. If our findings remain consistent with those of Experiments 1 and 2, we expect that explicit participants might still show a disadvantage on highly difficult problems relative to implicit participants. In contrast, if the diagnostic exemplars provide a better environment for rule induction, then we would expect participants in the explicit condition to be less likely to become fixated on the incorrect rule and therefore be more successful than implicit participants on highly difficult problems and on transfer items. We also predict that more explicit participants will discover the rule under these conditions.

Given that this new learning environment would be more diagnostic and that we expected more discoverers in this experiment, we took the opportunity to explore the differences between discoverers and nondiscoverers. To this end, we included a measure of cognitive style preference in this experiment. Specifically, we suspected that discoverers might be highly analytical, given their long response time across all problems in the first two experiments.

EXPERIMENT 3

Method

Participants. Participants were 89 undergraduate students (29 male, 56 female, 4 who did not report gender) at a Midwestern university. Participants had a mean age of 19.34 ($SD = 2.04$), with a mean of 2.05 ($SD = 1.26$) years of university education. Participants were recruited for credit via the departmental participant pool.

Materials and Procedure. This experiment was identical to that used in Experiment 2, except for the distribution of problems. Four blocks of 60 trials were presented. Each block contained a randomly selected order of 25 high-difficulty, 25 medium-difficulty, and 10 low-difficulty problems. High- and medium-difficulty problems are very challenging, and, based on pilot testing, we found it necessary to include low-difficulty problems to maintain the motivation of the participants.

After the computer task, participants completed the *Rational Experiential Inventory* (REI) (Pacini & Epstein, 1999) to estimate individual differences in preferences to rely on experiential/intuitive and rational/analytical thought. The measure consists of two subscales: rational (alpha = .87) and experiential (alpha = .89). Each subscale consists of 20 items to which participants responded on a 5-point Likert scale. The implicit and explicit instructions, the posttest and near transfer items, and posttask questionnaire were the same as used in Experiment 2.

Results and Discussion

Given more diagnostic exemplars, 18 participants discovered the rule (all explicit). These participants spent approximately 4620.2 ms (SD = 1456.7) on each prediction. Once again the time spent per problem by those who discovered the rule was greater when compared to implicit (M = 2035.7 ms, SD = 493.7) and explicit participants (M = 2511.7, SD = 892.7). The response times for explicit participants were not consistent with the previous two experiments in that performance was less variable, which may suggest that these participants became frustrated and "gave up" looking for the rule.

In this case, with a problem set that included mostly difficult problems, we expected that there might be differences between those who were looking for the rule and found it, and those who were unsuccessful. Consistent with the idea that the nondiscoverers may have "given up," an analysis of the accuracy on prediscovery trials did show differences as a function of discovery status (discoverer vs. nondiscoverer), $F(1,52) = 15.03, p < .001$. As would be expected, performance varied as a function of difficulty, Wilk's $\Lambda = .25, F(2, 51) = 76.3, p < .001$, without a significant interaction.

Analyses of individual differences in cognitive style allowed us to compare discoverers and nondiscoverers with respect to their preferred mode of processing. Results for the REI showed that discoverers were significantly more rational (M = 79.56, SD = 9.77) than either explicit (M = 67.39, SD = 10.55) or implicit (M = 69.06, SD = 9.31) nondiscoverers, $F(2, 83) = 9.53, p < .0005$. The three groups showed no differences in preference for experiential/intuitive processing, $F < 1.00$.

To make comparisons with the previous experiments, we analyzed nondiscoverers separately. A 2 (Strategy) × 4 (Block) × 3 (Difficulty) ANOVA revealed main effects of Difficulty, Wilk's $\Lambda = .14, F(2, 67) = 210.4, p < .001$, and Strategy $F(1, 68) = 5.01, p = .028$. These effects were qualified by a significant interaction between Strategy and Difficulty, Wilk's $\Lambda = .87, F(2, 67) = 5.13, p = .008$. An analysis of the simple effects showed that the accuracy of implicit participants was significantly greater for low-difficulty problems, $F(1, 68) = 4.25, p = .043, d = .49$ (see Figure 10.4). No differences were found for medium-difficulty problems, and as predicted, accuracy was greater for implicit participants for high-difficulty

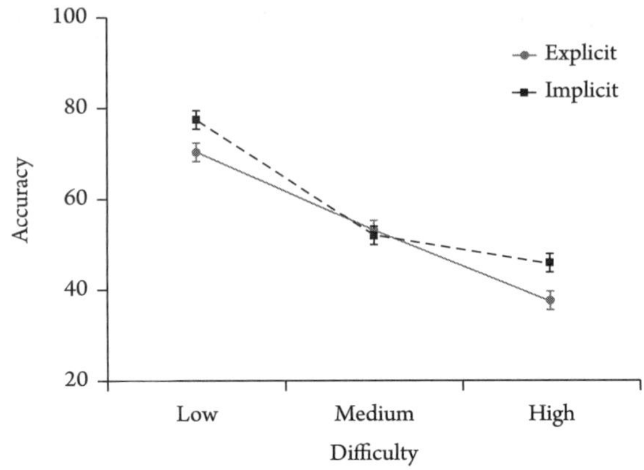

Figure 10.4 Accuracy as a function of problem difficulty and strategy type for Experiment 3. Experiment 3 used proportionally more medium- and high-difficulty problems.

problems, $F(1, 68)$, 9.24, $p = .003$, $d = .73$. The accuracy of explicit participants in this case (37.9%) was significantly greater than chance, $t(35) = 2.47$, $p = .018$, as was that of implicit participants (45.6%), $t(34) = 8.33$, $p < .001$.

On the posttest problems, the accuracy of implicit participants ($M = 7.5$) was significantly better than that of explicit participants ($M = 4.9$), $t(59) = 3.61$, $p = .001$, and was significantly greater than chance-level responding, $t(28) = 5.04$, $p < .001$. Again, there were no differences in mean accuracy or confidence for the multipeg transfer problems ($ps > .05$), but a similar number of explicit and implicit participants (38%) got two or more problems correct. Discoverers were at ceiling for all transfer measures.

These results show that under conditions of more diagnostic exemplars, participants who sought the rule were more likely to discover it. These participants were also unusually analytical in their general cognitive style. However, among those who did not discover the rule, the pattern of results parallels the findings in the first two experiments. For explicit participants who did not discover the rule, they performed relatively worse than in the first experiment on all types of problems. It appears that we were successful in encouraging explicit participants to test hypotheses because they did not appear to be fixated on an incorrect rule. Their performance on low- and medium-difficulty problems was no better than that of the implicit participants (and actually *worse*, in the case of low difficulty). However, they did continue to show significantly less accurate performance on difficult problems. It is possible that the participants who did not find the rule simply found the set of problems to be too challenging. This hypothesis is corroborated by their response time. Explicit participants did not spend nearly as much time on each trial (two to three seconds) as explicit participants did in Experiment 2 (three to four seconds). Implicit participants again performed remarkably similarly to their performance in Experiments 1 and 2. Their success on the posttest and transfer task again supports the hypothesis that the implicit processing led to unselective learning and thus greater flexibility in applying the knowledge to novel problems.

GENERAL DISCUSSION

In this series of experiments, we successfully replicated and extended the pattern of findings from the process-control paradigm (Berry & Broadbent, 1988). Using the balance-scale task, we were able to easily manipulate the distribution of problems of varying levels of difficulty and to manipulate the problem-solving approach taken by participants by asking them either to simply do their best while making predictions, or to deliberately try to crack the pattern. Our task involves a real-world physics problem, and in the case of our explicit strategy instructions, it simulates the inductive inference and hypothesis testing required in scientific discovery.

In our experiments, we found evidence for three primary conclusions: (a) The explicit approach was more likely to lead to successful rule induction than the implicit approach. (b) Among those who did not discover the rule, the explicit approach led to worse performance on difficult problems than did the implicit approach. (c) The implicit approach led to consistently better performance on difficult problems than did the explicit approach.

Rule Discovery Depends on Both Strategy and Problem Difficulty

In our first two experiments, only a few participants went on to successfully induce the correct rule ($n = 6$ for Experiment 1 and $n = 7$ for Experiment 2). What distinguished their performance from nondiscoverers was the additional amount of time they spent per problem, regardless of difficulty level. That is, rather than quickly respond to easy problems, they spent additional time—presumably assessing if a current rule under consideration could be applied to easy problems as well. With the exception of two participants in the implicit condition who clearly disregarded our instructions (as evidenced by their long response times), all of the discoverers were in the explicit condition and were following our instructions to deliberately try to find an algorithm that would allow flawless predictions. Therefore, successfully finding a rule did require one to be looking for it.

Under conditions designed to promote induction of the correct rule, more explicit participants discovered the rule. In our third experiment, we presented participants with a particularly challenging learning environment—but one that was diagnostic with respect to having the highest proportion of difficult problems (which can only be solved with the correct rule). Here, 18 participants discovered the correct rule (i.e., 33% of those receiving the explicit instructions). This experiment provides evidence that the most successful participants (rule discoverers) were highly analytical in cognitive style. These results confirm the value of the explicit approach to problem solving. We speculate that cognitive ability or motivational factors may have also played a role in the success of these participants.

When a Bad Strategy Is Worse Than No Strategy at All

Across all three experiments, the implicit approach was found to be more successful than the explicit approach on the most difficult problems among participants who did not discover the rule. One hypothesis we examined was that this pattern of results might have been due to fixation among participants taking the explicit approach. Evidence from Experiments 2 and 3 supports this hypothesis. Under the fixating conditions we presented in Experiment 2 (i.e., proportionally more easy and medium problems), fewer participants discovered the rule, and performance among explicit participants was higher on items for which the incorrect fixated solution yielded the correct response. This pattern supports the idea that they committed a rational error, likely by discovering and using one the various additive rules that have been identified (and will yield a correct prediction on the easier problems). Thus, in the case of a nondiagnostic learning environment, or one in which the entire problem space is poorly represented, a bad strategy can be worse than no strategy at all.

When It Helps to Think Less

Under conditions designed to promote induction of the correct rule, more explicit participants discovered the rule, yet among *nondiscoverers,* an implicit approach still led to better performance than the explicit approach in two respects. First, in all three experiments, the implicit approach resulted in better accuracy on the most difficult problems as compared to an explicit approach. The performance of those taking the explicit approach was typically no different than chance-level responding. Second, their superior performance on a posttest suggested that implicit participants had acquired relatively sophisticated and accurate knowledge of the task. It is remarkable that the participants taking an implicit approach not only outperformed explicit participants for difficult problems, but also had better overall accuracy in our third experiment, which presented a particularly challenging set of problems, and they did this in much less time as well. For our participants, the implicit approach was optimal when the correct rule was not discovered.

Based on findings from the process-control literature (e.g., Berry & Broadbent, 1988), we predicted that implicit processing might lead to more flexible knowledge of the torque rule. Our evidence was inconclusive. Although implicit participants outscored explicit nondiscoverers on a posttest of highly difficult problems, their performance on transfer problems (multipeg problems) was no better than those in the explicit condition. The format of the transfer problems may have prohibited participants from expressing implicit knowledge of the task. Previous research with process-control paradigms has shown a dissociation between implicit performance and later knowledge of the task (Berry & Broadbent, 1984, 1988; Geddes & Stevenson, 1997). In addition, the time elapsed between the initial task and the transfer task may have precluded some participants from reporting what they knew during the balance-scale task itself. Additional work should be done to test the limits of the transferability of knowledge of participants who take a less analytic approach, perhaps using perceptually dissimilar transfer tasks or those involving a more complex calculation of torque.

Implications for Real-World Scientific Problem Solving

The goal of the present volume is to examine the ways in which implicit and explicit processes contribute to the creation of scientific knowledge. As pointed out by Schunn and Anderson (1999), there are issues with studying the cognitive processes involved in scientific reasoning processes for a number of reasons, including the idea that many studies are done in laboratories with undergraduates because of issues of sample size and experimental control, and because practicing scientists are not often available to serve as participants in psychological studies (with some exceptions, e.g., Schunn & Dunbar, 1996). When it comes to studying implicit processes, in particular, real scientists working in real laboratories may afford, at best, post hoc recollections of previous problem-solving efforts. By using a real-world physics problem, we hope that we have found an experimental task that is more ecologically valid than some of the laboratory tasks that have been used previously, but which also allows the experimental control and reasonable sample size needed to allow us to generate reasonable hypotheses about how implicit processes might work in real-world scientific practice. If implicit processing and tacit knowledge are important in problem solving in general, we must do more research to elucidate these processes in scientific problem solving (Pozzali, 2008).

With respect to real scientific practice, our results suggest that an explicit, analytical approach may work best when working in familiar paradigms, or when proposing "next step" experiments that provide an incremental gain in a research program. The implicit approach, however, will be beneficial in instances when a researcher is at a crossroads, embarking on research in a new literature, faced with conflicting evidence, or simply when past approaches have failed. In such cases, the analytical approach may be unproductive or even fixating. When analysis fails, it may be better to step back, take a holistic approach, and trust a hunch about which direction to go. Scientists should acknowledge that such hunches are based on tacit knowledge developed via years of experience, even if that expertise remains implicit or nonverbalizable. In fact, breakthroughs or groundbreaking discoveries may be more likely to happen when following an implicit approach, precisely because this openness to new possibilities may result in discoveries that we cannot even anticipate.

Classic research by Bowers, Regehr, Balthazard, and Parker (1990) has shown that the mechanism for the expression of intuitive knowledge may be spreading activation. Scientists should trust that their tacit knowledge directs problem solving; even when they cannot articulate the source of their ideas, all of their knowledge is being brought to bear on a problem. As Polanyi (1966) argued, "If all knowledge is explicit, i.e., capable of being clearly stated, then we cannot know a problem or look for its solution. And... if problems nevertheless exist, and discoveries can be made by solving them, we can know things, and important things, that we cannot tell" (p. 22).

Limitations and Future Directions

One critique of our experimental design regards the strength of our instructional manipulation. Implicit participants may have actually adopted some type of strategy rather than passively making predictions. Participants may naturally engage with a task like this analytically, particularly because its surface features are scientific. However, response time data confirmed that implicit participants made predictions very quickly (1.5 to 2 seconds on average), which was consistently faster than the nondiscoverers in the explicit condition (2.5 to 3.5 seconds on average) or the discoverers (3.9 to 4.6 seconds on average). Even if implicit participants were testing hypotheses, their superior performance shows that their approach was substantially different than those explicitly told to test hypotheses. Implicit participants' predictions were significantly more accurate and at times twice as fast as explicit participants. Something about the implicit instructions worked for our participants on high-difficulty problems.

Notably, the balance-scale task used here did not permit participants to actively manipulate the balance. They were presented with instances, asked to make predictions, and received feedback on the accuracy of their predictions, in contrast to other paradigms (e.g., SDE) that require participants to propose an input or to manipulate a variable to observe its effect on the output.

Future studies should further explore the difference between participants in the explicit condition who eventually discover the rule and those who never do. Most previous work with process-control or artificial grammar paradigms has not included data from discoverers. The torque rule that underlies success in the balance-scale paradigm is difficult to discover, yet our experiments show that some undergraduate students are able to crack the pattern. In future studies, it would be ideal to create groups of equal sample sizes and formally compare performance of those who do and those who do not discover the rule. Preliminary evidence from Experiment 3 shows that cognitive style may explain the difference between those who do and do not discover the rule. Another potential mediating variable is cognitive ability. Follow-up studies could test these hypotheses, possibly by investigating discoverers' patterns of accuracy and reaction time in making predictions.

We found that an explicit approach is sometimes successful, yet at other times frustrating and/or leads to fixation on rules that work for some of the problems (but not all of them). These results imply that for individuals taking a reliably analytical approach, the explicit approach is more likely to be successful. Yet there will be problems for which the explicit approach leads to shortsighted solutions or, worse, an impasse in problem solving. For such cases, these results imply that the implicit approach should be employed. If the explicit approach is initially unsuccessful, problem solvers would do best to step away from the problem and attempt to get a fresh perspective, perhaps no longer directing their search for a solution and allowing less hypothesis-driven processing to occur. This hypothesis should be tested in follow-up studies by presenting participants with highly difficult problems and allowing for an incubation period in which a more implicit approach may show a benefit for participants who have reached an impasse. Another strategy would be to examine the effect of implicit versus explicit processing on later hypothesis generation. If tacit knowledge provides valuable hunches about fruitful hypotheses, participants who interact with the task under implicit processing instructions should have an advantage in generating hypotheses after switching to the explicit processing condition.

Finally, additional work should be done to test the limits of the transferability of knowledge of participants who take an implicit approach. The current results suggest that their knowledge of the balance-scale problems was more easily applied in the posttest of highly difficult problems and other, relatively similar, balance problems. However, their performance on the transfer task was no better than that of the explicit participants. It would be useful to test whether this knowledge would transfer to less perceptually similar problems or problems that required a yet more complex calculation of torque. Whereas implicit participants in these experiments performed very similarly despite changes in the distribution of items, it may be possible to observe the effect of the distribution of problems in performance on far transfer items.

Conclusions

In order to test the robustness of the interaction between processing mode and problem difficulty found in process-control tasks, we attempted to replicate it with a scientific discovery paradigm. The results of the present experiments are very provocative, as the pattern of results follows from current theory about dual processes. We found that the explicit approach can be successful for some individuals. At the same time, we also found that "thinking too hard" may be counterproductive, and that in some circumstances, thinking less may be beneficial for solving complex problems. We provide evidence that rule-seeking behavior can have a restrictive, fixating influence on hypothesis testing. Our evidence also shows that when problems are so difficult that a viable hypothesis does not come to mind, an implicit approach is more appropriate, allowing passively acquired knowledge to be expressed.

REFERENCES

Bargh, J. A., & Ferguson, M. J. (2000). Beyond behaviorism: On the automaticity of higher mental processes. *Psychological Bulletin, 126,* 925–945.

Ben-Zeev, T. (1995). The nature and origin of rational errors in arithmetic thinking: Induction from examples and prior knowledge. *Cognitive Science, 19,* 341–376.

Berry, D. C. (1997). *How implicit is implicit learning?* New York: Oxford University Press.

Berry, D. C., & Broadbent, D. E. (1984). On the relationship between task performance and associated verbalizable knowledge. *Quarterly Journal of Experimental Psychology, 36A,* 209–231.

Berry, D. C., & Broadbent, D. E. (1988). Interactive tasks and the implicit-explicit distinction. *British Journal of Psychology, 79,* 251–272.

Bowers, K. S., Regehr, G., Balthazard, C., & Parker, K. (1990). Intuition in the context of discovery. *Cognitive Psychology, 22,* 72–110.

Burns, B. D., & Vollmeyer, R. (2002). Goal specificity effects on hypothesis testing in problem solving. *Quarterly Journal of Experimental Psychology, 55A,* 241–261.

DeShon, R. P., & Alexander, R. A. (1996). Goal setting effects on implicit and explicit learning of complex tasks. *Organizational Behavior and Human Decision Processes, 65*, 18–36.

Dijksterhuis, A., & Nordgren, L. F. (2006). A theory of unconscious thought. *Perspectives on Psychological Science, 1*(2), 95–109.

Epstein, S., Pacini, R., Denes-Raj, V., & Heier, H. (1996). Individual differences in intuitive-experiential and analytical-rational thinking styles. *Journal of Personality and Social Psychology, 71*, 390–405.

Evans, J. St. B. T. (2008). Dual-processing accounts of reasoning, judgment, and social cognition. *Annual Review of Psychology, 59*, 255–278.

Evans, J. St. B. T., & Over, D. E. (1996). *Rationality and reasoning*. Hove, UK: Psychology Press.

Feist, G. J. (2006). *The psychology of science and the origins of the scientific mind*. New Haven, CT: Yale University Press.

Frensch, P. A. (1998). One concept, multiple meanings: On how to define the concept of implicit learning. In M. A. Stadler & P. A. Frensch (Eds.), *Handbook of implicit learning* (pp. 47–104). Thousand Oaks, CA: Sage.

Geddes, B. W., & Stevenson, R. J. (1997). Explicit learning of a dynamic system with a non-salient pattern. *Quarterly Journal of Experimental Psychology, 50A*(4), 742–765.

Hogarth, R. M. (2005). Deciding analytically or trusting your intuition? The advantages and disadvantages of analytic and intuitive thought. In T. Betsch & S. Haberstroh (Eds.), *The routines of decision making* (pp. 67–82). Mahwah, NJ: Lawrence Erlbaum.

Holyoak, K. J., & Spellman, B. A. (1993). Thinking. *Annual Review of Psychology, 44*, 265–315.

Inhelder, B., & Piaget, J. (1958). *The growth of logical thinking from childhood to adolescence*. New York: Basic Books.

Keren, G., & Schul, Y. (2009). Two is not always better than one: A critical evaluation of two-system theories. *Perspectives on Psychological Science, 4*, 533–550.

Keselman, A. (2003). Supporting inquiry learning by promoting normative understanding of multivariable causality. *Journal of Research in Science Teaching, 40*, 898–921.

Klaczynski, P. A. (2000). Motivated scientific reasoning biases, epistemological beliefs, and theory polarization: A two-process approach to adolescent cognition. *Child Development, 71*, 1347–1366.

Klahr, D., & Dunbar, K. (1988). Dual search space during scientific reasoning. *Cognitive Science, 12*, 1–48.

Klahr, D., & Siegler, R. S. (1978). The representation of children's knowledge. In H. W. Reese & L. P. Lipsitt (Eds.), *Advances in child development and behavior* (Vol. 12, pp. 61–116). New York: Academic Press.

Klein, G. (1998). *Sources of power: How people make decisions*. Cambridge, MA: MIT Press.

McMackin, J., & Slovic, P. (2000). When does explicit justification impair decision making? *Applied Cognitive Psychology, 14*, 527–541.

Newell, A., & Simon, H. A. (1972). *Human problem solving*. Englewood Cliffs, NJ: Prentice-Hall.

Pacini, R., & Epstein, S. (1999). The relation of rational and experiential information processing styles to personality, basic beliefs, and the ratio-bias phenomenon. *Journal of Personality and Social Psychology, 76*, 972–987.

Polanyi, M. (1966). *The tacit dimension*. Garden City, NY: Doubleday.

Pozzali, A. (2007). Tacit knowledge, implicit learning and scientific reasoning. *Mind & Society, 7,* 227–237.

Pretz, J. E. (2008). Intuition versus analysis: Strategy use and experience in complex everyday problem solving. *Memory & Cognition, 36,* 554–566.

Reber, A. S. (1976). Implicit learning of synthetic languages: The role of instructional set. *Journal of Experimental Psychology: Human Learning and Memory, 2,* 88–94.

Roediger, H. L. III (1990). Implicit memory: Retention without remembering. *American Psychologist, 45,* 1043–1056.

Schauble, L. (1996). The development of scientific reasoning in knowledge-rich contexts. *Developmental Psychology, 32,* 102–119.

Schauble, L., Klopfer, L. E., & Raghavan, K. (1991). Students' transition from an engineering model to a science model of experimentation. *Journal of Research in Science Teaching, 28,* 859–882.

Schooler, J. W., Ohlsson, S., and Brooks, K. (1993). Thoughts beyond words: When language overshadows insight. *Journal of Experimental Psychology: General, 122*(2), 166–183.

Schunn, C. D., & Anderson, J. R., (1999). The generality/specificity of expertise in scientific reasoning. *Cognitive Science, 23,* 337–370.

Schunn, C. D., & Dunbar, K. (1996). Priming, analogy, and awareness in complex reasoning. *Memory and Cognition, 24,* 271–284.

Siegler, R. S. (1976). Three aspects of cognitive development. *Cognitive Psychology, 8,* 481–520.

Simon, H. A., & Lea, G. (1974). Problem solving and rule induction: A unified view. In L. W. Gregg (Ed.), *Knowledge and cognition* (pp. 105–128). Hillsdale, NJ: Lawrence Erlbaum.

Sloman, S. A. (1996). The empirical case for two systems of reasoning. *Psychological Bulletin, 119,* 3–22.

Stanovich, K., & West, R. (2000). Individual differences in reasoning: Implications for the rationality debate? *Behavioral and Brain Sciences, 23,* 645–665.

Sternberg, R. J. (1986). *Intelligence applied: Understanding and increasing your intellectual skills.* San Diego, CA: Harcourt Brace Jovanovich.

Sun, R., Terry, C., & Slusarz, P. (2005). The interaction of the explicit and the implicit in skill learning: A dual-process approach. *Psychological Review, 112,* 159–192.

Sweller, J., Mawer, R. F., & Ward, M. R. (1983). Development of expertise in mathematical problem solving. *Journal of Experimental Psychology: General, 112,* 639–661.

Tversky, A. & Kahneman, D. (1974). Judgment under uncertainty: Heuristics and biases. *Science, 185,* 1124–1131.

VanLehn, K. (1986). Arithmetic procedures are induced from examples. In J. Hiebert (Eds.), *Conceptual and procedural knowledge: The case of mathematics* (pp. 133–179). Hillsdale, NJ: Lawrence Erlbaum.

Wilkening, F., & Anderson, N. H. (1982). Comparison of two rule-assessment methodologies for studying cognitive development and knowledge structures. *Psychological Bulletin, 92,* 215–237.

Wilson, T. D., & Schooler, J. W. (1991). Thinking too much: Introspection can reduce the quality of preferences and decisions. *Journal of Personality and Social Psychology, 60,* 181–192.

Zimmerman, C. (2000a). The development of scientific reasoning skills. *Developmental Review, 20,* 99–149.

Zimmerman, C. (2000b). *A network interpretation approach to the balance scale task* (Doctoral dissertation, University of Alberta, Edmonton, 1999). *Dissertation Abstracts International, 61*(3-B), 1667.

Zimmerman, C. (2007). The development of scientific thinking skills in elementary and middle school. *Developmental Review, 27,* 172–223.

Zimmerman, C., Raghavan, K., & Sartoris, M. L. (2003). The impact of the MARS curriculum on students' ability to coordinate theory and evidence. *International Journal of Science Education, 25,* 1247–1271.

PART FOUR
PSYCHOLOGICAL PERSPECTIVES: INFLUENCE ON SCIENCE

11 Scientists Are Human

Implicit Cognition and Researcher Conflict of Interest

Anthony G. Greenwald

Researchers have conflicts of interest to which they give little heed, but which can inappropriately influence their scientific judgment and their conduct of research. These often-unrecognized conflicts are problematic partly because they can operate outside of awareness, and also because they are generally unrecognized in scientific codes of research ethics. The American Psychological Association's code of research ethics presently gives almost no attention to researchers' conflicts of interest. The Association for Psychological Science has no code of research ethics at all (see Greenwald, 2009, pp. 32–33).

This chapter observes that scientists' unrecognized conflicts of interest stem from implicit affective–cognitive processes of types that have only recently received research attention. These conflicts occur in four professional roles that are found in virtually all scientific disciplines: (a) gatekeepers who decide on acceptance of articles submitted for publication or on award of research funds, (b) reviewers who provide advice to those gatekeepers, (c) experts who offer scientific opinions in court cases, and (d) researchers who decide when and how to report the findings of their research.

As will be shown, these conflicts bring self-interest into play. In business and legal settings, conflicts involving self-interest are well recognized when a decision has the potential to produce inappropriate financial gain to the decision maker or to the decision maker's relatives. Such financial self-interest possibilities are also recognized in ethical codes of many scientific societies. Scientists' ethical codes even extend the category of "relatives" beyond familial relatives to students, collaborators, and colleagues. Nevertheless, scientists' codes of ethics typically do not extend to two more abstract sets of relatives—preferred theories and published findings.

BEYOND CONFIRMATION BIAS

A pervasive stereotype of the laboratory scientist is that he is a seeker of empirical truth—an objective collector and appraiser of research evidence. The presumption that the scientist is male is not this stereotype's only problematic ingredient—the other is its presumption that the scientist is objective. One form of deviation from scientific objectivity that has been long recognized is *confirmation bias*. Confirmation bias is the tendency to give weight to hypothesis-confirming evidence while overlooking hypothesis-disconfirming evidence. Rather than being seen as an ethical problem, confirmation bias is seen more as an unfortunate deficit in analytic thinking, much as astigmatism is an unfortunate impairment of vision.[1]

Past discussions of confirmation bias in psychology have often characterized it as a deviation from objectivity that calls for corrective treatment. The proposed corrective treatments have focused mostly on problems that result from statistical significance testing practices. The criticized practices are ones that inflate Type I error rates, in turn producing dubious rejections of null hypotheses, and in further turn producing unjustified conclusions that theorized predictions have been confirmed. As pointed out by Greenwald et al. (1986)—and as remains true 25 years later—it is both interesting and troublesome that none of many proposed corrective treatments for confirmation bias has ever taken hold in psychology. As just the most obvious example, despite many urgings to correct confirmation bias by abandoning statistical significance testing, statistical significance testing remains psychology's dominant method for evaluating empirical findings.

IMPLICIT SOCIAL COGNITION AND BALANCED IDENTITY THEORY

The inability of proposed correctives for confirmation bias to gain traction suggests that the psychological factors underlying confirmation bias are not yet sufficiently understood. This apparent gap in understanding provides the opportunity to suggest here an approach that draws on recently emerging interest in implicit processes in social cognition.

Starting in the mid-1980s, psychology experienced what can now be identified as an *implicit revolution*. The implicit revolution was part successor to, and part continuation of, a cognitive revolution that started about 1960. Work of the last 25 years has strengthened psychologists' understanding that human memory, judgment, and behavior accomplish a great deal that is not revealed by research conducted exclusively with self-report methods. The vanguard of the implicit revolution consisted of memory researchers' demonstrations that brief experiences could influence a subject's later responses to perceptual and judgment tasks, even when subjects could neither identify the source of this influence nor remember its occurrence (e.g., Graf & Schacter 1985; Jacoby & Dallas, 1981). This revolution rapidly spread to social

[1] Some well-cited treatments of confirmation bias are those by Bakan (1966), Greenwald (1975), Selvin and Stuart (1966), and Sterling (1959). An overview of these and other analyses of confirmation bias can be found in Greenwald, Pratkanis, Leippe, and Baumgardner (1986).

psychology, where it brought recognition of the extent to which actors in many social situations remained unaware of influences that affected judgments and social behavior (Greenwald & Banaji, 1995).

Balanced identity theory (BIT) was offered by Greenwald et al. (2002) as a "unified theory of implicit social cognition" (see also Cvencek, Greenwald, & Meltzoff, 2012). BIT describes mental associative processes that, among other effects, cause people to like things that they associate themselves with and to identify themselves with things that they like. BIT's main theoretical proposition—the *balance–congruity* principle—was named to acknowledge its debt to core ideas of Heider's (1946, 1958) balance theory and Osgood and Tannenbaum's (1955) congruity theory, both of which provided theoretical accounts of the often-observed tie between liking something and having a personal connection to it.[2] Greenwald et al. (2002) described their balance–congruity principle as an organizing principle of human social knowledge within an associative *social knowledge structure*, a fragment of which is illustrated schematically in Figure 11.1. The description of a structural property (*shared first-order link*) of this associative network and the statement

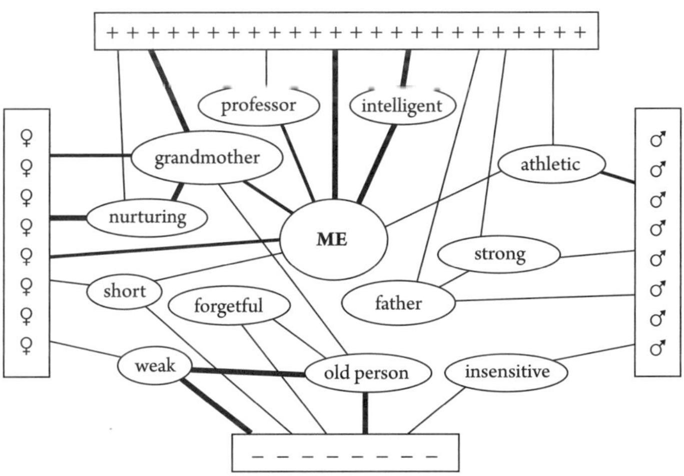

Figure 11.1 A social knowledge structure. This structure includes associations that correspond to social psychological constructs of self-concept, self-esteem, stereotype, and attitude in the psyche of an elderly female academic. Nodes (ovals) represent concepts and links (lines) represent associations. Line thickness represents strength of association. The self-concept includes links of the ME node to concepts that include roles (professor, grandmother) and trait attributes (intelligent, athletic); self-esteem is the collection of associations—either direct or mediated through components of the self-concept—of the ME node to valence (+ + + or – – –); stereotypes are associations of group concepts such as old person, grandmother, professor, male (♂♂♂), and female (♀♀♀) with attribute concepts; and attitude is the collection of links, either direct or mediated via components of a stereotype, that connect a social concept to valence. (Figure 11.1, including caption, is reproduced from Greenwald et al. [2002], p. 5)

[2] Perhaps the strongest of many early lines of empirical evidence for the connection between liking things and having a personal tie to them is the similarity–attraction effect that was extensively documented by Byrne (e.g., 1969). A variety of subsequent supporting evidence was reviewed by Greenwald and Banaji (1995, pp. 10–14).

of the balance–congruity principle, which uses this property, are reproduced here from Greenwald et al. (2002, p. 6):

Definition 1: Shared first-order link. When each of two nodes is linked to the same third node, the two are said to have a shared first-order link.
Principle 1: Balance–congruity. When two unlinked or weakly linked nodes share a first-order link, the association between these two should strengthen.

Two properties of this social knowledge structure—its central representation of self ("ME" in Figure 11.1) and its strong association of self with positive valence—both of which were stated as assumptions of BIT, are based on findings regarding self and valence established in more than three decades of research by personality and social psychologists (cf. Greenwald & Banaji, 1995; Greenwald et al., 2002). When combined with the balance–congruity principle, these two assumptions generate BIT's explanation of people coming to like (associate with positive valence with) things that they associate with self.

To extend BIT to scientific judgment, consider a possible addition to Figure 11.1—a node that represents a theory published by the researcher who possesses Figure 11.1's knowledge structure. Because Figure 11.1's node for representing this theory is (by assumption) associated with ME, there exists a shared first-order link configuration that strengthens association between the researcher's theory and positive valence. Further application of the balance–congruity principle will explain development of associations to positive valence for other nodes associated with the theory, such as ones that represent supporting empirical findings and ones that represent colleagues associated with the theory.

BIT's second principle, named *imbalance–dissonance* to acknowledge its debt both to balance theory and to Festinger's (1957) dissonance theory, explains how things not associated with self resist association with positive valence. Like the balance–congruity principle, the imbalance–dissonance principle also builds on a structural property of the social knowledge structure (Greenwald et al., 2002, p. 6):

Definition 2: Bipolar opposition of nodes. To the extent that two nodes have fewer shared first-order links than expected by chance, they can be described as bipolar-opposed. As diagramed in Figure 11.1, [the social knowledge structure] has two prominent pairs of bipolar-opposed nodes, those for valence (positive, negative) and sex or gender (male, female).
Principle 2: Imbalance–dissonance. The network resists forming new links that would result in a node having first-order links to both of two bipolar-opposed nodes.

To be applied to scientists' conflicts involving self-interest, the imbalance–dissonance principle needs an assumption that the knowledge structure has a prominent node that is bipolar-opposed to the ME node (a "THEM" node, not shown in Figure 11.1). The reasonableness of this assumption is suggested by findings of much recent research in which subjects have readily grouped non-self-associated people and concepts together with words representing *them* or *other* (e.g., Greenwald & Farnham, 2000). The imbalance–dissonance principle explains why non-self–associated concepts do not form associations with positive valence, and may instead form associations with THEM and/or with negative valence.

Greenwald et al. (2002) showed that BIT is well supported in research examining correlations among implicit measures of associations that correspond to group identities, self-concepts, attitudes, and stereotypes. A recent overview by Cvencek et al. (2012) combined those earlier balanced identity studies with subsequent ones in a meta-analysis that showed strong support for BIT's predictions in studies with implicit measures of association strengths. In unexpected contrast with these consistent BIT-supporting results, there has been essentially no support for BIT's predictions in analyses of data collected using *self-report* measures of association strengths. Although this was initially surprising, it is also understandable considering the long history of results that have shown distortion of self-report measures by extraneous influences such as demand characteristics (Orne, 1962), evaluation apprehension (Rosenberg, 1969), and various subject motivations (Weber & Cook, 1972).

R. K. MERTON AND T. C. CHAMBERLIN

Sociologist Robert K. Merton's essay, "The Normative Structure of Science" (Merton, 1973) identified the "ethos" of science with four *norms*, which he identified as *universalism, communism, disinterestedness*, and *organized skepticism*.[3] Social psychologists and sociologists distinguish two types of norms. Some norms are *descriptive* of what most people actually do. Other norms are *prescriptive*—they capture widely shared knowledge of what people *should* do.

Merton understood his four norms as being of the prescriptive type—he called them "institutional imperatives" (1973, p. 270). The norm of *universalism* demands that scientists remain uninfluenced by cultural factors—scientific knowledge should not be shaped by nationality, race, religion, gender, or other cultural values. *Communism* (relabeled "communalism" by later writers) demands that, rather than treating scientific knowledge as personal property, scientists should cooperate by freely sharing their findings. (Merton wrote, "Secrecy is the antithesis of this norm; full and open communication its enactment" [p. 274].) The norm of *disinterestedness* captures the importance of scientific objectivity—scientists should reject pressures to disguise findings or to commit fraud ("the activities of scientists are subject to rigorous policing, to a degree perhaps unparalleled in any other field of activity"; "disinterestedness has a firm basis in the public and testable character of science" [p. 276]). The norm representing scientists' questioning and challenging of authority, *organized skepticism*, encompassed not only readiness to challenge established lay opinion or religious belief, as when astronomers and biologists historically

[3] Merton's article originally appeared in 1942 with the title, "A Note on Science and Democracy," published in the *Journal of Legal and Political Sociology*. It was then reprinted in Merton's 1949 book, *Social Theory and Social Structure*, as a chapter titled "Science and Democratic Social Structure," before being reprinted again in 1973 as "The Normative Structure of Science."

questioned geocentric or creationist views, but also readiness to question existing scientific views.

Once the four norms are stated, an obvious question arises: Are Merton's *prescriptive* norms also *descriptive* norms? In other words, in addition to describing what scientists *should* do, do they accurately describe what scientists *do*? Given how admirable the four norms are, it is easy to assume that scientists largely adhere to them. But this chapter takes a quite different view, seeing Merton's norms serving more to describe a mask worn by many scientists than to describe the actual conduct of scientists. This skeptical view of Merton's norms has many antecedents, perhaps none more to the point than the characterization offered more than a century ago by geologist T. C. Chamberlin (1890/1965):

> The moment one has offered an original explanation for a phenomenon which seems satisfactory, that moment affection for his intellectual child springs into existence; and as the explanation grows into a definite theory, his parental affections cluster about his intellectual offspring, and it grows more and more dear to him, so that, while he holds it seemingly tentative, it is still lovingly tentative, and not impartially tentative.... There is an unconscious selection and magnifying of the phenomena that fall into harmony with the theory and support it, and an unconscious neglect of those that fail of coincidence. The mind lingers with pleasure upon the facts that fall happily into the embrace of the theory, and feels a natural coldness toward those that seem refractory. Instinctively, there is a special searching-out of phenomena that support it, for the mind is led by its desires. There springs up, also, an unconscious pressing of the theory to make it fit the facts, and a pressing of the facts to make them fit the theory. (p. 755)

In contrast to Merton's scientist—who is assumed to be free of self-interest—Chamberlin's scientist readily allows self-interest to override objectivity. Being biased in favor of personal theories and succumbing to pressures to disguise results, Chamberlin's scientist readily violates the norm of disinterestedness. It is interesting that Chamberlin explicitly identified "selection and magnifying of the phenomena that fall into harmony with the theory," "neglect of those that fail of coincidence," and "pressing of the facts to make them fit the theory" as "unconscious" processes. He did not need the 100-year-later findings of implicit social cognition research to form his view of scientists' affective–cognitive biases.

EVIDENCE?

Is there evidence available to decide whether scientists conform to Merton's prescription of disinterestedness or, alternatively, readily lapse in their objectivity, as Chamberlin supposed? Unfortunately, little direct evidence is available. With the relatively rare exceptions of well-publicized cases of fraud, scientists' norm violations may remain largely covert. There are nevertheless two forms of indirect evidence—self-observation and inference from properties of published research.

Self-observation. Drawing just on knowledge of their own behavior, many scientists may be able to conclude that they wander from Merton's norm of disinterestedness. Try this five-item test:[4]

1. You are reviewing a manuscript reporting research on a topic on which you have published. Assuming that your findings have no direct bearing on interpretation of the research you are reviewing, will it affect your judgment of this manuscript if, all else being equal, (a) the manuscript does not cite your work, (b) it cites your work, or (c) it cites your work approvingly?
2. In analyzing findings of the test of a prediction based on your original theoretical analysis, are you equally likely to scrupulously check your first statistical analysis for correctness if it (a) confirms your prediction or (b) disconfirms your prediction? And a follow-up question: Would you give the same answer if the prediction was one made by a theoretical rival?
3. When an initial empirical test disconfirms your theoretical prediction, are you more likely to (a) conduct an exact replication to allow a more conclusive indication of your prediction's incorrectness, or (b) change procedures in the hope that modified implementations of critical variables will produce a confirmation?
4. When you conduct a single study that includes several tests of your original theoretical prediction, and you obtain a mixture of confirming and disconfirming results, will your eventual report include (a) all of these prediction-relevant findings, or (b) just the theory-confirming ones?
5. When an initial empirical test confirms your hypothesis, but two subsequent exact replications fail to confirm, will you be more likely to (a) write up and submit a report of just the first (confirming) result, (b) submit a report that meta-analyzes the series of three studies to provide the most powerful hypothesis test, (c) change procedures to test again with different operationalizations, or (d) discontinue research on the problem?

The scenario of each question offers a confrontation between scientific objectivity and self-interest. In #1, citation of your work should (objectively) not influence your judgment. In #2, neither the outcome of the analysis nor whether the prediction is from your own or a rival's theory should affect the wisdom of checking its accuracy. In #3, the decision to change procedure suggests greater interest in obtaining the sought outcome than in providing a powerful test that might disconfirm it. In both #4 and #5, scientific objectivity should call for treating all tests as equally informative and therefore equally worthy of report.

There are occasional studies that have investigated what scientists do when initial findings disconfirm their predictions. As an example, Greenwald (1975) found that researchers reported that they were considerably more likely either to conduct a new test with modified procedures or to give up on a problem when an initial

[4] Readers who have not yet had the chance to observe themselves in the situations described in these questions (e.g., beginning PhD students) may believe that, when they encounter these situations, they will readily conform to Merton's norm of disinterestedness. Those readers may discover that their answers will change within a few years.

result disconfirmed (rather than confirmed) their predicted outcome (Greenwald, 1975, p. 4). However, there are not enough such studies to permit strong conclusions.

Indirect evidence can also be obtained by noticing what gets published and what does not. Consider the following four observations, for all of which multiple examples can be found in published research literature:

1. An intriguing result is published. No replication is ever published—not by the original investigator and not by others. It remains quite plausible that the original author and/or others tried to reproduce the result, but did not confirm it, and were content not to report those attempts in any public form.
2. An intriguing result is published. Some replications appear. As still more replications appear, increasingly these reveal some difficulty in replication (cf. Lehrer, 2010). Although the result may be regarded as established, early-wave researchers may have based claims on results selected to exaggerate evidence for confirmation, leading to overstatement of the finding's strength.
3. In an article that reports a series of studies, each confirming the authors' theoretical prediction, the series of studies inexplicably does not repeat, from one study to the next, procedures that confirmed the hypothesis. It is difficult to avoid believing that the initial study's procedure did not replicate and that the researchers persevered to find alternative occasionally successful procedures—but without reporting intervening nonreplications.
4. A published article gives obvious indications that the authors collected more hypothesis-relevant data than they reported. The clue may be that the research operations associated with variables central to the hypothesis test are not the operations that have been routinely and effectively used for the same variables in much prior research. It is plausible that the researchers are suppressing nonconfirming data.

REMEDIES

The four professional roles in which Greenwald (2009) identified conflicts of interest that are not treated in codes of scientific ethics were those of gatekeepers, reviewers, expert witnesses, and researchers. Although there has been little formal recognition of these self-interest conflicts, they certainly have widespread informal recognition. For example, the objective evaluation of a manuscript submitted for publication should, in principle, not vary with theoretical allegiances of editors or reviewers. Nevertheless, most publishing researchers are aware that evaluations of their manuscripts will vary as a function of reviewers' and editors' theoretical allegiances. This awareness is likely to guide a decision as to where to submit a new manuscript by considering who is on the editorial board and what is known about their theoretical predilections.

The starting point in considering remedies for implicit-self-interest-rooted conflicts of interest is to consider what is known about how to disrupt effects of implicit cognition on judgment and action. This summary was offered by Greenwald and Banaji (1995): "When attentional effort is directed to a weak stimulus or memory trace [i.e., one that might lead to an implicit effect], the reduced effectiveness of that cue is likely due to conscious strategies overriding and interfering with automatic cognitive effects" (p. 18). They observed further that "augmentation of an implicit effect [can be expected] if the direction of the effect is consistent with a conscious motive" (p. 18). Because scientists likely have conscious motives to promote their theories, this last statement suggests that awareness that the theory being tested is one's own is more likely to enhance than to disrupt objectivity-impairing effects of implicit self-interest.

A conclusion follows: Implementing remedies for implicit-self-interest conflicts will require not only providing some effective remedial strategies, but also giving scientists a conscious motive to block the effects of self-interest. To start with the first of these two needs, here is a brief summary of the remedies suggested by Greenwald (2009) for each of the four professional roles on which he focused.

Gatekeeper. Implicit self-interest may intrude when an editor or grant review coordinator is evaluating a submission that addresses a theoretical issue on which this gatekeeper has a pro or con position. An easily stated resolution is recusal—the gatekeeper can pass the decision to an alternate who does not have such a basis for conflict.

Reviewer. Implicit self-interest may intrude when a reviewer has an established theoretical position that agrees with or opposes that presented in a journal submission or grant proposal that he or she has just been invited to evaluate. Recusal is a possible remedy, but may be unsatisfactory because of the difficulty of locating other expert reviewers who lack any such basis for conflict. In this case a remedy is for the reviewer to forgo anonymity. Or, without forgoing anonymity, the reviewer can directly state the relevant theoretical allegiance (or other basis for possible self-interest) at the start of the review.

Expert witness. The conflict here is not the obvious financial one of being a paid advocate. Judges, attorneys, and jurors are well aware of the paid advocacy role and take it into account in considering any expert's testimony. The situation in which implicit self-interest may not be obvious is when the expert submits a manuscript or grant application, with the possible consequence that the publication or grant award will enhance the expert's basis for future paid advocacy. Here, a satisfactory remedy is for the expert to describe the basis for self-interest at the time of submitting the manuscript or grant proposal.

Researcher. Because researchers' conflicts involving self-interest occur in situations that lack demarcated decision points, the remedies for the other three roles—recusal or explicit acknowledgment of the basis for self-interest—do not apply. Researchers' conflicts could be brought into the open if researchers were obliged to maintain a public record of all empirical hypothesis tests that they undertake. Alas, this remedy—which would be an answer to meta-analysts' prayers—may be so difficult to enforce as to be impractical.

How to Implement Remedies?

As just demonstrated, straightforward remedies for implicit-self-interest conflict are relatively easy to state for three of the four roles—gatekeeper, reviewer, and expert witness. However, widespread voluntary adoption of these remedies seems unlikely because scientists are likely not only (a) to have conscious motives to nurture and protect their favored theories, but also (b) to see themselves as immune to this possible interference with their objectivity.

If effective remedy requires conscious motivation to adhere to the norm of disinterestedness, the solution may be to offer incentives to adhere to that norm. Although many forms of incentives are imaginable, the most obvious one is the formulation of standards for research conduct that include sanctions for deviation—an ethical code.

For an ethical code to improve adherence to ethical standards it must be possible to detect nonadherence. It would accomplish little to have a standard stated simply as an admonition to "be objective" in conducting reviews, making gatekeeper decisions, offering expert testimony, and evaluating one's own research results. However, deviations might be readily noticed—to those overseeing journal or grant review processes—for a standard that included a statement of appropriate behavior, such as "Reviewers having published theoretical views on questions at issue in a manuscript or proposal should decline to review or, alternately, make clear their theoretical perspective when providing an evaluation."

Researchers' conflicts, as already noted, are more problematic. It is difficult for editors or reviewers to detect deviations from an ethical guideline to present full (rather than partial, selective) report of findings bearing on research hypotheses. Nevertheless, existence of that guideline might provide the occasion for research collaborators to bring deviations to an author's attention. In an ideal world, such informal operation of ethical standards seems greatly preferable to enforcement by official sanctions from a professional association. Thus, if a reviewer failed to acknowledge having a published theoretical position relevant to a manuscript or grant submission, this could be corrected by an editor or review coordinator bringing the lapse to the reviewer's attention.

In summary, the desirability of Merton's four norms, coupled with their nonappearance in scientific codes of ethics, suggests wide belief that scientists voluntarily and faithfully adhere to them. However, that presumed belief lacks empirical support. This chapter suggests that the conduct of many or most scientists may be closer to Chamberlin's characterization of the scientist as an emotionally swayed nurturer of personal theoretical creations—one who will readily deviate from objectivity to protect and promote those creations. Scientists must be seen as being as human as are other parents when it comes to caring for their offspring. To the extent that it is desirable for scientists to adhere to Merton's norm of disinterestedness, it is inappropriate to assume that scientists are superior beings whose professional judgment is uninfluenced by the implicit cognitive–affective processes that pervasively characterize other humans.

REFERENCES

Bakan, D. (1966). The test of significance in psychological research. *Psychological Bulletin, 66*, 432–437.

Byrne, D. (1969). Attitudes and attraction. In L. Berkowitz (Ed.), *Advances in experimental social psychology* (Vol. 4, pp. 35–89). New York: Academic Press.

Chamberlin, T. C. (1965). The method of multiple working hypotheses. *Science, 148,* 754–759. (Original work published 1890).

Cvencek, D., Greenwald, A. G., & Meltzoff, A. N. (2012). Balanced identity theory: Review of evidence for implicit consistency in social cognition. In B. Gawronski & F. Strack, (Eds.), *Cognitive consistency: A fundamental principle in social cognition* (pp. 157–177). New York: Guilford Press.

Festinger, L. (1957). *A theory of cognitive dissonance.* Palo Alto, CA: Stanford University Press.

Graf, P., & Schacter, D. (1985). Implicit and explicit memory for new associations in normal and amnesic subjects. *Journal of Experimental Psychology: Learning, Memory, and Cognition, 11,* 501–518.

Greenwald, A. G. (1975). Consequences of prejudice against the null hypothesis. *Psychological Bulletin, 82,* 1–20.

Greenwald, A. G. (2009). What (and where) is the ethical code concerning researcher conflict of interest? *Perspectives on Psychological Science, 4,* 32–35.

Greenwald, A. G., & Banaji, M. R. (1995). Implicit social cognition: Attitudes, self-esteem, and stereotypes. *Psychological Review, 102,* 4–27.

Greenwald, A. G., Banaji, M. R., Rudman, L. A., Farnham, S. D., Nosek, B. A., & Mellott, D. S. (2002). A unified theory of implicit attitudes, stereotypes, self-esteem, and self-concept. *Psychological Review, 109,* 3–25.

Greenwald, A. G., & Farnham, S. D. (2000). Using the implicit association test to measure self-esteem and self-concept. *Journal of Personality and Social Psychology, 79,* 1022–1038.

Greenwald, A. G., Pratkanis, A. R., Leippe, M. R., & Baumgardner, M. H. (1986). Under what conditions does theory obstruct research progress? *Psychological Review, 93,* 216–229.

Heider, F. (1946). Attitudes and cognitive organization. *Journal of Psychology, 21,* 107–112.

Heider, F. (1958). *The psychology of interpersonal relations.* New York: John Wiley.

Jacoby, L. L., & Dallas, M. (1981). On the relationship between autobiographical memory and perceptual learning. *Journal of Experimental Psychology: General, 110,* 306–340.

Lehrer, J. (2010). The truth wears off. *New Yorker,* December 13, 52–57.

Merton, R. K. (1942/1973). The normative structure of science. In R. K. Merton (Ed.), *The sociology of science: Theoretical and empirical investigations* (pp. 267–278). Chicago: University of Chicago Press, 1973.

Orne, M. T. (1962). On the social psychology of the psychological experiment: With particular reference to demand characteristics and their implications. *American Psychologist, 17,* 776–783.

Osgood, C. E., & Tannenbaum, P. H. (1955). The principle of congruity in the prediction of attitude change. *Psychological Review, 62,* 42–55.

Rosenberg, M. J. (1969). The conditions and consequences of evaluation apprehension. In R. Rosenthal & R. L. Rosnow (Eds.), *Artifact in behavioral research* (pp. 279–349). New York: Academic Press.

Selvin, H. C., & Stuart, A. (1966). Data-dredging procedures in survey analysis. *American Statistician, 20,* 20–23.

Sterling, T. D. (1959). Publication decisions and their possible effects on inferences drawn from tests of significance—or vice versa. *Journal of the American Statistical Association, 54,* 30–34.

Weber, S. J., & Cook, T. D. (1972). Subject effects in laboratory research: An examination of subject roles, demand characteristics, and valid inference. *Psychological Bulletin, 77,* 273–295.

12 Bias, Feminism, and the Psychology of Investigating Gender

Alice H. Eagly

Scientists assume that science provides a better approach to understanding the world than alternative methods such as intuitive insights or prescientific analyses. One of the advantages of science is that its rules and methods build in considerable protection against bias through, for example, statistical hypothesis testing, peer review as a precondition for publication, the replication of studies, and measuring instruments that minimize human subjectivity. Although few contemporary scientists would claim that science is objective, these practices do give at least limited protection against a range of potential biases. However, despite some exploration of judgmental biases that affect the reasoning of scientists (e.g., Mahoney, 1979), this aspect of the psychology of science has received relatively little attention. Therefore, I offer an analysis of some of these biases in relation to research on the psychology of gender. I also consider whether the influence of feminism on the psychology of gender has countered or exacerbated the effects of the judgmental biases that I consider. To gain some insight about these issues, I analyze the history of science on the psychology of gender with respect to the probable role of scientists' judgmental biases.

In this chapter, I argue that scientists routinely produce findings and interpretations that are to some extent coerced by cognitive and affective biases that pervade human judgment and thinking. These biases are particularly relevant to psychology (more than, for example, physics or chemistry) because the interpretation of human behavior strongly engages them. These largely implicit biases are generally outside of the awareness of scientists as they make their everyday decisions about the design of studies and the reporting and interpretation of findings. In this chapter, I consider two types of bias that are particularly relevant to research and theory on the psychology of gender but likely affect other psychological research as well, especially research pertaining to social groups.

The first type of bias that I consider engages *correspondent inference* and the linked principle of *psychological essentialism*, which are phenomena of social cognition. The second type of bias is *ingroup bias*, which affects the practice of science through the linked principle of the *congeniality bias* in information processing, thus involving two basic phenomena of social psychology.

Correspondent inference is the tendency to ascribe human behaviors to psychological dispositions of persons, rather than to situational pressures that act on these persons. In everyday social cognition, these dispositions are generally personality traits and abilities that correspond to the characteristics of the observed behavior (Gilbert, 1998). This preference for traits as causes of behavior makes it difficult for people to take the power of situations into account. Also, when thinking about individuals as members of social categories, especially gender categories, social perceivers tend to essentialize their traits—that is, to see category members as having deep, hidden, and stable attributes that shape their behavior. To the extent that scientists think like people in general, they would be biased in favor of ascribing male and female behavior to their underlying traits, and especially to intrinsic, deeply embedded traits shared by most men and most women. This essentialist bias helps stabilize the social order because it supports existing social arrangements by making male-female and other group differences seem sensible and group hierarchies seem reasonable and difficult to reverse (Morton, Postmes, Haslam, & Hornsey, 2009).

Ingroup bias is the tendency to favor one's own group—that is, to have more positive attitudes toward ingroups than outgroups and to favor ingroups behaviorally (Brewer, 1999; Yzerbyt & Demoulin, 2010). The linked congeniality bias is the tendency to favor information that agrees with one's attitudes, at all stages of information processing (Eagly & Chaiken, 2005). To the extent that people favor their own sex as an ingroup, information congenial to their sex is experienced as better—more worth attending to and more plausible and valuable than information that favors the other sex. Because scientists are female or male, they would plausibly manifest an evaluative ingroup bias in favor of their own sex, a bias that could have multiple effects on gender research through congeniality biases in information processing.

In this speculative chapter, I consider the possible effects of these two types of bias on the scientific study of gender. To my knowledge, these biases are rarely discussed in texts on methods, nor is it likely that many scientists are aware that they can be biased in these senses. Nonetheless, these biases can reverberate through the production, interpretation, and dissemination of research. The culprit in producing science that is biased in these ways is scientists' basic humanity whereby the same tendencies that typify people's everyday thinking can contaminate scientific thinking.

CORRESPONDENCE AND ESSENTIALISM BIASES

Much of the research and theorizing about gender centers on interpreting sex differences and similarities in human behavior. Scientists seek not only to describe sex differences but also to explain them. So, for example, if women are more socially sensitive than men (Hall, 1984), what causes this difference? Relevant to understanding scientists' reasoning about causation is basic research in social cognition that has probed perceivers' explanations of behavior and the properties that they ascribe to members of social categories. Scientific reasoning may reflect the same tendencies that social cognitive research has found characterize everyday thinking about people and social categories.

One key generalization from social cognitive research is that in general people infer psychological traits from observations of others' behavior, a tendency known as *correspondent inference* or *correspondence bias* (Gilbert, 1998). For example, upon observing a person giving very careful directions to a stranger who appears to be lost, social perceivers automatically identify the behavior in trait terms and characterize the actor by the traits that are implied—that is, as kind and helpful. When explaining why such acts occur, trait explanations usually win over situational explanations.

Observers typically match traits to behaviors even in the presence of clear-cut information about situational constraints that could plausibly influence the behavior. For example, in a classic experiment by Ross, Amabile, and Steinmetz (1977), participants in a laboratory quiz game were randomly assigned to the role of questioner or contestant; the questioner then prepared questions to which the answerer had to respond. Even though the questioner possessed clear and unambiguous advantage from the privilege of making up difficult questions from her or his own knowledge, observers of the quiz game interaction and its participants judged the person asking the questions as considerably more knowledgeable than the person answering the questions. Participants thus inferred dispositional differences between persons in the questioner and contestant roles and ignored the power of the social roles that structured this situation, a phenomenon that Ross (1977) labeled the *fundamental attribution error*. Subsequent research has confirmed that correspondent inference is widespread (Gawronski, 2003) and that social perceivers infer traits from observed behavior by a largely spontaneous process (Uleman, Saribay, & Gonzalez, 2008). It therefore seems likely that scientists, just as nonscientists, would favor trait explanations of behaviors more than situational explanations and, more specifically, ascribe differences in female and male behavior to underlying differences in psychological traits.

What kinds of traits might be favored in explanations of sex differences? Because male and female are human social categories, it is relevant to understand how people think about social categories in general. Social categories can be "natural kind" in that their members are thought to have stable, natural qualities that set them apart from other categories, or they can be arbitrary categories that do not acquire such properties, such as a category consisting of the people standing on a corner waiting for a signal light to change. Distinctive enduring dispositions are not inferred for individuals arbitrarily grouped together but for other, more stable social categorizations, dispositional traits are readily inferred.

Gender categories produce especially clear-cut dispositional inferences about women and men in general because male and female are discrete social groupings with relatively fixed boundaries. Moreover, membership in one or the other of these categories is involuntary. Research by Haslam, Rothschild, and Ernst (2000) thus found that, among 40 social categories presented to research participants, female-male social categories were judged as the most natural, necessary, immutable, discrete, and stable, even compared with categories of ethnicity and race. The sex classification is thus viewed as the inevitable result of sex following from sex chromosomes.

Given the naturalness of classifying people by sex, it is not surprising that perceivers tend to essentialize men and women—that is, to view category members as having underlying essences that account for their behavior (Prentice & Miller, 2007). These essences need not be genetic or biological because essentialist reasoning can emphasize nurture or nature (Rangel & Keller, 2011). In fact, research supports the conclusion that people usually recognize that both nature and nurture contribute to the essences of male and female (e.g., Martin & Parker, 1995). Nonetheless, essentialist reasoning dominates in everyday understanding of female and male behavior (Wood & Eagly, in press). Therefore, if scientists think like other people, essentialist theorizing about gender would have an edge over other types of scientific theorizing. Later in this chapter, I discuss scientists' explanations of sex differences to see whether essentialism predominates and what influence feminism might have on such interpretations.

INGROUP AND CONGENIALITY BIASES

To consider another type of judgmental bias that taints science, I consider the ways in which a bias toward favoring one's own gender likely affects science. This bias would arise from social identity considerations (Turner, 2000), in conjunction with a general tendency to favor information congenial to one's attitudes.

Sex-of-Author Effects in Gender Research

To begin this discussion of ingroup bias, I will share a personal story. In the late 1970s, I began my first meta-analysis, which summarized scientific research on the question of whether there is a sex difference in how easily people are influenced by others (Eagly & Carli, 1981). The project produced a small but significant overall tendency toward greater female influenceability that was somewhat larger in group settings. The surprise result in this project was that the sex of the authors of the articles correlated with the studies' findings, $r\ (88) = .41, p < .001$, with male authors more likely than female authors to find women more influenceable than men. Desiring to understand this unexpected relationship, my coauthor, Linda Carli, and I performed a secondary analysis on the findings of another early meta-analysis on sex differences—Hall's (1978) review of the ability to correctly decode nonverbal cues, a form of social sensitivity. The project had produced a moderate tendency toward greater decoding skill in women than men. Our secondary analysis showed that the sex of the authors of the articles correlated with the studies' findings, $r\ (45) = .36, p = .013$, with female authors more likely than male authors to find women more skilled at decoding. We concluded that "at least in these two areas, both male and female researchers portray their own gender more favorably than researchers of the opposite sex do" (Eagly & Carli, 1981, p. 17).

A subsequent meta-analysis of mine, on female and male leadership styles (Eagly & Johnson, 1990), established that female authors found larger tendencies for women to manifest more democratic and interpersonally oriented styles than men. To the extent that these styles are culturally valued, perhaps especially by women, this finding also suggests that authors portray their own sex more favorably than the other sex. In addition, Wood's (1987) meta-analysis on the effects of groups' sex composition on group performance found that the overall tendency for groups with more men to be more productive was stronger in the male-authored studies—here producing a more favorable portrayal of men by male authors. Additionally, Anderson and Leaper's (1998) meta-analysis of interruptions in conversations found that the overall tendency for men to intrusively interrupt conversations more than women do was more pronounced in female-authored studies. Female authors thus portrayed men more unfavorably (and women more favorably) than male authors did. Yet Thomas and French (1985) found that on some, but not all, tests of motor performance, female authors found *larger* sex differences in the direction of superior male performance than did male authors. Finally, quite a few other meta-analyses of sex differences have not found sex-of-author effects although author sex has not been routinely examined as a moderator.

It is troubling that sex-of-author effects have been obtained and have usually remained difficult to explain by moderators or mediators available in the meta-analyses. It would make sense that meta-analyses would reveal that male and female investigators had conducted studies with different attributes, thus producing sex-of-author effects. One effort to explore this type of mediation was partially successful: Becker's (1986) secondary analysis of the Eagly and Carli (1981) influenceability data found that, for one of the three categories of influence studies, the percentage of male authors was positively associated with the number of stimulus items presented to subjects, presumably yielding a more reliable estimate of the sex difference. However, this mediator was not appropriate to explain the sex-of-author effects in the other categories of the Eagly and Carli (1981) data or in most other meta-analyses. Also, Thomas and French (1985) speculated that female authors, often collecting their own data, inspired males to perform especially well on motor performance tasks to impress this female audience. Given such possibilities, it seems appropriate to explore a wide range of mediators by which authors' gender could influence research. Moreover, attention to author characteristics reveals other provocative results, such as the tendency for originators of paradigms to produce more positive findings than subsequent researchers who are ostensibly replicating aspects of the known findings (e.g., Johnson & Eagly, 1989; Strube & Garcia, 1981; Wood, Lundgren, Ouellette, Busceme, & Blackstone, 1994).

Ingroup Bias as an Influence on Research

Regardless of what process mediates sex-of-author effects, most of these effects are consistent with ingroup bias. In social psychology, bias in favor of the groups in which people have membership is a well-established principle (Yzerbyt & Demoulin, 2010). This phenomenon presumably emerges from a need for positive social identity. Because social identifications are an important aspect of the self-concept, people are motivated to evaluate their own social groups more favorably than other groups (Ellemers & Haslam, 2012; Turner, 2000), to maintain a positive self-concept. Empirically, ingroup bias is well established across a range of experimental paradigms (see meta-analyses by Bettencourt, Charlton, Dorr, & Hume, 2001; Mullen, Brown, & Smith, 1992).

People are thus motivated to favor the groups to which they belong, perhaps especially their gender group (Hogg, 2006). Research suggests that people, presumably including scientists, are profoundly identified with their gender (Wood & Eagly, 2009). A sense of oneself as female or male develops early in childhood and is elaborated throughout development. Gender, along with race or ethnicity and age, is generally a primary identity, which is relevant across all social situations (Brewer, 1988, 1999). Nevertheless, the principle that people favor their own gender does not necessarily imply that they are negative toward the other gender. Merely a more positive evaluation of one's own than the other gender predicts acting more favorably toward one's own group.

Ingroup bias is no doubt relevant not only to sex-of-author effects but more generally to many aspects of researching and theorizing about gender. Scientists who compare women and men in their studies are inevitably researching their own gender group's status or performance relative to that of the other gender group. These scientists are thus in some sense studying themselves. Their gender identities likely are engaged by relevant aspects of the research.

Attitudinal Selectivity

How might ingroup gender bias affect science? This evaluative bias could produce selectivity in processing scientific information pertaining to gender issues. Knowledge of attitudinal selectivity has emerged from a long-term research tradition in psychology (see Eagly & Chaiken, 1993, 2005). In classic theorizing (see Eagly, 1998), selectivity operates at all stages of information processing. People might avoid exposing themselves to challenging information in the first place. If exposed to such information, they might avoid attending to it, distort its meaning, and have difficulty storing and retrieving it. In addition, they may selectively evaluate it so that congenial information seems more truthful and valid than uncongenial information. In the 20th century, selectivity research became especially prominent in cognitive dissonance theorists' claims that people avoid, distort, or repress information that is incongruent with their attitudes or behaviors (e.g., Festinger, 1957, 1964), thereby manifesting what has been termed a *congeniality* bias (e.g., Eagly & Chaiken, 1993, 2005).

There is ample research on these selectivity processes. Concerning selective exposure, a recent meta-analysis by Hart et al. (2009) found a moderate preference for congenial over uncongenial information ($d = 0.36$). That is, in situations that provide people with choices between congenial information, which supports their attitudes, beliefs, or behaviors, and uncongenial information, which challenges these tendencies, they generally prefer the congenial information. If people do receive uncongenial information, they generally evaluate it relatively negatively. They tend to counterargue it (e.g., Edwards & Smith, 1996), and they rate it as less convincing and less valid than congenial information (Lord, Ross, & Lepper, 1979). However, with respect to selective memory, it is not clear that congenial information has a memorability advantage (see Eagly, Chen, Chaiken, & Shaw-Barnes, 1999). The principle underlying this seeming deviation from the congeniality bias is that uncongenial information becomes more memorable because it is often processed more deeply as message recipients counterargue it, whereas congenial information becomes more memorable because of its advantages of pleasantness and fit with pre-existing schemas (Eagly, Kulesa, Brannon, Shaw, & Hutson-Comeaux, 2000). Often these countervailing forces are of fairly equal strength, thus producing null effects of information congeniality on memory.

How might attitudinal selectivity produce sex-of-author effects? Assuming that gender is an important identity that produces ingroup bias, researchers would be pleased or displeased by the results of their comparisons between male and female research participants, depending on whether the comparison made their own sex look better or worse than the other sex. In evidence that such effects occur, experiments manipulating the favorability of research findings to each sex found that participants of both sexes were more positive about findings that portrayed their own sex favorably and presumably affirmed their gender identity (Morton, Haslam, Postmes, & Ryan, 2006). This finding was *more* pronounced among participants who possessed at least some scientific training, thus failing to support the idea that scientific training or knowledge mitigates this form of attitudinal selectivity.

Concerning sex-of-author effects appearing in meta-analyses, it is important that gender hypotheses were often subsidiary to studies' focal hypotheses, thus allowing authors a choice when it came to reporting sex comparisons. The female-male statistical comparison could to varying degrees support difference or similarity, and the author could include or exclude this information from his or her report. For example, in the influenceability literature (Eagly & Carli, 1981), female authors might have favored reporting null or counterstereotypical findings, which would counter the negative stereotype that women are easily influenced. Male authors might have preferred to report comparisons that confirm the stereotypical difference, which would affirm men's identities as strong, independent, and not easily swayed in laboratory conformity experiments. Along comes the meta-analyst who gathers up the information from the available studies, and lo and behold, there is a sex-of-author effect in the direction of favoring authors' own sex.

This explanation of sex-of-author effects remains speculative. Proving this explanation would require additional research into researchers' decision making about including sex comparisons in their articles. Of course, authors would not necessarily partition their data by sex to examine whether a difference occurred. Those interested in gender would be more likely to do so and then would have the opportunity to decide whether to include this finding in their report. Because of the potential for this type of selectivity, I have argued that scientists should always include reports of sex comparisons in their research reports, at least tucked away in a footnote (Eagly, 1995). Such a practice would prevent researchers from presenting these comparisons only if they favored their own sex. If the practice of routine reporting were accepted by the community of researchers, meta-analysts of sex differences and similarities would have a wealth of much more representative information to aggregate and integrate.

HISTORY OF THE PSYCHOLOGY OF GENDER CONSIDERED IN RELATION TO SCIENTISTS' JUDGMENTAL BIASES

Early Scientific Theory

Now I turn to exploring the history of research on the psychology of gender in terms of the possible influence of the judgmental biases considered in this chapter. In considering scientifically derived interpretations of gender, it is appropriate to start with Charles Darwin (1882), whose views have been extremely influential. He was an astute observer of nature who proposed the theory of evolution, a brilliant exposition of the development of species that is widely acknowledged as one of the all-time greatest scientific achievements. He eventually extended his theory to explain human sex differences as well as racial and social class differences. This endeavor proved to be far more difficult than explaining the development of species, and Darwin's ideas about differences between human groupings have met with at least as much disfavor as favor. Here I consider only his interpretations of human sex differences. Using the same types of observational methods that he applied to understanding nonhuman species, Darwin took into account what he observed about women and men in Victorian England. What were these observations (Vicinus, 1972)? In England, as in America, women led very restricted lives. Englishwomen could not keep their earnings or inherit property or money or vote in elections. Education for women, beyond the basics, was thought to be unnecessary except insofar as it prepared them for the domestic role. Higher education was inaccessible to women and regarded as a threat to their reproductive potential. In the middle and upper classes, women's role was to bear children (Queen Victoria had nine) and to serve as loyal helpmates to their husbands. Working-class women generally undertook menial jobs, often as servants in the homes of wealthier families.

Given the conditions of women's lives, it is not surprising that Darwin's theory of human sex differences emphasized that women are inferior to men: "The chief distinction in the intellectual powers of the two sexes is shewn by man's attaining to a higher eminence, in whatever he takes up, than can woman—whether requiring deep thought, reason, or imagination, or merely the use of the senses and hands" (Darwin, 1882, p. 564). Darwin also wrote that the distinctive psychological characteristics of women, which he identified as tenderness, intuition, rapid perception, and perhaps imitation, are "characteristic of the lower races, and therefore of a past and lower state of civilization" (Darwin, 1882, pp. 563–564).

Were Darwin's disparaging interpretations of the psychology of women valid in any sense for Victorian England? Absent ingroup bias favoring men, he might have given women more credit for managing the challenges of domestic life shaped by high birthrates and a clear-cut male-female division of labor. Nevertheless, his generalizations to some extent fit his observations. Given their lack of education, women did not possess nearly as much understanding of the world outside of the home as men did and displayed little accomplishment in any sphere of the humanities or sciences. Men gained knowledge of the wider world and, if privileged, had access to higher education and intellectually demanding occupations. Within the limits of their social class, men could follow their ambitions and aspirations, whereas women could not, unless they narrowed their scope to childbearing and domestic pursuits.

Taking such observations into account, Darwin extended evolutionary theory, derived to explain the traits of nonhuman species, to human sex differences (see Shields & Bhatia, 2009). He regarded these sex differences as built in through evolution especially by means of sexual selection processes of male-male competition for mates and female choice. Specifically, the argument was that women have a high investment in parenting due to producing relatively few large, metabolically expensive eggs and their gestating and nursing of infants. Men's large supply of less costly sperm and freedom from gestation and lactation would yield a potentially faster reproductive rate. Consequently, men would compete with one another for access to fertile women, gaining fitness by sexual promiscuity. In contrast, women, bearing the greater cost of reproduction, would gain fitness by selectively choosing mates who provide resources and protection for them and their offspring (see Kenrick, Maner, & Li, 2005).

Darwin thus argued that the sexes differ in their reproductive decisions, with men sexually promiscuous and women choosy. Also, because men can gain fitness benefits from shared sexual access to women, men engage in activities such as provisioning that can yield control over resources and women's sexuality. Darwin further argued that men's superior qualities are preferentially transmitted to their sons. He also reasoned that human females' earlier maturation rendered them incomplete, placing their existence somewhere between that of a child and adult man, thus with stunted intellectual and emotional attributes. He further proposed that men are more variable in their traits than women; thus, persons of superior achievement would be predominantly male. In Darwin's reasoning, these attributes of women and men were built in through human evolution and thus are relatively fixed and unalterable.

Would it have been possible for Darwin to propose a different type of theory? As 21st-century scientists look back on Victorian England, they of course readily take into account the constraints on women's lives. Confined to the home or to menial work, depending on social class, women lived within the narrowly defined womanhood of that era. These constraints of social context stand out for modern observers because many have been eliminated (e.g., inability to vote or obtain higher education) and others considerably reduced (e.g., employment discrimination, lack of reproductive freedom). As a result, women's accomplishments beyond the domestic role have escalated. Darwin's lack of familiarity with societies that allowed women more rights and access to resources made it unlikely that he would consider social context as a cause of women's attributes. Moreover, in not considering social context, Darwin's reasoning was consistent with the overriding human judgmental tendency to infer traits from behaviors and to essentialize gender categories. Furthermore, Darwin's evolutionary theory endowed all species and the sexes within species with intrinsic qualities that reflected their evolutionary history. It would have been surprising had he thought differently about the human species.

Given the cognitive edge that essentialist explanations of sex differences have over situational explanations in everyday thinking, most 19th-century and early 20th-century psychologists also viewed human nature as the cause of sex differences (see review by Shields, 1975). For example, psychologists who studied the brain regarded the smaller size of the female brain as causing female intellectual inferiority. In addition, the female brain was thought to be particularly underdeveloped in key regions such as the parietal lobes or the frontal lobes—wherever intelligence was thought to reside according to the science of the day. Psychologists believed that the more "primitive" parts of brains dominated in women, accounting for their being ruled by their instincts and emotions.

Critiques of Essentialist Explanations of Sex Differences

The only prominent 19th-century exception to essentialist thinking about sex differences was John Stuart Mill (1869), who challenged consensus beliefs in his famous essay "On the Subjugation of Women." Mill (1869, p. 324) wrote, "I deny that any one knows or can know, the nature of the two sexes, as long as they have only been seen in their present relation to one another. Until conditions of equality exist, no one can possibly assess the natural differences between women and men, distorted as they have been. What is natural to the two sexes can only be found out by allowing both to develop and use their faculties freely." Mill's bold analysis complemented his activism on behalf of women's suffrage and gender equality.

A few psychologists associated with the early 20th-century, *first-wave* feminist movement challenged essentialist explanations of sex differences by questioning the existence of the differences themselves (Shields, 1975). Helen Thompson Woolley was the first psychologist to put intellectual sex differences to empirical test (Thompson, 1903). She examined 25 women and 25 men in a series of tests of their intellectual, motor, and sensory abilities. Any sex differences were negligible. Also, insofar as men attained greater eminence than women, Thompson pointed to environmental causes, especially women's limited opportunities. A few other psychologists—Leta Stetter Hollingworth (1914) in particular—continued to use empirical methods to undermine the hypothesis of female inferiority. However, the influence of these pioneering feminist psychologists was lessened by the then existing restrictions on women's access to desirable careers in science.

The discipline of psychology sputtered along in the first half of the 20th century, slowed down by two world wars and one major depression. Although developmental psychologists made some progress in studying the behaviors of girls and boys, systematic empirical research on gender was limited. Adding more essentialist concepts to psychologists' repertoire, Sigmund Freud (1933) brought forth metaphoric concepts such as "penis envy" and the Oedipus and Electra complexes that endowed women with the moral inferiority of a lesser superego.

Among many other prominent psychologists, essentialist ideas akin to maternal instinct continued to hold sway. G. Stanley Hall (1918), one of the pioneers of

American psychology, was only slightly less essentialist than most other psychologists in acknowledging that women might want to enter male spheres of activity. He wrote that allowing women to compete with men would cause women then to neglect their instinctive urges to mother and so would bring about "race suicide." Later in the century, Bruno Bettelheim (1965, p. 15) wrote, in a discussion of women's careers in science, "We must start with the realization that, as much as we want women to be good scientists or engineers, they want first and foremost to be womanly companions of men and to be mothers." Similarly, Erik Erikson (1964, p. 586) maintained that women have a "somatic design that harbors an 'inner space' destined to bear the offspring of chosen men, and with it, a biological, psychological, and ethical commitment to take care of human infancy."

Such highly essentialist analyses endowed women with an intrinsic desire to center their lives on marriage and childrearing. To be sure, social science of the 20th century offered a few very different voices. Famously, in her study *Sex and Temperament in Three Primitive Cultures*, anthropologist Margaret Mead (1935, p. 280) espoused a radically social constructionist view of gender summarized by the statement, "many, if not all, of the personality traits which we have called masculine or feminine are as lightly lined to sex as are the clothing, the manners, and the form of head-dress that a society at a given period assigns to either sex." Psychology, in contrast, lacked a similarly prominent mid-20th-century advocate of cultural influences on gender.

The *second wave* of feminist activism, which began in the 1960s in the United States, was well supplied with critics of essentialist theorists of sex differences. These feminists regarded psychologists' theories of gender as promoting women's disadvantage. Betty Friedan's best-selling book, *The Feminine Mystique* (1963), laid some of the blame for women's discontent on the influence of Sigmund Freud and his followers. Feminist psychologists went further by offering broad-brush critiques of the psychological science of the day. For example, in an early statement, Naomi Weisstein (1968, 1971) ridiculed prominent psychologists' characterizations of women as childlike, dependent, unassertive, and interested only in finding a husband and bearing children. A few years later, Stephanie Shields (1975) wrote about the "social myths" such as maternal instinct promulgated by psychologists.

Some feminist critics escaped the biases of correspondent inference and essentialism by arguing that the most important causes of sex differences derived from the social context, which psychological theories of gender had usually not endowed with causal force. In Weisstein's (1968, p. 75) words, "One must understand the social conditions under which women live if one is going to attempt to explain the behavior of women. And to understand the social conditions under which women live, one must be cognizant of the social expectations about women." Why did social expectations become such a prominent causal category in feminists' writing, thus overcoming the usual human tendency to favor essentialist explanations? The importance of social expectations increased because the woman's movement demanded change in these social conditions. To activists, it was obvious that causes of women's disadvantage were located in the social environment, and not in unchanging, unalterable traits of women. Feminist activists of the day sought to change this environment, especially with respect to social patterns such as violence against women, sexual harassment, discrimination in employment and education, and restrictions on women's reproductive freedom.

The Psychology of Gender in Fuller Flower

Gender research became a major endeavor in psychology in the second half of the 20th century (Eagly, Eaton, Rose, Riger, & McHugh, in press). In illustration, Figure 12.1 displays over time the number of journal articles pertaining to human population groups and listed in PsycINFO with the words "sex" or "gender" in their title. In the mid-1960s, the incidence of "sex" in titles began to rise, followed later by a sharp rise in articles with "gender" in their titles. As the term "sex" declined somewhat in popularity, "gender" rose, but "sex" has gained again from the mid-1990s onward (see also Haig, 2004). Whether framed by sex or gender concepts, the psychology of gender has become a popular topic within psychological science.

It might be expected that most of this research, especially insofar as it was inspired by feminist activism, would emphasize social context, thereby pointing to the rules, norms, and behavior patterns that activists sought to change. Not necessarily. When second-wave feminist psychologists first turned to studying gender, the emphasis of most of this research was on women's intrinsic nature rather than the social context of women's lives. These feminism-influenced psychologists did not escape the tendency to locate the causes of sex differences in the essential qualities of women and men. The gender research that attracted the most attention in these years placed causes of sex differences inside the person. Although this new research helped reverse the neglect of women in psychological theories and typically portrayed women more favorably than earlier research and theory, these early second-wave feminist psychologists did not fundamentally change psychologists' essentialist thinking about sex differences.

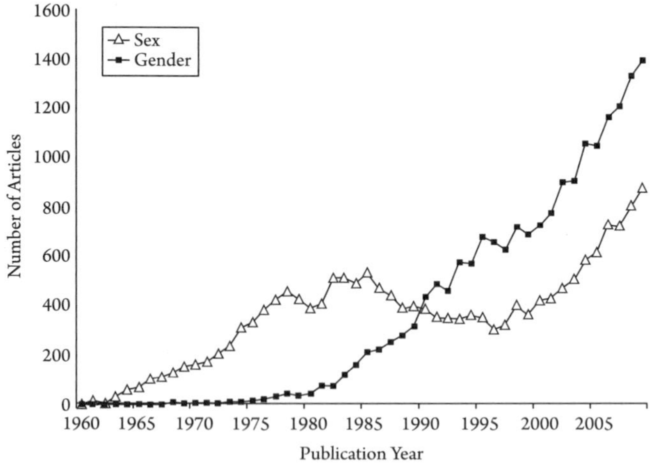

Figure 12.1. Annual frequency of journal articles with sex or gender in their title listed in PsycINFO 1960–2010, addressing human population groups.

Examples of this trait-oriented approach to understanding gender include Horner's (1972) work on fear of success, Gilligan's (1982) research on the caring and relational quality of women's moral reasoning, and Bem's (1974) and Spence's (Spence & Helmreich, 1978) research on psychological masculinity and femininity. Other feminists revised psychoanalysis by banishing its phallocentric bent and instead focusing on mothering as producing near-universal personality differences in women and men (e.g., Chodorow, 1978). In such research, psychologists continued to locate causes of masculine and feminine traits in essential qualities of men and women although not necessarily in traits that are inherent in human nature.

Why did these feminist psychologists also advocate essentialist ideas about gender when the feminist social movement pointed to women's social context as the engine of female disadvantage? Two reasons are plausible: (a) These feminist ideas were not entirely new but linked to prior psychological theories that relied on personal qualities to explain behavior, and (b) essentialist thinking is scientists' path of least resistance in theorizing about social categories. Concerning links to prior theories, the feminist psychologists who proposed the new theories during the early phases of the second wave feminism had been trained in the earlier, prefeminist traditions of psychology but desired to correct and elaborate these theories to take account of women. Thus, Horner's (1972) research emerged from Atkinson and McClelland's work on achievement motivation (McClelland, Atkinson, Clark, & Lowell, 1953), Gilligan's (1982) research from Kohlberg's (1969) research on moral reasoning, Bem's (1974) and Spence's (Spence & Helmreich, 1978) research from the personality psychology tradition of assessing individual differences in traits, and feminist psychoanalytic theories (e.g., Chodorow, 1978) from Freudian personality theory. Much of this early feminist research was thus consistent with the general tendency to regard men and women as possessing personal qualities that determine their behavior. Although this new research was temporarily influential and popular, it also met with robust feminist critique for what was deemed its insufficient emphasis on social context (e.g., Mednick, 1989)—that is, for not following the lead of activists who had hoped that psychology might counter the oppression of women by placing emphasis on the social context.

Where would situationalist theories arise? Perhaps mainly in social psychology because it possesses a theoretical vocabulary for understanding the social context of behavior. Because the power of the situation is the overriding theme of social psychology, the advice to look to the social context fit better into this subfield of psychology than into most others. Therefore, in the 1970s and beyond, research flourished on gender stereotypes and prejudice against women in its many manifestations, including backlash against women's assertiveness and the objectification of women's bodies (see Wood & Eagly, 2010, for review). Aspects of the social environment that had not been labeled or studied by psychologists emerged as active research areas among social, industrial/organizational, and clinical psychologists—for example, sexual harassment, sexual abuse and coercion, intimate partner violence, and restrictions on reproductive freedom. This phenomenon-oriented research was intended to contribute to the amelioration of social problems that oppressed women. This research is regarded as decidedly feminist, precisely because of its switch to placing causes in the environment rather than in the person, consistent with the feminist movement's emphasis on changing roles and social structures to ameliorate women's disadvantage.

In concert with these research areas, feminist social psychologists located in both psychology and sociology departments offered new theories that emphasized social expectations and the social structures within which gender relations are embedded. For example, Eagly's (1987; Eagly & Wood, 2012) social role theory argues that membership in the female or male social category subjects people to social expectations that affect their social behavior. Deaux and Major's (1987) theory argued that gender is enacted in dyadic interactions as a function of gendered beliefs about the self, others' expectations, and contextual influences that make gender more or less salient. Yet this trend to emphasize social context was even stronger in sociology departments. Ridgeway (e.g., 1991, 2011), a sociological social psychologist, argued that sex functions as a status cue because of people's past experience in observing hierarchical relations between men and women—for example, male managers and female secretaries and clerks. The higher status of men than women then accounts for many differences in female and male behavior. Other sociologists offered strongly situationalist theories by, for example, viewing gender as a performance induced by social expectations, with both men and women "doing gender" in response to these expectations (West & Zimmerman, 1987).

In addition to these developments, the core of Darwin's ideas about sex differences reemerged as evolutionary psychology in the 1980s (e.g., Buss, 1984). Like Darwin, contemporary evolutionary psychologists offer explanations of female and male behavior that are highly essentialist. They explain men's greater size and strength along with their aggressiveness and competitiveness as due to sexual selection mechanisms of male competition and female choice, assumed to be built in through evolutionary pressures. The idea of female intellectual inferiority is still present in this tradition, but for the most part narrowed to focus on skills, especially in mathematics, that lead to excellence in science (e.g., Geary, 1996). The greater male variability hypothesis is a continuing focus of interest (Archer & Mehdikhani, 2003), as are many other hypotheses derived from Darwin's evolutionary theory.

The contrast between social psychological theories emphasizing social context and evolutionary psychology theories emphasizing essential, evolved sex-differentiated attributes is apparent in contemporary science (Eagly & Wood, 2011). The match of evolutionary psychology to the essentialist naïve psychology of sex differences gives it a certain explanatory edge, especially with the general public. Moreover, the rise in evolutionary psychology in the 1980s during the period when gender roles were becoming destabilized fits the generalization that the threat of change in gender relations increases the attractiveness of essentialist explanations of sex differences, at least in the minds of men (Morton et al., 2009).

Emphasis on social context is also influential but, arrayed as it is with feminist influence, it is particularly welcome in feminist-influenced audiences and perhaps for women in general, who tend to be less essentialist than men in relation to gender (Smiler & Gelman, 2008). For other audiences, an emphasis on social context can appear tainted by association with a progressive politics deployed to foster gender equality. In fact, research has demonstrated that liberals are more likely than conservatives to rely on situational explanations for social and personal problems, evidently because they are motivated to correct their first reactions favoring personal causes by reaching for situational explanations more consistent with their values (Skitka, Mullen, Griffin, Hutchinson, & Chamberlin, 2002). Many feminist researchers may be similarly motivated to overcome correspondent inference and thereby identify and test situational attributions for female-male differences.

Science does offer means for determining which theory or theories are more valid or, more productively, for melding the valid portions of each theory into a more complex nature-nurture theory. Progress in this direction is not fast, however, because scientists have entrenched positions on issues of sex differences and similarities, just as they have entrenched positions on many other scientific issues. In the psychology of gender as in other domains, scientists' defensive processes allow them to cling to their own theories, and, to some extent, to produce tainted evidence (Mahoney, 1979). Similarly, in daily life people engage in selective evaluation of scientific evidence by reacting more positively to information that fits their preferences (e.g., Bastardi, Uhlmann, & Ross, 2011).

CONCLUSION

I have presented two types of judgmental biases that sometimes contaminate research and have illustrated them in relation to research on the psychology of gender. One bias is correspondent inference, allied with psychological essentialism, which locates causes in the essences of women and men. The other bias is ingroup favoritism, allied with the attitudinal congeniality bias, which produces a tendency to favor information favorable to one's own group, including scientific information. Both types of biases are important in everyday life, and it is plausible that they act on scientists as well, including those who study gender. An implicit preference to favor men and their dominance in society may have fueled earlier scientific depictions of women as unintelligent and ruled by maternal instinct, given that scientists were almost exclusively male. With women's large-scale entry into science, especially psychology, coinciding with and following after the second-wave feminist movement (Kite et al., 2001), feminist ideology encouraged scientists, especially feminists, to overcome the favoring of essentialist causes of sex differences and to instead consider situational forces as responsible for women's disadvantaged social position.

The presence of more female researchers also would counter men's ingroup bias by introducing women's ingroup bias. It is thus unlikely that women are any freer than men from ingroup bias, but their inclusion among researchers lessens the imbalance toward male ingroup bias that existed when most scientists were men. Women's inclusion can also raise uncomfortable issues about biases to the extent that women produce different findings than men or advocate different theories. Ethnic, racial, and social class diversity among scientists would similarly add new ingroup biases that could counter traditional biases that thrived when psychological science was pursued primarily by white men.

Ingroup bias can also foster male scientists' emphasis on essentialist causes of sex differences because locating women's disadvantage in deeply rooted dispositions, especially dispositions inherent in human nature, favors the patriarchal status quo and decreases belief in the malleability of sex differences (Brescoll & LaFrance, 2004; Morton et al., 2009). For example, if women are less assertive and dominant than men, and this sex difference is built into humans' nature, women would be forever subordinated in society, never gaining equal access to power and authority. If women naturally possess nurturing tendencies and men do not, there would be little hope that care of children would ever be equally shared between women and men.

Once scientists accept a theory, that acceptance further sets in motion processes influenced by the congeniality bias in information processing. Scientists often design tests that subtly or not so subtly favor their own theory and are unaware of such biases. For example, a widely accepted finding frequently cited by evolutionary psychologists—that men are more likely than women to interpret friendliness as sexual interest—seemed well established through many replications of a design that failed to also test whether men are more likely than women to interpret sexual interest as friendliness (see Farris, Treat, Viken, & McFall, 2008b, for review). An experimental design that allowed a test of this two-way linkage between friendliness and sexual interest supported this reciprocal model and produced the revised generalization that men are less able than women to discriminate between different emotions (Farris, Treat, Viken, & McFall, 2008a). Thus, a one-sided hypothesis had led to one-sided experimentation that was eventually corrected by researchers who were skeptical about a widely accepted generalization.

Scientists not only design biased tests of hypotheses but also are less critical about evidence that favors their own theory than evidence that challenges it (Mahoney, 1977). Scientists go out of their way to read and cite studies that favor their own theories and ruthlessly critique unwelcome findings (e.g., Chapman, Ragg, & McGeechan, 2009; Kelley & Blashfield, 2009).

For an understanding of the endpoint of science that is biased by such processes, I offer a statement by George Miller (1963, p. 150), "[A scientist] is looking for something that matches up to his image of what the world must be, something that meets a test he himself imposes, something that has meaning only in terms of the standards he lives by. In that sense, the scientist is Everyman, looking just as you and I. We go and look for the things we want, and when we find them we find part of ourselves."

As Miller (1963) suggested, scientists impose frameworks that make sense in terms of their own lives and cultural contexts. Scientists' ideas for explaining human behavior are influenced by implicit biases that are largely invisible to the scientists themselves. Therefore, gender biases are usually unarticulated or so cloaked in scientific terminology that they are unrecognizable as bias. However, these biases often become salient from the perspective of other theories, ideologies, social groups, cultures, and historical periods. Researchers committed to alternative theories are likely to raise issues about theories that oppose or differ from their own perspectives. By such means, science is to some extent self-correcting. Yet, at least in the short run scientists usually do find what they are looking for, very often compromising the objectivity of science.

Is there any hope for a science of human behavior that is less compromised by these bias-producing processes? Perhaps. In addition to a motive to defend one's attitudes, beliefs, and behaviors from challenges, known as *defense motivation*, social psychologists have also proposed a contrasting motive known as *accuracy motivation*, a general propensity to favor accurate appraisals of reality (Chaiken & Legerwood, 2012; Chaiken, Liberman, & Eagly, 1989; Eagly et al., 1999). Accuracy motivation and defense motivation can oppose one another because uncongenial information can be accurate and congenial information can be inaccurate. Overcoming defense motivation in such situations does not necessarily come easily, given that congenial information is more pleasant and usually has a schematic advantage over uncongenial information. Congenial information just "makes sense" because it fits in readily with preexisting attitudes and beliefs. To counter such bias, scientific training ought to include study of judgmental biases. Also, professors who teach graduate students ought to convey that science is a search for the truth, not a competitive sport.

Meta-analytic projects can be helpful in adjudicating the findings produced in the context of alternative theories. Meta-analysts can bring disparate findings together in a common framework and represent methodological differences between studies that may be responsible for inconsistent findings. Meta-analysts can also classify research findings by researchers' theoretical positions. However, biases can run deep into primary research and be undetectable to meta-analysts. In the case of sex-of-author effects, the ease of determining author sex has brought one probable source of bias into the open. Similarly, author bias in the form of favoring investigators who share one's ethnicity has been detected in citation patterns (Greenwald & Schuh, 1994). However, in such cases, meta-analyses and studies of citation patterns can only document bias, not remove it from the scientific literature.

AUTHOR NOTE

I thank Wendy Wood, Joan Linsenmeier, and the editors of this book for comments on a draft of this chapter.

REFERENCES

Anderson, K. J., & Leaper, C. (1998). Meta-analyses of gender effects on conversational interruption: Who, what, when, where, and how. *Sex Roles, 39*, 225–252. doi:10.1023/A:1018802521676

Archer, J., & Mehdikhani, M. (2003). Variability among males in sexually selected attributes. *Review of General Psychology, 7*, 219–236. doi:10.1037/1089-2680.7.3.219

Bastardi, A., Uhlmann, E. L., & Ross, L. (2011). Wishful thinking: Belief, desire, and the motivated evaluation of scientific evidence. *Psychological Science. 22*, 731–732. doi:10.1177/0956707611406447

Becker, B. J. (1986). Influence again: Another look at studies of gender differences in social influence. In J. S. Hyde & M. Linn (Eds.), *The psychology of gender: Advances through meta-analysis* (pp. 178–209). Baltimore, MD: Johns Hopkins University Press.

Bem, S. L. (1974). The measurement of psychological androgyny. *Journal of Consulting and Clinical Psychology, 42,* 155–162. doi:10.1037/h0036215

Bettelheim, B. (1965). The commitment required of a woman entering a scientific profession in present-day American society. In J. A. Mattfield & C. G. Van Aken (Eds.), *Women and the scientific professions* (pp. 3–19). Cambridge, MA: MIT Press.

Bettencourt, B. A., Charlton, K., Dorr, N., & Hume, D. L. (2001). Status differences and in-group bias: A meta-analytic examination of the effects of status stability, status legitimacy, and group permeability. *Psychological Bulletin, 127,* 520–542. doi:10.1037/0033-2909.127.4.520

Brescoll, V., & LaFrance, M. (2004). The correlates and consequences of newspaper reports of research on sex differences. *Psychological Science, 15,* 515–520. doi:10.1111/j.0956-7976.2004.00712.x

Brewer, M. B. (1988). A dual process model of impression formation. In T. Srull & R. S. Wyer Jr. (Eds.), *Advances in social cognition* (Vol.1, pp. 1–36). Hillsdale, NJ: Lawrence Erlbaum.

Brewer, M. B. (1999). The psychology of prejudice: Ingroup love or outgroup hate? *Journal of Social Issues, 55,* 429–444.

Buss, D. M. (1984). Evolutionary biology and personality psychology: Toward a conception of human nature and individual differences. *American Psychologist, 39,* 1135–1147. doi:10.1037/0003-066X.39.10.1135

Chaiken, S., & Legerwood, A. (2012). A theory of heuristic and systematic information processing. In P. van Lange, A. Kruglanski, & E. T. Higgins (Eds.), *Handbook of theories in social psychology* (Vol. 1, pp. 246–266). London, England: Sage.

Chaiken, S., Liberman, A., & Eagly, A. H. (1989). Heuristic and systematic processing within and beyond the persuasion context. In J. S. Uleman & J. A. Bargh (Eds.), *Unintended thought: Limits of awareness, intention, and control* (pp. 212–252). New York, NY: Guilford Press.

Chapman, S., Ragg, M., & McGeechan, K. (2009). Citation bias in reported smoking prevalence in people with schizophrenia. *Australian and New Zealand Journal of Psychiatry, 4,* 277–282. doi:10.1080/00048670802653372

Chodorow, N. J. (1978). *The reproduction of mothering: Psychoanalysis and the sociology of gender.* Berkeley, CA: University of California Press.

Darwin, C. (1882). *The descent of man, and selection in relation to sex* (new ed.). New York, NY: D. Appleton and Company,

Deaux, K., & Major, B. (1987). Putting gender into context: An interactive model of gender-related behavior. *Psychological Review, 94,* 369–389. doi:10.1037/0033-295X.94.3.369

Eagly, A. H. (1987). *Sex differences in social behavior: Asocial-role interpretation.* Hillsdale, NJ: Lawrence Erlbaum.

Eagly, A. H. (1995). The science and politics of comparing women and men. *American Psychologist, 50,* 145–158. doi:10.1037/0003-066X.50.3.145

Eagly, A. H. (1998). Attitudes and the processing of attitude-relevant information. In J. G. Adair, D. Bélanger, & K. L. Dion (Eds.), *Advances in psychological science: Social, personal and cultural aspects* (Proceedings of the 26th International Congress of Psychology, Vol. 1, pp. 185–201). East Sussex, England: Psychology Press.

Eagly, A. H., & Carli, L. L. (1981). Sex of researchers and sex-typed communications as determinants of sex differences in influenceability: A meta-analysis of social influence studies. *Psychological Bulletin, 90,* 1–20. doi:10.1037/0033-2909.90.1.1

Eagly, A. H., & Chaiken, S. (1993). *The psychology of attitudes*. Fort Worth, TX: Harcourt Brace Jovanovich.

Eagly, A. H., & Chaiken, S. (2005). Attitude research in the 21st century: The current state of knowledge. In D. Albarracín, B. T. Johnson, & M. P. Zanna (Eds.), *The handbook of attitudes* (pp. 743–767). Mahwah, NJ: Lawrence Erlbaum.

Eagly, A. H., Chen, S., Chaiken, S., & Shaw-Barnes, K. (1999). The impact of attitudes on memory: An affair to remember. *Psychological Bulletin, 125,* 64–89. doi:10.1037/0033-2909.125.1.64

Eagly, A. H., Eaton, A., Rose, S., Riger, S., & McHugh, M. (in press). Feminism and psychology: Analysis of a half-century of research on women and gender. *American Psychologist.*

Eagly, A. H., & Johnson, B. T. (1990). Gender and leadership style: A meta-analysis. *Psychological Bulletin, 108,* 233–256. doi:10.1037/0033-2909.108.2.233

Eagly, A. H., Kulesa, P., Brannon, L. A., Shaw, K., & Hutson-Comeaux, S. (2000). Why counterattitudinal messages are as memorable as proattitudinal messages: The importance of active defense against attack. *Personality and Social Psychology Bulletin, 26,* 1392–1408. doi:10.1177/0146167200263007

Eagly, A. H., & Wood, W. (2011). Feminism and the evolution of sex differences and similarities. *Sex Roles, 64,* 758–767. doi: 10.1007/s11199-011-9949-9

Eagly, A. H., & Wood, W. (2012). Social role theory: A biosocial analysis of sex differences and similarities. In P. van Lange, A. Kruglanski, & E. T. Higgins (Eds.), *Handbook of theories in social psychology* (Vol. 2, pp. 458–476). London, England: Sage.

Edwards, K., & Smith, E. E. (1996). A disconfirmation bias in the evaluation of arguments. *Journal of Personality and Social Psychology, 71,* 5–24. doi:10.1037/0022-3514.71.1.5

Ellemers, N., & Haslam, S. A. (2012). Social identity theory. In P. van Lange, A. Kruglanski, & E. T. Higgins (Eds.), *Handbook of theories in social psychology* (Vol. 2, pp. 379–398). London, England: Sage.

Erikson, E. (1964). Inner and outer space: Reflections on womanhood. *Daedalus, 93,* 582–606.

Farris, C., Treat, T. A., Viken, R. J., & McFall, R. M. (2008a). Perceptual mechanisms that characterize gender differences in decoding women's sexual intent. *Psychological Science, 19,* 348–354. doi:10.1111/j.1467–9280.2008.02092.x

Farris, C., Treat, T. A., Viken, R. J., & McFall, R. M. (2008b). Sexual coercion and the misperception of sexual intent. *Clinical Psychology Review, 28,* 48–66. doi:10.1016/j.cpr.2007.03.002

Festinger, L. (1957). *A theory of cognitive dissonance.* Evanston, IL: Row, Peterson.

Festinger, L. (1964). *Conflict, decision, and dissonance.* Palo Alto, CA: Stanford University Press.

Freud, S. (1933). *New introductory lectures on psycho-analysis.* New York, NY: Penguin Books.

Friedan, B. (1963). *The feminine mystique.* New York, NY: W. W. Norton.

Gawronski, B. (2003). On difficult questions and evident answers: Dispositional inference from role-constrained behavior. *Personality and Social Psychology Bulletin, 29,* 1459–1475. doi:10.1177/0146167203256375

Geary, D. C. (1996). Sexual selection and sex differences in mathematical abilities. *Behavioral and Brain Sciences, 19,* 229–284. doi:10.1017/S0140525X00042400

Gilbert, D. T. (1998). Ordinary personology. In D. T. Gilbert, S. T. Fiske, & G. Lindzey (Eds.), *The handbook of social psychology* (4th ed., Vol. 2, pp. 89–150). Boston, MA: McGraw-Hill.

Gilligan, C. (1982). *In a different voice: Psychological theory and women's development.* Cambridge, MA: Harvard University Press.

Greenwald, A. G., & Schuh, E. S. (1994). An ethnic bias in scientific citations. *European Journal of Social Psychology, 24,* 623–639. doi:10.1002/ejsp.2420240602

Haig, D. (2004). The inexorable rise of gender and the decline of sex: Social change in academic titles, 1945–2001. *Archives of Sexual Behavior, 33,* 87–96. doi:10.1023/B:ASEB.0000014323.56281.0d

Hall, G. S. (1918). *Youth, its education, regimen and hygiene.* New York, NY: Appleton.

Hall, J. A. (1978). Gender effects in decoding nonverbal cues. *Psychological Bulletin, 85,* 845–857. doi:10.1037/0033-2909.85.4.845

Hall, J. A. (1984). *Nonverbal sex differences: Communication accuracy and expressive style.* Baltimore, MD: Johns Hopkins University Press.

Hart, W., Albarracin, D., Eagly, A. H., Lindberg, M., Merrill, L., Brechan, I., & Lee, K. H. (2009). Feeling validated versus being correct: A meta-analysis of selective exposure to information. *Psychological Bulletin, 135,* 555–588. doi:10.1037/a0015701

Haslam, N., Rothschild, L., & Ernst, D. (2000). Essentialist beliefs about social categories. *British Journal of Social Psychology, 39,* 113–127. doi: 10.1348/014466600164363

Hogg, M. A. (2006). Social identity theory. In P. J. Burke (Ed.), *Contemporary social psychological theories* (pp. 111–136). Palo Alto, CA: Stanford University Press.

Hollingworth, L. S. (1914). Variability as related to sex differences in achievement. *American Journal of Sociology, 19,* 510–530.

Horner, M. S. (1972). Toward an understanding of achievement-related conflicts in women. *Journal of Social Issues, 28*(2), 157–175.

Johnson, B. T., & Eagly, A. H. (1989). Effects of involvement on persuasion: A meta-analysis. *Psychological Bulletin, 106,* 290–314. doi:10.1037/0033-2909.106.2.290

Kelley, L. P., & Blashfield, R. K. (2009). An example of psychological science's failure to self-correct. *Review of General Psychology, 13,* 122–129. doi:10.1037/a0015287

Kenrick, D. T., Maner, J. K., & Li, N. P. (2005). Evolutionary social psychology. In D. M. Buss (Ed.), *The handbook of evolutionary psychology* (pp. 803–827). Hoboken, NJ: John Wiley.

Kite, M. E., Russo, N. F., Brehm, S. S., Fouad, N. A., Hall, C. C. I., Hyde, J. S., & Keita, G. P. (2001). Women psychologists in academe: Mixed progress, unwarranted complacency. *American Psychologist, 56,* 1080–1098. doi:10.1037/0003-066X.56.12.1080

Kohlberg, L. (1969). Stage and sequence: The cognitive-developmental approach to socialization. In D. Goslin (Ed.), *Handbook of socialization theory and research* (pp. 347–480). Chicago, IL: Rand McNally.

Lord, C. G., Ross, L., & Lepper, M. R. (1979). Biased assimilation and attitude polarization: The effects of prior theories on subsequently considered evidence. *Journal of Personality and Social Psychology, 37,* 2098–2109. doi:10.1037/0022-3514.37.11.2098

Mahoney, M. J. (1977). Publication prejudices: An experimental study of confirmatory bias in the peer review system. *Cognitive Therapy and Research, 2,* 69–70. doi:10.1007/BF01173636

Mahoney, M. J. (1979). Psychology of the scientist: An evaluative review. *Social Studies of Science, 9,* 349–375. doi:10.1177/030631277900900304

Martin, C. L., & Parker, S. (1995). Folk theories about sex and race differences. *Personality and Social Psychology Bulletin, 21,* 45–57. doi: 10.1177/0146167295211006

McClelland, D. C., Atkinson, J. W., Clark, R. A., & Lowell, E. L. (1953). *The achievement motive.* Century psychology series. New York, NY: Appleton-Century-Crofts.

Mead, M. (1935). *Sex and temperament in three primitive societies.* New York, NY: William Morrow.

Mednick, M. T. (1989). On the politics of psychological constructs: Stop the bandwagon, I want to get off. *American Psychologist, 44,* 1118–1123. doi:10.1037/0003-066X.44.8.1118

Mill, J. S. (1869). *The subjection of women.* New York, NY: Appleton.

Miller, G. A. (1963). Thinking, cognition, and learning. In B. Berelson (Ed.), *The behavioral sciences today* (pp. 139–150). New York, NY: Basic Books.

Morton, T. A., Haslam, S. A., Postmes, T., & Ryan, M. K. (2006). We value what values us: The appeal of identity-affirming science. *Political Psychology, 27,* 823–838. doi:10.1111/j.1467-9221.2006.00539.x

Morton, T. A., Postmes, T., Haslam, S. A., & Hornsey, M. J. (2009). Theorizing gender in the face of social change: Is there anything essential about essentialism? *Journal of Personality and Social Psychology, 96,* 653–664. doi:10.1037/a0012966

Mullen, B., Brown, R., & Smith, C. (1992). Ingroup bias as a function of salience, relevance, and status: An integration. *European Journal of Social Psychology, 22,* 103–122. doi:10.1002/ejsp.2420220202

Prentice, D. A., & Miller, D. T. (2007). Psychological essentialism of human categories. *Current Directions in Psychological Science, 16,* 202–206. doi:10.1111/j.1467-8721.2007.00504.x

Rangel, U., & Keller, J. (2011). Essentialism goes social: Belief in social determinism as a component of psychological essentialism. *Journal of Personality and Social Psychology, 100,* 1056–1078. doi: 10.1037/a0022401

Ridgeway, C. (1991). The social construction of status value: Gender and other nominal characteristics. *Social Forces, 70,* 367–386. doi:10.2307/2580244

Ridgeway, C. (2011). *Framed by gender: How gender inequality persists in the modern world.* New York, NY: Oxford University Press.

Ross, L. (1977). The intuitive psychologist and his shortcomings: Distortions in the attribution process. In L. Berkowitz (Ed.), *Advances in experimental social psychology* (Vol. 10, pp. 173–220). New York, NY: Academic Press.

Ross, L. D., Amabile, T. M., & Steinmetz, J. L. (1977). Social roles, social control, and biases in social-perception processes. *Journal of Personality and Social Psychology, 35,* 485–494. doi:10.1037//0022–3514.35.7.485

Shields, S. A. (1975). Functionalism, Darwinism, and the psychology of women: A study in social myth. *American Psychologist, 30,* 739–754. doi:10.1037/h0076948

Shields, S. A., & Bhatia, S. (2009). Darwin on race, gender, and culture. *American Psychologist, 64,* 111–119. doi:10.1037/a0013502

Skitka, L. J., Mullen, E., Griffin, T., Hutchinson, S., & Chamberlin, B. (2002). Dispositions, scripts, or motivated correction? Understanding ideological differences in explanations for social problems. *Journal of Personality and Social Psychology, 83,* 470–487. doi:10.1037/0022-3514.83.2.470

Smiler, A. P., & Gelman, S. A. (2008). Determinants of gender essentialism in college students. *Sex Roles, 58,* 864–874. doi:10.1007/s11199-008-9402-x

Spence, J. T., & Helmreich, R. L. (1978). *Masculinity and femininity: Their psychological dimensions, correlates, and antecedents.* Austin, TX: University of Texas Press.

Strube, M. J., & Garcia, J. E. (1981). A meta-analytic investigation of Fiedler's contingency model of leadership effectiveness. *Psychological Bulletin, 90,* 307–321. doi:10.1037/0033-2909.90.2.307

Thomas, J. R., & French, K. E. (1985). Gender differences across age in motor performance: A meta-analysis. *Psychological Bulletin, 98,* 260–282. doi:10.1037/0033-2909.98.2.260

Thompson, H. B. (1903). *The mental traits of sex.* Chicago, IL: University of Chicago Press.

Turner, J. C. (2000). Social identity. In A. E. Kazdin (Ed.), *Encyclopedia of psychology* (Vol. 7, pp. 341–343). Washington, DC: American Psychological Association.

Uleman, J. S., Saribay, S. A., & Gonzalez, C. M. (2008). Spontaneous inferences, implicit impressions, and implicit theories. *Annual Review of Psychology, 59,* 329–360. doi:10.1146/annurev.psych.59.103006.093707

Vicinus, M. (Ed.). (1972). *Suffer and be still: Women in the Victorian age.* Bloomington, IN: Indiana University Press.

Weisstein, N. (1968). *Kinder, Kirche, Kuche as scientific law: Psychology constructs the female.* Boston, MA: New England Press.

Weisstein, N. (1971). Psychology constructs the female; or, the fantasy life of the male psychologist (with some attention to the fantasies of his friends, the male biologist and the male anthropologist). *Journal of Social Education, 35,* 362–373.

West, C., & Zimmerman, D. H. (1987). Doing gender. *Gender and Society, 1,* 125–151.

Wood, W. (1987). Meta-analytic review of sex differences in group performance. *Psychological Bulletin, 102,* 53–71. doi:10.1037/0033-2909.102.1.53

Wood, W., & Eagly, A. H. (2009). Gender identity. In M. R. Leary & R. H. Hoyle (Eds.), *Handbook of individual differences in social behavior* (pp. 109–125). New York, NY: Guilford Press.

Wood, W., & Eagly, A. H. (2010). Gender. In S. Fiske, D. Gilbert, & G. Lindzey (Eds.), *Handbook of social psychology* (Vol. 1, pp. 629–667). New York, NY: John Wiley.

Wood, W., & Eagly, A. H. (in press). Biosocial construction of sex differences and similarities in behavior. In J. M. Olson & M. P. Zanna (Eds.), *Advances in experimental social* psychology (Vol. 46). London: Elsevier.

Wood, W., Lundgren, S., Ouellette, J. A., Busceme, S., & Blackstone, T. (1994). Minority influence: A meta-analytic review of social influence processes. *Psychological Bulletin, 11,* 323–345. doi:10.1037/0033-2909.115.3.323

Yzerbyt, V., & Demoulin, S. (2010). Intergroup relations. In S. Fiske, D. Gilbert, & G. Lindzey (Eds.), *Handbook of social psychology* (Vol. 2, pp. 1024–1083). New York, NY: John Wiley.

13 The Theory Ladenness of the Mental Processes Used in the Scientific Enterprise[1]

Evidence from Cognitive Psychology and the History of Science

William F. Brewer

NATURALIZED PHILOSOPHY OF SCIENCE

This chapter is an exercise in naturalized philosophy of science—the use of evidence from science to address problems in the philosophy of science. In the late 19th century psychology broke away from philosophy, and for a period both sides were comfortable with this division of intellectual labor. However, in the middle of the 20th century there were two trends which led to a reconsideration of the division. In the course of criticizing certain aspects of logical positivism, Quine (1969) argued that there is a strong role for the use of evidence from experimental psychology in epistemology (see Kornblith, 1994 for more details). The second trend arose from the work of Norwood Hanson (1958) and Thomas Kuhn (1962) in the philosophy of science. Both of these writers made use of evidence from psychology and from the

[1] Some of the ideas in this chapter were initially worked out with Bruce Lambert in a wonderful collaboration in 1990–91. Lester Loschky helped me work through the research on perception and attention in a series of discussions in 1994-96 and again in 2003. Others who have, over the years, provided feedback about the ideas in this chapter are: Fouad Abd-El-Khalick, Clark Chinn, Don Dulany, Allan Franklin, Dan Reisberg, Brian Ross, Dan Simons, Ryan Tweney, and Corinne Zimmerman. I thank Ellen Brewer, Clark Chinn, Jason Finley, and Bruce Lambert for comments on an earlier draft of this chapter.

history of science in their work. Hanson often reviewed findings from experimental psychology in his writings, but he was a little hesitant about the status of this evidence in making philosophical arguments. In one place Hanson (1969) makes use of results from psychology, but then states, "A thorough examination of that, however, is a task for the experimental psychologist, a title to which I can lay no claim whatever. We have been concerned here with a conceptual inquiry, and that is the province of philosophy" (p. 104). Kuhn was much more explicit about the use of naturalized evidence and evidence from the history of science. The very first sentence in *The Structure of Scientific Revolutions* (1962) states, "History, if viewed as a repository for more than anecdote or chronology, could produce a decisive transformation in the image of science by which we are now possessed" (p. 1).

In somewhat more recent times the philosopher Ronald Giere (1988) has made strong arguments for the central role of evidence from cognitive psychology in the philosophy of science. I think the emerging view is that a mature philosophy of science will need to make use of evidence from a variety of disciplines, for example, psychology, sociology, biology, and history of science (cf. Giere, 2006; Proctor & Capaldi, 2006). The philosopher Peter Godfrey-Smith (2003) gives a clear and simple account of the naturalist approach. He stated, "For me, naturalism holds that the best way to address many philosophical problems is to approach them within our best current scientific picture of the world" (p. 238).

STRUCTURE OF THE CHAPTER

In this chapter, I first describe the problem of the theory ladenness of perception as originally posed by the philosophers Hanson and Kuhn. Then I provide a general framework from cognitive psychology that is used to organize and evaluate the evidence for theory ladenness in the scientific process. The core of the chapter outlines the evidence from cognitive psychology and the history of science on the issue of theory ladenness. The chapter concludes with a discussion of the conscious/unconscious nature of the top-down theory-laden psychological processes and with a short section on the normative implications of the material in this chapter for carrying out science.

THE PROBLEM OF THEORY LADENNESS

In this section of the chapter, I outline the claim that perception is *theory-laden*, that is, that one's theories/beliefs can influence one's conscious visual experience.

Norwood Hanson

Hanson's book, *Patterns of Discovery* (1958), made creative and thoughtful arguments for the theory ladenness of observation in science. One of his arguments was an extended thought experiment involving the perceptual experiences of Tycho Brahe (who held a geocentric theory of the earth/sun relation) and Johannes Kepler (who held a heliocentric theory of the earth/sun relation). Hanson wrote, "Let us consider Johannes Kepler; imagine him on a hill watching the dawn. With him is Tycho Brahe" (p. 5). Hanson asked, "Do Kepler and Tycho see the same thing in the east at dawn?" (p. 5). He asserted, "Tycho [sees] a mobile sun, Kepler [sees] a static sun" (p. 17) and "for... Kepler the horizon drops; for... Tycho the sun rises" (p. 182). Many philosophers have used their intuitions about this example, decided that the outcome of the thought experiment is that both Kepler and Brahe see the same (geocentric) view, and concluded that Hanson was wrong. However, a careful reading of Hanson's discussion of this example in his book *Perception and Discovery* (1969) shows that Hanson was making an expertise argument; he assumed that individuals who hold the heliocentric view but are not professional astronomers would, in fact, have a geocentric perception. Hanson states, "Unlike most of us who still speak of *sunrise* and *sunset* and who treat the Nautical Almanac as a kind of textbook of ptolemaic astronomy, the 20th Century astronomer will *not* see the sun as up over the horizon ready to vault to its zenith position" (p. 119). In his writings on the theory-laden issue, Hanson did not typically refer to actual experimental findings from psychology (though he did sometimes give footnotes that show he had read this literature widely). However, Hanson made thoughtful arguments based on stimuli that had been used in psychological experiments. He pointed out that there is a class of ambiguous figures (see Figure 13.1) which can give rise to two qualitatively different perceptual experiences. He argued that these two different experiences were due to different "conceptual organizations" of the incoming information (1969, p. 108). He chose these cases carefully because he realized that they provided strong evidence against the standard philosophical account—that this type of figure leads to one sensory/perceptual experience and then that single experience is *interpreted* in two different ways (cf. 1958, p. 5). Hanson made an explicit comparison between the Brahe/Kepler thought experiment and the perceptions of ambiguous picture stimuli. He concluded that "seeing is a 'theory-laden' undertaking" (1958, p. 19) and that "we usually *see* through spectacles made of our past experience [and] our knowledge" (1969, p. 149).

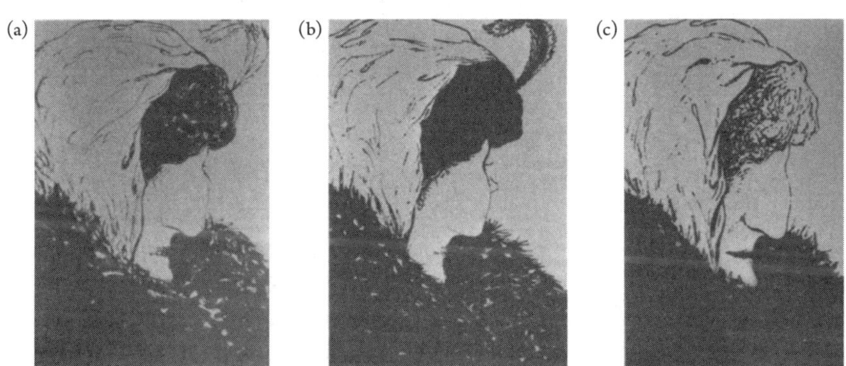

Figure 13.1 Old woman/young woman ambiguous figure used in Leeper, 1935. (Figure a is ambiguous; Figure b is an unambiguous young woman; Figure c is an unambiguous old woman) (used with permission of the author from Neisser, 1967, p. 142).

Thomas Kuhn

Kuhn's famous book, *The Structure of Scientific Revolutions* (1962), made very strong claims about the theory ladenness of perception. Kuhn used data from psychological experiments and placed much more weight on cases from the history of science than did Hanson. Kuhn described an experiment in which participants were asked to describe what they saw when presented with playing cards that were normal (e.g., a red four of hearts) and anomalous (e.g., a black four of hearts) exposed for very short time intervals (Bruner & Postman, 1949). He emphasized the fact that many participants shown anomalous cards reported seeing nonanomalous cards (e.g., reported seeing a standard black four of spades when presented with the anomalous black four of hearts). He also presented some of the qualitative reports of the participants which seemed to indicate that the anomalous cards were actually being misperceived (e.g., "I can't make the suit out, whatever it is. It didn't even look like a card that time. I don't know what color it is now or whether it's a spade or a heart," Kuhn, 1962, p. 63). The results of this experiment will be discussed in more detail later in this chapter. Kuhn also described Hanson's (1958) analysis of ambiguous figures. Kuhn found the idea of a visual figure that could give rise to two different visual experiences to be a powerful metaphor for his own thinking about theory change. He noted that "the switch of gestalt . . . is a useful elementary prototype for what occurs in full-scale paradigm shift" (p. 85).

Kuhn was clearly aware of the potential epistemological consequences of these psychological findings. If a scientist's theories impacted his or her observations, then how could the observations be used to decide between two competing theories? Many readers of Kuhn have interpreted his work as leading to some form of relativism (e.g., Barnes, 1974; Woolgar, 1988). Kuhn claimed that this was a misreading of his ideas (Baltas, Gavroglu, & Kindi, 2000, p. 307) and in places in his writing Kuhn seems to take a more moderate view. For example in his discussion of the Bruner and Postman (1949) experiment, Kuhn (1962) pointed out that with long exposures most participants could accurately report both the anomalous and nonanomalous cards, suggesting that the misperceptions were a somewhat limited phenomenon. In a more general discussion of the implications of the theory-laden position, Kuhn referred to the case of competing scientific paradigms and stated, "Practicing in different worlds, the two groups of scientists see different things when they look from the same point in the same direction. Again, that is not to say that they can see anything they please. Both are looking at the world" (1962, p. 149).

Nevertheless, it is obvious that Kuhn found the argument for the theory ladenness of observations to be very compelling and very troubling. In reviewing his account of the chemical revolution, Kuhn noted, "Lavoisier, we said, saw oxygen where Priestly had seen dephlogisticated air and where others had seen nothing at all" (p. 117). In discussing the different theories of Aristotle and Galileo, Kuhn stated, "I am, for example, acutely aware of the difficulties created by saying that when Aristotle and Galileo looked at swinging stones, the first saw constrained fall, the second a pendulum. . . . Nevertheless, I am convinced that we must learn to make sense of statements that at least resemble these" (p. 120). Probably the most extreme statement Kuhn ever made about the implications of theory ladenness was his claim that "after a revolution scientists are responding to a different world" (p. 110).

Conclusions

It is obvious that Hanson and Kuhn were practicing a naturalized form of philosophy of science with the focus on evidence from psychology and from the history of science. They both concluded that the evidence supported a strong degree of theory ladenness. These ideas have had an enormous impact on thinking in the field of philosophy of science (see Suppe, 1977, for a review and discussion).

It has now been quite some time since Hanson and Kuhn wrote their important works on the theory-ladenness issue. At the time they were writing, the vast majority of research in experimental psychology was focused on issues such as memory for nonsense syllables or maze learning in rats that did not seem to be relevant to the issues Hanson and Kuhn were interested in understanding. In the intervening period the field of cognitive psychology has flourished, and so a major goal of this chapter is to analyze the theory-ladenness issue in the light of current research and current thinking in cognitive psychology.

A FRAMEWORK FOR UNDERSTANDING THEORY LADENNESS

In this section of the chapter, I lay out the general framework that I have used to select and analyze the relevant work in cognitive psychology.

Theory Ladenness across Mental Processes

In attempting to understand the theory-ladenness issue in the context of current cognitive psychology (Brewer & Lambert, 1993, 2001), we came to believe that the issue of theory-ladenness in the philosophy of science had been focused too narrowly on the topic of visual perception. In retrospect it is clear that this emphasis derives from an intellectual tradition in philosophy that gives a prominent place in epistemology to individual perceptual experience. The tradition starts with the British Empiricists, continues with the Logical Positivists, and occurs unabated in the writings of the postpositivists such as Hanson and Kuhn. When we tried to analyze some of the historical examples discussed by the philosophers of science, it seemed to us that these episodes showed that there was theory ladenness across a variety of psychological processes. In attempting to work out these ideas, we found ourselves looking for a way to select and organize the mental processes that scientists use to carry out the scientific endeavor. One heuristic I have used to develop this typology is to think about the types of content that might make up the chapters in a textbook of cognitive psychology, in the hope that those topics might be more likely to be natural kinds. This is still a work in progress. In the current chapter I discuss the issue of theory ladenness in perception, attention, thinking, experimenting, memory, and communication. I am not satisfied with the organization of the subtopics under the heading of *thinking*; this topic clearly needs additional work. More detail will be given in this chapter to the topics of perception and attention because it is the possibility of the theory ladenness of these two psychological processes that has been thought to have major epistemological and metaphysical ramifications.

Data Are Frequently Not the Scientist's Individual Perceptual Experience

There is another reason to study cognitive processes other than perception. When thinking about science, it seems likely that the prototype most philosophers of science have had in mind was a scientist making an individual perceptual observation (e.g., Galileo rolling balls down an inclined plane or Newton looking at spectra generated by prisms). A number of scholars (e.g., Bogen & Woodward, 1992; Fodor, 1991) have pointed out that in current science the individual scientist is frequently part of a large group, and that with modern computer-based data acquisition and analysis, the actual data used in science are rarely the direct perceptual observations of an individual scientist. Bogen and Woodward (1992) put the point very forcefully: "It is data rather than perceptual beliefs that play a central evidential role in science and data are typically not descriptions of perceptual appearances or reports of perceptual belief at all" (p. 599). The implications of these arguments are that we need to look not just for the occurrence of theory ladenness of perceptual observation, but for the possible theory ladenness of data taken more broadly (e.g., in tables of numbers, statistical output, etc.). So we need to explore the role of theory in gathering data, evaluating data, selecting data, interpreting data, and in the carrying out of experiments.

The Term "Theory" Used Broadly

What is the *theory* in theory ladenness? Examination of the philosophy of science literature on theory ladenness shows that the term "theory" is used very broadly. Within cognitive psychology there has been much discussion of different forms of mental representation—images, concepts, schemas, mental models, naïve theories (Brewer, 1987, 1999, 2003; Johnson-Laird, 1983; Kosslyn, 1980; Markman, 1999; Rumelhart & Norman, 1988). While discussions of these different forms of mental representation are directly related to a naturalized discussion of the nature of laws and theories in science, it seems to me that these distinctions are of limited importance for discussions of the theory-ladenness issue. Thus for the purposes of this chapter I will use the term "theory" to mean roughly any form of top-down conceptual information.

The Use of Laboratory Data Derived from Nonscientist Participants

In an ideal world one would like the psychological evidence presented in this chapter to be based on data from scientists. In practice, almost all the psychological data are from nonscientist participants. It seems to me that for most of the topics covered in this chapter, one can safely make the inferential leap. Clearly, there are differences between scientists and nonscientists; however, it often seems possible to study analogous tasks. To investigate the role of visual knowledge/skill on attention, one might study nonscientists who were experts at some computer game, while with scientists one might study experts at the use of microscopes. If one is going to study the role of knowledge on memory with undergraduates, one might use materials related to contemporary rock groups, while with scientists one might use materials related to their specialty (e.g., the Higgs boson). While recognizing that there are differences between scientists and nonscientists, it seems reasonable to assume that the underlying mental processes in these cases would be the same. There may be some aspects of the psychology of scientists (e.g., the carrying out of experiments) where becoming a member of the culture of science leads to qualitative differences between scientists and nonscientists, but it will require a much improved knowledge base in the field of the psychology of science to resolve such issues. The small number of studies that actually provide data from working scientists are included in the sections on the history of science.

The Top-Down/Bottom-Up Synthesis in Perception

In analyzing the data on perception from cognitive psychology and from the history of science, I am going to work within the modal view in current cognitive psychology (e.g., Lindsay & Norman, 1977, p. 251; McClelland & Rumelhart, 1981). The modal view assumes that perception/observation is the joint product of top-down information (*theory* in the terms of this chapter) and bottom-up information (information from the world). Note that adopting this framework has powerful metaphysical/epistemological consequences. As I pointed out above many scholars have taken Hanson and Kuhn's arguments for the theory ladenness of observation to lead to epistemological relativism. The top-down/bottom-up synthesis allows one to accept evidence for top-down effects in observation while the bottom-up information can serve to avoid the slippery slope to relativism.

Jerry Fodor (1983, 1984) along with several other philosophers (Gilman, 1990; Raftopoulos, 2001) has rejected the top-down/bottom-up synthesis and proposed a theory of perception based on special purpose modules that are insulated from top-down information. There is not space in this chapter to discuss these issues; however, see Brewer and Loschky (2005) for an analysis and critique of these views.

In Perception the Strongest Top-Down Effects Occur with Degraded Input

Our previous analysis of perception (Brewer & Lambert, 1993, 2001, and particularly Brewer & Loschky, 2005) suggested that the proportion of information that comes from top-down processes versus bottom-up processes varies across perceptual tasks. Hallucinations and dreams are almost entirely driven by top-down processes. The perception of a vivid, unexpected object that one is attending to would include much bottom-up information. Our review of the literature in experimental psychology and in the history of science suggested that, in cases where theory had a strong impact on perception, the stimuli tended to be degraded, presented for a short time, or ambiguous. This finding has obvious implications for the methodology used by scientists when the data consist of personal perceptual observations (see the examples from the history of science and the discussion of the normative implications that occur later in this chapter). In addition, this finding reduces the impact that the occurrence of theory ladenness has on the epistemological status of observation in science.

The Facilitatory and Inhibitory Role of Theory

In retrospect, in our initial papers on theory ladenness (Brewer & Lambert, 1993, 2001), we may have focused too strongly on the negative role of top-down processes. We were eager to show the existence of top-down processes, so we pointed out many examples from cognitive psychology and the history of science where there was evidence for theory distorting perception, biasing attention, shifting interpretations, distorting memory, and so forth.

In this chapter I want to emphasize that having a theory is a two-edged sword. In psychological experiments we typically know what the input stimulus is, and so we can empirically determine the facilitatory or inhibitory role of theory. Examination of the evidence from cognitive psychology shows that having a correct theory about the information being presented can reduce perceptual thresholds, direct attention to the appropriate stimuli, facilitate the interpretation of information, and improve memory. It is probably the case that most of the time the inhibitory role of theory is a side effect of a theory-based process that normally functions to facilitate a particular mental process. For example, when attention is successfully directed at some theory-related object, it reduces attention to non-theory-related objects.

I think the same argument works for the history of science examples—theory can facilitate or inhibit the work of the scientist. Thus, by analogy with the psychological data, we can see that the scientist lucky enough to have a theory that reflects the world will find that the theory improves perception of the world, directs attention to appropriate aspects of the world, facilitates interpretation of the world, improves memory of information about the world, and so forth. For a scientist who has an incorrect theory of the world, the same psychological processes will inhibit finding the true state of the world. I realize this analysis is going to raise some philosophical hackles, but for a realist such as myself this gives a very nice account of the psychology of many of the historical episodes to be discussed in this chapter. Note that the working scientist does not know if his or her current theory correctly reflects the world or not. This is what makes science so hard. However, by looking at historical cases, where we now better understand the state of the world, we can see how theory facilitated or inhibited the scientist's understanding of the world. Note that this psychological/historical argument rejects the view held by relativists in the sociology of science that practicing Whig history is a conceptual mistake and instead embraces the view that in the history of *science* we often know who was right so we can use that to our advantage in studying the psychology of episodes from the history of science.

A classic example (almost a counterbalanced design!) can be found in the history of searches for planets in our solar system. In 1846 Urbain Le Verrier, a French mathematician, used Newtonian theory and the discrepancies in Uranus's orbit to predict the location of an unknown planet. He sent his predictions to the German astronomer Johann Galle who discovered Neptune within hours (Grosser, 1979). Clearly, theory enabled Galle to direct his attention to the right segment of the sky and make the discovery. Le Verrier also noticed that there were discrepancies in Mercury's orbit and soon a number of astronomers "discovered" the planet Vulcan in an orbit between Mercury and the sun (see Baum & Sheehan, 1997). We now know there is no planet Vulcan and that the discrepancies in Mercury's orbit can be accounted for by Einstein's theory of general relativity. So in the case of the planet Vulcan the scientists' beliefs in Newtonian theory played an inhibitory role in understanding the nature of the world and led to observations of an object that does not exist. The Neptune/Vulcan story serves as a good exemplar of my general claim that the mental processes of scientists working with a theory that correctly maps the world facilitate the ability of the scientist to understand the world, while the mental processes of the scientists working with a theory that incorrectly maps the world will inhibit the ability of the scientist to understand the nature of the world.

THE ORGANIZATION OF THE MATERIAL FROM COGNITIVE PSYCHOLOGY AND HISTORY OF SCIENCE

In the next sections of this chapter I examine the impact of theory on some of the mental processes used by scientists in carrying out science. I review the evidence for theory ladenness in perception, attention, thinking, experimenting, memory, and communication. For each of these processes I first cover material from cognitive psychology and then from the history of science. I try to include examples of both the facilitatory and inhibitory role of theory. However, it is much easier to find inhibitory examples in the history literature since when things go wrong it often exposes the psychological processes at work, whereas, when things go well, there is less visible evidence of the psychological processes.

PERCEPTION

In philosophy and psychology the term "perception" is sometimes used very broadly. In this chapter I use "perception" more narrowly to mean (for visual input) the characteristics of the phenomenally experienced visual world.

Cognitive Psychology

Ambiguous Figures

There have been a number of studies of the impact of perceptual and conceptual information on the perception of ambiguous figures. The classic study in this area is the early work of Leeper (1935). He used the old woman/young woman ambiguous figure (see Figure 13.1a) in his experiment. This figure gives rise to two different perceptual experiences, and experiencing one is relatively incompatible with simultaneously experiencing the other. Leeper presented an unambiguous version of this figure (e.g., an unambiguous old woman) to participants (see Figure 13.1c) and then showed them the ambiguous figure. This visual priming was extremely powerful. Of those seeing the unambiguous old woman first, 97% saw the ambiguous figure as an old woman; and of those seeing the unambiguous young woman first, 100% saw the exact same figure as a young woman. Note that this experimental manipulation seems somewhat analogous to a novice scientist looking at an idealized textbook drawing of a cell and then having that visual information facilitate his or her identification of an actual slide containing cells. Once participants perceived the ambiguous figure as either the old woman or the young woman, Leeper found it hard to get them to see the other option. This is consistent with a personal anecdote told by the philosopher of science William Bechtel in making the point that figures of this type involve perceptual experience rather than cognitive inference. He stated, "For many years I could not see the young woman, although numerous people had tried to point out the features. Then in frustration I threw a copy of the diagram upside down in front of a colleague and demanded to know again where this obscure image was. As the drawing hit the table, the picture of the young woman clearly came into view—I simply saw her, I did not infer anything" (Bechtel, 1988, p. 46). Several other studies have investigated the impact of top-down information of a more conceptual nature. My favorite of these is a study (Liu, 1976) that shows that listening to scary music (Disney's "Chilling, Thrilling Sounds of the Haunted House") shifts the percentage of the (witchlike?) *old woman* responses from the 14% given by a control group up to 43%.

Similar findings have been reported with the ambiguous rat/man drawing (see Figure 13.2). Bugelski and Alampay (1961) found that if participants first saw a set of pictures of animals (not including a picture of a rat) before seeing the

Figure 13.2 Rat/man: ambiguous figure (used with permission of the CPA from Bugelski & Alampay, 1961, p. 206).

ambiguous rat/man figure, 97% saw the rat, whereas in a group shown a set of human pictures first only 12% saw the rat. Reese (1963) replicated these findings with children. In terms of the top-down/bottom-up synthesis these studies focus on the impact of top-down information. Experimental evidence for the influence of bottom-up information (e.g., presenting a clear unambiguous picture and asking what the picture shows) is somewhat hard to find since it is so obvious that no one is likely to do the experiment. However, Bugelski and Alampay were kind enough to give us the appropriate data. They point out that in naming the conceptual set pictures (e.g., fish, duck, dog) no participant made an error.

Now let me try to relate these experiments to my general arguments for the facilitatory or inhibitory role of theory in science. Imagine one of the options in the ambiguous figures (say the *rat* of the *rat/man* figure) to be the state of nature. Then a scientist who has conceptual information about rats or previous perceptual experience similar to the rat will be facilitated in perceiving information related to the true state of nature, while the scientist who holds the *man theory* will be inhibited in perceiving information related to the true state of nature. The control groups in these studies can be thought of as similar to the scientist without a theory, whose perception of information from the world will neither be facilitated nor inhibited by specific theories, but will likely be determined by more general background beliefs. This analogy between the laboratory experiments and episodes in the history of science may sound a little strong, but the examples of the theory ladenness of perception from the history of science that come later in this chapter seem to fit this account very well.

Fragmented Figures

There have been several relevant studies using fragmented figures (see Figure 13.3). These stimuli have the property that at first they often appear to be a set of random shapes, but they can also be seen as meaningful objects. Perceiving the fragmented figures as meaningful objects typically takes some time and is experienced as a sudden shift in the phenomenal experience, not as an interpretation. (You can judge for yourself by first looking at the figures, then reading the caption, then looking back at the fragmented figures.)

Once again Leeper (1935) carried out the classic study. He showed participants a set of fragmented figures. One group was simply told that the figures could be seen as "coherent pictures" while another group was given top-down information in the form of the name of the superordinate conceptual class (e.g., "a means of transportation" for the *airplane*; "a common tool" for the *saw*) before seeing each fragmented figure. The top-down conceptual information strongly facilitated the ability of the participants to see the underlying meaningful objects. Steinfeld (1967) carried out a similar experiment with a single fragmented picture of a ship and found strong facilitation by top-down conceptual information. A control group with no prior knowledge about the picture took on average 24 seconds to see the meaningful object. A group that read a story about a woman (an interference condition) took 38 seconds to see the meaningful object; whereas a group that read a story about

Figure 13.3 Fragmented figures used in Leeper, 1935. (man, airplane, typewriter, dog, bus, man on horse, elephant) (used with permission of Pion Ltd, London from Reynolds, 1985, p. 49).

the sinking of an ocean liner (a facilitation condition) took only 4.5 seconds to see the meaningful object. Reynolds (1985) also found strong facilitation by top-down conceptual information. He used a control group that was not even informed that the fragments could be perceived as a meaningful object. These participants saw the meaningful objects for only 9% of the fragmented figures, whereas a group given superordinate conceptual class information saw the meaningful objects for 74% of the same fragmented figures. As applied to the case of the working scientist, the results of these experiments with fragmented figures suggest the power of having a correct theory when making observations in situations with very noisy data.

Anomalous Stimuli

Bruner and Postman (1949) carried out a very interesting study on the perception of anomalous information (discussed briefly earlier in this chapter). The design of this study leaves open several alternative explanations of the findings, but I am going to analyze it in detail because this is the psychology experiment that Thomas Kuhn gives the most space to in *The Structure of Scientific Revolutions*. Kuhn liked the study because it seemed to show the theory ladenness of perception and the problems that individuals have in dealing with anomalous information. Bruner and Postman presented participants with playing cards for short time intervals. Some of the cards were ordinary (e.g., a red five of hearts), some were anomalous (e.g., a black five of hearts). They found that the exposure times needed to identify the anomalous cards was much longer than for ordinary cards. Bruner and Postman appear to interpret the results as showing an impact of a violation of expectation on *perception* (the title of the paper was *On the Perception of Incongruity: A Paradigm*). It seems to me that the higher thresholds might just be due to reluctance of the participants to report seeing something as unlikely as a black five of hearts (cf. Neisser, 1967, pp. 118–134). Probably Bruner, Postman, and Kuhn were all aware of this methodological problem. Kuhn implicitly argued for a perceptual account by quoting some of the responses of the participants that certainly suggest that there was a perceptual impact (see the earlier discussion of the experiment). Bruner and Postman made similar arguments by presenting data on what they called "compromise reactions." For example, when shown a red spade or a red club card some participants gave responses such as "brown," "purple," and "grayish red." These responses certainly suggest an actual impact on perception; in the terms of this chapter, they seem to be a mixture of top-down and bottom-up information. Overall, it seems to me that the increased thresholds to anomalous information may be due to a mixture of true perceptual effects and reporting effects. These reporting effects can be thought of as examples for which theories about the nature of normal cards influenced the choice of what to communicate—a topic to be discussed later in this chapter. Also note that in their discussions of this experiment both Kuhn and Bruner and Postman emphasize the impact of *top-down* beliefs on perception, yet the data show clear evidence for *bottom-up* effects. For example, as the exposure times were increased (up to a maximum of 1 second) eventually the participants were correct for all of the normal cards and for most of the anomalous cards. The Bruner and Postman experiment is a little less analytic than the other studies reviewed above, but nevertheless it still fits well within the framework I am using to organize the results of these studies on the theory ladenness of perception.

Studies of Scientific Observation

While the data from these experiments showing the impact of top-down information on perception appear quite convincing, it seems to me that a person interested in the psychology of science might use an ecological validity argument and say that it is possible that these lab-based results do not generalize to typical tasks that occur in doing science. It turns out that there is a literature in science education that nicely generalizes the work from laboratory tasks to tasks more directly like those of working scientists. Gunstone and White (1981) studied first-year physics undergraduates carrying out a classic falling-bodies experiment. First they asked the participants what they predicted would happen if two identical spheres of different weights were dropped and found that roughly three-quarters of the participants thought they would reach the ground at the same time and one-quarter thought the heavy sphere would reach the ground first. Then the experimenters carried out the falling-bodies experiment three times and asked the participants to record their observations of what occurred. This was a difficult perceptual task, but the balls hit the ground at roughly the same moment. The participants who held the *equal-speed theory* observed the balls to hit the ground at the same moment in 98% of their observations, whereas those who held the *heavier-first theory* only observed the balls to hit the ground simultaneously in 67% of their observations. Note that the data from those with the *heavier-first theory* fits the top-down/bottom-up synthesis very nicely. For 33% of the observations the top-down information was stronger while for 67% of the observations the bottom-up information appeared to dominate.

Chinn and Malhotra (2002) carried out another Galilean experiment with falling objects. The participants were elementary school children and the falling bodies were two rocks of the same size, but different weights. Chinn and Malhotra used a pretest questionnaire to establish what the children believed about the relative speed of heavy and light falling objects. One group of children did not get to feel the rocks and so did not know that they were of different weights. This control group allows one to see the degree of theory-based facilitation or inhibition in the other children's observations. The children in this study saw the rocks fall three times and were asked to indicate if the rocks hit at the same time or if one or the other rock hit first (a difficult perceptual judgment). The control group (who did not know the rocks were different in weight) reported that the rocks hit at the same time in 35% of their observations. The children who held a *heaver-first theory* reported that the rocks hit at the same time in only 25% of their observations, thus showing theory-based inhibition, whereas the children with a *simultaneous-hit theory* showed strong theory-based facilitation, reporting that the rocks hit at the same moment in 72% of their observations. In these studies, designed to be close analogues to observation in science, the designs do not provide ironclad evidence that the children's theories influenced their actual *perception* of the falling bodies, but they certainly provide evidence that these children's theories influenced their *scientific observations*, which is the important issue in establishing the relevance of the psychological data to the larger issues of theory ladenness in science.

Conclusions

Overall, these studies provide support for the top-down/bottom-up approach to perception and show that theories can have both a facilitatory role and an inhibitory role in the perceptual process. One important empirical finding from these studies is that the strongest impact of theory ladenness on perception tends to occur when the stimuli are weak, degraded, ambiguous, presented for a short time, or otherwise require a difficult perceptual judgment.

History of Science

Early Astronomy

The field of astronomy is rich with examples of theory-laden perception (Hetherington, 1988; Sheehan, 1988). This is not because astronomers are, as individuals, more susceptible to top-down influences—it is because for most of the history of the field the working astronomer gathered data by direct perceptual observation. The astronomers at the leading edge of astronomy were pushing their instruments to the limits, and thus they frequently had to deal with relatively degraded bottom-up information. I have argued above that these are the conditions in which theory ladenness of perception is most likely to occur.

The discovery of the rings of Saturn is a classic example. In 1610 Galileo used his telescope to discover that Jupiter had four moons and to work out their periods. This spectacular discovery set the stage for his first telescopic observations of Saturn about six months later. He described what he saw as, “the star of Saturn is not a single star, but is a composite of three, which almost touch each other, never change or move relative to each other and are arranged in a row along the zodiac, the middle one being three times larger than the two lateral ones” (Van Helden, 1974, p. 105). Clearly, Galileo now held a belief that planets were likely to have moons, and he saw the ring around Saturn as two satellite moons. Galileo’s observations of Saturn were quickly followed up by other scientists, and for over 40 years astronomers drew Saturn as having moons or handles or mixtures of the two (cf. Figure 13.4). Note that during this period none of Saturn’s actual moons had been discovered, so all the moons in the drawings were due to theory-laden perception. Also notice the delicate interplay in the drawings between top-down information about moons and

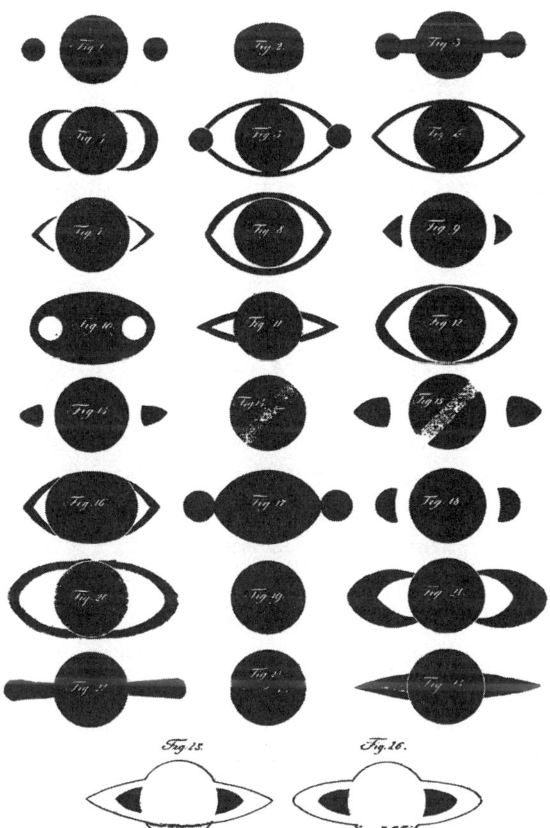

Figure 13.4 Drawings of the rings of Saturn (used with permission of the University of Chicago Press from Shapley, 1949, p. 13).

handles and the bottom-up information from the actual ring that they were observing. Soon after Huygens announced his theory that what they were looking at was a planet with a ring around it, the drawings rapidly shifted to the view that we now know to be correct.

Another example of theory-laden perception can be seen in William Herschel's discovery of Uranus in 1781. When Herschel first observed Uranus he assumed that it was a comet. Austin (1967) has analyzed Herschel's data and has shown that the data indicate that the apparent size of the "comet" was increasing. This is the data that would be expected if a comet was approaching the earth, but Austin's calculations show that Uranus was actually moving *away* from earth during the period of the observations. Austin ruled out a number of possible physical explanations for the data and turned to a psychological account. He argued that the observations of the object's diameter were extremely "difficult" and "delicate" and concluded, "It is in just these sorts of cases that preconceptions and expectations can exercise their greatest influence on measurements" (p. 284).

Later Astronomy

A case of theory-laden perception played a large role in delaying the acceptance of our current understanding that spiral nebulae are not part of our own Milky Way galaxy but are independent galaxies far removed from our own galaxy. In the early 1920s Adriaan Van Maanen measured the internal motions of seven spiral nebulae and found that all showed rotation and all rotated in the same direction (a direction that would lead to an unwinding of the spiral) (Hart, 1993; Hetherington, 1988). If the spiral nebulae were at large distances from our galaxy, then Van Maanen's data would require the spiral nebulae to be rotating at speeds close to the speed of light; thus his data were strong evidence against the multiple galaxies theory. During the 1920s a number of findings were reported using a variety of different procedures that seemed to support the theory that the spiral nebulae were actually distant galaxies. However, for about a decade Van Maanen's important measurements blocked the general acceptance of this view.

The technique Van Maanen used to study the internal motions of nebulae involved making extremely difficult perceptual decisions about whether given points in the nebula had moved in photographs taken 5 or 10 years apart. Van Maanen was known as a very careful observer. He used a wide variety of controls to ensure accuracy of the bottom-up information (Hart, 1993; Hetherington, 1988): He only compared photographs taken with the same telescope. He used sets of photographs from different observatories. He maintained the measuring device at a relatively constant temperature. He also had another astronomer measure a small subset of the data, so clearly he was aware of the possibility of top-down theory-based biases. Nevertheless, despite all of these controls, it is almost certain that his measurements of the internal motions of the nebulae were theory-laden. Edwin Hubble developed a technique using Cepheid variable stars that provided powerful evidence that the spiral nebulae were at enormous distances, so Van Maanen's data had to be in error. Eventually, Hubble and several other astronomers attempted to replicate Van

Maanen's findings using the same photographs and the same equipment and they found no evidence for internal motions. So in making a perceptual judgment that was at the limits of the human sensory system Van Maanen must have allowed his beliefs about the rotations of the nebulae to influence his measurements.

A number of astronomers have become sensitive to the psychological factors involved in doing astronomy. Their understanding of the issues is so close to my own that I can use their conclusions to summarize the issues. On the theory-laden issue in astronomy Sheehan (1988) stated, "Once a definite expectation is established, it is inevitable that one will see something of what one expects; this reinforces and refines one's expectations in a continuing process until finally one is seeing an exact and detailed—but ultimately fictitious—picture" (p. 85). Sheehan also noted the role of theory in facilitating astronomical observation. He stated, "It is often wondered, after an object is once discovered, that it was not found long before. Faint objects are often reported as detectable in telescopes two or even three times smaller than those used for their discovery. Following the discovery of Saturn's inner ring (the crêpe ring)... for instance, the new ring was seen by a number of observers with small instruments" (p. 63). Hetherington (1983) drew similar conclusions from the history of astronomy and concluded that having a theory is a two-edged sword. He pointed out that "scientists are guided by theory in selecting those observations that may be important" (p. 729) and that "the warping of judgments by knowledge, the influence on observational reports of preconceived opinion, is inevitable" (p. 729).

N-rays

One of the most well-known historical episodes involving the theory ladenness of perception occurred in the "discovery" of N-rays by the French physicist René Blondlot (Klotz, 1980; Nye, 1980). In the early part of the 20th century, many new forms of radiation were discovered, so it was reasonable to expect that more were waiting to be revealed. In 1903 Blondlot announced the discovery of N-rays (named after his university—the University of Nancy). These rays were emitted by a wide variety of sources—cathode ray tubes, Nernst lamps, and the sun. The rays could penetrate wood, iron, and gold; they were blocked by water and rock salt, and they could be refracted by aluminum prisms. N-rays could be detected by their producing a small increase in the brightness of a spark gap or by an increase in the brightness of a phosphorescent powder that had been exposed to the sun. This discovery caused a sensation in the world of physics. From 1903 to 1906 N-rays were discussed in 300 papers by 100 scientists, and during this period roughly 15% of the articles in a major French science journal dealt with N-rays (Price, 1975).

However, there was a problem. Almost no physicists in the United States, England, or Germany could replicate Blondlot's work. An American physicist, R. W. Wood, was delegated to go to Blondlot's lab and try to uncover the difficulty. (It is probably not a coincidence that Wood had devoted some energy to uncovering the tricks used by spiritualists and mediums.) In an initial series of experiments, Wood discovered he could not see any changes in the N-ray detectors when Blondlot and his assistants could. Wood carried out several covert control experiments in which he manipulated the physical set up so that the physical situation was not in agreement with the psychological expectations of the University of Nancy scientists. The most dramatic test was one in which a Nernst lamp was supposed to be emitting N-rays that were then refracted by an aluminum prism and detected by the increase in brightness of a phosphorescent chemical. In his two page report of his visit to Blondlot's lab, Wood described the outcome of this experiment—"I subsequently found that the removal of the prism (we were in a dark room) did not seem to interfere in any way with the location of the maxima and minima in the deviated (!) ray bundle" (Wood, 1904). Wood's removal of the prism had modified the physical conditions in a way that would eliminate the N-ray effects and yet the observer, not knowing about the modification, continued to report N-ray observations.

It seems highly unlikely that any fraud was involved in the N-ray affair. A large number of scientists using Blondlot's methodology were able to replicate his results and Blondlot was open to having a skeptic such as Wood come into his lab. Wood clearly favored a theory-laden psychological account of the N-ray observations. In one section of his paper he referred to the experimenters as being "unconsciously influenced" (1904, p. 530). Note that the perceptual task was quite difficult in both of the procedures used to observe N-rays. Blondlot argued that to obtain reliable results it was necessary for the "electric spark to be unstable ('short and feeble')" (Nye, 1980, p. 137). Blondlot published a set of instructions dealing with the need to be concerned with psychological factors. The observers had to be trained; they had to be dark adapted; they were to look to the side of the detector so the observation would be made with the more light sensitive rods in the peripheral visual field. Thus, Blondlot was concerned with the *bottom-up* effects in the detection task. When someone proposed that he carry out a control experiment similar to the one Wood carried out, Blondlot argued that the observer must be able to regulate the time and occurrence of the N-rays in order to avoid fatigue (Nye, 1980, p. 144). The N-ray affair appears to be a classic case of an experiment in which a scientist with strong theoretical beliefs has to make a very difficult perceptual judgment, and the results show that in these circumstances scientific observations are highly theory-laden.

Cambridge versus Vienna

In the 1920s there was a major disagreement between researchers from Cambridge, England, and those in Vienna about data relating to the artificial disintegration of the elements. Stuewer (1985) has written a careful account of this episode in physics and shown that the discrepancy in the results from Vienna was due to theory-based effects on perception. Ernest Rutherford (leader of the Cambridge group and one of the world's outstanding experimental physicists) had discovered artificial nuclear disintegration by bombarding nitrogen with alpha particles and finding that some of the atoms of nitrogen had transmutated into oxygen. He and James Chadwick then began a systematic program of research, which involved firing alpha particles at a wide range of elements and looking for protons caused by the impact of the alpha particles. The technique for observing the protons involved looking through a microscope at small flashes of light that occurred on a scintillation screen. This was known to be a very difficult perceptual task, and in a letter to Niels Bohr, Rutherford said, "The experiments look easy, when they are really very difficult and full of pitfalls for the inexperienced. So much is this so that I have decided not to get any other work done except under my personal eye" (Stuewer, 1985, p. 255). Using the scintillation method Rutherford and Chadwick found that some elements produced protons and others did not.

At about this same time two young researchers, Hans Pettersson and Gerhard Kirsch, at the Institute for Radium Research in Vienna, began to repeat the Cambridge experiments with very different results—they found that, when bombarded with alpha particles, essentially all the elements emitted protons. After a while the two groups reached an impasse; each stating that its methodology was superior. Chadwick was dispatched to Vienna to examine the experimental procedures used by Pettersson and Kirsch. He decided their experimental apparatus was sound and focused on the procedures used to count the scintillations. He discovered that the actual counting was done by three young women who knew that it was good to see as many particles as possible. Chadwick then used shielding to manipulate the number of the alpha particles striking the scintillation screen *without* informing the counters. He found that they continued to report scintillations even when he removed the alpha particle source. So essentially he manipulated the physical parameters while holding the human counter's expectations constant, and the data followed the counter's expectations, not the physical manipulations. In Chadwick's account to Rutherford of the procedures used in Vienna he stated, "There was no question of cheating. Rather, they were deluding themselves. They had been informed of the nature of the experiments being conducted... they saw what they were expected to see" (Stuewer, 1985, p. 288). The top-down/bottom-up synthesis casts an interesting light on this episode. The Vienna researchers were quite aware that scintillation counting was a very difficult perceptual task and there were possible psychological factors that needed to be addressed. They required that the counters be dark adapted for a half hour before the session; they were only allowed to count for 20 seconds at a time; they took rest breaks between counting sessions; they were not to count when they were fatigued. Note that (as with the N-ray experiments) all these careful and appropriate methodological procedures were directed at possible problems with *bottom-up* factors. What Chadwick did was apply a methodological control for *top-down* factors in the counting process, and unfortunately for the Vienna researchers he found that for the procedures used in Vienna the top-down factors were very strong.

Expertise in Visual Perception

Charles Babbage (1830) provided a wonderful account of the facilitatory contribution of theory-based experience on perception:

A striking illustration of the fact that an object is frequently not seen, *from not knowing how to see it*, rather than from any defect in the organ of vision, occurred to me some years since, when on a visit at Sough. Conversing with Mr. [John] Herschel on the dark lines seen in the solar spectrum by Fraunhofer, he inquired whether I had seen them; and on my replying in the negative, and expressing a great desire to see them, he mentioned the extreme difficulty he had had even with Fraunhofer's description in his hand and the long time which it had cost him in detecting them. My friend then added, "I will prepare the apparatus, and put you in such a position that they shall be visible, and yet you shall look for them and not find them: after which, while you remain in the same position, I will instruct you *how to see them*, and you shall see them, and not merely wonder you did not see them before, but shall find it impossible to look at the spectrum without seeing them." On looking as I was directed, notwithstanding the previous warning, I did *not* see them; and after some time I inquired how they might be seen, when the prediction of Mr. Herschel was completely fulfilled. (p. 211)

Conclusions

Overall, it seems to me that these cases from the history of science are in close agreement with the laboratory data from cognitive psychology: Perception is theory-laden. The top-down effects due to theory are much stronger when the bottom-up stimuli are weak or degraded. Researchers in the physical sciences may tend to underestimate the potential problems that can arise due to theory ladenness. When a scientist's theories map the world, the psychological processes facilitate the ability of the scientist to gain an understanding of the world, but when a scientist's theories do not map the world, the same psychological processes inhibit the scientist's attempt to understand the world.

ATTENTION

In this section I will use *attention* broadly to cover cases where the eyes remain fixed and mental attention is deployed to an object of interest, cases where the eyes shift to the object, and cases where there is an actual physical search for the object. As with perception, theory-directed attention is a two-edged sword. Focused attention facilitates performance, whereas information that is not attended to is inhibited. Chun and Marois (2002) refer to this as "the bright side of visual attention" and "the dark side of visual attention." The top-down, bottom-up framework is often applied to attention (e.g., Connor, Egeth, & Yantis, 2004). Visually salient stimuli are thought to capture attention in a bottom-up fashion, whereas top-down processes can direct attention to any goal-relevant stimuli.

Cognitive Psychology

Laboratory Tasks

Mack and Rock (1998) introduced an important new laboratory procedure for studying attention. In a programmatic series of experiments, participants were asked to make a perceptual judgment such as reporting which arm of a cross is longer. The stimuli were typically displayed for 200 milliseconds on a computer screen, which does not provide enough time for a shift in eye fixation. After a number of trials an unexpected object, such as a small black square, appeared within one of the quadrants formed by the cross. Right after the participants had made their perceptual judgment on this critical trial they were asked, "Did you see anything on the screen on this trial that had not been there on previous trials?" The number of participants who reported not seeing the unexpected object ranged from 25% to 80% depending on the details of the experimental procedure. Mack and Rock refer to this dramatic finding as "inattentional blindness." Because of the need for participants to have no expectations about an unexpected object appearing in the stimulus display, this crucial experimental manipulation was carried out only once with each participant. As a control procedure, after the critical trial, the participants were given some additional trials and asked to ignore the cross and report whatever else they saw on the screen. Under this condition of more focused attention, the participants were essentially always successful at reporting the object that had previously been displayed in the critical unexpected trial. The series of experiments carried out by Mack and Rock shows that objects in the focus of attention are highly likely to be observed, yet a significant proportion of unexpected objects in the visual field are not observed (i.e., they show inattentional blindness). Notice that this task from the experimental psychology laboratory is very similar to many search tasks carried out by astronomers using telescopes and biologists using microscopes. It seems highly likely that the findings generalize to those situations, thus supporting the theory ladenness of attention in scientific observation.

More Molar Tasks

Shinoda, Hayhoe, and Shrivastava (2001) carried out a study of the role of expectations on attention using a driving simulator. Participants were asked to drive around a virtual town following normal traffic rules. In one condition when a stop sign appeared on the side of the street at an intersection, 100% of the participants detected the sign, whereas when the stop sign appeared at an unexpected location (on the side of the street at mid block) only about 20% of the participants detected the sign. These results seem to provide strong evidence for the theory ladenness of attention.

Simons and Chabris (1999) carried out a study designed to see if inattentional blindness could be shown for even very salient stimuli. These researchers asked participants to focus on one aspect of a complex and dynamic event recorded on videotape. In this event there were two teams (one in white shirts; one in black shirts) of three players. The members of each team were passing a basketball back and forth in a game that lasted 75 seconds. In one condition, to establish focused attention, participants were asked to mentally count the number of aerial and bounce passes made by the team in white shirts. During the game a woman in a full-body gorilla suit (see Figure 13.5) walked through the scene (visible for about 5 seconds). The gorilla actually briefly occluded the basketball or a relevant player. When asked, "Did you notice anything unusual?" only about 50% of the participants reported seeing the gorilla. Thus, with goal-directed attention focused on the white team, the participants showed strong inattentional blindness for very salient non-goal-related information from the scene. Clearly, this is a striking case of the theory ladenness of attention.

Figure 13.5 Gorilla passing through ball game (used with permission from Simons & Chabris,1999, p. 1070; figure provided by Daniel Simons).

In a follow-up study Simons (2010) made use of the fact that the gorilla study has become widely known. Simons repeated the standard gorilla experiment, but added two new unexpected events (the color of a curtain changed from red to gold and one player on the nonattended team left the game). For those participants who stated that they had never heard of the gorilla experiment, Simons found that about 50% did not report seeing the gorilla, thus replicating the finding of inattentional blindness from the earlier experiment. Of those participants who stated that they knew about the gorilla experiment, 100% reported seeing the gorilla. Clearly, these participants were using some of their finite attentional resources to search for the gorilla in addition to counting basketball passes. However, this deployment of attention came with the usual trade off—only 17% of those who knew about the gorilla reported seeing even one of the two new unexpected events. So overall these experiments show that theory-directed focusing of attention facilitates theory-directed observation at the expense of non-theory-based observations, and that even very salient stimuli can show inattentional blindness.

History of Science

Earlier in the chapter I discussed the history of the searches for Neptune and Vulcan. Both of those episodes in the history of science provide convincing evidence that theory plays a strong role in directing the attention of scientists.

Prediscovery Observations

Another set of data from astronomy that provides evidence for the theory ladenness of attention is the astronomers' use of *prediscovery observations*. Prediscovery observations are observations that occurred before the "official" discovery of some object. They are particularly important for establishing the orbits of objects in the solar system, since if one can find an observation of the newly discovered object that had been made many years earlier, it becomes possible to establish the orbit with much improved accuracy. After a new asteroid is discovered, it is now standard to use theory to predict where it should have been in the past and then to use computers to search earlier photographs taken of that general location (Boattini, D'Abramo, Forti, & Gal, 2001). This is an example of attention in the broad sense that theory is directing the astronomer to look at a particular region on a photograph that probably no human had ever previously examined.

However, when prediscovery observations were used to refine the orbits of Uranus and Neptune they revealed the impact of a more immediate form of attention. These observations were made before photography, so the astronomers were actually looking at the (undiscovered) planet and then drawing or otherwise indicating where in the sky it was seen. There were at least 22 prediscovery observations of Uranus (Forbes, 1983); however, the observers interpreted them as stars and thus did not attend to them carefully enough to discover that they moved with respect to the other background stars. In two cases there is evidence that the observers actually did attend to and observe the movement of the undiscovered planet. In 1795 Jérôme Lalande observed Neptune (officially discovered in 1846) several times and noticed that it was not in the same location (Price, 2000, p. 352), and in a spectacular case of prediscovery in 1613, in the very earliest years of the use of the telescope, Galileo also observed Neptune over several nights and noted on his star map that on the second night this "star" seemed to have moved with respect to the other stars (Kowal & Drake, 1980). In both of these cases the astronomers chose to retain the belief that they were seeing a fixed star and assume the apparent motion was due to observational errors on their part (an example of data *rejection* in the Chinn-Brewer framework, Brewer & Chinn, 1994a).

The Piltdown Forgery

The Piltdown forgery provides another example of theory-directed attention. The Piltdown skull and jawbone were discovered in 1912. The British paleontological community considered them to be the remains of ancient hominids and a major find in the evolution of modern humans. However, over time, as other fossils were discovered, the Piltdown skull did not fit the emerging picture of early human evolution. In 1953 Weiner had the insight that the fossil record would be much more orderly if one did not have to include the Piltdown skull, and one way that this could occur would be if the skull was a hoax (Weiner, 1980). With this new theory in mind, he and others were allowed to examine the original skull in the British Museum of Natural History, and they were immediately able to see the obvious evidence that it was a forgery. La Gros Clark stated, "The evidences of artificial abrasion immediately sprang to the eye. Indeed so obvious did they seem it may well be asked—how was it that they had escaped notice before" (Gould, 1980, p. 112). One book on the subject included a picture of the Piltdown jaw with the following description, "The Piltdown molars showing the non-alignment of the flat crowns. This unnatural placement cried out for notice in 1912 but oddly escaped attention for some forty years" (Millar, 1972, p. 8). Thus the uncovering of the Piltdown hoax provides additional evidence for the theory ladenness of attention in science.

Bottom-up Cases

In addition to the evidence for top-down processes operating in attention, the history of science also provides good examples where salient bottom-up information attracted attention in spite of strong top-down beliefs. Tycho Brahe's account of his discovery of the Supernova of 1572 makes the point well, "I was contemplating the stars in a clear sky, I noticed that a new and unusual star, surpassing the other stars in brilliancy, was shining almost directly above my head; and since I had, almost from boyhood, known all the stars of the heavens perfectly... it was quite evident to me that there had never been any star in that place in the sky, even the smallest, to say nothing of a star so conspicuously bright as this. I was so astonished at this sight that I was not ashamed to doubt the trustworthiness of my own eyes" (Brahe, 1929).

A more recent example of bottom-up attention occurred in the discovery of the cosmic background radiation by Penzias and Wilson (Bernstein, 1993). These two researchers found that a very sensitive antenna they were working with produced too much background noise. They spent most of a year trying to locate the source of the noise (even at one point entertaining the hypothesis that it was due to a buildup of pigeon droppings!). Eventually they discussed their problem with a group of physicists working at Princeton who explained that their noise was not a methodological artifact, but that they had, in fact, discovered the radiation left over after the Big Bang.

Discovery of Charon

The discovery of Pluto's moon, Charon, is a well-documented case in the history of science (Christy, 1990; Christy & Harrington, 1978; Harrington & Harrington, 1979; Stern & Mitton, 1999) and provides good evidence for the role of attention in science. In 1978 James Christy was working at the United States Naval Observatory in Washington, DC. He was looking at some new photographs of Pluto and trying to make very careful measurements of Pluto's location in order to improve its orbit. He noticed that the images of Pluto were asymmetric, with a small bump off to one side (see Figure 13.6), and that the location of the bump changed in photographs taken on different days. Crucially, he also noticed that the stars in the photograph were clear points and that suggested that the asymmetric character of Pluto was real and not a methodological artifact due to some movement of the telescope. He told his supervisor that he thought he had discovered a moon orbiting Pluto, and his boss simply said, "Jim, you're crazy." His supervisor's dismissal of the finding was quite reasonable given that there had been a number of searches for a moon around Pluto with larger telescopes and they had all come up empty-handed. However, the next day Christy went looking for prediscovery pictures of Pluto in the Observatory's archives and found photographs of Pluto taken in 1965 and 1970. He could see the bump on these photographs and was able to do a calculation giving the period of the

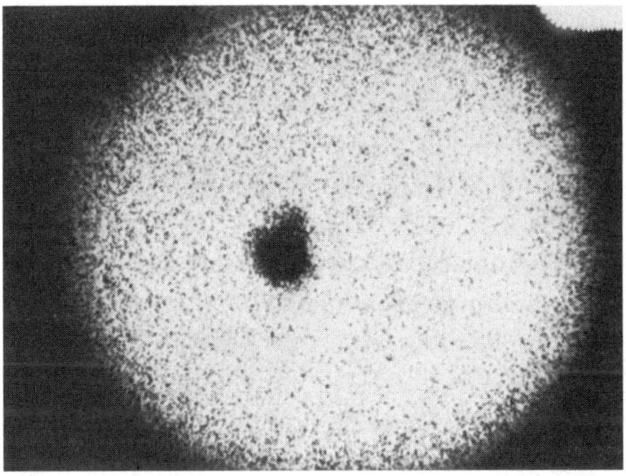

Figure 13.6 Charon discovery picture (bump is top right of dark central image of Pluto) (used with permission of the AAS from Christy & Harrington, 1978, p. 1005).

bump to four decimal places! After a number of cross-checks the announcement of the discovery of Charon was made several weeks later.

In an account of his discovery, Christy (1990) explained his success in terms of top-down and bottom-up factors. He pointed out a number of bottom-up problems (e.g., overexposed or underexposed film) with the observations made by the previous investigators. However, note that bottom-up factors cannot account for the core of this episode, since a number of prediscovery photographs were used to cinch the discovery of Charon; and this shows that all the necessary bottom-up information was available in the set of 10 photographs that he used. Yet many astronomers, including Christy himself, had looked at those plates very carefully over the years, and had not noticed the bump. However, several astronomers must have become aware of the asymmetrical images (i.e., bottom-up information) since someone had written on the 1965 photographs "Pluto image elongated" (Christy, 1990, p. 176) and someone had written "image defective" on the cover of the 1978 discovery photographs (Stern & Mitton, 1999, p. 52). However, in each case the astronomer must have interpreted the bump on the image as resulting from a methodological problem. In fact, the astronomer who took the 1965 photographs said he had noticed the elongated image of Pluto, but attributed it to atmospheric turbulence (Stern & Mitton, 1999, p. 56). Christy provided a nice account of the top-down factors that led to the failure of so many astronomers to notice the crucial bumps on the photographs of Pluto. Christy stated, "The elongation on the images of Pluto was also hidden by the bias in the minds of most astronomers, including myself, that moons did not exist that close to planets and, because many attempts to find a moon of Pluto had failed, that Pluto did not have a moon" (1990, p. 175).

Conclusions

These cases from the history of science provide evidence on a number of different issues. They show the close interplay between attention and theory-based interpretation (a topic we will return to in the next section of this chapter). Note that some of the examples discussed above (e.g., the discovery of Charon) are cases where the interpretation shifts from something random/meaningless to a relevant theory (as with fragmented figures in the psychological experiments), whereas in other cases (e.g., the Piltdown skull) the interpretation shifts from one theory (the skull is of an ancient human) to another theory (the skull is a hoax) as in the psychological experiments with ambiguous figures. More generally, these cases confirm the findings from cognitive psychology and provide evidence that attention is theory-laden, that attention has facilitatory and inhibitory effects, and that attention is directed by both bottom-up and top-down factors.

THINKING

I am using "thinking" as the term to cover a broad range of central mental processes involved in carrying out the scientific enterprise (understanding/interpretation, explanation, cognitive styles, data gathering, data evaluation, data selection). I have found the psychological processes in this category difficult to deal with, and while I feel I have captured a number of important subcategories of scientific thinking, I am not confident that this organization cuts nature at her joints. Also my relativist friends should pay attention to this topic since if I were going to try to defend a relativist position I would not focus on perception and attention, but on the dramatic theory ladenness that can sometimes be found in the domain of scientific thinking.

Cognitive Psychology

Understanding/Interpretation

One of the major roles of theories is in leading to understanding and new interpretations. Bransford and Johnson (1972) carried out a classic series of demonstration experiments designed to show that having a theory about something increased comprehension and memory for the information. In one experiment the theory was background knowledge about an everyday activity. Participants heard an obtusely written passage such as "First you arrange things into different groups depending on their makeup.... It is important not to overdo any particular endeavor. That is, it is better to do too few things at once than too many..." Before hearing the passage one group was told the passage was about "washing clothes." A control group heard the passage without the topic (or with the topic given after the passage). Comprehension rating scores and memory scores on the passage for the theory group were about double those for the control group, thus showing the role of theory in facilitating understanding.

Vosniadou and Skopeliti (2005) carried out a study which shows that the conceptual shift involved in a new interpretation can be very deep. These researchers compared a sample of 10-year-old children, most of whom held spherical, heliocentric theories of the earth, with a sample of 6-year-old children, most of whom held various versions of flat earth theories (see Vosniadou & Brewer, 1992 for a discussion of these types of naïve models). The children were given a set of cards that included the names of ordinary terrestrial objects (e.g., house, rock, tree) and names for solar objects (e.g., sun, moon, earth). The children were asked to sort the cards into two piles. Vosniadou and Skopeliti found that the majority of the older children put *earth* with the solar objects, while the majority of the younger children put *earth* with the terrestrial objects. These data suggest that as the children shift from flat models of the earth to heliocentric, spherical models of the earth, there is an ontological shift in the status of the earth (as there was in the history of astronomy). Clearly, children's interpretations of the world are highly theory-laden.

Explanation

Naïve theories are a subclass of mental representation that provide explanations (Brewer, Chinn, & Samarapungavan, 1998). For example, Vosniadou and Brewer (1994) studied children's understanding of the day/night cycle. They found that children held a variety of naïve theories and that the different theories led to different explanations for the alternation of day and night. Some children explained the occurrence of night by saying the sun moves out into space; others said the sun revolves around the earth; others said the earth rotates. These findings show that children's explanations are deeply theory-laden.

Cognitive Styles

In the course of carrying out scientific reasoning, both scientists and nonscientists employ different sets of mental tools. For example, Reiner, Chi, and Resnick (1988) have argued that many young children prefer to explain phenomena in terms of *touch causality* (e.g., pushing, lifting) instead of in terms of invisible forces (e.g., magnetism, heat). So this seems to be a case where the cognitive representations brought to bear on the task are theory-laden.

Data Gathering

The process of gathering data is another scientific activity that appears to be theory-laden. This process overlaps somewhat with the theory ladenness of attention; however in the present case I want to focus on data gathering in less directly perceptual cases. Klahr and Dunbar (1988) conducted a study of undergraduates carrying out "experiments" to understand how to program a programmable toy vehicle (Big Trak). They discovered that some participants tended to just push keys and see what happened (inductivists?), while others developed theories and then pushed the keys that would allow them to test that theory. In this particular task the theorists were faster at finding the solution than were the inductivists. So for the theorists in this task, the choice of which data to gather was theory-laden and facilitated discovery and understanding.

Data Evaluation

In the course of carrying out the scientific process, scientists spend much of their time *evaluating* their own data and the data of other scientists.

Brewer and Chinn (1994b) carried out a study with undergraduates to investigate the possibility that the evaluation of data is theory-laden. One group of participants read passages supporting one theory (dinosaurs were warm-blooded), while another group of participants read passages supporting another theory (dinosaurs were cold-blooded). Evidence from a belief-rating task showed that most of the undergrads came to believe the theory they had studied. When asked to evaluate the quality of the data consistent (or inconsistent) with the theories they had studied, the participants gave higher ratings to theory-consistent data. Thus, the quality of the exact same piece of data was evaluated differently depending on the participant's theory. This study directly shows that data evaluation is theory-laden.

History of Science

Understanding/Interpretation

The role of theory in leading to differences in understanding and interpretation has been well documented in the history and philosophy of science since Kuhn emphasized its importance in his 1962 book. For example, scientists holding a geocentric theory interpreted the day/night cycle as due to the sun going around the earth, while scientists holding a heliocentric theory interpreted the day/night cycle as due to the earth rotating on its axis (Kuhn, 1957).

The paleontologists are sometimes in the unenviable situation of having their theory-laden interpretations available for public display. Shipman (1986) provided a wonderful account of how the Iguanodon has been interpreted over time. In the earliest drawing of the Iguanodon by Gideon Mantell (who first described it in 1825), the dinosaur looks somewhat like a living iguana. The model made for the Crystal Palace exhibition in 1851 appears to have been based on a rhinoceros. The models displayed in the Royal Museum of Natural History in Brussels appear to have been influenced by kangaroos. The current representations appear to be based on raptors (fast-moving predators with a rigid horizontal tail). Note that the top-down/bottom-up framework gives a good account of what has been going on as paleontologists try to understand what Iguanodons looked like. At first there were only a few bones and so most of the representation had to come from top-down information (e.g., Mantell found that the teeth of the Iguanodon were similar to those of living iguanas so he represented them as giant iguanas). Then as more bones became available, they provided bottom-up evidence that constrained the interpretations (e.g., specimens were found showing that Iguanodons have small forelegs). These historical episodes make it clear that scientific understanding and interpretation is highly theory-laden.

Explanation

Studies of revolutions in the history of science (Kuhn, 1962; Thagard, 1992) give many examples of the theory ladenness of explanation. For example, a burning candle was explained by proponents of the phlogiston theory as due to the release of phlogiston, whereas Lavoisier explained the combustion as due to the candle combining with oxygen.

Cognitive Styles

Malley (1979) has given an intriguing example of the impact of theory on the cognitive style used in carrying out science. She points out that the Curies had a considerable lead time and better access to radioactive material than did Rutherford and Soddy, yet it was Rutherford and Soddy who developed the transmutation theory to account for spontaneous radioactive decay. Malley argued that Rutherford and Soddy thought in terms of mechanical physical models, while the Curies preferred not to move too far from the data, and when they did, they preferred to explain phenomena in terms of energy. Malley concluded that Rutherford and Soddy's mechanical models led to the discovery because in this particular domain "nature was working against the French" (p. 223).

The physicist Weinberg (1992) made an argument for differences in cognitive styles leading to Thomson's discovering the electron. Thomson is considered to have made the discovery with a series of experiments involving cathode-ray tubes. Weinberg pointed out, "The same experiment was done in Berlin at just about the same time by Walter Kaufmann.... Yet Kaufmann is never listed as a discoverer of the electron, because he did not think that he had discovered a new particle. Thomson was working in an English tradition going back to Newton, Dalton, and Prout—a tradition of speculation about atoms and their constituents. But Kaufmann was a positivist; he did not believe that it was the business of physicists to speculate about things that they could not observe" (p. 178). Both of these historical examples seem to show that different top-down cognitive styles provide different cognitive tools for thinking about scientific problems.

Data Gathering

The history of science provides much evidence that theory directs what data a scientist collects. For example, when Darwin was in the Galápagos he collected a number of birds from the different islands. However, at this point he had not developed his theory of natural selection, so he did not realize that he would need information about the geography and ecology of the birds. Later when he was developing his theory and needed to know the locations of the different species of finches, he had to borrow specimens from other members of the crew who had happened to label their specimens by island (Browne, 1995).

Apparently, Darwin learned his lesson well. In a letter he wrote in 1861 to Fawcett he stated, "About thirty years ago there was much talk that geologists ought only to observe and not theorise; and I well remember some one saying that at this rate a man might as well go into a gravel-pit and count the pebbles and describe the colours. How odd it is that anyone should not see that all observation must be for or against some view if it is to be of any service!" (Darwin to C. R. Fawcett, September 18, 1861).

Data Evaluation

Mahoney (1977) carried out a study of the manuscript review process that showed that data evaluation is theory-laden. Mahoney sent fictitious manuscripts to the reviewers of a journal that was known for having a particular theoretical orientation (behaviorist psychology). In one manuscript the data provided evidence for a behaviorist theoretical position, and in the other manuscript the data were reversed so the evidence went against a behaviorist position. Some of the reviewers received the manuscript with probehaviorist data and others the manuscript with antibehaviorist data. He found that the antibehaviorist manuscript was more likely to be rejected by the probehaviorist reviewers than the probehaviorist manuscript and that the ratings for the quality of the methodology were particularly theory-laden, with the quality of the (identical) methodology in the probehaviorist manuscript being rated almost twice as high as that in the antibehaviorist manuscript.

Kim and Ahn (2002) showed that theory influences the weight a sample of experienced clinical psychologists give to a symptom in making a diagnosis. Kim and Ahn determined the psychologists' theories for a number of mental disorders (e.g., anorexia nervosa) by having them draw a causal map of the symptoms for each disorder. The psychologists were also asked how important a given symptom was for making a diagnosis of each disorder. Kim and Ahn found that causal centrality in the psychologists' theories was strongly correlated with their rating of diagnostic importance. This example of strong theory-based data evaluation is particularly impressive because clinical psychologists make their diagnoses based on the *Diagnostic and Statistical Manual of Mental Disorders* and, for most disorders, that system explicitly gives equal weight to each symptom in making a diagnosis.

Data Selection

Often the scientist has to choose from large sets of data (their own and that gathered by others) which data to use in working out a scientific problem. This process is theory-laden. A very well-studied (but complex) case is the data used by Robert Millikan to determine the charge on the electron. Holton (1978) carried out a very careful analysis of Millikan's lab notebooks and Millikan's published articles. It turns out that it was very important for Millikan to use only very clean data, since the logic of a dispute he was having with the physicist Felix Ehrenhaft was such that including methodologically bad data tended to support Ehrenhaft's position and go against Millikan's position that the results should be in simple multiples of a basic charge. Holton's analysis shows that Millikan used his methodological beliefs about the quality of individual observations to include or exclude them in his analysis. Holton argued that this selection of data was rational and ethical and pointed out that Ehrenhaft, who included all of his data, came to a conclusion that is now known to be incorrect. Note that, in addition to this selection of the data, Millikan also made some clear ethical violations (e.g., he stated in his published paper that it included all the data when it did not); however, ethical issues are not the focus of the current chapter and require independent treatment.

Another well-studied case is Newton's calculation of the velocity of sound and the precession of the equinoxes. Westfall (1973) has shown that Newton's revisions from the first edition of the *Principia* to the second edition were highly theory-laden. Newton was certain his theory was correct and was trying to show that he could calculate the velocity of sound and the precession of the equinoxes to a very high degree of accuracy. He tended to include in his calculations those measurements of the speed of sound and measurements of the tides that were more favorable to his theory. Note, however, that there were limits to his selectivity. A new measurement of the speed of sound had taken place between the two editions and Newton decided it was more carefully done than any of the previous work; thus he based his calculations on this number even though it went against his initial theoretical calculations.

EXPERIMENTING

Experimenting is a crucial aspect of what scientists do and so needs to be in this analysis. Carrying out experiments is a way of understanding the world that has been developed by the social institution of science. Given this, one has to be careful in comparing the data in cognitive psychology from nonscientists with those of professional scientists, but it seems to me that there is enough overlap to make this possible (cf. Brewer, 2008; Brewer & Samarapungavan, 1991).

Cognitive Psychology

Karmiloff-Smith and Inhelder (1975) carried out a study that shows the theory ladenness of experimentation. They asked children to balance wooden blocks on a narrow metal rail. A number of the blocks had their weight evenly distributed so that they balanced at their geometric center. Other blocks had visible or invisible weights on one end so that they balanced far off center. The youngest children did not appear to have a theory; they simply moved the blocks around until they balanced. At about age six or seven most of the children appear to have adopted a strong theory—things balance in the center. These children placed the blocks on the rail on their centers. For ordinary blocks the data came in well for this theory; however, the theory did not work for the unbalanced blocks. Most of these children held to the *geometric center theory* and put the unbalanced blocks aside as impossible to balance. Thus we appear to have a case where the six- and seven-year-old children's experiments were directed by theory and when the data did not fit the theory, the children adjusted the data by not including the unbalanced blocks (see Brewer & Chinn, 1994a for a more detailed psychological account of how individuals deal with anomalous data).

History of Science

There are many individual episodes in the history of science that show theory ladenness in carrying out experiments. However, there is a literature in physics on *intellectual phase locking* (Cohen & DuMond, 1965, p. 551) or the *bandwagon effect* (Franklin, 1984, p. 789) that makes the point in a more general way. Intellectual phase locking occurs when the theory ladenness of carrying out experiments causes the experiments that follow an inaccurate experiment to show a tendency to agree with the earlier incorrect result. This effect can be detected by plotting the results of measuring a particular physical constant by the date of the experiment. Figure 13.7 shows some examples of these plots. If there were no effect of intellectual phase locking, then in an idealized world one would expect early experiments to show means (with large error bars) that varied unsystematically and then the later experiments would home in on a particular mean with much smaller error bars. It is the case that the measurements of some physical properties follow that ideal pattern. However, there are other patterns. One common pattern is for the earliest measurement to be quite far (e.g., too low) from the current value; then the means of subsequent experiments move up over time (typically staying within the error bars of the most recent experiment) until at some point they level off. Another pattern is to have the measurements cluster over a period of time, then suddenly an experiment (that must have been hard to get published!) gives a measurement that is many standard deviations outside the previous means, and the later measurements cluster around the new measurement.

These plots reporting the measurements of physical constants over time capture the data from hundreds of experiments, so they clearly establish the generality of intellectual phase locking; however, they do not give an account of *how* the effect occurs. Luckily a number of physicists have provided accounts of how they think the scientists' beliefs in the data from the earlier experiments influence the way the later experiments were carried out. Birge (1957) wrote that E. O. Lawrence stated that "in any highly precise experimental arrangement there are initially many instrumental difficulties that lead to numerical results far from the accepted value of the quantity being measured. It is, in fact, just such wide divergences that are the best indication of instrumental errors of one kind or another. Accordingly, the investigator searches for the source or sources of such errors, and continues to search until he gets a result close to the accepted value. *Then he stops*" (p. 51). Franklin quoted an ironic comment made within the physics community about a scientist "who fights the systematics until he or she gets the 'right' answer (read 'agrees with previous experiment') and then publishes" (1984, p. 791). Richard Feynman (1985), in his unique style, described the intellectual phase lock effect as revealed in measurement of the charge of the electron:

> Millikan measured the charge on an electron by an experiment with falling oil drops, and got an answer which we now know not to be quite right. It's a little bit off, because he had the incorrect value for the viscosity of air. It's interesting to look at the history of measurements of the charge of the electron, after Millikan. If you plot them as a function of time, you find that one is a little bit bigger than Millikan's, and the next one's a little bit bigger than that, and the next one's a little bit bigger than that, until finally they settle down to a number which is higher. Why didn't they discover that the

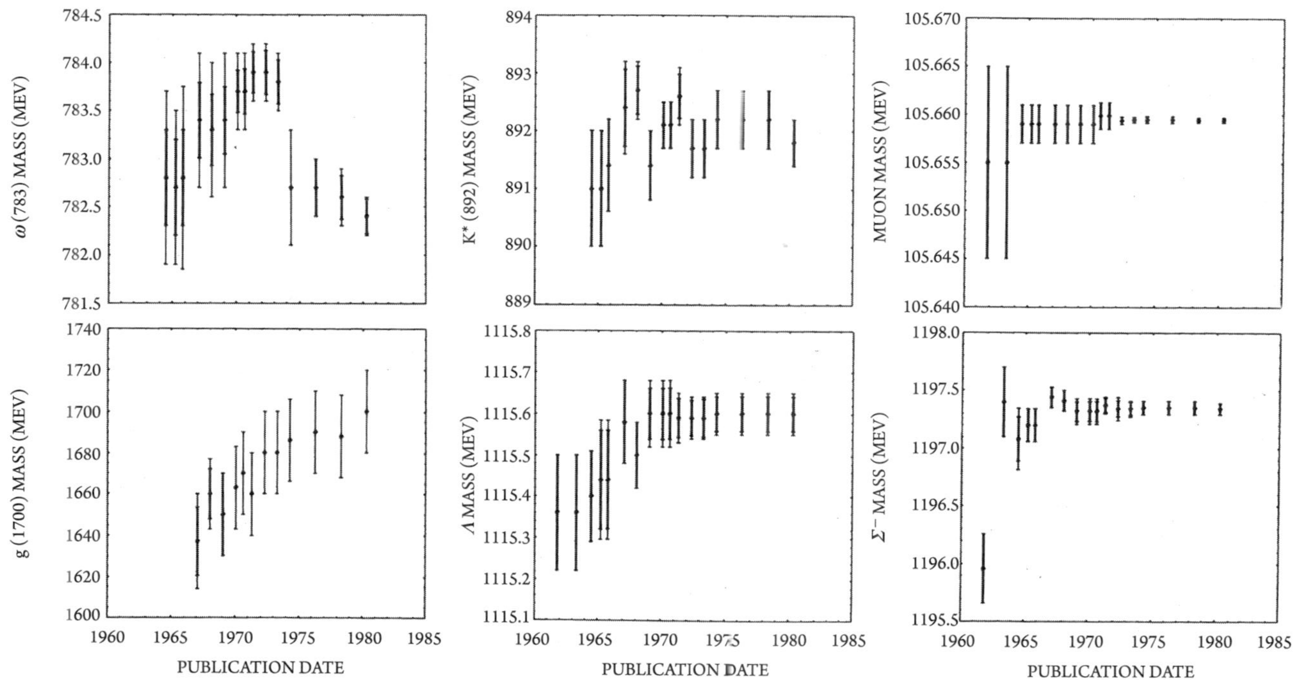

Figure 13.7 Intellectual phase lock: physics measures (used with permission from Particle Data Group, 1980, p. S284).

new number was higher right away? It's a thing that scientists are ashamed of—this history—because it's apparent that people did things like this: When they got a number that was too high above Millikan's, they thought something must be wrong—they would look for and find a reason why something might be wrong. When they got a number closer to Millikan's value they didn't look so hard. And so they eliminated the numbers that were too far off, and did other things like that. (Feynman,1985, p. 342)

MEMORY

Memory for data and theories is the psychological material that scientific reasoning is based on. Hoddeson and Brewer (2004) have examined the oral histories given by scientists about their work and found a number of interesting types of memory distortions, suggesting that this may be a psychological process in scientists that is open to theory ladenness.

Cognitive Psychology

There is a broad literature in cognitive psychology focused on the way that theory influences human memory (e.g., Bartlett, 1932; Schacter, 1995). This literature shows that memory is clearly theory-laden. Theory can facilitate memory. Mishra and Brewer (2003) studied the facilitatory role of theory in memory for scientific text. They gave two groups of participants a passage with data (e.g., facts about the Hawaiian Islands) relevant to a scientific theory (the plate tectonic theory of island chains). Before reading the data text, one group read a text that outlined the plate tectonic theory that provided an explanatory account for the data; the other group read a passage, equally long, that did not provide an explanation for the data. When asked to recall the text describing the data, the group that had been provided with the explanation showed higher recall for theory-relevant information from the data text.

However, the impact of theory on human memory is a two-edged sword in this area, as it is in other psychological processes. There are clear examples of the *inhibitory* role of theory-laden memory processes. If an individual has a theory and is attempting to recall information that diverges from the theory, there will be a tendency for recall of that information to be distorted in the direction of the theory. This can be seen in a study by Zangwill (1937). In this study participants were shown a series of inkblots. One set of inkblots looked somewhat like animals (and the participants were given explicit instructions that the shapes were intended to suggest animals). The other set of inkblots looked somewhat like mountains (and participants were given explicit instructions that the blots were designed to suggest mountains). Unknown to the participants there was one ambiguous inkblot that appeared in *both* sets (see Figure 13.8). After viewing each inkblot briefly the participants were asked to draw them from memory. The data displayed in Figure 13.8 show that in recall the exact same stimulus figure was recalled in animal-like form when it had been presented with other animal-like blots, but was recalled in mountain-like form when it had been presented with other mountain-like blots. Both of these experiments show memory to be theory-laden; in one case it is facilitatory and in the other case it is inhibitory.

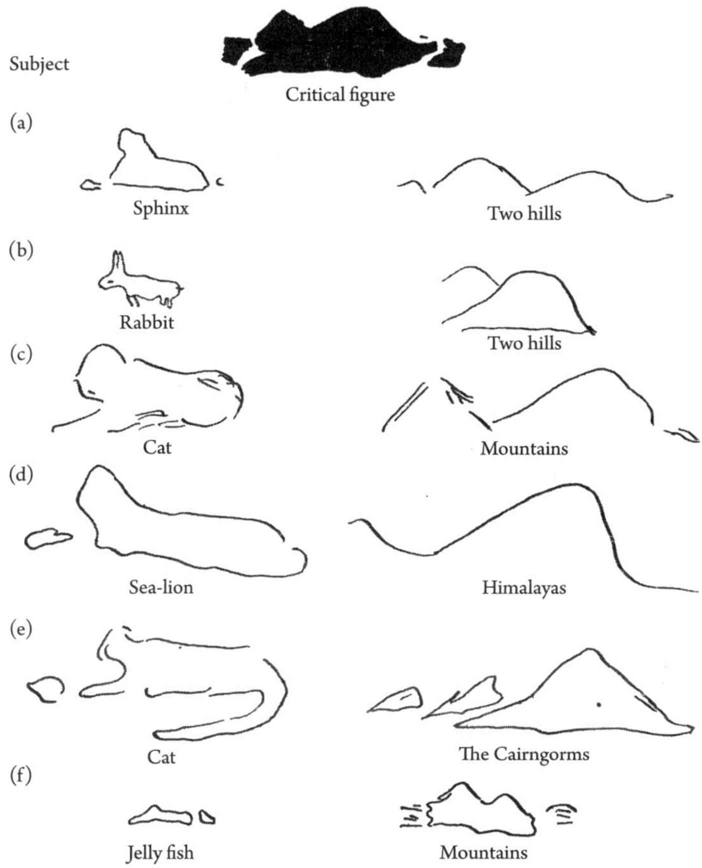

Figure 13.8 Zangwill inkblot recalls from six participants (critical figure is at top, animal context recalls to the left, mountain context recalls to the right) (used with permission from Zangwill, 1937, p. 15).

History of Science

Vicente and Brewer (1993) reviewed a number of cases in which the scientific literature distorts the historical record (e.g., that Galileo carried out experiments on falling bodies from the Tower of Pisa). They argued that many of these distortions may be theory-based memory errors. They supported their claim by examining a case in which it was known that the published scientific literature very frequently did not contain an accurate account of a particular set of experiments dealing with memory for chess pieces by chess experts. Vincente and Brewer asked undergraduate participants to read accurate accounts of the set of experiments that is often misdescribed in the literature and then later asked the participants to recall the experiments. They found that the participants' recalls of the experiments contained most of the same errors that had been found frequently in the scientific literature. Overall, these and other findings on the theory-laden nature of memory suggest that when two scientists with different theoretical orientations are thinking about a scientific problem, they may be working with quite different memory representations of the information in the area.

COMMUNICATION

Once scientific work is completed, the norms of the scientific community require that it be published and communicated to other scientists. Typically, the goal of the scientist writing up work for publication is to persuade other scientists of the value of their data or the correctness of their theories; thus the communication process is obviously open to theory ladenness. Note that some of the theory-laden processes in communication may operate in a more explicit and conscious fashion than those in most of the other psychological domains I have discussed.

Cognitive Psychology

Van Dijk and Kintsch (1978) carried out a series of studies investigating how participants summarize text. They accounted for the data with a set of "macrostructures," which are used to reduce and organize the information that is in the complete text. For example, one of their macrorules deleted nonthematically related information. Thus van Dijk and Kintsch's approach to summarization assumes that a writer's summary is generated in a theory-laden fashion.

History of Science

I suspect every working scientist believes that reviews of the literature by scientists with deep commitments to a particular scientific position are strongly theory-laden. It turns out that there are some empirical studies of review articles that provide data to support these beliefs. Joyce, Rabe-Hesketh, and Wessely (1998) studied 89 "comprehensive reviews" of chronic fatigue syndrome. They found that a given review's sample of research articles was clearly driven by the theoretical backgrounds of the authors. Writers from microbiology and immunology included twice as many infectious disease articles as did psychiatrists, whereas psychiatrists included twice as many psychiatric articles as did the writers from microbiology and immunology. So here is clear evidence of theory ladenness in the *selection* of articles to be included in reviews. Chalmers, Frank, and Reitman (1990) showed that the *evaluation* of the quality of articles in reviews is also theory-laden. In one case they examined 63 reviews about the effectiveness of a form of radiotherapy after radical mastectomy. Of the reviews written by radiotherapists and surgeons, 72% were classified as "enthusiastic," whereas of the reviews written by individuals from other medical specialists, only 15% were enthusiastic. Thus there is clear evidence that in communicating scientific information to other scientists, writers show theory ladenness in both the selection of the articles to be discussed and in evaluating those articles.

In Chapter 11 of *The Structure of Scientific Revolutions*, Kuhn asserted that scientific textbooks are theory-laden. He argued that textbook writers want to produce a textbook that shows that the science has progressed in a steady, cumulative fashion. Kuhn pointed out that, in fields that have undergone major paradigmatic shifts in their history, this means the writer has to rework the actual historical record ("Whig history"). Kuhn stated, "Partly by selection and partly by distortion, the scientists of earlier ages are implicitly represented as having worked upon the same set of fixed problems" (1962, p. 137).

MULTIPLE PROCESSES

Cognitive Psychology

The experiments from cognitive psychology in the earlier sections of this chapter were chosen to be as analytic and specific to a given mental process as possible. However, in real life, for a given scientific episode, it seems likely that top-down theories influence multiple mental processes. The classic study by Hastorf and Cantril (1954) is a good example. These investigators studied a particularly controversial football game between Princeton and Dartmouth. Undergraduates from each institution were shown the same film of the game and as they watched they were asked to indicate any infraction of the rules that they noticed. The Princeton undergraduate participants reported over twice as many rule infractions by the Dartmouth players as did the Dartmouth undergraduate participants, thus showing the strong impact of top-down factors. This study is not analytic with respect to the mental processes that were influenced, but they might well include perception, attention, interpretation, evaluation, memory, and communication.

History of Science

All of the historical examples used earlier in this chapter were chosen to be examples of the theory ladenness of a particular mental process (as well as this can be done with uncontrolled episodes from the history of science). However, as mentioned above, in actual practice many examples of theory ladenness in science probably involve the operation of multiple top-down processes. I end this section of the chapter with one of my favorite examples of theory ladenness in science—an account by Charles Darwin. This example almost certainly includes multiple theory-laden psychological processes—perception, attention, and interpretation. In 1831 while Darwin was finishing up his undergraduate work at Cambridge, he went on a geology field trip to Wales with the famous geologist Adam Sedgwick (Browne, 1995, pp. 140–143). At that time neither Sedgwick nor Darwin knew about the ice ages. In 1842, when Darwin had learned about the ice age theory, but remained unconvinced by it, he decided to repeat his 1831 tour and look for evidence relevant to the ice age theory. "[In 1831] we spent many hours in Cwm Idwal, examining all the rocks with extreme care, as Sedwick was anxious to find fossils in them; but neither of us saw a trace of the wonderful glacial phenomena all around us; we did not notice the plainly scored rocks, the perched boulders, the lateral and terminal moraines. Yet these phenomena are so conspicuous that . . . a house burnt down by fire did not tell its story more plainly than did this valley" (Darwin, 1887/1958, p. 70). One of the glacial features that Darwin observed at Cwn Idwal was a group of supraglacial erratics—slabs of rock transported across the top of the glacier

Figure 13.9 Darwin's supraglacial erratics, Cwm Idwal, Wales (picture taken by author August 14, 2011).

that fell down crevasses when the glacier began to melt and imbedded themselves in the underlying moraine (gravel). The supraglacial erratics that Darwin did not pay attention to in 1831, but described in 1842 are still very visible at Cwn Idwal, Wales (see Figure 13.9).

CONSCIOUS AND UNCONSCIOUS MENTAL PROCESSES IN CARRYING OUT SCIENCE

It is difficult to find explicit studies on the conscious/unconscious nature of theory-driven top-down effects. However, the more informal evidence is fairly clear. The information reviewed in this chapter shows top-down influences occurring across a very wide variety of psychological processes. In many cases the top-down influence was negative (inhibitory). Given that most participants in psychological experiments wish to be seen as performing well and that scientists do not want psychological factors to distort their attempts to understand the world, it seems almost certain that the participants in these studies and the scientists carrying out their work were not aware of the impact the top-down processes were having on their performance. For example, the participants in the studies using the old woman/young woman ambiguous figure who have been primed to see a young woman simply see the young woman and are not aware that there are top-down processes at work in determining their perception. Similarly, the astronomers who were observing the rings around Saturn and drawing them as moons were not aware that their beliefs that planets have moons around them were distorting their observations.

It turns out that trying to explicate the role of conscious/unconscious processes in science leads to issues of scientific ethics. If a scientist deliberately and consciously reports an increase in the brightness of a phosphorescent powder when none is actually occurring or deliberately and consciously tampers with an experiment to obtain the expected data, most scientists would consider this to be a clear violation of the ethical norms of science (Committee on Science, Engineering, and Public Policy, 1995). However when a top-down process is operating unconsciously to inhibit or distort the scientist's behavior, we would not typically consider the scientist to have violated the ethical norms of science. (Not all observers of the scientific scene appear to see this distinction. In their book, *Betrayers of the Truth*, Broad and Wade (1982) include several of the historical episodes I have discussed in this chapter as examples of scientific fraud.) Overall, I think the examples in this chapter from the history of science provide much evidence to support the view that the theory-driven top-down processes in science typically operate outside of awareness.

NORMATIVE IMPLICATIONS FOR SCIENTIFIC PRACTICE

Up to this point this chapter has been a descriptive/analytic account of science (how science is done). However, several readers of an earlier version of this chapter wanted me to discuss the normative implications (how science *ought* to be done) of the psychological findings outlined in the previous sections of this chapter. I am a little uncomfortable in this role, but will briefly discuss these issues.

It is interesting that these topics were discussed by Francis Bacon in *Novum Organum* in 1620 in one of the very first discussions of the philosophy of science. Bacon (1620/1952) outlined a number of "idols of the tribe" which are errors in scientific reasoning that he thought are characteristics of the human mind. Several of his examples are similar to those discussed in this chapter. For example he stated, "Man always believes more readily that which he prefers. He, therefore, rejects difficulties for want of patience in investigation" (Bacon, 1620/1952, p. 111). He also stated, "The human understanding, from its peculiar nature, easily supposes a greater degree of order and equality in things than it really finds" (p. 110). Bacon provided a number of normative suggestions and argued that these and other problems in scientific thinking will cause trouble "unless mankind when forewarned guard themselves with all possible care against them" (p. 109).

It seems to me that the scientific community has already adopted a number of procedures designed to reduce some of the negative consequences of theory ladenness. For example, the introduction of the keeping of laboratory notebooks increases the retention of low-level bottom-up information and reduces the potential impact of the theory ladenness of memory. Some individual scientists have developed their own personal defenses against theory-based memory distortions. Darwin noted, "Whenever a published fact, a new observation or thought came across me, which was opposed to my general results, [I made]....a memorandum of it without fail and at once; for I had found by experience that such facts and thoughts were far more apt to escape from memory than favourable ones" (1887/1958, p. 123).

Another approach that is widely practiced in science is to develop ways to obtain a stronger bottom-up input from nature. This strategy has driven the development of larger telescopes, more powerful microscopes, more powerful particle accelerators, and so forth. Developments such as these enable very strong bottom-up information to override incorrect theory-based beliefs. For example, imagine that we could travel in time and show a modern photograph of Saturn taken by the *Voyager* or *Cassini* spacecraft to the early astronomers who were having trouble understanding what they were seeing when viewing Saturn. If we asked them what they saw in the picture, I think they would say that that it was a planet (or some other heavenly object) with a ring around it (and with a little work we could probably convince them it was a picture of Saturn).

The examples from the history of science described in this chapter suggest that physical scientists do not apply control procedures for the possibility of error due to top-down processes with the same vigor that they do in controlling for error due to bottom-up processes (Hetherington, 1983, made a similar point). The examples of the N-ray episode, the Cambridge/Vienna controversy, and the Van Maanen case show that there are circumstances where the results of experiments can be determined almost totally by top-down factors and not by bottom-up information from the world. Independently manipulating the psychological and physical factors, as was eventually done in each of these cases, can provide controls for theory-based top-down effects on experiments. In a recent book Chabris and Simons (2010) argue that in a wide range of domains (including attention and memory) certain aspects of individuals' metacognitive intuitions are not accurate (e.g., almost everyone believes they would notice a gorilla walking across a scene they were viewing). Thus, the undervaluing of the potential role of top-down factors by scientists may reflect a larger aspect of how the mind works.

However, note that some disciplines have become aware of the potential problems posed by theory-based effects and developed methodological procedures to reduce them. For example, Stuewer (1985, p. 292) noted that after the resolution of the Cambridge/Vienna controversy the Cambridge experimenters became worried about the potential for theory ladenness in the perception of human observers in particle physics and devoted resources into developing electrical methods of particle counting. In the medical and psychological communities, the use of double-blind experimental designs is very common. Franklin (2002) pointed out that the particle physics community has become concerned enough about the possibility of theory-based data selection that some groups have adopted a policy of "blind analysis" in which various procedures (e.g., adding random numbers to the data) are used to mask the incoming data from the experimenters until the experiment is over.

The evidence reviewed in this chapter supports the view that top-down processes can have both facilitatory and inhibitory consequences for understanding nature. This finding makes the normative implications somewhat complex. Clearly, one wants to establish top-down controls to avoid situations such as the N-ray episode. However, if one wants the facilitatory aspects of top-down processes, one has to accept some potential negative consequences. At the time the science is being done, the scientist does not know if his or her theory maps the world and thus does not know if the theory-based processes are facilitating or inhibiting their progress in understanding nature. Perhaps pluralism within the community of scientists is our best option. Some scientists should practice a strong form of theory-based science (cf. Kuhn, 1963 on the virtues of dogmatism in science) while other scientists should try to reduce theory ladenness. Then we can hope that the social institutions of science (e.g., the review process, organized skepticism) will be able to even out some of the difficulties produced by the theory ladenness of the psychological processes of individual scientists who are engaged in the scientific endeavor.

CONCLUSIONS

This chapter is intended to be an example of naturalized philosophy of science. A number of philosophers of science have claimed that there is no evidence for the theory ladenness of perception. For example, Couvalis stated, "There is no evidence that theory problematically permeates experience, as opposed to merely helping us focus on some aspect of the world" (1997, p. 14). It seems to me that the naturalized philosophy of science evidence presented in this chapter argues strongly for the theory ladenness of perception and thus refutes philosophical claims to the contrary.

The reason why many philosophers of science have not wanted to accept the view that perception/observation is theory-laden is that they believe that this position leads to epistemological relativism. For example, Klee stated that "it is important to understand what is at stake in the dispute over whether or not all scientific terms are theory-laden. What is at stake is nothing less than the... 'objectivity' of science" (1997, p. 59). The evidence in this chapter supports a view in which perception is the product of bottom-up *and* top-down information. Both the historical record and the evidence from cognitive psychology suggest that the top-down factors have their greatest impact when bottom-up information is weak or degraded. If the experimental situation is such that the bottom-up information is clear and salient (e.g., a reading of 10 on a meter that goes from 1 to 10) then the issue of theory penetrating perception is much less epistemologically troublesome.

REFERENCES

Austin, R. H. (1967). Uranus observed. *British Journal for the History of Science, 3,* 275–284.

Babbage, C. (1830). *Reflections on the decline of science in England, and on some of its causes.* London: B. Fellowes.

Bacon, F. (1952). *Novum organum. Great books of the Western world.* (Vol. 30). Chicago: Encyclopaedia Britannica. (Original work published 1620).

Baltas, A., Gavroglu, K., & Kindi, V. (2000). A discussion with Thomas S. Kuhn. In J. Conant & J. Haugeland (Eds.), *The road since structure* (pp. 255–323). Chicago: University of Chicago Press.

Barnes, B. (1974). *Scientific knowledge and sociological theory.* London: Routledge & Kegan Paul.

Bartlett, F. C. (1932). *Remembering.* Cambridge: Cambridge University Press.

Baum, R., & Sheehan, W. (1997). *In search of planet Vulcan: The ghost in Newton's clockwork universe.* New York: Plenum.

Bechtel, W. (1988). *Philosophy of science: An overview for cognitive science.* Hillsdale, NJ: Lawrence Erlbaum.

Bernstein, J. (1993). Three degrees above zero. In J. Bernstein (Ed.), *Cranks, quarks, and the cosmos* (pp. 65–81). New York: Basic Books.

Birge, R. T. (1957). A survey of the systematic evaluation of the Universal Physical Constants. *Nuovo Cimento, Supplement, 6,* 39–67.

Boattini, A., D'Abramo, G., Forti, G., & Gal, R. (2001). The Arcetri NEO precovery program. *Astronomy & Astrophysics, 375,* 293–307.

Bogen, J., & Woodward, J. (1992). Observations, theories and the evolution of the human spirit. *Philosophy of Science, 59,* 590–611.

Brahe, T. (1929). On a new star, not previously seen within the memory of any age since the beginning of the world. In H. Shapley & H. E. Howarth (Eds.), *A source book in astronomy* (pp. 13–19). New York: McGraw-Hill.

Bransford, J. D., & Johnson, M. K. (1972). Contextual prerequisites for understanding: Some investigations of comprehension and recall. *Journal of Verbal Learning and Verbal Behavior, 11,* 717–726.

Brewer, W. F. (1987). Schemas versus mental models in human memory. In P. Morris (Ed.), *Modelling cognition* (pp. 187–197). Chichester, UK: John Wiley.

Brewer, W. F. (1999). Scientific theories and naive theories as forms of mental representation: Psychologism revived. *Science & Education, 8,* 489–505.

Brewer, W. F. (2003). Mental models. In L. Nadel (Ed.), *Encyclopedia of cognitive science* (Vol. 3, pp. 1–6). London: Nature.

Brewer, W. F. (2008). In what sense can the child be considered to be a "little scientist"? In R. A. Duschl & R. E. Grandy (Eds.), *Teaching scientific inquiry: Recommendations for research and implementation* (pp. 38–49). Rotterdam, Netherlands: Sense.

Brewer, W. F., & Chinn, C. A. (1994a). Scientists' responses to anomalous data: Evidence from psychology, history, and philosophy of science. *Philosophy of Science Association, 1,* 304–313.

Brewer, W. F., & Chinn, C. A. (1994b). The theory-ladenness of data: An experimental demonstration. In A. Ram, & K. Eiselt (Eds.), *Proceedings of the 16th Annual Conference of the Cognitive Science Society* (pp. 61–65). Hillsdale, NJ: Lawrence Erlbaum.

Brewer, W. F., Chinn, C. A., & Samarapungavan, A. (1998). Explanation in scientists and children. *Minds and Machines, 8,* 119–136.

Brewer, W. F., & Lambert, B. L. (1993). The theory-ladenness of observation: Evidence from cognitive psychology. In *Proceedings of the 15th Annual Conference of the Cognitive Science Society* (pp. 254–259). Hillsdale, NJ: Erlbaum.

Brewer, W. F., & Lambert, B. L. (2001). The theory-ladenness of observation and the theory-ladenness of the rest of the scientific process. *Philosophy of Science, 68,* S176–S186.

Brewer, W. F., & Loschky, L. (2005). Top-down and bottom-up influences on observation: Evidence from cognitive psychology and the history of science. In A. Raftopoulos (Ed.), *Cognitive penetrability of perception: Attention, action, strategies, and bottom-up constraints* (pp. 31–47). Hauppauge, NY: Nova Science.

Brewer, W. F., & Samarapungavan, A. (1991). Children's theories vs. scientific theories: Differences in reasoning or differences in knowledge? In R. R. Hoffman & D. S. Palermo (Eds.), *Cognition and the symbolic processes: Applied and ecological perspectives* (pp. 209–232). Hillsdale, NJ: Lawrence Erlbaum.

Broad, W., & Wade, N. (1982). *Betrayers of the truth.* New York: Simon & Schuster.

Browne, J. (1995). *Charles Darwin voyaging: A biography.* Princeton, NJ: Princeton University Press.

Bruner, J. S., & Postman, L. (1949). On the perception of incongruity: A paradigm. *Journal of Personality, 18,* 206–223.

Bugelski, B. R., & Alampay, D. A. (1961). The role of frequency in developing perceptual sets. *Canadian Journal of Psychology, 15,* 205–211.

Chabris, C., & Simons, D. (2010). *The invisible gorilla and other ways our intuitions deceive us.* New York: Crown.

Chalmers, T. C., Frank, C. S., & Reitman, D. (1990). Minimizing the three stages of publication bias. *Journal of American Medical Association, 263,* 1392–1395.

Chinn, C. A., & Malhotra, B. A. (2002). Children's responses to anomalous scientific data: How is conceptual change impeded? *Journal of Educational Psychology, 94,* 327–343.

Christy, J. W. (1990). A moment of perception. In M. Littmann (Ed.), *Planets beyond: Discovering the outer solar system* (Updated and Revised) (pp. 175–176). New York: John Wiley.

Christy, J. W., & Harrington, R. S. (1978). The satellite of Pluto. *Astronomical Journal, 83,* 1005–1008.

Chun, M. M., & Marois, R. (2002). The dark side of visual attention. *Current Opinion in Neurobiology, 12,* 184–189.

Cohen, E. R., & DuMond, J. W. M. (1965). Our knowledge of the fundamental constants of physics and chemistry in 1965. *Reviews of Modern Physics, 37,* 537–594.

Committee on Science, Engineering, and Public Policy, National Academy of Sciences. (1995). *On being a scientist: Responsible conduct in research* (2nd ed.). Washington, DC: National Academy Press.

Connor, C. E., Egeth, H. E., & Yantis, S. (2004). Visual attention: Bottom-up versus top-down. *Current Biology, 14,* R850–R852.

Couvalis, G. (1997). *The philosophy of science: Science and objectivity.* London: Sage.

Darwin, C. (1861). Letter, C. R. Darwin to H. Fawcett, September 18, 1861. Darwin Correspondence Project. www.darwinproject.ac.uk/entry-3257

Darwin, C. (1958). *The autobiography of Charles Darwin 1809–1882.* (N. Barlow, Ed.). New York: W. W. Norton. (Original work published 1887).

Feynman, R. P. (1985). *"Surely you're joking, Mr. Feynman!"* New York: W. W. Norton.

Fodor, J. A. (1983). *The modularity of mind: An essay on faculty psychology.* Cambridge, MA: MIT Press.

Fodor, J. A. (1984). Observation reconsidered. *Philosophy of Science, 51,* 23–43.

Fodor, J. A. (1991). The dogma that didn't bark (A fragment of a naturalized epistemology). *Mind, 100,* 201–220.

Forbes, E. G. (1983). The pre-discovery observations of Uranus. In G. Hunt (Ed.), *Uranus and the outer planets* (pp. 67–79). Cambridge: Cambridge University Press.

Franklin, A. (1984). Forging, cooking, trimming, and riding on the bandwagon. *American Journal of Physics, 52,* 786–793.

Franklin, A. (2002). *Selectivity and discord.* Pittsburgh, PA: University of Pittsburgh Press.

Giere, R. N. (1988). *Explaining science: A cognitive approach.* Chicago: University of Chicago Press.

Giere, R. N. (2006). Naturalized philosophy of science. In D. M. Borchert (Ed.), *Encyclopedia of philosophy* (2nd ed., Vol. 6, pp. 500–503). Detroit, MI: Thomson Gale.

Gilman, D. (1990). Observation: An empirical discussion. *Philosophy of Science Association, 1,* 355–364.

Godfrey-Smith, P. (2003). *Theory and reality: An introduction to the philosophy of science.* Chicago: University of Chicago Press.

Gould, S. J. (1980). *The panda's thumb.* New York: W. W. Norton.

Grosser, M. (1979). *The discovery of Neptune.* New York: Dover.

Gunstone, R. F., & White, R. T. (1981). Understanding of gravity. *Science Education, 65,* 291–299.

Hanson, N. R. (1958). *Patterns of discovery.* Cambridge: Cambridge University Press.

Hanson, N. R. (1969). *Perception and discovery: An introduction to scientific inquiry.* San Francisco, CA: Freeman, Cooper.

Harrington, R. S., & Harrington, B. J. (1979). The discovery of Pluto's moon. *Mercury, 8,* 1–3, 6, 17.

Hart, R. C. (1993). Van Maanen's internal motions in spiral nebulae. In N. S. Hetherington (Ed.), *Encyclopedia of cosmology* (pp. 646–656). New York: Garland.

Hastorf, A. H., & Cantril, H. (1954). They saw a game: A case study. *Journal of Abnormal and Social Psychology, 49,* 129–134.

Hetherington, N. S. (1983). Just how objective is science? *Nature, 306,* 727–730.

Hetherington, N. S. (1988). *Science and objectivity: Episodes in the history of astronomy.* Ames: Iowa State University Press.

Hoddeson, L., & Brewer, W. F. (2004, July). *Oral history and the psychology of human memory.* Paper presented at the 12th Triennial Conference of the International Society for the Study of Time: Time and Memory. Clare College, Cambridge, UK.

Holton, G. (1978). Subelectrons, presuppositions, and the Millikan-Ehrenhaft dispute. *Historical Studies in the Physical Sciences, 9,* 161–224.

Johnson-Laird, P. N. (1983). *Mental models.* Cambridge, MA: Harvard University Press.

Joyce, J., Rabe-Hesketh, S., & Wessely, S. (1998). Reviewing the reviews: The example of chronic fatigue syndrome. *Journal of the American Medical Association, 280,* 264–266.

Karmiloff-Smith, A., & Inhelder, B. (1975). "If you want to get ahead, get a theory." *Cognition, 3,* 195–212.

Kim, N. S., & Ahn, W-k. (2002). Clinical psychologists' theory-based representations of mental disorders predict their diagnostic reasoning and memory. *Journal of Experimental Psychology: General, 131,* 451–476.

Klahr, D., & Dunbar, K. (1988). Dual space search during scientific reasoning. *Cognitive Science, 12,* 1–48.

Klee, R. (1997). *Introduction to the philosophy of science: Cutting nature at its seams.* New York: Oxford University Press.

Klotz, I. M. (1980, May). The N-Ray affair. *Scientific American, 242,* 168–170, 173–175.

Kornblith, H. (Ed.). (1994). *Naturalizing epistemology* (2nd ed.). Cambridge, MA: MIT Press.

Kosslyn, S. M. (1980). *Image and mind.* Cambridge, MA: Harvard University Press.

Kowal, C. T., & Drake, S. (1980). Galileo's observations of Neptune. *Nature, 287,* 311–313.

Kuhn, T. S. (1957). *The Copernican revolution.* Cambridge, MA: Harvard University Press.

Kuhn, T. S. (1962). *The structure of scientific revolutions.* Chicago: University of Chicago Press.

Kuhn, T. S. (1963). The function of dogma in scientific research. In A. C. Crombie (Ed.), *Scientific change* (pp. 347–369). New York: Basic Books.

Leeper, R. (1935). A study of a neglected portion of the field of learning—The development of sensory organization. *Journal of Genetic Psychology, 46,* 41–75.

Lindsay, P. H., & Norman, D. A. (1977). *Human information processing: An introduction to psychology* (2nd ed.). New York: Academic Press.

Liu, A.-Y. (1976). Cross-modality set effect on the perception of ambiguous pictures. *Bulletin of the Psychonomic Society, 7,* 331–333.

Mack, A., & Rock, I. (1998). *Inattentional blindness.* Cambridge, MA: MIT Press.

Mahoney, M. J. (1977). Publication prejudices: An experimental study of confirmatory bias in the peer review system. *Cognitive Therapy and Research, 1,* 161–175.

Malley, M. (1979). The discovery of atomic transmutation: Scientific styles and philosophies in France and Britain. *Isis, 70,* 213–223.

Markman, A. B. (1999). *Knowledge representation.* Mahwah, NJ: Lawrence Erlbaum.

McClelland, J. L., & Rumelhart, D. E. (1981). An interactive activation model of context effects in letter perception: Part 1. An account of basic findings. *Psychological Review, 88,* 375–407.

Millar, R. (1972). *The Piltdown men.* London: Victor Gollancz.

Mishra, P., & Brewer, W. F. (2003). Theories as a form of mental representation and their role in the recall of text information. *Contemporary Educational Psychology, 28,* 277–303.

Neisser, U. (1967). *Cognitive psychology.* New York: Appleton-Century-Crofts.

Nye, M. J. (1980). N-rays: An episode in the history and psychology of science. *Historical Studies in the Physical Sciences, 11*(Pt. 1), 125–156.

Particle Data Group. (1980). Review of particle properties. *Reviews of Modern Physics, 52,* S1–S286.

Price, D. D. S. (1975). *Science since Babylon* (Enlarged ed.). New Haven, CT: Yale University Press.

Price, F. W. (2000). *The planet observer's handbook* (2nd ed.). Cambridge: Cambridge University Press.

Proctor, R. W., & Capaldi, E. J. (2006). *Why science matters: Understanding the methods of psychological research.* Malden, MA: Blackwell.

Quine, W. V. (1969). Epistemology naturalized. In W. V. Quine (Ed.), *Ontological relativity and other essays* (pp. 69–90). New York: Columbia University Press.

Raftopoulos, A. (2001). Is perception informationally encapsulated? The issue of the theory-ladenness of perception. *Cognitive Science, 25,* 423–451.

Reese, H. W. (1963). "Perceptual set" in young children. *Child Development, 34,* 151–159.

Reiner, M., Chi, M. T. H., & Resnick, L. (1988). Naive materialistic belief: An underlying epistemological commitment. In *Proceedings of the 10th Annual Conference of the Cognitive Science Society* (pp. 544–551). Hillsdale, NJ: Lawrence Erlbaum.

Reynolds, R. I. (1985). The role of object-hypotheses in the organization of fragmented figures. *Perception, 14,* 49–52.

Rumelhart, D. E., & Norman, D. A. (1988). Representation in memory. In R. C. Atkinson, R. J. Herrnstein, G. Lindzey, & R. D. Luce (Eds.), *Learning and cognition: Vol. 2. Stevens' handbook of experimental psychology* (2nd ed.) (pp. 511–587). New York: John Wiley.

Schacter, D. L. (1995). Memory distortion: History and current status. In D. L. Schacter, J. T. Coyle, G. D. Fischbach, M.-M. Mesulam, & L. E. Sullivan (Eds.), *Memory distortion: How minds, brains, and societies reconstruct the past* (pp. 1–43). Cambridge, MA: Harvard University Press.

Shapley, D. (1949). Pre-Huygenian obervations of Saturn's ring. *Isis, 40*(1), 12–17.

Sheehan, W. (1988). *Planets & perception: Telescopic views and interpretations, 1609–1909.* Tucson: University of Arizona Press.

Shinoda, H., Hayhoe, M. M., & Shrivastava, A. (2001). What controls attention in natural environments? *Vision Research, 41,* 3535–3545.

Shipman, P. (1986, October). How a 125 million-year-old dinosaur evolved in 160 years. *Discover, 7,* 94–102.

Simons, D. J. (2010). Monkeying around with the gorillas in our midst: Familiarity with an inattentional-blindness task does not improve the detection of unexpected events. *i-Perception, 1,* 3–6.

Simons, D. J., & Chabris, C. F. (1999). Gorillas in our midst: Sustained inattentional blindness for dynamic events. *Perception, 28,* 1059–1074.

Steinfeld, G. J. (1967). Concepts of set and availability and their relation to the reorganization of ambiguous pictorial stimuli. *Psychological Review, 74,* 505–522.

Stern, A., & Mitton, J. (1999). *Pluto and Charon: Ice worlds on the ragged edge of the solar system.* New York: John Wiley.

Stuewer, R. H. (1985). Artificial disintegration and the Cambridge-Vienna controversy. In P. Achinstein & O. Hannaway (Eds.), *Observation, experiment, and hypothesis in modern physical science* (pp. 239–307). Cambridge, MA: MIT Press.

Suppe, F. (Ed.). (1977). *The structure of scientific theories* (2nd ed.). Urbana: University of Illinois Press.

Thagard, P. (1992). *Conceptual revolutions.* Princeton, NJ: Princeton University Press.

van Dijk, T. A., & Kintsch, W. (1978). Cognitive psychology and discourse: Recalling and summarizing stories. In W. U. Dressler (Ed.), *Current trends in textlinguistics* (pp. 61–80). Berlin: Walter de Gruyter.

Van Helden, A. (1974). Saturn and his anses. *Journal for the History of Astronomy, 5,* 105–121.

Vicente, K. J., & Brewer, W. F. (1993). Reconstructive remembering of the scientific literature. *Cognition, 46,* 101–128.

Vosniadou, S., & Brewer, W. F. (1992). Mental models of the earth: A study of conceptual change in childhood. *Cognitive Psychology, 24,* 535–585.

Vosniadou, S., & Brewer, W. F. (1994). Mental models of the day/night cycle. *Cognitive Science, 18,* 123–183.

Vosniadou, S., & Skopeliti, I. (2005). Developmental shifts in children's categorizations of the earth. In B.G. Bara, L.W. Barsalou, & M. Bucciarelli (Eds.), *Proceedings of the 27th Annual Conference of the Cognitive Science Society* (pp. 2325–2330). Mahwah, NJ: Lawrence Erlbaum.

Weinberg, S. (1992). *Dreams of a final theory.* New York: Pantheon Books.

Weiner, J. S. (1980). *The Piltdown forgery.* New York: Dover.

Westfall, R. S. (1973). Newton and the fudge factor. *Science, 179,* 751–758.

Wood, R. W. (1904). The *n*-rays. *Nature, 70,* 530–531.

Woolgar, S. (1988). *Science: The very idea.* Chichester, UK: Ellis Horwood.

Zangwill, O. L. (1937). A study of the significance of attitude in recognition. *British Journal of Psychology, 28,* 12–17.

14 The Practice of Psychological Science in Social-Personality Research

Are We Still a Science of Two Disciplines?

Jessica L. Tracy, Richard W. Robins, and Jeffrey W. Sherman

> I shall discuss the past and future place within psychology of two historic streams of method, thought, and affiliation which run through the last century of our science. One stream is *experimental psychology*; the other, *correlational psychology*. Dashiell optimistically forecast a confluence of these two streams, but that confluence is still in the making. Psychology continues to this day to be limited by the dedication of its investigators to one or the other method of inquiry rather than to scientific psychology as a whole.
> (Cronbach, *1957, p. 671)*

In his 1957 American Psychological Association presidential address, the eminent educational psychologist Lee Cronbach made a distinction between "two streams" of scientific psychology, *experimental* and *correlational*. Cronbach's use of his presidential address to target this issue reflects the importance with which it was imbued at the time. Many researchers, across the various domains of psychological science, were grappling with questions about the relative merits of these two streams or "disciplines," and about whether they should become more integrated.

Fifty years later, Cronbach's distinction still appears to represent the state of the field today. In his address accepting the 2007 APA Award for Distinguished Scientific Applications of Psychology, Peter Bentler (2007) commented that "Cronbach's hope has not progressed much." Indeed, the split between Cronbach's streams transcends methodological preferences and is considerably broader than a simple division between researchers who conduct experiments by manipulating variables, and those who measure variables and search for correlations. Rather, the two streams characterize almost every aspect of the research endeavor. According to Cronbach, the two approaches differ in their "philosophical underpinnings, methods of inquiry, topical interests, and loci of application" (p. 671). To clarify, Cronbach was likely referring to the research designs, measures, and statistical analyses a researcher uses (these may be what Cronbach meant by "methods of inquiry"); the processes and causal factors he or she views as responsible for effects sought and found; the content areas studied (i.e., Cronbach's "topical interests"); the general philosophical or theoretical framework underlying a researcher's goals (i.e., Cronbach's "philosophical underpinnings"), and the ways in which he or she evaluates findings—for example, whether internal or external validity is emphasized (see Table 14.1).

Most researchers today would likely agree that the concepts, methods, and approaches of both the *experimental* and the *correlational* stream of thought are important and, in fact, essential to a complete program of research; both have contributed enormously to the current state of knowledge in psychological science.[1] Yet the distinction between these two streams permeates all domains of psychological science, albeit to different degrees. To give a few prominent examples, within clinical psychology, *experimental* stream researchers tend to conduct experimental or quasi-experimental studies on humans and animals, exploring basic processes (e.g., emotion, motivation, neurobiology) by manipulating the presumed causal influences on the etiology and maintenance of clinical disorders. In contrast, *correlational* stream clinical researchers are more likely to conduct longitudinal and epidemiological studies aimed at identifying predictors of psychiatric disorders and their consequences. Researchers in clinical psychology have also debated the utility of studying psychopathologies such as depression by seeking predictors in actual patient populations (a *correlational* stream approach), versus using "analogue" studies in which experiments are conducted on individuals (typically college students) who score high but in the normal range on measures of depressive affect (an *experimental* stream approach). Some argue that research on nonclinical samples cannot be used to make inferences about the causes and consequences of mental illness (Kazdin, 1978; Vredenburg, Flett, & Krames, 1993; Westen, Novotny, & Thompson-Brenner, 2004), whereas others maintain that the internal validity of controlled experiments, even those relying on college student samples, provides important insights into the processes underlying mental illness that offset any potential limitations in external validity (Crits-Christoph, Wilson, & Hollon, 2005).

The split between Cronbach's streams is also evident in health psychology, where there is a distinction between researchers who examine the way chronic

[1] We have adopted Cronbach's (1957) labels of "experimental" and "correlational" to identify the two streams, and to refer to the full range of methods, statistics, research designs, and philosophical approaches that define each of the streams. Table 1 provides a complete portrait of each stream, based on Cronbach's original definition. It is important to note that although we use Cronbach's labels, we, like him, do not intend to imply that either stream can be characterized solely, or even primarily, as research guided by the correlational approach or research guided by the experimental approach.

Table 14.1 Overview of the two streams of psychology, based on Cronbach (1957)

	Correlational stream	*Experimental stream*
Research design/approach	• Correlational • Longitudinal • Psychobiographic/case study • Twin/adoption study	• Dyadic/group interaction • Experimental • Quasi-experimental
Statistical/data analytic procedures	• Correlation • Convergent/discriminant function analyses • Cluster analyses • Factor/principal component analyses • Growth curve modeling • Item-response theory (IRT) • Multidimensional scaling • Multiple regression • Partial correlation • Reliability analysis • Structural equation modeling (SEM) • Time series analysis	• Analyses of variance (ANOVA) • *t*-test

(*continued*)

Table 14.1 (Continued)

	Correlational stream	*Experimental stream*
Assessment methods/measures	• Autonomic nervous system assessment • Experience sampling • Hormone levels • Informant report • Molecular genetics • Narrative/open-ended questionnaires • Self-report scales • Structured interviews	• Behavioral observation • Behavioral response • Implicit measures • Judgments of groups/nations/cultures • Memory tasks • Reaction time measures
Type of validity emphasized	• External (generalizability and mundane realism)	• Internal
Theoretical approach to research	• Effects that are consistent with common sense are most informative • Emphasis on consistency of behaviors, thoughts, and feelings	• Counter-intuitive effects are most informative • Emphasis on influence of situations on behaviors, thoughts, and feelings

dispositional variables influence long-term health outcomes in real-world contexts (e.g., Miller, Chen, & Cole, 2009), and those who assess online physiological or neural responses to experimental manipulations (e.g., Dickerson & Kemeny, 2004). In the field of biopsychology, researchers reflecting the *correlational stream* approach typically use observational methods to study nonhuman animal behaviors over time, often in naturalistic contexts (e.g., Weinstein, Capitanio, & Gosling, 2008); whereas researchers reflecting the *experimental stream* conduct experiments on nonhuman animals, manipulating genetic (e.g., through gene knockout studies), physiological (e.g., through brain lesions and psychopharmacological interventions), and social factors (e.g., exposure to dominant vs. submissive conspecifics, social isolation), and observing the behavioral and physiological consequences (e.g., Francis, Champagne, Liu, & Meaney, 1999; Winstanley, 2007). Within the field of developmental psychology, two main subareas—cognitive development and socioemotional development—map fairly closely onto the two streams, with cognitive developmentalists generally favoring experimental studies of basic processes and social developmentalists generally favoring correlational (e.g., longitudinal) studies of phenomena that can only be observed and measured outside the lab, such as parenting, antisocial behavior, and peer relations.

A split between the two streams is also apparent, although perhaps less obviously so, in cognitive psychology. Although the field is generally dominated by researchers working within the experimental stream, several researchers have argued for a "cognitive ethology" that openly addresses the distinction between mental processes that are operationalized as responses to laboratory manipulations, and mental processes that occur in everyday life (Kingstone, Smilek, & Eastwood, 2008). Along with earlier cognitive researchers such as Broadbent (1991), Neisser (1976, 1991), and Bruner (1990), cognitive ethologists assert that researchers should conduct complementary studies on ecologically valid behaviors, everyday life, or "acts of meaning." At the same time, others note that laboratory studies produce essential knowledge about mental processes such as attention and memory, regardless of their external generalizability (Mook, 1983).

In some disciplines the two streams coexist without competing, and researchers who identify with one stream faithfully support and respect researchers who represent the other. In other disciplines, researchers from the two streams may clash over scarce resources, such as academic positions, grant funding, power within a department, and top graduate students who, in turn, often feel that they must choose which stream to align with (Swann & Seyle, 2005). In general, the presence of a split between the two streams is a fact taken almost for granted across most areas of psychological science practiced today. Yet an important question underlies this apparent fact: To what extent is the split *real*? Is the distinction between the two streams an accurate representation of ongoing research in today's psychological scientific climate, or is it a stereotyped, mythologized distinction that allows for quick and easy conceptualizations of different kinds of research, but does not characterize actual researchers in terms of the work they do? To what extent do psychological scientists truly embody one stream versus the other? And to what extent are they in fact more likely to represent a middle-ground, hybrid perspective, making use of whatever methods, approaches, and theoretical principles best apply to their research, regardless of any "official" stream associations? In other words, have the streams finally merged, as Broadbent, Cronbach, and Neisser hoped, or do researchers still tend to conduct research within one stream or the other?

To address these questions, we conducted a study examining the research practices of a group of prominent psychological scientists: editors and editorial board members of seven leading social-personality psychology journals (Tracy, Robins, & Sherman, 2009a). We chose social-personality researchers as a test case for examining whether psychological science remains a discipline of two streams for several reasons. First, a recent study of the interconnections among various subfields of psychology (i.e., examining the extent to which the flagship journals of each subfield cite articles from flagship journals of other subfields) found that social-personality psychology was the most central "broker" subfield, or "mediating hub of knowledge," across the past three decades (Yang & Chiu, 2009). In other words, social-personality is both the largest provider and the largest consumer of research within psychology as a whole. Given this finding, of the widespread dissemination of social-personality research to other subfields, it is important that social-personality methods and research approaches be widely accessible and comprehensible. Furthermore, researchers across disciplines may have a vested interest in understanding the basic structure of social-personality research practices (e.g., whether they can be characterized as having two streams).

Second, social-personality is characterized by the use of a particularly wide range of methods, and the research it produces seems to well represent both the *correlational* and *experimental* streams. As a broad generalization, personality psychologists are often assumed to work within the *correlational stream* (defined broadly as described above and in Table 14.1), whereas social psychologists are often assumed to work within the *experimental stream* (again defined broadly as in Table 14.1). There are, of course, many exceptions, and many researchers are likely to best be considered hybrids, more aptly characterized as "social-personality psychologists" than "social" or "personality" psychologists, and correspondingly conducting research that reflects both approaches. However, the split between the streams in social-personality research seems obvious; in fact, to some extent, social-personality psychology can be viewed as encompassing two separate areas of psychology, rather than as a single area with two general subemphases. Furthermore, each of the two "subareas" has major connections with other prominent areas of psychology (e.g., personality with clinical, social with cognitive), and for some researchers these connections may be more relevant or self-identifying than their connections with the other subarea within social-personality.

Thus, in our study of the structure of contemporary psychology research practices, we asked prominent social-personality researchers to complete an extensive survey about the methods, statistics, and research designs they use, as well as the content areas they seek to understand. By classifying these individuals as personality or social psychologists (which we did based on the journals they were affiliated with, as well as their own self-reported affiliations), we were then able to quantify the differences between the research practices of the two areas, and test whether these differences fit with Cronbach's distinction. We were also interested in researchers' own explicit beliefs about these differences, and whether these beliefs were accurate. Thus, we also asked these individuals to report on the research practices they believed to be typical of personality and social psychologists. In sum, this survey allowed us to directly address three important questions: (1) How do personality and social psychologists differ in research methods, designs, analyses, and general approaches to research? (2) Do the differences between personality and social psychologists converge with these researchers' own explicit beliefs about the two groups? and (3) To what extent do the various methods and statistical procedures used by these researchers reflect the two streams identified by Cronbach (i.e., do the two columns of Table 14.1 "hang together" empirically to form two distinct factors)?

A META-SOCIAL-PERSONALITY STUDY

In 2006, we attempted to recruit all individuals who were serving as editors and editorial board members of the following leading journals in social-personality psychology: *European Journal of Personality* (*EJP*), *European Journal of Social Psychology* (*EJSP*), *Journal of Experimental Social Psychology* (*JESP*), *Journal of Personality* (*JP*), *Journal of Personality and Social Psychology* (*JPSP*), *Personality and Social Psychology Bulletin* (*PSPB*), and *Personality and Social Psychology Review* (*PSPR*). We chose to recruit editors and editorial board members of these journals for several reasons. First, they are very likely to conduct personality and social research and to perceive themselves as personality or social psychologists (or both). Second, these individuals are typically among the most productive researchers working in the field, so they are collectively responsible for a large body of social-personality research. Third, members of editorial boards cover a broad range of career stages, providing a sample that includes individuals who are at the early, middle, and late stages of their scientific careers. Fourth, members of editorial boards decide what is (and is not) accepted for publication in social-personality journals, and thus are the "gatekeepers" of social-personality psychology. These individuals are highly knowledgeable about what constitutes social-personality research; in fact, one could argue that they set the standards for the field. Fifth, including editors from these particular journals allowed us to equate the sample across personality and social psychology, and ensure that both groups were fairly equally represented.

Of the 407 individuals contacted, 39% ($N = 159$; 29% female; median age = 45 years, range = 30–70) agreed to participate, which involved completing a detailed survey asking about their research practices.[2] Seventy-four percent of these individuals were classified as either personality or social psychologists, based on the journals with which they were affiliated. The remaining participants ($n = 42$) could not

[2] This response rate is comparable to rates typically found in survey research relying on mail (20%) and telephone responding (60%; Visser, Krosnick, & Lavrakas, 2000). Given that Internet responses are more convenient than mail-in responses but that it is easier to refuse to participate over e-mail than phone, it is not surprising that the response rate found here falls in between these typical rates. It is also noteworthy that surveys with lower response rates may be more predictive of outcomes than surveys with greater response rates (Visser, Krosnick, Marquette, & Curtin, 1996).

be classified on this basis because they either served on both social and personality journals, or only on a journal that is explicitly a mixture of social and personality research. Twenty-two of these unclassified participants were subsequently classified based on the extent to which they reported studying "issues and topics related to personality psychology" and "issues and topics related to social psychology"; those with scores greater than the midpoint of the scale (4) on the "social psychology" variable and less than the midpoint (4) on the "personality psychology" variable were classified as social psychologists, and those with the reverse pattern were classified as personality psychologists. We decided not to apply further criteria to classify the remaining 20 respondents (13% of the sample), in order to maintain the distinctiveness of the two categories. Overall, 46% of the sample ($n = 74$) was classified as social psychologists and 41% ($n = 65$) was classified as personality psychologists.

A full description of the sample, all items included in the survey, and all results, are reported in Tracy et al. (2009a; see also Tracy, Robins, & Sherman, 2009b). Here, we briefly describe several key results from the study, which address our three core research questions.

What Is the Difference between Personality and Social Researchers? What Do These Researchers Think Is the Difference?

Research designs and approaches. Figure 14.1, Panel A, displays mean frequencies of each of the 12 research designs included in the survey, separately for respondents classified as personality versus social psychologists. As can be seen, the two groups differ in the majority of approaches asked about, but these differences are not absolute. For example, social researchers more frequently use experimental designs and personality researchers more frequently use correlational designs, but both groups use both designs fairly frequently. Furthermore, there are certain designs, such as correlational, that all researchers use at least occasionally. In addition, the mean frequency of use for several designs—cross-cultural, cross-species, field study, quasi-experimental, and psychobiographical—did not differ between the groups.

As can be seen from Figure 14.1, Panel B, personality and social researchers' explicit beliefs about the research designs used by each group were quite accurate. Participants' accurately predicted that personality researchers use more correlational, cross-sectional, longitudinal, patient study, and twin/adoption study designs; whereas social researchers use more dyadic/group and experimental designs. However, not all beliefs were on target. Participants mistakenly expected that personality psychologists would use more cross-cultural, cross-species, and psychobiographical designs, but this was not the case. For cross-species and psychobiographical approaches, the inaccuracy may be due to very low frequency of these designs, by both groups—at least among those researchers who serve on major social-personality editorial boards. For cross-cultural designs, the absence of a predicted difference suggests that cultural psychology spans across personality and social psychology. Interestingly, the split

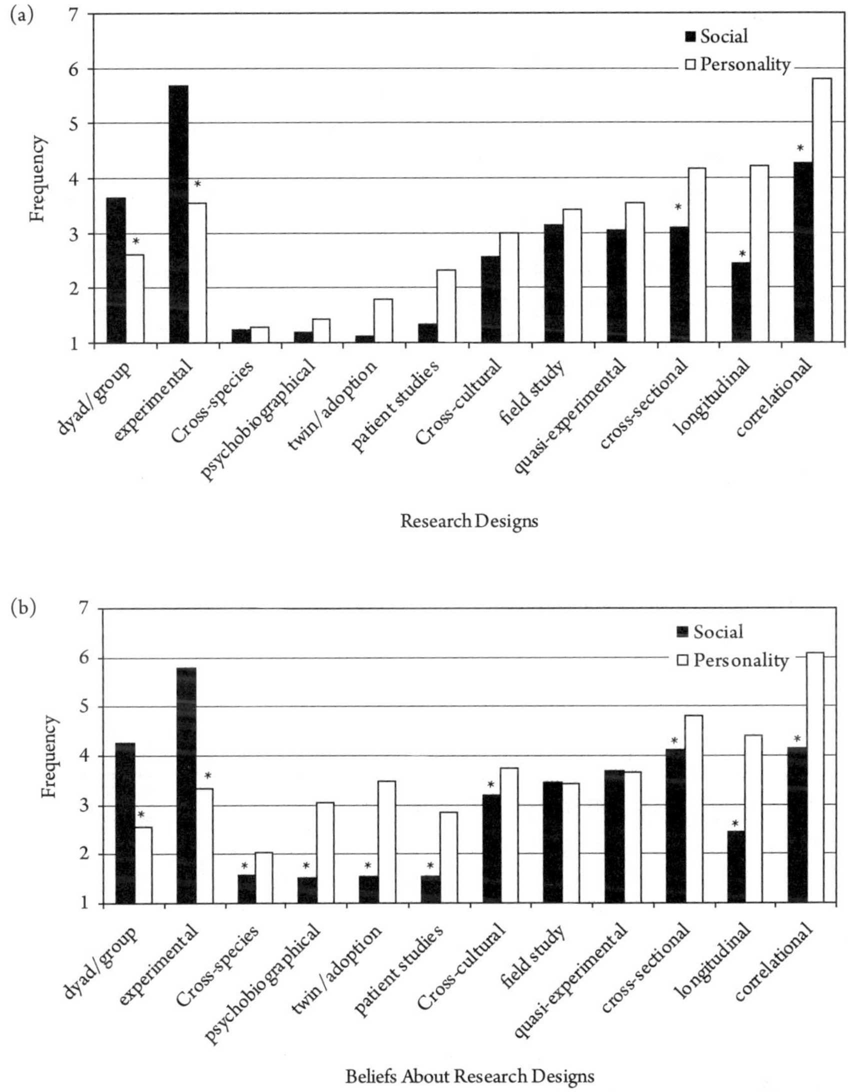

Figure 14.1 Research designs used by personality and social psychologists (Panel A) and beliefs about research designs used by personality and social psychologists (Panel B).
Note. N = 139. *p < .05.

between the two streams seems to exist within this more narrow subfield as well: *correlational*-stream, personality-oriented cultural psychologists—referred to as "cross-cultural psychologists"—tend to focus on cross-cultural consistency (i.e., human universals), whereas *experimental*-stream, social-oriented cultural psychologists—referred to as "cultural psychologists"—tend to focus on cultural differences (i.e., viewing culture as a situational variable—or meaning system—that determines behavior; Heine, 2001).

Statistical procedures and data analytic strategies. As Figure 14.2, Panel A, shows, personality and social psychologists differ in the frequency with which they use most statistical procedures. Although social researchers use Analysis of Variance

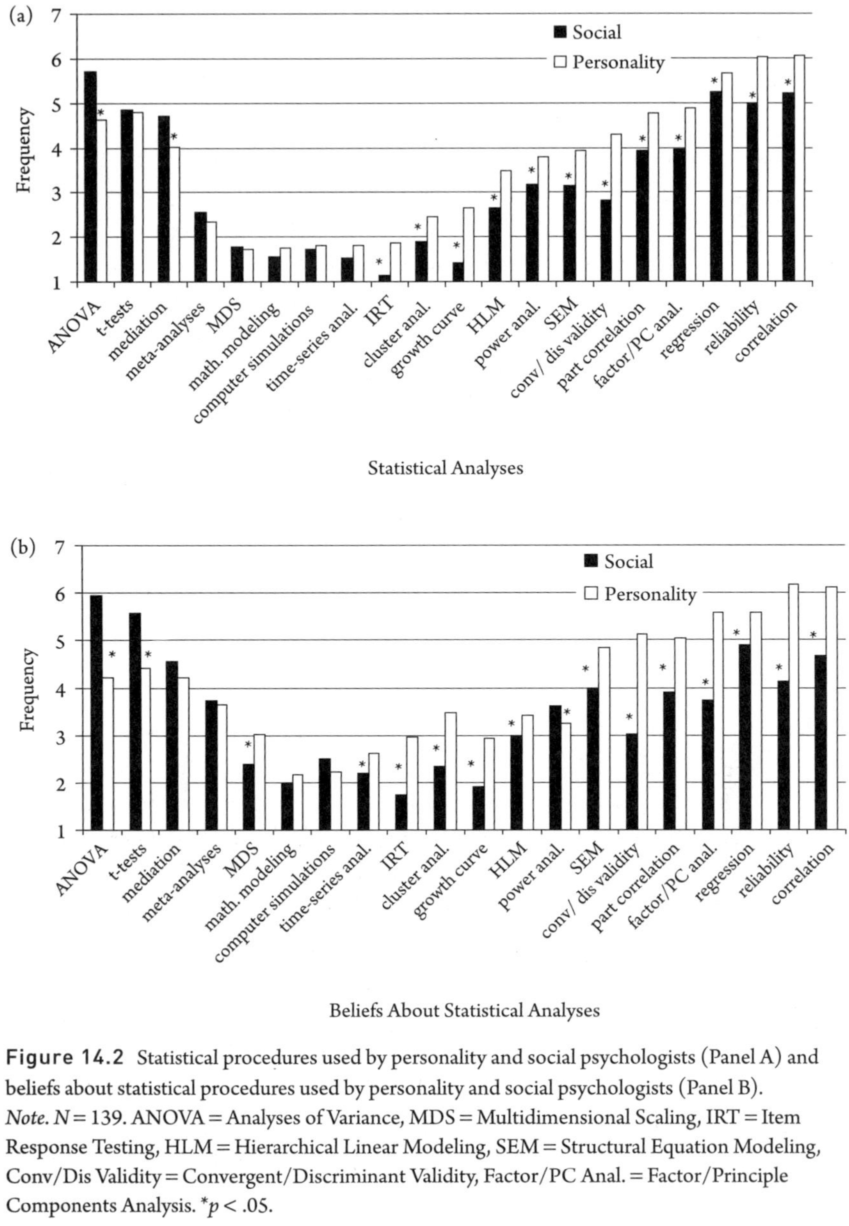

Figure 14.2 Statistical procedures used by personality and social psychologists (Panel A) and beliefs about statistical procedures used by personality and social psychologists (Panel B).
Note. $N = 139$. ANOVA = Analyses of Variance, MDS = Multidimensional Scaling, IRT = Item Response Testing, HLM = Hierarchical Linear Modeling, SEM = Structural Equation Modeling, Conv/Dis Validity = Convergent/Discriminant Validity, Factor/PC Anal. = Factor/Principle Components Analysis. $^{*}p < .05$.

(ANOVA) and tests of mediation more frequently, personality researchers use almost every other procedure included in the survey more frequently, suggesting that personality researchers use a wider range of statistical techniques to analyze their data. This may go hand in hand with their greater reliance on correlational, rather than experimental, approaches to research design; nonexperimental studies tend to require more sophisticated quantitative techniques to infer causal relations. However, in most cases these differences were again relative; both groups very frequently use the same statistical procedures, including *t*-tests, multiple regression, ANOVA, tests of mediation, correlations, factor or principal components analyses, partial correlations, and reliability analyses. In fact, over 90% of respondents in both groups reported using each of these procedures at least some of the time.

As with research designs, explicit beliefs about statistical procedures largely converged with the statistics actually used (see Figure 14.2, Panel B); personality researchers were accurately expected to use more correlations, tests of convergent/discriminant validity, cluster analyses, discriminant function analyses, factor/principal components analyses, growth curve modeling, hierarchical linear modeling, item-response theory, multiple regression, partial correlations, reliability analyses, and structural equation modeling; whereas social researchers were correctly expected to use more ANOVA. However, participants incorrectly believed that personality researchers would more frequently use time series analyses and multidimensional scaling; in fact, these two very infrequently used procedures are used equally infrequently by both groups.

Assessment methods/measures. As Figure 14.3, Panel A, shows, personality and social psychologists differ in the frequency with which they use most assessment methods. Once again though, most of these differences were relative, and there are certain methods that both groups use very frequently and do not differ in; for example, self-report and judgments of self/other. These two methods are, by far, the most frequently used forms of assessment; over 98% of individuals in both groups reported using each of these methods at least some of the time.

Once again, explicit beliefs about assessment methods were generally on target (see Figure 14.3, Panel B). Personality researchers were correctly expected to use more emotion sampling methods, hormone level assessment, informant reports, molecular genetics, and structured interviews; whereas social researchers were correctly expected to use more behavioral response assessment, implicit measures, memory tasks, judgments of groups/nations/cultures, reaction times, and other judgment tasks. However, there were several inaccuracies; personality researchers were expected to more frequently use narrative/open-ended and self-report assessments; these methods—particularly self-report—are in fact commonly used by both groups. This discrepancy between belief and actual practice has a noteworthy implication: Personality researchers may be more frequently critiqued for an overreliance on self-report, but this critique may be based more on a perception of overreliance than actual overreliance. Or, to the extent that the critique is applicable, it is equally applicable to both groups.

Types of validity emphasized. We also asked participants to rate the importance they placed on each of the following forms of validity: construct, internal, external defined in terms of generalizability (i.e., do findings generalize to other studies?), and external defined in terms of mundane realism (i.e., do findings generalize to processes/behaviors that occur outside the lab?). As can be seen from Figure 14.4, Panel A, although both groups equally value construct validity, social researchers place greater importance on internal validity than do personality researchers, whereas personality researchers place greater importance on both kinds of external validity than do social researchers.

Consistent with these differences, participants' beliefs reflected social researchers' greater emphasis on internal validity and personality researchers' greater emphasis

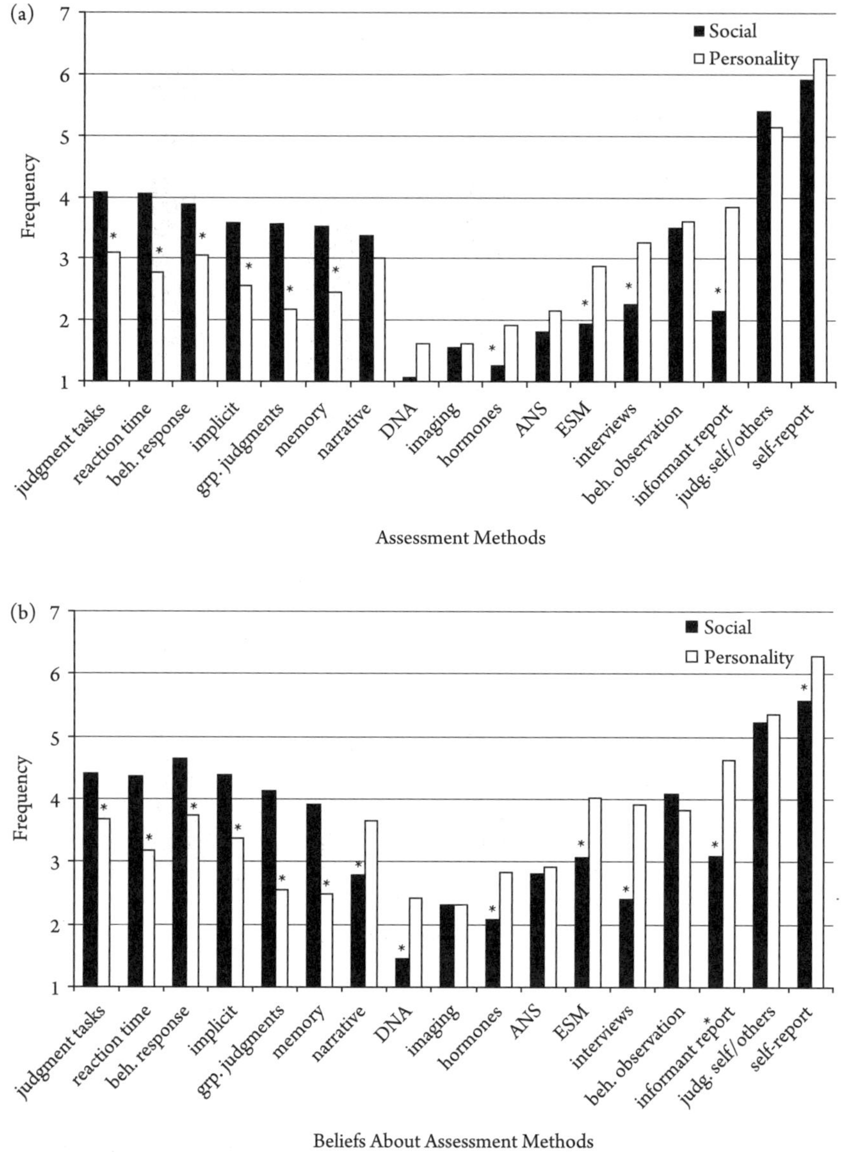

Figure 14.3 Assessment methods used by personality and social researchers (Panel A) and beliefs about assessment methods used by personality and social researchers (Panel B).
Note. $N = 139$. ANS = Autonomic Nervous System/physiological response methods, ESM = Experience Sampling Methods. $^*p < .05$.

on external validity. However, participants expected a difference for construct validity, which did not emerge. This inaccuracy may be due to an implicit association between personality research and measurement, and a tendency to conceptualize construct validity as largely about the valid measurement of constructs. This belief was belied, however, by the high level of importance all researchers place on ensuring that their constructs of interest—be they manipulated or measured—are

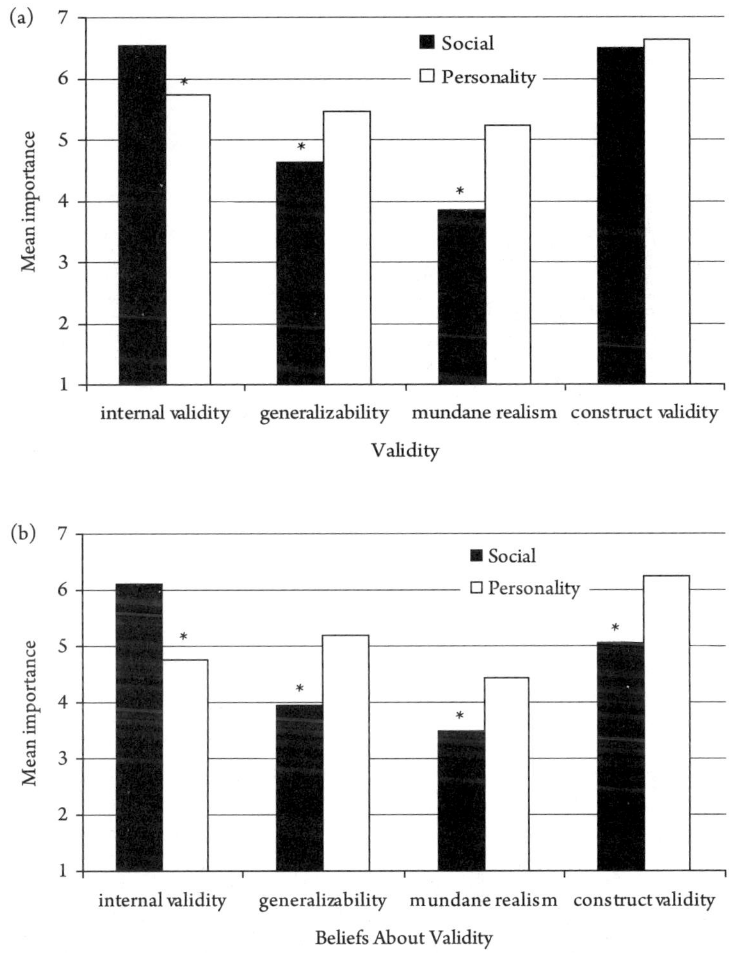

Figure 14.4 Forms of validity emphasized by personality and social psychologists (Panel A) and beliefs about forms of validity emphasized by personality and social psychologists (Panel B). *Note.* $N = 139$. $^{*}p < .05$.

operationalized in a valid manner. In addition, although participants correctly predicted that social researchers would place greater emphasis on internal validity than would personality researchers, they did not recognize that, in fact, both groups value internal validity very highly, and even personality researchers rate it no less important than external validity.

Philosophical and theoretical approaches. Some of the sharpest differences between personality and social researchers emerged in the context of broad philosophical approaches to science and research—though it is noteworthy that the items we included to assess these issues were developed specifically to distinguish between the two groups, and between the *experimental* and *correlational* streams. First, we found that 64% of social researchers reported seeking counterintuitive effects rather than effects that are consistent with common sense, whereas only 35% of personality researchers reported seeking counterintuitive effects over commonsensical ones, $p < .05$. This difference is consistent with a 2004 debate in *Brain and Behavioral Sciences* about the merits of seeking counterintuitive effects, which largely featured personality and social researchers on opposite sides of the spectrum (see Krueger & Funder, 2004). Given the prominence of this debate, it is not surprising that researchers' explicit beliefs about personality and social researchers precisely mirrored this finding; 60% of respondents expected social researchers to seek more counterintuitive than commonsensical effects, whereas only 10% expected personality researchers to show this preference, $p < .05$.

This distinction may reflect these researchers' differing emphases on describing a phenomenon and understanding its genetic/developmental/interpersonal history, versus seeking to understand the cognitive mechanisms that underlie the phenomenon. For example, in the self-judgment literature—an area of research populated by both personality and social psychologists—personality researchers typically seek to show that self-reports are relatively accurate, converge with reports made by others, and predict important outcomes; whereas social researchers tend to document errors and biases in self-judgments that inform on the ways in which the self-evaluative system functions (Dunning, 2005; Funder, 1987). Thus, social psychologists seek counterintuitive effects because they believe that these effects reveal something about the workings of the system, whereas personality psychologists seek commonsensical effects because these effects describe more general patterns of behavior (Darley & Todorov, 2004; Epley, van Boven, & Caruso, 2004; Krueger & Funder, 2004).

Second, two of the strongest differences that emerged between the groups reflected orientations toward the long-standing person-situation debate, suggesting that this issue continues to play a role in determining whether an individual is a personality or social researcher (in fact, a recent special issue of the *Journal of Research in Personality* featured articles from prominent researchers on both sides of the debate, discussing current perspectives on this topic; see Donnellan, Lucas, & Fleeson, 2009). Specifically, personality researchers were more likely to characterize their overarching theoretical approach with the statement, "Individuals' behaviors, thoughts, and feelings tend to be consistent across situations and over time" ($Ms = 5.09$ vs. 3.48, $d = 1.15$, $p < .05$); whereas social researchers were more likely to characterize their theoretical approach with the statement, "Situations drive most behaviors, thoughts, and feelings" ($Ms = 5.30$ vs. 3.65, $d = 1.25$, $p < .05$). Despite these large-size differences between groups, it is important to note that 5.1 and 5.3 are not at the highest end of the scale, and 3.5 and 3.6 are closer to the midpoint than low end of the scale. Thus, given that these were some of the largest differences that emerged between the two groups, the most accurate characterization of these findings is that most researchers believe in the importance of both dispositional and situational influences on behavior, but there is a relative difference between the two groups. In terms of explicit beliefs, researchers were accurate, but they expected a considerably larger difference than the one that emerged; $ds = 2.77, 2.62$; $ps < .05$. Thus, although the person-situation debate no longer entirely drives the distinction between personality and social researchers, in the minds of researchers it is still a major factor underlying the split between the two groups.

Does the Structure of Social-Personality Research Conform to Cronbach's Two Streams?

Despite the differences we found between personality and social research practices, it seems that there are a set of research practices, within each domain, that are frequently used by both groups, as well as other practices that are used infrequently by both. Thus, despite between-group differences that appear to be consistent with the distinction between Cronbach's *correlation* and *experimental* streams, it remains possible that the underlying structure of social-personality research is *not* best characterized by the two-stream approach. That is, we do not know whether researchers who use experimental designs also tend to use assessment methods such as reaction times and judgments tasks, and emphasize internal validity; and whether those who use correlational designs also tend to assess their variables of interest with structured interviews and informant reports, and value external validity. It is possible that there is in fact only one central stream, with the differences we found representing tendencies toward particular approaches. To address this question, we conducted a factor analysis, using varimax rotation, on the items assessing research designs, statistics, assessment methods, and validity.

As can be seen from Table 14.2, a two-factor solution provided a good fit for the data (for greater detail, see Tracy et al., 2009a). Furthermore, the first factor seemed to clearly represent the *correlational* stream. Most of the highest positive loading items were statistical procedures (e.g., "convergent/discriminant validity," "HLM," "partial correlation"), research designs (e.g., "correlational approach," "longitudinal study approach"), and assessment methods (e.g., "informant report," "structured interviews") associated with the *correlational* stream and with personality research, both in terms of actual research practice and researchers' explicit beliefs. The second factor, in contrast, seemed to clearly represent the *experimental* stream, with the highest loading items tending to be assessment methods (e.g., "reaction time measures," "memory tasks," "implicit measures"), research designs (e.g., "experimental"), and statistical procedures (e.g., "ANOVA") associated with the *experimental* stream, and with the actual practices of, and explicit beliefs about, social researchers. Returning to our predictions in Table 14.1, 39 out of 42 items (i.e., 93%) relevant to methodology had higher loadings on the predicted factor, assuming the two-factor solution represents Cronbach's two streams.

To empirically determine the extent to which these two factors map onto the split between personality and social research, we saved factor scores for the two factors and correlated them with the personality and social research topics variables. Results showed that the split between personality and social research was fairly clear-cut; personality researchers scored higher on the first (*correlational*) factor ($Ms = .62$ vs. $-.53$, $d = 1.47$, $p < .05$); and social researchers scored higher on the second (*experimental*) factor ($Ms = .33$ vs. $-.44$, $d = .88$, $p < .05$). Similarly, the *correlational* factor was strongly positively correlated with personality research identity

Table 14.2 Two-factor varimax-rotated solution of methodology items

Items	*Factor 1*	*Factor 2*
Correlational designs	.63	
Longitudinal designs	.63	–.28
Hierarchical Linear Modeling	.63	
Convergent/discriminant validity	.62	
Informant reports	.59	
Growth curve modeling	.58	
Item-Response Theory	.58	
Structural Equation Modeling	.55	
Factor/Principal Components analyses	.54	
Molecular genetics/DNA testing	.54	
Twin/adoption designs	.53	
Cluster analyses	.52	
Partial/part correlations	.52	
Correlations	.49	
Discriminant function analyses	.47	
Structured interviews	.47	
Patient studies	.46	
Cross-sectional designs	.46	
External validity (generalizability)	.45	
Multiple regression	.42	
External validity (mundane realism)	.41	
Time series analyses	.41	
Reliability analyses	.39	
Power analyses	.39	
Hormone assessment	.37	
Experience sampling method	.36	

Self-report assessment	.34	
Meta-analyses	.34	.28
Field study designs	.34	
Cross-cultural designs	.31	
Mathematical modeling	.30	
Quasi-experimental designs	.26	
Computer simulations	.26	
Reaction time assessment		.77
Memory tests		.76
Implicit measures		.64
Experimental designs	–.57	.60
"Other" judgment tasks		.59
ANOVA	–.38	.54
Behavioral response measurement		.47
Internal validity		.43
t-tests		.41
Multidimensional scaling		.34
Tests of mediation/path analyses		.35
Judgments of groups		.34
fMRI		.28
ANS response assessment	.25	.26

Note. $N = 155$. Factor loadings below.25 were suppressed.

($r = .58$), and strongly negatively correlated with social research identity ($r = -.56$); for the *experimental* factor, this pattern reversed (*rs* = –.29,.36; all *ps* < .05). To the extent that the two-factor solution represents the structure of social-personality research, then, the split between the two streams seems to fairly accurately characterize the distinction between personality and social researchers. This finding raises an important, and potentially troubling, question, suggested by Cronbach (1957): is social-personality, and, perhaps, all of psychology, a field divided?

Evidence for a Merging of the Streams

In fact, other findings from our survey suggest a more optimistic conclusion, for those who share Cronbach's wish for the field. First, we found that almost half the sample (44%; $n = 68$) was best classified as *hybrids*, in that these researchers rated themselves as "4" (the midpoint of the scale, anchored with "sometimes") or higher on *both* the personality and social research topics questions. Furthermore, only 10% of the sample ($n = 15$) reported never studying topics related to one of the two fields and always studying topics related to the other. Thus, the large majority (90%) of the leading researchers in the field view themselves as studying topics from both areas.

As for the hybrid researchers, they were found to score higher on the *correlational* factor than social researchers (i.e., those who rated themselves as 4 or higher on the social research scale and below 4 on the personality scale), $d = .84$, $p < .05$, and lower than personality researchers, $d = .72$, $p < .05$. Hybrid researchers also scored higher on the *experimental* factor than personality researchers, $d = .92$, $p < .05$, and lower than social researchers, $d = .31$, $p < .05$ (one-tailed). This pattern suggests that for these hybrids, the two streams have, in essence, merged. In fact, factor scores on the *correlational* and *experimental* factors did not significantly differ for these individuals.

Other evidence for a possible merging of the streams comes from the results we obtained on the content areas researchers reported studying. In contrast to our findings on research methods, for content areas far more similarities emerged than differences. Specifically, personality and social researchers were equally likely to report studying most of the topics that constitute the core of social-personality research: aggression, attribution, brain functioning, creativity, clinical disorders, cultural psychology, education/achievement, emotion, evolutionary psychology, gender/sexuality, intelligence/cognitive ability, implicit processes, interpersonal attraction, judgment and decision making, motivation, nonverbal behavior, political psychology, positive psychology, relationships, personality traits, persuasion, self-concept, self-esteem, self-regulation, social development, and statistics. Furthermore, many of the topics that did show a difference between groups were those that, by definition, are explicitly personality or social research topics (e.g., personality traits, personality development, social cognition, social influence, social roles; see Tracy et al., 2009a). These findings suggest that, personality and social researchers study largely the same topics; but, given the results on the methodology items, they do so, to some extent, in different ways.

To more directly examine whether this is the case, we identified subsamples of personality and social researchers who reported studying the same topic, and directly tested whether they differed in how frequently they used correlational versus experimental designs, and correlations versus ANOVA statistical analyses. We conducted these analyses on all research topics that showed no group differences and were listed as a content area by at least 25% of each group: attribution, cultural psychology, emotion, gender/sexuality, implicit processes, motivation, relationships, self-concept, self-esteem, and self-regulation. In all cases except two, social and personality researchers differed significantly in all four methodological variables examined, with personality researchers more frequently using correlational design and correlational statistics, and social researchers more frequently using experimental design and ANOVA, all $ps < .05$. Thus, the two groups differ more in *how* than *what* they study.

IMPLICATIONS AND FOR THE FIELD OF SOCIAL-PERSONALITY RESEARCH

In summary, our research suggests that Cronbach's two streams are alive and well in psychological science. Our findings also suggest that social-personality researchers hold accurate beliefs about the research practices of members of their field, suggesting that at least these most prominent individuals within the field know themselves and their field quite well. It is noteworthy, however, that beliefs about differences tended to be larger (in most cases by at least a standard deviation) than actual differences. That is, researchers tend to overemphasize the magnitude of the differences between the two groups, although they are accurate in their assessment of where these differences lie.

Importantly, despite the clear distinction that emerged between personality and social researchers, there was also a tremendous amount of overlap between the two groups. Differences were typically moderate to large in size, but almost all of these differences were relative, as many approaches, designs, methods, and statistics were used by both personality and social researchers. Furthermore, for content areas, there was a great deal of similarity between the two groups, suggesting that personality and social psychologists study largely the same topics, but they do so in different ways. This finding has an important implication: If the two groups are interested in understanding the same psychological constructs, yet use distinct approaches, methods, and analyses to do so, then the field as a whole may benefit from greater adoption of an interactionist, or symbiotic approach (Swann & Seyle, 2005).

To take a prominent research example, both personality and social researchers study aggression, but they do so in different ways. Social psychologists have documented a broad range of situational factors that can lead to frustration and anger (e.g., Anderson et al., 2010; DeWall & Anderson, 2011), demonstrating how certain conditions or contexts can influence aggressive behaviors, and how seemingly nonaggressive individuals can, under extreme circumstances, be influenced to commit atrocities (e.g., Zimbardo, 2007). From this perspective, the prison abuse that occurred at Abu Ghraib reflects a "rotten barrel" rather than a few "rotten apples." In contrast, personality researchers have documented stable individual differences in aggressive tendencies and shown that these aggressive tendencies are highly heritable and consistent across situations and over long periods of time (e.g., Lahey, Moffitt, & Caspi, 2003). For example, even in bad situations, not everyone becomes "evil," and in real-life contexts such as at Abu Ghraib, we can predict which individuals are likely to commit criminal acts on the basis of previous antisocial behavior, and from scores on variables such as authoritarianism, negative emotionality, impulsivity, (low) intelligence, and (low) self-esteem (Altemeyer, 1996; Caspi et al., 1994; Donnellan, Trzesniewski, Robins, Moffitt, & Caspi, 2005; Miller & Lynam, 2001).

Of course, both situations and stable individual differences are likely to influence behaviors, and it is noteworthy that the basic effects in personality psychology (i.e., correlations of individual differences) and the basic effects in social psychology (i.e., differences between conditions/situations) are independent, such that large main effects of situations do not imply an absence of stable individual differences, and large correlations between person factors do not imply an absence of situation effects on the behaviors predicted by these factors (Funder, 2006). In other words, an individual like former army reserve specialist Charles Graner, an Abu Ghraib prison guard who was convicted of abusing prisoners and had a previous history of spousal and coworker abuse, is probably more dispositionally aggressive than an individual like the Dalai Lama, and would likely show higher levels of aggression than the Dalai Lama across a wide range of situations. Yet, given evidence from Zimbardo's famous prison experiment, both individuals are probably more likely to engage in aggressive behaviors in the Abu Ghraib prison than in a monastery. Thus, the most fruitful approach may be to simultaneously study both dispositions and situations, by conducting experiments that compare the effects of different situational factors on behavior, while also correlating relevant predictors and outcomes and testing which situational conditions moderate the effects of which personality dispositions. Indeed, using precisely this approach, Bushman (1995) demonstrated that both person and situation factors—as well as the interaction between the two—predict aggression. Our finding that many social-personality researchers are best considered hybrids suggests that this combined approach is, in fact, taken with some regularity.

In other areas of psychology, considerable debate has emerged over how best to approach and study topics of wide interest, such as the emotional phenomenon of fear. The neuroscientific view holds that fear should be studied through its neuroanatomy and neurochemistry, using MRI, brain-lesioned patients and animals, and single-cell recording in rats (e.g., Amaral, 2002; Damasio, 1999; Davidson, 2001; LeDoux, 1996; Panksepp, 1998). More traditionally, psychoanalysts have studied fear as displaced anxiety, typically due to repressed sexuality (Freud, 1909/1955). Behaviorists later reduced fear to a behavioral response to conditioned stimuli (Wolpe & Rachman, 1960), and cognitive psychologists viewed fear as a cognitive appraisal of threat and, in fact, viewed all emotions as cognitions paired with undifferentiated arousal (Schachter & Singer, 1962). Yet, as Miller and Keller (2000) argued, "Researchers are learning a great deal about the biology of fear—and the psychology of fear—from studies of the amygdala (e.g., Lang, Davis, & Öhman, 2000), but this does not mean that fear *is* activity in the amygdala. That is simply not the meaning of the term. 'Fear' is not reducible to biology" (p. 212). In fact, this holds for almost any important psychological phenomenon: It cannot be completely explained from any single perspective. Fear cannot be reduced to amygdala activity, nor to repressed castration anxiety, nor to a conditioned response to a stimulus, nor to a cognitive appraisal of threat. Fear may be all of these things, and it can best be understood when the various ideas are viewed as complementary, rather than antagonistic.

Cronbach (1957) wrote that "correlational psychology studies only variance among organisms; experimental psychology studies only variance among treatments. A united discipline will study both of these, but it will also be concerned with the otherwise neglected interactions between organismic and treatment variables" (p. 681). In fact, modern researchers have lived up to Cronbach's hope for the field, and not only his worry. Many researchers have adopted a symbiotic perspective and examine interactions between person and experimental variables. The hybrids in our sample represent this trend, but as Swann and Seyle (2005) noted, a larger number of researchers (who may nonetheless identify with the personality or social group) have also done precisely this, integrating correlational and experimental approaches in their research, using assessment methods and statistics from either stream as necessary, and producing bodies of work that have greatly added to the field's knowledge. Examples of the interactionist approach pervade the field, and many are summarized by Swann and Seyle (2005), but to name just a few prominent exemplars: In Carver and Scheier's (1998) research on self-focused attention, these researchers both manipulated self-focus and measured dispositional "self-consciousness"; in Higgins's (1987) model of actual, ought, and ideal selves, Higgins outlined a causal process that could be manipulated and laid the groundwork for measuring individual differences in self-views; and in Dweck's (1999) model of implicit self-theories, Dweck proposed scales for assessing self-theories as stable individual differences, but also directly influenced and manipulated them through interventions. The lines of research that resulted from each of these initial programs are impressive, and each has shaped the field in important ways. Many more examples exist, and the fact that researchers can fruitfully merge the two streams in this most literal way makes the streams' overlap quite clear and demonstrates their compatibility as research endeavors. At the same time, it is worth noting that both personality and social psychology have flourished as somewhat separate subfields, and both maintain several distinct high-impact journals that largely publish research emphasizing one stream or the other. Thus, the current structure of the field also may be a fruitful division of labor, allowing for a high level of productivity and efficiency within both subfields.

BROADER IMPLICATIONS: WHAT DOES IT ALL MEAN FOR PSYCHOLOGICAL SCIENCE?

The fact that the split between the *correlational* and the *experimental* streams of thought remains prominent in social-personality psychology suggests that this split is likely to be prominent in other areas of psychological science as well, and thus that there is no single paradigm for conducting psychological research. It would be surprising if a general paradigm had been adopted across the rest of psychological science, but, for whatever reason, had not yet taken hold within the area of social-personality. Thus, from a Kuhnian perspective, psychological science is still in the preparadigmatic stage, characterized by multiple, coexisting research practices and schools of thought. According to Kuhn (1962, 1970), the extent to which psychology is a paradigmatic science is reflected in its adoption of an integrated set of methods and approaches. If there are still two separate streams, relying upon divergent sets of methods and approaches, it would suggest that psychology has yet to reach the paradigmatic stage.

Kuhn portrayed the development of a science as moving from a *preparadigmatic* stage, characterized by multiple, competing schools, to a *paradigmatic* (or *normal science*) stage, characterized by a single dominant paradigm of shared assumptions and methodologies. Once a field has attained paradigmatic status, further scientific development involves a succession of revolutions in which the dominant paradigm within the field is displaced by a new one. The question of whether psychology has reached the paradigmatic stage of science has long been a source of fierce debate (e.g., Briskman, 1972; Leahey, 1991; Robins, Gosling, & Craik, 1999; Schultz, 1981; Warren, 1971; Watson, 1977). For example, Masterman (1970) argued that psychology may be characterized by a *multiparadigmatic* stage, which precedes the eventual emergence of a single dominant paradigm.

In contrast to Kuhn and Masterman, Lakatos's (1970) theory of scientific development allows for multiple, competing schools at any stage of a science's historical development. These schools ("research programmes" in Lakatos's terminology) may coexist indefinitely, and there is no assumption that one school will eventually emerge as the dominant paradigm. Lakatos viewed scientific progress as akin to a horse race in which competing schools progress, degenerate, are revived, and so on, depending on the ability of the school to generate new hypotheses that lead to empirical discoveries. Lakatos (1970) argued that a school becomes "progressive" when it yields new predictions that lead to empirical successes (Gholson & Barker, 1985, p. 757). Latour (1987) also suggested that multiple schools of thought within a single discipline might coexist, but in his view each school's prominence is determined by sociological, not scientific, factors. Specifically, schools rise in prominence when they successfully disseminate their scientific products to the rest of the field through communication networks that determine what becomes attended to and widely known.

Given that the two streams have coexisted now for over a century, and have ebbed and flowed in their relative prominence within the field as a whole, psychology may simply be a Lakatosian science that is constituted of two paradigms. Indeed, many of the most important variables that define human behavior and real-world outcomes cannot be directly manipulated, yet when manipulation is possible most scientists would, at some point, want to find a way to do that. Thus, it seems likely that psychology will always rely upon both *correlational* and *experimental* stream research practices, and psychological science as a whole may be better off for not being uniparadigmatic. Future research might address this issue by comparing indices of scientific progress among sciences that have a single paradigm and those, like psychology, which are multiparadigmatic. Despite Kuhn's arguments, it is not clear that psychology has been disadvantaged by the absence of a single consensual view on the best way to practice research.

Nonetheless, the success of multiple paradigms in psychology may, in part, lie in the fact that the two streams are not entirely separate. Although most researchers subscribe more fully to one stream or the other, our research demonstrates that the majority of researchers make use of designs, methods, and statistics from both streams, and (for the most part) view both as valid ways of conducting research. Thus, although psychology can be said to be multiparadigmatic, the two paradigms are not paradigms in the traditional sense. Rather, they might be more accurately viewed as two perspectives, or general approaches, that shape an individual's research but do not constrict it, in the way that, for example, the "standard model" does for physicists, or Darwin's theory of evolution does for biologists. Indeed, there may be other divides, such as that between researchers who conduct quantitative research and those who conduct qualitative research, that better represent distinct paradigms than the *correlational/experimental* split.

It is also noteworthy that Kuhn developed the concept of the scientific paradigm partly to distinguish the natural sciences from the social sciences. In Kuhn's (1962) view, paradigms cannot exist in the social sciences, where multiple theories are always competing for attention and empirical support. Given this background, the extent to which the approaches, methods, processes, content areas, and philosophical perspectives in social-personality research tend to be shared and endorsed by most researchers across both sides of the divide is, at some level, quite remarkable. It would seem unfair to relegate psychological science to a perpetual nonparadigmatic state; rather, at worst it may be preparadigmatic until a full merging of the streams occurs. Attaining this state would require that a truly integrative interactionist approach becomes dominant, both in terms of the underlying theoretical assumptions and the research methods employed to study psychological phenomenon.

REFERENCES

Altemeyer, B. (1996). *The authoritarian specter.* Cambridge, MA: Harvard University Press.

Amaral, D. G. (2002). The primate amygdala and the neurobiology of social behavior: Implications for understanding social anxiety. *Biological Psychiatry, 51,* 11–17.

Anderson, C. A., Shibuya, A., Ihori, N., Swing, E. L., Bushman, B. J., Sakamoto, A.,… Saleem, M. (2010). Violent video game effects on aggression, empathy, and prosocial behavior in Eastern and Western countries. *Psychological Bulletin,136,* 151–173.

Bentler, P. M. (2007). Can scientifically useful hypotheses be tested with correlations? *American Psychologist, 62,* 772–782.

Briskman, L. B. (1972). Is a Kuhnian analysis applicable to psychology? *Science Studies, 2,* 87–97.

Broadbent, D. E. (1991). A word before leaving. In D. E. Meyer & S. Kornblum (Eds.), *Attention and performance, 14* (pp. 863–879). Cambridge, MA: Bradford Books / MIT Press.

Bruner, J. (1990). *Acts of meaning.* Cambridge, MA: Harvard University Press.

Bushman, B. J. (1995). Moderating role of trait aggressiveness in the effects of violent media on aggression. *Journal of Personality and Social Psychology, 69,* 950–960.

Carver, C. S., & Scheier, M. F. (1998). *On the self-regulation of behavior.* New York: Cambridge University Press.

Caspi, A., Moffitt, T. E., Silva, P. A., Stouthamer-Loeber, M., Krueger, R. F., & Schmutte, P. S. (1994). Are some people crime-prone? Replications of the personality-crime relationships across countries, genders, races, and methods. *Criminology, 32,* 163–195.

Crits-Christoph, P., Wilson, G. T., & Hollon, S. D. (2005). Empirically supported psychotherapies: Comment on Westen, Novotny, and Thompson-Brenner (2004). *Psychological Bulletin, 131,* 412–417.

Cronbach, L. J. (1957). The two disciplines of scientific psychology. *American Psychologist, 12,* 671–684.

Damasio, A. R. (1999). *The feeling of what happens: Body and emotion in the making of consciousness.* New York: Harcourt, Brace.

Darley, J., & Todorov, A. (2004). Psychologists seek the unexpected, not the negative, to provoke innovative theory construction. *Brain and Behavioral Sciences, 27,* 331–332.

Davidson, R. J. (2001). The neural circuitry of emotion and affective style: Prefrontal cortex and amygdala contributions. *Social Science Information, 40,* 11–37.

DeWall, C. N., & Anderson, C. A. (2011). The general aggression model. In M. Mikulincer & P. Shaver (Eds.), *Understanding and reducing aggression, violence, and their consequences* (pp. 15–33). Washington, DC: American Psychological Association.

Dickerson, S. S., & Kemeny, M. E. (2004). Acute stressors and cortisol responses: A theoretical integration and synthesis of laboratory research. *Psychological Bulletin, 130,* 355–391.

Donnellan, M. B., Lucas, R. E., & Fleeson, W. (2009). Personality and assessment at age 40: Reflections on the past person-situation debate and emerging directions of future person-situation integration. *Journal of Research in Personality, 43* (special issue), 117–290.

Donnellan, M. B., Trzesniewski, K., Robins, R. W., Moffitt, T. E., & Caspi, A. (2005). Low self-esteem is related to aggression, antisocial behavior, and delinquency. *Psychological Science, 16,* 328–335.

Dunning, D. (2005). Self-insight: Roadblocks and detours on the path to knowing thyself. New York: Psychology Press.

Dweck, C. S. (1999). *Self-theories: Their role in motivation, personality, and development.* Philadelphia, PA: Psychology Press / Taylor & Francis.

Epley, N., van Boven, L., & Caruso, E. M. (2004). Balance where it really counts. *Brain and Behavioral Sciences, 27,* 333.

Francis, D. D., Champagne, F. A., Liu, D., & Meaney, M. J. (1999). Maternal care, gene expression, and the development of individual differences in stress reactivity. *Annals of the New York Academy of Sciences, 896,* 66–84.

Freud, S. (1909/1977). Analysis of a phobia in a five-year-old boy. In The Pelican Freud Library (Vol. 8, Case Histories 1, pp. 169–306). London: Hogarth Press.

Funder, D. C. (1987). Errors and mistakes: Evaluating the accuracy of social judgment. *Psychological Bulletin, 101,* 75–90.

Funder, D. C. (2006). Toward a resolution of the personality triad: Persons, situations, and behaviors. *Journal of Research in Personality, 40,* 21–34.

Gholson, B., & Barker, P. (1985). Kuhn, Lakatos, and Laudan: Applications in the history of physics and psychology. *American Psychologist, 40,* 755–769.

Heine, S. J. (2001). Self as cultural product: An examination of East Asian and North American selves. *Journal of Personality, 69,* 881–906.

Higgins, E. T. (1987). Self-discrepancy: A theory relating self and affect. *Psychological Review, 94,* 319–340.

Kazdin, A. E. (1978). Evaluating the generality of findings in analogue therapy research. *Journal of Consulting and Clinical Psychology, 46,* 673–686.

Kingstone, A., Smilek, D., & Eastwood, J. (2008). Cognitive ethology: A new approach for studying human cognition. *British Journal of Psychology, 99,* 317–340.

Krueger, J. I., & Funder, D. C. (2004). Toward a balanced social psychology: Causes, consequence, and cures for the problem-seeking approach to social behavior and cognition. *Brain and Behavioral Sciences, 27,* 313–376.

Kuhn, T. S. (1962). *The structure of scientific revolutions.* Chicago: University of Chicago Press.

Kuhn, T. S. (1970). *The structure of scientific revolutions* (2nd ed.). Chicago: University of Chicago Press.

Lahey, B. B., Moffitt, T. E., & Caspi, A. (2003). *Causes of conduct disorder and juvenile delinquency.* New York: Guilford Press.

Lakatos, I. (1970). Falsification and the methodology of scientific research programmes. In I. Lakatos & A. Musgrave (Eds.), *Criticism and the growth of knowledge* (pp. 91–196). Cambridge: Cambridge University Press.

Lang, P. J., Davis, M., & Öhman, A. (2000). Fear and anxiety: Animal models and human cognitive psychophysiology. *Journal of Affective Disorders, 61,* 137–159.

Latour, B. (1987). *Science in action.* Cambridge, MA: Harvard University Press.

Leahey, T. H. (1991). *A history of modern psychology.* Englewood Cliffs, NJ: Prentice-Hall.

LeDoux, J. E. (1996). *The emotional brain: The mysterious underpinnings of emotional life.* New York: Simon & Schuster.

Masterman, M. (1970). The nature of a paradigm. In I. Lakatos & A. Musgrave (Eds.), *Criticism and the growth of knowledge* (pp. 59–89). Cambridge: Cambridge University Press.

Miller, G. A., & Keller, J. (2000). Psychology and neuroscience: Making peace. *Current Directions in Psychological Science, 9,* 212–215.

Miller, G. E., Chen, E., & Cole., S. W. (2009). Health psychology: Developing biologically plausible models linking the social world and physical health. *Annual Review of Psychology, 60,* 501–524.

Miller, J. D., & Lynam, D. (2001). Structural models of personality and their relation to antisocial behavior: A meta-analytic review. *Criminology, 39,* 765–798.

Mook, D. G. (1983). In defense of external invalidity. *American Psychologist, 38,* 379–387.

Neisser, U. (1976). *Cognition and reality.* San Francisco, CA: Freeman.

Neisser, U. (1991). A case of misplaced nostalgia. *American Psychologist, 46,* 34–36.

Panksepp, J. (1998). *Affective neuroscience: The foundations of human and animal emotions.* New York: Oxford University Press.

Robins, R. W., Gosling, S. D., & Craik, K. H. (1999). An empirical analysis of trends in psychology. *American Psychologist, 54,* 117–128.

Schachter, S., & Singer, J. E. (1962). Cognitive, social and physiological determinants of emotional state. *Psychological Review, 69,* 379–399.

Schultz, D. (1981). *A history of modern psychology* (3rd ed.). New York: Academic Press.

Swann, W. B. & Seyle, C. (2005). Personality psychology's comeback and its emerging symbiosis with social psychology. *Personality and Social Psychology Bulletin, 31,* 155–165.

Tracy, J. L., Robins, R. W., & Sherman, J. W. (2009a). The practice of psychological science: Searching for Cronbach's two streams in social-personality psychology. *Journal of Personality and Social Psychology, 96,* 1206–1225.

Tracy, J. L., Robins, R. W., & Sherman, J. (2009b). Will the real personality psychologist, and the real social psychologist, please stand up? *Journal of Research in Personality, 43,* 272–273.

Visser, P. S., Krosnick, J. A, & Lavrakas, P. J. (2000). Survey research. In H. T. Reis & C. M. Judd (Eds.), *Handbook of research methods in social and personality psychology* (pp. 223–252). Cambridge: Cambridge University Press.

Visser, P. S., Krosnick, J. A., Marquette, J., & Curtin, M. (1996). Mail surveys for election forecasting? An evaluation of the Columbus Dispatch poll. *Public Opinion Quarterly, 60,* 181–227.

Vredenburg, K., Flett, G. L., & Krames, L. (1993). Analogue versus clinical depression: A critical reappraisal. *Psychological Bulletin, 113,* 327–344.

Warren, N. (1971). Is a scientific revolution taking place in psychology? Doubts and reservations. *Science Studies, 4,* 407–413.

Watson, R. I. (1977). The historical background for national trends in psychology: The United States. In J. Brozek, & R. B. Evans (Eds.), *R. I. Watson's selected papers on the history of psychology* (pp. 83–94). Hanover, NH: University Press of New England.

Weinstein, T. A. R., Capitanio, J., & Gosling, S. D. (2008). Personality in animals. In O. P. John, R. W. Robins, & L. A. Pervin, (Eds.), *Handbook of personality: Theory and research* (3rd ed., pp. 328–350). New York: Guilford Press.

Westen, D., Novotny, C. M., & Thompson-Brenner, H. (2004). The empirical status of empirically supported psychotherapies: Assumptions, findings, and reporting in controlled clinical trials. *Psychological Bulletin, 130,* 631–663.

Winstanley, C. A. (2007). The orbitofrontal cortex, impulsivity and addiction: Probing orbitofrontal dysfunction at the neural, neurochemical and molecular level. *Annals of the New York Academy of Sciences, 1121,* 639–655.

Wolpe, J., & Rachman, S. (1960). Psychoanalytic "evidence": A critique based on Freud's case of Little Hans. *Journal of Nervous and Mental Disease, 130,* 135–148.

Yang, Y., & Chiu, C. (2009). Mapping the structure and dynamics of psychological knowledge: Forty years of APA journal citations (1970–2009). *Review of General Psychology, 13,* 349–356.

Zimbardo, P. (2007). *The Lucifer effect.* New York: Random House.

PART FIVE
SCIENTIFIC CREATIVITY

15 Scientific Creativity as *Blind* Variation

Explicit and Implicit Procedures, Mechanisms, and Processes

Dean Keith Simonton

A half century ago, Donald Campbell (1960) published a classic article in which he argued that scientific creativity entails a process of blind variation and selective retention, or BVSR. However, Campbell never elaborated BVSR into a comprehensive psychological theory. Instead, he became more interested in extending his basic position to sociocultural evolution and evolutionary epistemology (e.g., Campbell, 1965, 1974). This shift may help explain why Campbell's article at first appeared to exert the biggest influence on the philosophical research, especially in the philosophy of science (e.g., Bradie, 2001; Briskman, 1981/2009; Kantorovich, 1993; Stein & Lipton, 1989). Nonetheless, a handful of years before Campbell's death in 1996, his theory began to receive increased attention from researchers specializing in the psychology of creativity, including scientific discovery and innovation (e.g., Perkins, 1994; Simonton, 1993; Staw, 1990). Of these psychological publications, a majority were probably written by this chapter's author. My BVSR elaborations and extensions span over two decades and appeared in numerous journal articles, book chapters, and entire books, as well as a special issue of the *Journal of Creative Behavior* devoted to Campbell's theory (e.g., Simonton, 1998, 1999c, 2005, 2009a, 2010a). In particular, *Scientific Genius: A Psychology of Science* represented the first book-length treatment of BVSR with respect to exceptional scientific creativity (Simonton, 1988).

My efforts have provoked considerable criticism from those psychologists who do not believe that creativity (scientific or otherwise) can be explicated in BVSR terms (e.g., Gabora, 2005, 2010; Sternberg, 1998; Weisberg, 2000). Many of these anti-BVSR critiques were published as commentaries to three target articles

(Simonton, 1999b, 2007a, 2010a). My developments of BVSR theory have also drawn criticism from outside psychology, most notably from Kronfeldner (2010), a philosopher, and Dasgupta (2004, 2010), a computer scientist. Although I have attempted to respond to these critiques by diverse corrections, clarifications, and refinements (e.g., Simonton, 1999a, 2005, 2007b, 2010b), it also must be admitted that my work has not always made a positive contribution. Most notably, unlike Campbell (1960), very early I identified BVSR as a "Darwinian" process (Simonton, 1995), an identification that left the unfortunate and false impression that the BVSR theory was based on an analogy with biological evolution (Kronfeldner, 2010). That implication is contrary to Campbell's own position; he maintained that BVSR was a generic process with several specific manifestations, including not just evolution, but also perception, learning, and creative thought (see also Hull, 2001). In fact, Campbell made the point early in his 1960 article that the proper historical anticipation of BVSR was not Charles Darwin's 1859 *Origin of Species* but rather Alexander Bain's (1855/1977) earlier *The Senses and the Intellect* (see also Simonton, 2010b). Furthermore, Campbell drew heavily on the work of the mathematician Henri Poincaré (1921) and the physicist Ernst Mach (1896), both of whom offered proto-BVSR theories of scientific creativity that were free of any Darwinian influences.

Given this and other departures from Campbell's (1960) theory of scientific creativity, this chapter will go back to the original publication with the aim of updating the BVSR position. Moreover, this update will try not to violate Campbell's initial views. I believe that it is essential to be faithful as much as possible to those views because Campbell's initial position was fundamentally correct (see also Cziko, 1998; Kim, 2001; Nickles, 2003). I begin by defining the central terms. From those definitions, I develop some criteria useful for identifying the operation of blind variations in scientific creativity and discovery. I then can deal with the most common criticisms of the BVSR theory.

CONCEPTUAL DEFINITIONS

For Campbell (1960), the variants in the creative process are "thought trials," or what we can more simply refer to as "ideas," so that we can speak of "ideational variations." In the specific case of scientific creativity, these thought trials will often constitute explicit hypotheses, conjectures, explanations, and the like. Still, under certain circumstances, these ideas can include behaviors operating at a more implicit level, as will be shown later. Despite this flexibility in usage, it is inconsistent with Campbell's original position to consider a completed creative product, such as a scientific journal article, as an ideational variant (cf. Simonton, 1999c). Indeed, the creative product should be viewed as a composition of ideas that are the result of BVSR. That should render the published outcome more sighted than the contributing thought trials (but, for counterviews, see Kantorovich, 1993; Kim, 2001). This distinction in mind, it is now necessary to define the variation and selection components of BVSR.

BLIND VARIATION

According to Campbell (1960), blind variations have two features. First, "the variations emitted [must] be independent of the environmental conditions of the occasion of their occurrence" (p. 381). Second, "the occurrence of trials individually [must] be uncorrelated with the solution, in that specific correct trials are no more likely to occur at anyone point in a series of trials than another, nor than specific incorrect trials" (p. 381). Although this twofold definition seemed satisfactory for Campbell's purposes, critics deem it less so (e.g., Kronfeldner, 2010). Therefore, I will apply a formal definition here that constitutes an elaboration of the definition used in evolutionary biology (cf. Simonton, 2010a; Sober, 1992). Because BVSR seems to require a minimum of two variants, the definition will be specified in those terms, but the extension to three or more variants entails nothing new conceptually. I start with the definition of sighted variants to place blind variants in proper context.

Sighted Variants

Let X and Y represent two *possible* ideational variants, that is, we can assume that they have the expectancies or subjective probabilities $0 < p(X) < 1$ and $0 < p(Y) < 1$. Corresponding to each variant is a utility value u (Simonton, 2010a; cf. fitness value w in Sober, 1992). These values will be notated here as $u(X)$ and $u(Y)$, respectively. For our purposes, these u's will also be taken as probabilities, namely the respective probabilities that the variants will be selected and retained in the final creative product: $0 \leq u(X) \leq 1$ and $0 \leq u(Y) \leq 1$ (cf. Simonton, 2010a).

Now suppose that $p(X) > p(Y)$. Then the two variants are *sighted* if $u(X) > u(Y) \rightarrow p(X) > p(Y)$, where the symbol "$\rightarrow$" means "implies" (i.e., in the sense of subjective conditional probability). That is, one variant has a higher probability than the other variant exactly *because* it has a higher utility than does the other. The relative probabilities are derived from or informed by the relative utilities, obliging the p's to be isomorphic with the u's.

I have deliberately chosen the term "sighted" (e.g., Sternberg, 1998) over the more common "directed" or "guided" (cf. Stein & Lipton, 1989) because the former term avoids any reference to evolutionary theory (cf. Lenski & Mittler, 1993) and because it is prima facie opposed to "blind."

Blind Variants

In contrast, competing variant pairs are unequivocally *blind* when they meet one of the two following conditions:

First, $p(X) = p(Y)$ although $u(X) \neq u(Y)$: The two possible variants have equal probabilities even though their utilities are quite unequal. This condition is witnessed in fork-in-the-road choices where one path is decidedly better (shorter, straighter, more flat, safer, etc.) and the other path is unquestionably worse (longer, more tortuous, steeper, more dangerous, etc.), but the poor traveler has no idea which is which—and so must choose by flipping a coin. Blind ideational variants are often precisely of this character. They occur whenever at least two ideas are subjectively deemed virtually equiprobable despite the fact that their actual prospects of actually being true are extremely unequal. One is right, the other wrong.

Second, $p(X) > p(Y)$ although $u(X) < u(Y)$: The relative magnitudes of the two variant probabilities and their utilities are reversed. An illustration may be found in Maier's (1931) classic two-strings problem. Suppose two strings are hanging from the laboratory ceiling and the research participant must tie the ends together even though it is impossible to reach one end while holding the other with arms fully outstretched. Nonetheless, say that the participant is given the opportunity to use any of a number of objects, including a pair of pliers, placed on a nearby table. Presumably, the response variant with the highest probability would be to use the pliers to extend the arm's reach even though the pliers are still not long enough to do the trick. In contrast, the response with the lower probability will be to use the piers as a pendulum, tying the tool to the end of one string, setting it in motion, and catching it while holding the other string. This would represent an "unusual use" in the parlance of one very common creativity test (Guilford, 1967), where "unusual" implies low probability, that is, $p(\text{pendulum/pliers}) \approx 0$.

Although the above two conditions provide unequivocal evidence of blindness, they do not exhaust all conceivable manifestations of blind variations. In fact, it is feasible for the two variants to be blind even when their probabilities correspond perfectly with their utilities, with the continued proviso that the latter do not imply the former. Variant probabilities are then still ignorant of variant utility, any correspondence being coincidental. This may be considered a formal definition of the "lucky guess." For example, suppose a student taking a true-false test has a response bias toward selecting "true" and that the instructor constructing the test has a parallel bias in making up the questions. Then the student would do better than chance on the test even if she or he guessed on every single question. Of course, if the student actually knew in advance that $u(\text{true}) > u(\text{false})$, then the guesswork would be sighted (and even more so if the student answered true on every question rather than engaging in suboptimal probability matching). The point of this example is that sometimes blindness requires an inference from additional information. Even if the p's correspond perfectly with the u's, blind variation may obtain.

Blind-Sighted Continuum

The distinction between sighted and blind variations is equivalent to the distinction that Toulmin (1972) advanced between variations that are *coupled* to selection and those that are *decoupled* from selection. If "the factors responsible for the selective perpetuation of variants are entirely unrelated to those responsible for the original generation of those same variants" (p. 337), then variation and selection are decoupled, and hence the variants are emitted blindly. One advantage of Toulmin's terminology is that it avoids some of the unfortunate but irrelevant associations that some BVSR critics have to the term "blind" (e.g., Doyle, 2008; Sternberg, 1998).

Nevertheless, it proves useful to make explicit what is only implicit in Campbell (1960), namely, that the distinction between blind and sighted variations is quantitative rather than qualitative.[1] That is, thought trials can display variable degrees of blindness and sightedness. This variability comes from two sources.

The first source involves *imperfect preselection*. The definition provided earlier assumes that the thought trials are confined to the permissible or conceivable rather than the impermissible, inconceivable, or impossible. In the sciences, either a priori (logical) or a posteriori (factual) reasoning may specify that $p(Z) = 0$ because $u(Z) = 0$, and so variant Z is excluded from the "problem space." However, because the process of defining the problem space is fallible, preselection may admit false positives and omit false negatives. The former are variants of the type $p(Z) > 0$ but $u(Z) = 0$, the latter of the type $p(Z) = 0$ but $u(Z) > 0$.

The second source of variability involves the *residual decoupling* in the variants that survive preselection. For some problems, the variant probabilities may be partially implied by the variant utilities even if the utilities do not suffice to determine those probabilities. The implication is no more than a weak bias in the proper direction. To illustrate, if $u(X) = 1$ and $u(Y) = 0$ leads to the weak expectation or "hunch" that $p(X) > p(Y)$ but not that $p(X) = 1$ and $p(Y) = 0$, then some degree of coupling is evinced. The greater the magnitude of the inequality relating the two probabilities, then the more sighted the variants can be said to be. That is, as the difference $p(X) - p(Y)$ approaches 1, then the two variants become increasingly coupled.

In theory, these two sources of partial sightedness are orthogonal. Preselection might produce just two completely decoupled variants, or instead preselection might yield a dozen variants with some degree of coupling. In addition, preselection might favor just a single variant, but that variant may or may not have a positive utility, a contingency that I will return to at the end of this chapter. Even so, in practice it would seem likely that the two sources are positively associated with each other. The more variants surviving preselection, the higher the odds are that their respective probabilities are decoupled from their utilities.

SELECTIVE RETENTION

Although utility has been defined as the probability of eventual selection and retention, it must be noted that the selection process may adopt distinct forms. In particular, we must recognize two contrasts (cf. Dennett, 1995):

First, selection may be simultaneous or sequential. In simultaneous selection, variants compete in an ideational "struggle for existence." In sequential selection, the variants are selected in a sequence, where that sequence is not initially dictated by the utilities.

Second, selection may be external or internal. In the former, the variant is assessed directly against the environment, as occurs in an experimental test. In the latter, the

[1] Kronfeldner (2010) argued that blindness cannot be a quantitative variable, but her argument assumes that BVSR creativity is predicated on an analogy with biological evolution where the variation-generating mechanisms are presumed random. Her assumption is therefore false even if we accept that blindness in biology is qualitative rather than quantitative.

variant is tested against an internal representation of the environment, such as a cognitive map or mental model. Dennett (1995) called creatures that use internal selection "Popperian," quoting Popper's statement that such internalization "permits our hypotheses to die in our stead" (p. 375).

Although ideational variants are often selected sequentially and internally, such variants can also be subjected to simultaneous or external selection or both. An example of simultaneous and external selection would be whenever a scientist generates two alternative but mutually exclusive hypotheses (e.g., $X \rightarrow$ not-Y and $Y \rightarrow$ not-X) and designs a laboratory experiment that provides a critical test confirming either one or the other but not both.

IDENTIFICATION CRITERIA

Most opponents of BVSR object not to selective retention but rather to the blind variation or BV component (e.g., Kronfeldner, 2010; Sternberg, 1998; Thagard, 1988). Yet given the formal definition presented earlier, thought trials display blindness if we can identify two or more possible variants whose probabilities are at least partially decoupled from their utilities. Such decoupling may be either explicit or implicit. In the former case, the ideational variants are blind by design. That is, scientists will explicitly adopt procedures or methods that generate variants with generation likelihoods that do not and cannot correspond to their likelihoods of success. In the latter case, that of implicit BV, the scientist either generates variants that inadvertently exhibit certain properties of blindness or else relies on processes that must necessarily function with some amount of blindness. This explicit/implicit distinction will become more explicit in the two sections that follow.

Explicit BV

A straightforward example of explicit BV is that cited by Campbell (1960): radar sweeps. Radar on top of an airport control tower will scan the horizon 360° or 2π radians. At any time t the radar will be sending and receiving long-wave radiation at angle θ_t, so it is obvious that $p(\theta_t)$ is identical for all angles (i.e., all possible polar coordinates to the horizon are equiprobable). Even so, the utilities $u(\theta_t)$ will not be identical for all t. On the contrary, perhaps an echo indicating a "discovered" object is only returned within a very narrow arc. The variants are patently decoupled from their utility, and hence explicitly blind.

This may seem like a trivial example given that airport radars are seldom if ever used to make discoveries (except specious reports of UFOs). Yet radars can be used to conduct scientific research according to the same blind-by-design principle (e.g., in the earth and planetary sciences). More important, the exact same principle applies for other a priori scans, such as search grids. Astronomers can conduct searches of the heavens that are blind by design, pointing their optical or radio telescopes toward arbitrary spots in the sky without any way of knowing whether they will come across something interesting. Archeologists, paleontologists, and physical anthropologists can impose comparable search grids on their digs with similar purposeful blindness, a blindness that ensures that no potential discovery is ever overlooked within the confines of the search.

A second example comes from the computer programs that purport to make scientific discoveries (e.g., Langley, Simon, Bradshaw, & Zythow, 1987; Shrager & Langley, 1990). Even though the details differ from program to program, all can be said to operate according to the principle of a systematic search through a problem space. Yet close examination reveals that this search is at least partly blind. To see how this is so, consider BACON, a program devised to make discoveries using Baconian induction (e.g., Bradshaw, Langley, & Simon, 1983). A frequently reported example is BACON'S discovery of Kepler's Third Law of planetary motion. This law relates the planet's period of rotation around the sun, P, to its distance from the sun, D (actually the semimajor axis of the ellipse describing the orbit). The law says that $P^2 = kD^3$, that is, the square of the period is proportional to the cube of the distance. When given the raw data for the planets, BACON was able to arrive at the same law. Employing sequential selection, the program applied three specific heuristics to generate and test the following solutions: $P/D = k$, $P/D^2 = k$, and, finally, $P^2/D^3 = k$ (the product of the previous two quotients), the latter giving the correct answer.[2] These three solutions may be assigned the utilities $u(P/D) = 0$, $u(P/D^2) = 0$, and $u(P^2/D^3) = 1$. Because the generate-and-test order can be taken as indicating a set of sequentially changing solution probabilities, the three-step cycle still displays some degree of decoupling. The decoupling is not perfect given that the heuristics permitted BACON to "sightfully" bypass two solutions of the same type, namely, $P^2/D = k$ and $P^2/D^2 = k$. Although Simon (1983) was correct in claiming that BACON's simulated discovery "certainly does not rely on a brute-force trial-and-error exploration of the data" (p. 4571), only the "brute-force" part is wrong. The heuristic procedures reduced the number of "trial and errors" by about half.

Radar scans, search grids, and discovery programs can all be viewed as generating *systematic* blind variants. Moreover, radar scans, search grids, and discovery programs can also be classed as *combinatorial* as well as systematic. The first explores possible polar coordinates in an encircled region; the second examines possible Cartesian coordinates in a bounded space; and the third generates and tests possible mathematical functions of a given type (e.g., products and quotients of two variables or their powers). Far more interesting are combinatorial variants that are *stochastic* rather than systematic. These are especially worthy of note because all participants in the debate about BVSR creativity concur that random combinations are necessarily blind (e.g., Kronfeldner, 2010; Thagard, 1988). Accordingly, computer programs that prove to be creative via random combinatorial procedures provide evidence for the importance of BV (see also Simonton, 2010a).

[2] I ignore whether these heuristics reflect some benefit from more than three centuries' worth of hindsight. It is possible that BACON's three heuristics were not available back then, requiring Kepler to engage in more extensive BVSR, such as considering $P^2/D = k$ and $P^2/D^2 = k$ as possible answers. We forget that nobody prior to Kepler was even in the position to conceive a mathematical relation of this kind. The idea's importance even escaped Galileo Galilei, his great (and certainly mathematically competent) contemporary!

Happily, we do not have to look very far for such evidence: Computer scientists have already created numerous problem-solving programs that operate explicitly on BVSR principles, using random combinatorial operations. These programs include evolutionary algorithms, genetic algorithms, evolutionary programming, and genetic programming (Goldberg, 1989; Holland, 1975, 1992; Koza, 1992). Despite the explicitly random combinatorial procedures, these programs are effective problem solvers. They can make forecasts in currency trading, plan fiber-optic telecommunication networks, design stream and gas turbines, improve oil exploration and mining operations, enhance the efficiency of jet engines (Holland, 1992), and solve thorny problems in algebraic equations, the design of electrical circuits, the determination of animal foraging behavior, and the identification of optimal game-playing strategies (Koza, 1994; Koza, Bennett III, Andre, & Keane, 1999). These BVSR programs can even re-create scientific results as discovery programs have been able to do. As a case in point, Kepler's Third Law was rediscovered using genetic programming (Koza, 1992). Blind stochastic combinatorial models yield the same creative solutions as blind systematic combinatorial models.

Not all computer programs that evince creativity or discovery are explicitly BVSR in construction. Nevertheless, it may be impossible to construct a computer program that simulates these phenomena without introducing some blind component, and often this component will be stochastic rather than systematic. After examining the inner workings of highly creative programs, Boden (2004) concluded, "What is useful for creativity in minds and evolution is useful for creative computers too. A convincing computer model of creativity would need some capacity for making random associations and/or transformations," a requirement that would often be met "by reference to lists of random numbers" (p. 226). Scientific creativity and discovery can only be successfully simulated by introducing some BV mechanism, a stipulation that most often leads to stochastic combinatorial methods (see also Simonton, 2003, 2004, 2010a).

Even so, it is apparent that systematic and stochastic BV mechanisms are in crucial respects equivalent. Both systematic BACON and stochastic genetic programming were able to arrive at Kepler's Third Law. Although the former relied on a heuristic generate-and-test search while the latter depended on a more haphazard generate-and-test procedure, they both converged on the same discovery. The only procedural overlap between the two operations was the intrusion of some degree of blindness. Blindness may be systematic or stochastic, but it remains blindness in either case. A little thought reveals that the same operational equivalence holds for other instances of explicit BV. As an example, rather than conduct a systematic radar scan, equivalent outcomes could be obtained by randomly selecting all angles (without replacement until all directions have been exhausted). Evidently, it is most vital for a variant to be blind. It does not matter whether the process that generated the blindness is systematic or stochastic.

Implicit BV

The previous section raises a very intriguing puzzle. If the only way computers can simulate creativity or discovery is to incorporate some amount of BV—whether systematic or stochastic—does that fact in and of itself necessitate that human creativity must indeed function by parallel means? Does it really make sense that computers need blindness to accomplish what humans do using total sightedness? Alternatively, is it the case that human creativity and discovery are also dependent on some intrusion of variation blindness? If the latter holds, is it reasonable to infer blind variation from the facts of human creativity?

Admittedly, to infer blind variation a posteriori is more difficult and tenuous than to note merely when blind variation has been implemented a priori. Even so, a plausible case can be made for BVSR even by this post hoc approach. The argument is twofold. First, ideational variations may have telltale signs of blindness. Second, the ideational variations may be generated by processes that operate with some degree of blindness.

Variations with Properties of Blindness

The ideational variants emitted during the process of creativity have certain characteristic signs that they were generated by a BV process. Two related attributes are especially telling: superfluity and backtracking.

1. *Variation superfluity*—Bain (1855/1977) said "the greatest practical inventions [are] so much dependent upon chance [that] the only hope of success is to multiply the chances by multiplying the experiments" (p. 597). Accordingly, the variants must be blind to the extent that they are too numerous than would be expected from a sighted process. This inference is strengthened when two or more variants are extremely diverse, particularly when they are demonstrably incommensurate or contradictory (e.g., $X \rightarrow$ not-Y and $Y \rightarrow$ not-X).

In line with this criterion, the physicist Michael Faraday once admitted, "The world little knows how many of the thoughts and theories which have passed through the mind of a scientific investigator have been crushed in silence and secrecy by his own severe criticism and adverse examinations; that in the most successful instances not a tenth of the suggestions, the hopes, the wishes, the preliminary conclusions have been realized" (quoted in Beveridge, 1957, p. 79; for indirect evidence, see Tweney, 1989). Darwin (1892/1958) similarly said, "I have steadily endeavoured to keep my mind free so as to give up any hypothesis, however much believed (and I cannot resist forming one on every subject) as soon as facts are shown to be opposed to it. Indeed, I have had no choice but to act in this manner, for with the exception of the Coral Reefs, I cannot remember a single first-formed hypothesis which had not after a time to be given up or greatly modified" (pp. 55–56). The logician and economist Jevons (1877/1900) noted that

> it would be an error to suppose that the great discoverer seizes at once upon the truth, or has any unerring method of divining it. In all probability the errors of the great mind exceed in number those of the less vigorous one. Fertility of imagination and abundance of guesses at truth are among the first requisites of discovery; but the erroneous guesses must be many times as numerous as those that prove well founded. The weakest analogies, the most whimsical notions, the most apparently

absurd theories, may pass through the teeming brain, and no record remain of more than the hundredth part. (p. 577)

Hence, estimates of variant superfluity range from 90% (Faraday) to 99% (Jevons). Thomas Edison's "drag hunts" had a notoriously higher rate of superfluity, literally hundreds of experiments having to be conducted before he obtained a workable filament for his incandescent lamp (Perkins, 1994). Fortunately, Edison had a huge laboratory staff to carry out this blind search.

Although superfluous variants provide strong evidence of an underlying BV process, we have to take care not to assume that the converse is true, namely that a BV process will necessarily produce superfluity (Simonton, 2005). This is a common inferential mistake made by both proponents and opponents of BVSR (e.g., Dasgupta, 2004; Simonton, 1999c). BVSR only requires that there be a minimum of two possible variants satisfying the decoupling conditions in any given problem-solving condition. The inferential importance of superfluity is that the greater the number and diversity of variants, the lower the odds that they all have equal utilities. Furthermore, the power of this inference becomes reinforced by the second clue.

2. *Variation backtracking*—As observed before, creativity often operates under sequential rather than simultaneous selection. If the consecutive thought trials are sighted, then we would expect them to approach the final form asymptotically (if not instantaneously). Each generation-and-test would provide the necessary feedback to inch closer to the desired goal. By comparison, if the variants were inherently blind, then sometimes they would get "warmer" and sometimes "colder"—without any consistent pattern. Expressed differently, if variant utility was plotted as a function of variant order in the series of thought trials, then the values would be positive monotonic for the sighted case and nonmonotonic for the blind case (Simonton, 2007a). Each time a variant gets "colder" it represents a downward turn in the plot, introducing a zig-zag.

Helmholtz (1898) published an introspective report about his own thought trials that indicates that backtracking occurs in the sciences:

> I only succeeded in solving such problems after many devious ways, by the gradually increasing generalisation of favourable examples, and by a series of fortunate guesses. I had to compare myself with an Alpine climber, who, not knowing the way, ascends slowly and with toil, and is often compelled to retrace his steps because his progress is stopped; sometimes by reasoning, and sometimes by accident, he hits upon traces of a fresh path, which again leads him a little further; and finally, when he has reached the goal, he finds to his annoyance a royal road on which he might have ridden up if he had been clever enough to find the right starting-point at the outset. (p. 282)

The higher the proportion of backtracks, and the greater their magnitude, the stronger is the corresponding inference that the variants may have relied on blindness rather than sightedness. Indeed, it is reasonable to make a stronger deduction: The more frequent and more conspicuous the backtracking, the higher the likelihood that even the more successful variants lacked sightedness. Every "you're getting colder" implies that every "you're getting warmer" might have been nothing more than fortuitous. In the classic scenario of the drunkard engaged in a random walk, there will of necessity be occasions where the sad chap walks in the correct direction (e.g., toward home and a warm bed). Yet the drunkard should not be credited with sightedness if the next step goes off in the opposite direction. To engage in such asymmetrical attributions of blindness versus sightedness is to fall victim to the hindsight bias. Many if not most attributions of creator sightedness are probably no more than instances of observer hindsightedness—mere post hoc inferences of prescience.

Processes That Should Yield Blindness

There is no such thing as a single creative process (Simonton & Damian, in press). Instead, there are a great many different processes, mechanisms, or procedures that all can generate ideational variants in diverse circumstances. Any given problem will almost invariably be solved via a small subset of these operations, the specific composition of that subset varying from problem to problem in an erratic manner. However, among the most common processes involve associative richness, defocused attention, behavioral tinkering, and heuristic search.

1. *Associative richness*—James (1880) described the psychological processes that provided the basis of individual originality. These processes were chaotic rather than systematic, combinatorial or analogical rather than logical. In particular, he said that

> instead of thoughts of concrete things patiently following one another in a beaten track of habitual suggestion, we have the most abrupt cross-cuts and transitions from one idea to another, the most rarefied abstractions and discriminations, the most unheard of combination of elements, the subtlest associations of analogy; in a word, we seem suddenly introduced into a seething cauldron of ideas, where everything is fizzling and bobbling about in a state of bewildering activity, where partnerships can be joined or loosened in an instant, treadmill routine is unknown, and the unexpected seems only law. (p. 456)

Although contemporary research would not dare to describe creative thought in such an effusive, even rhapsodic manner, it is has become patent that creativity often requires the ability to engage in somewhat unconstrained associations. This associative richness may be obtained by some mixture of remote associations (Mednick, 1962), rare associations (Gough, 1976), divergent thinking of various kinds (Guilford, 1967), primordial or primary process cognition (Martindale, 1990), Janusian associations and homospatial imagery (Rothenberg, 1990), allusive or over-inclusive thought (Eysenck, 1995), and even clang associations (Hadamard, 1945). It is not necessary to describe each of these associative processes in detail to make two general points.

First, none of these processes can be claimed to yield associations having response probabilities that are coupled with their corresponding utilities. For example, when taking the Remote Associates Test, the associations that have the highest utilities relative to the target will have among the lowest probabilities of retrieval; that is precisely what renders them "remote" and thereby increases the time to solution (Mednick, 1962). Similarly, divergent thinking only has utility in problem solving insofar as it generates possibilities that would not be normally considered, such as using pliers to make a pendulum. Interestingly, even when Maier (1931) specifically instructed participants to use the pliers to solve the two-strings problem, their first response was to use them as tongs, and most participants were not able to conceive using the pliers as a weight until after they were given a hint (viz. the experimenter "accidentally" causing one cord to sway). Indeed, were the thinking process to yield alternative uses whose probabilities were coupled to their utilities as solutions, then we should more properly identify the process as convergent thinking. If the experimental task were to hammer in a nail and the participants are given the choice between a hammer and a screwdriver, then convergent thinking suffices to solve the problem. Divergent thought is useless.

Second, in the case of highly difficult or novel problems, it is unlikely that the initial responses will be generated by a single associative process. But if two or more processes are elicited, then these processes are often distinct enough to activate variable associative responses to a given stimulus situation. This divergence has repercussions insofar as creativity often depends on "spreading activation" proceeding from problem input to solution output (see, e.g., Langley & Jones, 1988; Mandler, 1995). This activation will then follow two or more associative pathways resulting from two or more of these associative mechanisms. As this activation spreads, some pathways may converge on certain nodes in the semantic network, nodes that may or may not provide a solution, whereas other pathways may continue their solitary advance, sometimes ending on a solution but most of the time not. Hence, we can conceive of variable utilities for two alternative associative pathways: Activation paths X and Y might have values $u(X) > u(Y)$. At the same time, it is very likely that these alternative pathways would have different associative strengths. In the extreme case, they might be $p(X) < p(Y)$ in a kind of first-will-be-last and last-will-be-first situation.

Hence, not only are the separate associative processes largely blind, but their joint operation when two or more processes are evoked should also be considered blind. This underlying process must oblige the resulting ideational variations to be at least partially blind.

2. *Defocused attention*—The former argument says that once the mind is stimulated with a given problem, spreading activation may expand along two or more routes with associative strengths that are decoupled from their probable success in leading to a solution. Yet it also must be recognized that the creative intellect is vulnerable to the vicissitudes of continued stimulation from the external world. More specifically, it has long been known that creativity is positively associated with defocused attention (Eysenck, 1995; Mendelsohn, 1976), and this correlation has received additional support from recent research on the relation between creativity and reduced

latent inhibition (Carson, Peterson, & Higgins, 2003; Peterson & Carson, 2000). Creative persons are more likely to respond to extraneous stimuli even when those stimuli have been proven irrelevant. This attentional proclivity departs diametrically from what is displayed by domain experts who are very proficient at ignoring information deemed irrelevant to the problem at hand (see also Ansburg & Hill, 2003). Furthermore, the extraneous sensory input resulting from defocused attention is likely to prime the ongoing associative process in unpredictable ways, leading it down new pathways that might not otherwise be pursued. This would enhance rather than diminish the decoupling of the associative strengths from their corresponding utilities.

This complex and volatile interplay between associative richness and defocused attention is probably especially important during what Wallas (1926) famously identified as the incubation phase of the creative process. After realizing that a problem cannot be readily solved despite repeated attacks, the creator will frequently move to other tasks, including the activities of everyday life ("the bath, the bed, and the bus" of Boden, 2004, p. 25). During this interval, the individual is involuntarily exposed to a much greater variety of unrelated stimuli that can prime various associations. Although most of these associations will still lead nowhere, and thus can be counted as blind variants emerging at the unconscious level, one accidental event may lead the associative process in a more fruitful direction. The archetypal example is the Eureka experience that Archimedes had taking a bath: The mathematician had no a priori reason to believe that the overflowing water would solve the problem, so he did not bathe to seek a solution. Laboratory experiments on insight have lent some support to the involvement of such "opportunistic assimilation" (Seifert, Meyer, Davidson, Patalano & Yaniv, 1995). Nonetheless, it must be emphasized that the cognitive susceptibility to extraneous input would be enhanced for individuals who simultaneously feature both associative richness and defocused attention.

3. *Behavioral tinkering*—Sometimes the ideational trials assume overt form. In the sciences, these blind variations may emerge in the guise of tinkering (Kantorovich, 1993). A famous illustration is Watson's (1968) route to discovering the DNA code:

> I spent the rest of the afternoon cutting accurate representations of the bases out of stiff cardboard.... [Later] I quickly cleared away the papers from my desk top so that I would have a large, flat surface on which to form pairs of bases held together by hydrogen bonds. Though I initially went back to my like-with-like prejudices, I saw all too well that they led nowhere.... [I] began shifting the bases in and out of various other pairing possibilities. Suddenly I became aware that an adenine-thymine pair held together by two hydrogen bonds was identical in shape to a guanine-cytosine pair held together by at least two hydrogen bonds. (p. 123)

Only after fumbling around with various possible hydrogen bonds was he able to discover that (a) adenine joined with thymine and guanine with cytocine and (b) the two sets of combined molecules had the same overall form, permitting them to unite the two spines of the double helix with uniform spacing. What is particularly instructive in Watson's report is that he began his tinkering by testing the pairings of adenine-adenine, cytocine-cytocine, and so forth, even though he should have known better (viz. because of Chargaff's rule). Watson's first variants, which he unwisely reckoned the most probable, truly had the lowest utilities (i.e., false positives).

For scientists with sufficient capacity for vivid visual imagery, such tinkering can become internalized, as in Albert Einstein's famed *Gedanken* ("thought") experiments.

4. *Heuristic search*—Newell and Simon (1972) introduced a classic model of human problem solving that has been developed into a more general theory of scientific creativity and discovery (Klahr, 2000; Klahr & Simon, 1999). Indeed, the Newell-Simon tradition helped provide the basis for the discovery programs discussed earlier in this chapter. This perspective makes a critical distinction between two approaches to problem solving: algorithms and heuristics. Algorithms represent precise step-by-step procedures that pretty much guarantee a solution, whereas heuristics are more indefinite "rules of thumb" that may or may not lead to a solution. Given this contrast, algorithms can be referred to as "strong" methods, whereas heuristics represent "weak" methods (Klahr, 2000). Moreover, whereas algorithms are often domain specific in application, heuristic methods can be applied to a wide range of problems. Such heuristic methods include hill climbing (steepest ascent), means-end analysis, working backward, analogy, and trial and error. Although the very last heuristic method is the only one that explicitly represents a generate-and-test or BV procedure, all weak methods have implicit blindness associated with their application (Simonton, 2003). That's what renders them weak; unlike algorithmic methods, none can ensure a solution. For example, although analogical thinking can prove a very useful method, it is not always apparent which of the possible analogies will work best, thereby requiring some kind of generate-and-test procedure, including where correspondences do and do not exist (cf., the "MAC/FAC" model; Gentner, 1998). Of course, scientists will endeavor to use algorithmic methods whenever feasible, but frequently they will have to fall back on a more heuristic search through the various possibilities. The reason for the latter course of action is that problems themselves may vary from well-defined problems with clear-cut means and goals to ill-defined problems where both the goals and the means are vague and insufficiently determined. The latter problems are far more "difficult."

The foregoing can be recast in BVSR terms. For well-defined problems, the algorithmic methods are strongly coupled to the most likely solution. Given the equation $2x^2 + 5x - 3 = 0$ and the goal to find the two roots, the answer is supplied without fail via the quadratic formula. The coupling is perfect. However, as the problems become more difficult, not only must the person increasingly rely on weak heuristic methods that do not promise a proper solution, but it also becomes even less evident which heuristic methods to try out first. That ambiguity elevates the trial and error heuristic to the status of a superordinate metaheuristic (Simonton, 2004). The creator may have no other option than to generate and test all conceivable heuristics without any assurance that even one will actually work. Because heuristic search has thereby become decoupled from the utilities for various heuristics with respect to a given problem, each applied heuristic generates one or more blind ideational variants. This interpretation is consistent with Newell, Shaw, and Simon's (1962) statement that "in spite of the primitive character of trial-and-error processes, they bulk very large in highly creative problem-solving; in fact, at the upper end of the range of problem difficulty there is likely to be a positive correlation between creativity and the use of trial-and-error generators" (pp. 72–73).

One of the beauties of Campbell's (1960) BVSR theory is that it does not require that creativity and discovery be produced by any single process or even predetermined set of processes. The theory only requires that the ideational variants thus produced be to some extent blind. The blind variants may be produced by systematic or stochastic combinatorial procedures, and they may be unconscious or conscious, implicit or explicit, cognitive or behavioral, and so forth—so long as the variant probabilities are at least partially decoupled from variant utilities. This conceptual inclusiveness permits researchers to avoid fruitless debates over what represents "the" creative process. Because BVSR does not represent a single procedure, it also circumvents the "No Free Lunch Theorem" that denies the possibility of a "one-size fits all" optimal problem-solving algorithm (Nickles, 2003). BVSR is not chosen as the optimal process per se, but only as the set of available processes that have any chance of working when confronted with extremely novel and complex problems. Indeed, as the earlier Helmholtz (1898) quotation illustrates, the optimal approach for a given problem becomes known only by means of hindsight. It takes many crooked and obstructed paths to discover the straight and direct path.

COMMON CRITICISMS

A half century has transpired since Campbell (1960) first put forward his BVSR model. During that period, the model has undergone considerable theoretical and empirical development. In the current chapter, I have tried to strengthen the argument even more by providing a more precise definition of what constitutes a *blind* variation. This definition then was reinforced by a series of illustrations that showed how blind ideational variants appear in scientific creativity and discovery. Of course, creativity appears in the arts as well as the sciences, so it is important to point out that BVSR can be shown to operate in artistic creativity as well. For instance, a detailed analysis of Picasso's *Guernica* sketches established that these particular variants display the predicted superfluity and backtracking (Doyle, 2008; Simonton, 2007a). In fact, BVSR creativity tends to be more conspicuous in the arts than in the sciences (Simonton, 2009a), albeit highly creative scientists probably rely more on BVSR than their less creative colleagues (Simonton, 2009b). Thus, the inclusiveness mentioned at the end of the previous section becomes all the more conspicuous. BVSR can not only embrace all processes involved in creativity, but also embrace all forms of creativity regardless of domain.

Even so, Campbell's (1960) theory continues to draw abundant and diverse criticisms (e.g., Kronfeldner, 2010; Sternberg, 1998). These complaints can be somewhat crudely divided into simple misunderstandings and profound misconceptions. The former can be dismissed in fewer words than the latter.

SIMPLE MISUNDERSTANDINGS

Because many critics do not understand Campbell's (1960) fundamental tenets, they will sometimes put forward irrelevant objections. Consider the following two examples.

First, it is simply untrue that the BVSR theory requires that ideational variants be random (cf. Russ, 1999; Schooler & Dougal, 1999). Although all stochastic variants are perforce blind, even systematic variants can be blind if generated in a blind fashion (i.e., if decoupled from the outcome). Furthermore, adherents of BVSR can reasonably claim that blind ideational variations are sometimes generated by processes that appear *as if* they are random even if not strictly so. This weakened claim then permits BVSR advocates to recruit on behalf of the theory a large number of stochastic combinatorial models of the creative process (e.g., Simonton, 2003, 2004, 2010a). In addition, openness to such quasi randomness helps creativity researchers comprehend why creativity might be enhanced when individuals undergo exposure to novel, unpredictable, or incongruous stimuli (e.g., Finke, Ward, & Smith, 1992; Proctor, 1993; Rothenberg, 1986; Sobel & Rothenberg, 1980; Wan & Chiu, 2002). Such input facilitates creativity by enabling the resulting spreading activation to act as if it were random even though the resulting associations actually represent merely blind variants.

Second, it is just incorrect to say that the BVSR theory of human creativity depends on a tight analogy with biological evolution (e.g., Kronfeldner, 2010). As a result, it is useless to debate whether certain specific aspects of creativity are analogous to characteristics of biological evolution (e.g., Stein & Lipton, 1989). The issue is irrelevant. According to the definition of blindness, creativity and evolution are connected by a formal equivalence rather than some loose and fuzzy metaphor (cf. Doyle, 2008; Sternberg, 1999). Indeed, both phenomena, along with many others ranging from antibody formation to neurological development, can be subsumed under a single universal selection theory (Cziko, 1995, 2001). What all such phenomena share is a core dependence on blind variations. This dependence holds even though the specifics, such as the modes of variation generation and selection, may differ dramatically.[3]

Once BVSR opponents recognize that Campbell's (1960) position is not contingent on an analogy with Darwin's theory of biological evolution then other criticisms based on presumed disanalogies become superfluous. For example, some critics argue that creative individuals are engaged in highly purposeful behavior, unlike the mindless organisms in Darwinian theory (e.g., Doyle, 2008; Sternberg, 1999). As one opponent expressed it, "Creativity is forward-looking and intentional, while evolution is not" (Sternberg, 1999, p. 357). However, if the BVSR

[3] Some BVSR critics argue that creativity cannot function using BVSR because biological evolution uses BVSR, and these two phenomena do not work in the same way (e.g., Gabora, 2010). This argument is logically fallacious. We can speak of "sound waves" and "light waves" even though gas molecules and photons do not "wave" in precisely the same manner (e.g., transverse versus longitudinal vibrations). Contrasts be what they may, both still generate phenomena distinctively characteristic of waves (e.g., interference). Similarly, both biological evolution and creative thought produce adaptive originality by selecting and retaining variations having some blindness, whether qualitative or quantitative.

account of creativity is conceived as originating with Bain (1855/1977) rather than with Darwin, then this complaint is well off the mark. In fact, Campbell specifically quoted Bain at length to emphasize the extreme importance of interest, curiosity, drive, energy, and determination in the creative process. "The number of trials necessary to arrive at a new construction is commonly so great that without something of an affection or fascination for the subject one grows weary of the task. This is the *emotional* condition of originality of mind in any department" (Bain, 1855/1977, p. 593). Later Bain stressed the requisite of "an Active turn, or profuseness of energy, put forth in trials of all kinds on the chance of making lucky hits" (p. 595). Hence, as Bain has it, the creator's dependence on trial and error requires more volition, more purpose and determination, than is needed for routine problem solving, which can almost function on cognitive autopilot (i.e., automaticity). As a consequence, there is no inconsistency between BVSR and the volitional nature of scientific creativity.

PROFOUND MISCONCEPTIONS

Although the previous objections are easily dismissed as being founded on basic misunderstandings of Campbell's (1960) theoretical argument, the next objection requires a more detailed response. Critics often assert that ideational variations must be sighted, guided, or directed (e.g., Kronfeldner, 2010; Sternberg, 1998; Thagard, 1988). In particular, creativity must be guided because scientists possess appreciable domain-specific expertise that they can bring to bear upon any given problem (e.g., Ericsson, 1999). Creative scientists know what they are doing. This contrast is frequently stated in either-or terms as if any application of expertise to solve a problem automatically excludes the operation of blindness (e.g., Kronfeldner, 2010). However, according to the definitions of sighted and blind variations given earlier, this dichotomy is specious. For creativity, sightedness and blindness are endpoints on a continuum.

Consider the impact of preselection: In addition to bona fide variants X and Y there might exist purely hypothetical variants W and Z, but the latter pair might have been altogether excluded from consideration by previously acquired knowledge. That is, $p(W) = p(Z) = 0$ precisely because $u(W) = u(Z) = 0$. Even so, the exclusion of W and Z does not instantly affect the coupling or decoupling of the respective probabilities and utilities for X and Y. This lack of an effect holds even if it should happen that the absolute probabilities for X and Y are changed after explicitly excluding W and Z. The expertise-guided exclusion of variants W and Z still does not instantaneously cause variants X and Y to be coupled.

This point can be most simply illustrated via a mundane analogy: Envision a student answering a multiple-choice test item that has four alternative answers, namely, W, X, Y, and Z, where X is the sole correct response. Stated formally, the conditions are $u(X) = 1$ and $u(W) = u(Y) = u(Z) = 0$. Now consider two extreme cases. On the one hand, the student has not studied for the exam and thus has zero expertise in the domain being tested. In that case, $p(W) = p(X) = p(Y) = p(Z) = .25$, and the student will have no option other than to make a 100% "blind" guess. The decoupling holds for all four variants. On the other hand, suppose that the student worked very diligently the night before and thus walked into the room having mastered all of the assigned material. Now the four answer probabilities are perfectly coupled to their utilities, that is, $p(X) = 1$ and $p(W) = p(Y) = p(Z) = 0$. In this instance, guessing was unnecessary because expertise whittled the possibilities down to just one. In the absence of two possible variants, blind variation becomes not just impossible but also pointless.

Yet what about circumstances that fall between the two extremes? For example, imagine a more mediocre student who studied just hard enough to deduce that *W* and *Z* are definitely incorrect but cannot exclude *Y* with equal confidence, so that $p(X) = p(Y) = .5$ even though $u(X) = 1$ and $u(Y) = 0$. Then we still have a decoupling of the probabilities and the utilities. To be sure, one might want to say that the guess is more "educated" (aka more sighted) in this third scenario than in the first. Indeed, that is what the acquired expertise is used for: to narrow the range of possibilities. By restricting the number of possible variants, the student doubled the probability of getting the right answer (viz. from .25 to.5). Nevertheless, an educated guess remains a guess. The student still has a 50–50 chance of being correct. To answer either *X* or *Y* would still count as a (partially) blind variant according to the definition given earlier.

A priori reductions in possible variants are usually explicit in BV mechanisms or processes that are explicitly blind. A radar scan is never spherical (i.e., all possible directions) but rather is restricted to a circular band along the horizon, or perhaps even just to a sector of the horizon, as in air-defense systems. A search grid is not dropped down anywhere randomly, but rather is imposed over an area that is most likely to yield discoveries (e.g., the earliest hominids are found in specific geological formations of Africa, never anywhere in the Americas). Even the rediscovery of Kepler's Third Law apparently confined the mathematical functions to just that subset that would prove most compatible with Kepler's First and Second Laws (which, after all, concerned ellipses and areas, respectively). In spite of these restrictions, the subsequent systematic combinatorial variants remain partially decoupled and hence partly blind. Similarly, in genuine acts of scientific creativity, domain-specific expertise is invariably used to narrow the range of possible variants. When Watson (1968) tinkered with his cardboard models of the DNA bases, he not only knew the overall helix configuration and the specific structures of the four bases (in their proper tautomeric forms only after a false start), but also realized the need to focus his attention on the viable hydrogen bonds that might connect them in congruent pairs. Despite all that initial expertise, Watson still had to blindly tinker to convert his residual ignorance to outright knowledge.

Recall what was said earlier about heuristic searches. Whenever possible, we try to solve problems using algorithmic methods. Those methods constitute an integral part of our domain-specific expertise. Still, those same methods often fail us when we are confronted with ill-defined, novel, or difficult problems, obliging us to fall back on unreliable heuristic methods—including the all-inclusive trial and error metaheuristic. We may even have to enter an incubation period in which spreading activation and defocused attention can do more unconscious or at least involuntary work. Yet we do not turn to these weak methods because we prefer them to the strong methods. Rather, we resort to them because our domain-specific knowledge and skill was not able to reduce the number of possible variants to just one. Blind variation begins where domain expertise ends. At least in the sciences, BV enters by default, not necessarily by preference (cf. Nickles, 2003).

Nonetheless, as intimated earlier, prior knowledge can also impose too many restrictions on the BV process. Some assumed impossible variants should actually be considered possible. Perhaps $p(Z) = 0$ even though $u(Z) = 1$ (e.g., the incorrect exclusion of noncircular planetary orbits by Ptolemy, Copernicus, and Galileo, but not Kepler). This possibility provides another reason why reduced latent inhibition can prove so useful to creativity. If expertise excludes the supposedly irrelevant according to accumulated experience, but the irrelevant is actually relevant in this circumstance, then defocused attention should prove an asset rather than a deficit. Augmenting this dispositional advantage, highly creative individuals often will have developmental backgrounds that encourage them to "think outside the box." Campbell (1960) observed that "persons who have been uprooted from traditional cultures, or who have been thoroughly exposed to two or more cultures, seem to have the advantage in the range of hypotheses they are apt to consider, and through this means, in the frequency of creative innovation" (p. 391; for empirical support, see Leung, Maddux, Galinsky, & Chiu, 2008). Likewise, Kuhn (1970) asserted that highly innovative scientists are often "very new to the field whose paradigm they change" because these very individuals "are particularly likely to see that these rules no longer define a playable game and to conceive another set that can replace them" (p. 90; for support, see Jeppesen & Lakhani, 2010; Simonton, 1984). It seems that whereas blind variation begins where expertise ends, excessive expertise can harm exceptional creativity if it prematurely truncates the number of permitted variants. Sometimes the thought trial with the lowest a priori probability of success will be the one that enjoys the highest utilities.

The need to go beyond the constraints imposed by domain-specific expertise is illustrated in a recent study of the relation between scientific eminence and psychopathology (Ko & Kim, 2008). On the one hand, scientists who worked within a given paradigm were more likely to attain eminence if they exhibited low levels of psychopathology. On the other hand, scientists who endeavored to overthrow the old paradigm were more prone to achieve eminence if they displayed high levels of psychopathology. It is no accident that psychopathology, like creativity, is also associated with reduced latent inhibition (Eysenck, 1995).

CONCLUSION

Creative ideas are often defined as having two qualities: (a) novelty or originality and (b) utility or adaptiveness (Simonton, 2000). In the specific case of scientific creativity, a solution to a problem must both be novel and work (e.g., be logically coherent and consistent with the facts).[4] This definition has a natural connection

[4] The criterion of utility or adaptiveness also applies to artistic creativity, albeit the specific manifestations of this criterion differ from the sciences. For example, an artistic idea has utility if it satisfies some aesthetic criterion, such as "beauty," which is most often subjectively determined.

with Campbell's (1960) BVSR theory (or what Nickles, 2003, explicitly called BV+SR): The BV step is responsible for generating the novelty while the SR step is responsible for determining utility. Nevertheless, an opponent of BVSR theory can readily argue that there is no intrinsic rationale for stipulating the "B" part of BV. Why cannot ideational variations be completely sighted rather than blind? Why not SVSR instead? All variants might be satisfactory, and selection becomes reduced to finding the optimal variant.

At this point, it becomes edifying to add a third component to the definition, namely the very criterion enforced by the United States Patent Office when evaluating applications (http://www.uspto.gov/inventors/patents.jsp). To succeed, inventors must not only prove that their inventions are new and useful, as already specified, but also "nonobvious." If the invention is an obvious extension of an already patented invention, then the application is denied. Moreover, the obviousness is gauged not by the knowledge of the average person on the street but rather by someone who has the relevant domain-specific expertise. An expert should be able to look at the invention and respond with an emphatic "why didn't I think of that?" Presumably, the inventor experienced the same surprise much earlier, having screamed "Eureka!" upon making the discovery (cf. "mind popping" in Mandler, 1995).

This third criterion probably should be universally adopted by creativity researchers as well. To be creative, an idea should be novel, useful, and surprising, even to those in the know (cf. "new, surprising, and valuable" in Boden, 2004, p. 1). With this additional criterion, the connection between BVSR and creativity becomes even more obvious—perhaps too obvious to be patentable. In fact, sometimes BVSR and nonobvious-novel-utility seem so tightly bound conceptually as to represent a virtual tautology. To be sure, BVSR advocates can shoot themselves in the foot by modifying Campbell's position to render it less defensible. I already mentioned my tactical error in calling creativity "Darwinian" (rather than "Popperian"), and another culpa mea was my arguing that the units of selection are products rather than thought trials (e.g., Simonton, 1999c). But if these unwise innovations are stripped away, then I believe Campbell's BVSR provides the most tenable theory for understanding creativity, scientific or otherwise. When properly presented and explained, even previous BVSR opponents have been known to become converts (e.g., Weisberg & Hass, 2007).

Any such presentation and explanation must take two propositions, namely A = "creative thought requires blind variations" and B = "creative thought is sighted," and then show that A does not imply not-B (cf. Kronfeldner, 2010). On the contrary, the noncontradictory nature of these two statements should be clear from the formal definitions given earlier. BVSR theory assumes that blindness is a quantitative property of two or more *possible* variants. Naturally, for some problems, only one variant is possible, in which case the variant is completely sighted. This situation would mean that the single preselected variant X has the properties $p(X) = u(X) = 1$. Even so, a BVSR theorist would then infer that it would be extremely unlikely that the resulting idea could be considered creative. Although the idea would be most likely useful given that, by assumption, $u(X) = 1$, the idea would probably be deficient in novelty, and would most definitely fall short in surprise. The idea is precisely what is expected in routine, even algorithmic thinking. Its very supreme sightedness renders it obvious rather than nonobvious.

Indeed, under these conditions of extreme preselection, the thought trial could only prove surprising if it turned out to be *wrong*. That is, we were utterly unjustified to suppose $p(X) = u(X) = 1$ because $u(X) = 0$. We would then have to confront an *anomalous* outcome. If that anomaly proves actually useful, and not just novel and surprising, then it can be considered a serendipitous discovery. These serendipitous discoveries often play a very dramatic role in the history of science. Furthermore, it is not difficult to broaden BVSR's formal definition to make it include these events as special or degenerate cases (Kantorovich, 1993). In particular, it is obvious (I hope) that the single expectation was blind (or decoupled) with respect to the outcome. That blindness may have ensued from the incorrect exclusion of Y during preselection; i.e., $p(Y) = 0$ but $u(Y) = 1$. This is a (blind) error of omission.

I would like to close this chapter with one final assertion. In line with Newell et al.'s (1962) suggestion that I quoted earlier, we might conjecture that the degree of blind variation—with respect to both preselection and coupling—is associated with the magnitude of creativity a scientist displays. Support for this association is to be found in the empirical research indicating that highly creative persons tend to have dispositional traits and developmental experiences that would provide more support for the production of blind ideational variations (Simonton, 2010a). At the same time, it is probable that this association would be curvilinear rather than linear, each domain having an optimal amount of ideational blindness for maximal creativity (Simonton, 2009a). This optimum would be lowest in the "hard sciences," moderate in the "soft sciences," and highest in the arts (see Simonton, 2009b). BVSR theory thus sheds light on all forms of creativity, whether scientific or artistic. That feature allows Campbell's (1960) conception to be far more inclusive than any alternative theoretical framework.

REFERENCES

Ansburg, P. I., & Hill, K. (2003). Creative and analytic thinkers differ in their use of attentional resources. *Personality and Individual Differences, 34,* 1141–1152.

Bain, A. (1977). *The senses and the intellect* (D. N. Robinson, Ed.). Washington, DC: University Publications of America. (Original work published 1855).

Beveridge, W. I. B. (1957). *The art of scientific investigation* (3rd ed.). New York: Vintage.

Boden, M. A. (2004). *The creative mind: Myths & mechanisms* (2nd ed.). New York: Routledge.

Bradie, M. (2001). The metaphysical foundation of Campbell's selectionist epistemology. In C. Heyes, & D. L. Hull (Eds.), *Selection theory and social construction: The evolutionary naturalistic epistemology of Donald T. Campbell* (pp. 35–52). Albany: State University of New York Press.

Bradshaw, G. F., Langley, P. W., & Simon, H. A. (1983). Studying scientific discovery by computer simulation. *Science, 222,* 971–975.

Briskman, L. (2009). Creative product and creative process in science and art. In M. Krausz, D. Dutton, & K. Bardsley (Eds.), *The idea of creativity* (2nd ed., pp. 17–41). Leiden, Netherlands: Brill. (Original work published 1980).

Campbell, D. T. (1960). Blind variation and selective retention in creative thought as in other knowledge processes. *Psychological Review, 67,* 380–400.

Campbell, D. T. (1965). Variation and selective retention in socio-cultural evolution. In H. R. Barringer, G. I. Blanksten, & R. W. Mack (Eds.), *Social change in developing areas* (pp. 19–49). Cambridge, MA: Schenkman.

Campbell, D. T. (1974). Evolutionary epistemology. In P. A. Schlipp (Ed.), *The philosophy of Karl Popper* (pp. 413–463). La Salle, IL: Open Court.

Carson, S., Peterson, J. B., & Higgins, D. M. (2003). Decreased latent inhibition is associated with increased creative achievement in high-functioning individuals. *Journal of Personality and Social Psychology, 85,* 499–506.

Cziko, G. A. (1995). *Without miracles: Universal selection theory and the second Darwinian revolution.* Cambridge, MA: MIT Press.

Cziko, G. A. (1998). From blind to creative: In defense of Donald Campbell's selectionist theory of human creativity. *Journal of Creative Behavior, 32,* 192–208.

Cziko, G. A. (2001). Universal selection theory and the complementarity of different types of blind variation and selective retention. In C. Heyes, & D. L. Hull (Eds.), *Selection theory and social construction: The evolutionary naturalistic epistemology of Donald T. Campbell* (pp. 15–34). Albany: State University of New York Press.

Darwin, F. (Ed.). (1958). *The autobiography of Charles Darwin and selected letters.* New York: Dover. (Original work published 1892).

Dasgupta, S. (2004). Is creativity a Darwinian process? *Creativity Research Journal, 16,* 403–413.

Dasgupta, S. (2010). On the blind-mindedness of creative thought. *Physics of Life Reviews, 7,* 188–189.

Dennett, D. C. (1995). *Darwin's dangerous idea: Evolution and the meanings of life.* New York: Simon & Schuster.

Doyle, C. (2008). Exploring the creation of Picasso's *Guernica*: Composition studies, chance, metaphors, and expertise. *Creativity Research Bulletin, 20,* 445–450.

Ericsson, K. A. (1999). Creative expertise as superior reproducible performance: Innovative and flexible aspects of expert performance. *Psychological Inquiry, 10,* 329–333.

Eysenck, H. J. (1995). *Genius: The natural history of creativity.* Cambridge: Cambridge University Press.

Finke, R. A., Ward, T. B., & Smith, S. M. (1992). *Creative cognition: Theory, research, applications.* Cambridge, MA: MIT Press.

Gabora, L. (2005). Creative thought as a non-Darwinian evolutionary process. *Journal of Creative Behavior, 39,* 262–283.

Gabora, L. (2010). Why blind-variation and selective-retention is an inappropriate explanatory framework for creativity. *Physics of Life Reviews, 7,* 182–183.

Gentner, D. (1998). Analogy. In W. Bechtel & G. Graham (Eds.), *A companion to cognitive science* (pp. 107–113). Oxford, UK: Blackwell.

Goldberg, D. E. (1989). *Genetic algorithms in search, optimization, and machine learning.* Reading, MA: Addison-Wesley.

Gough, H. G. (1976). Studying creativity by means of word association tests. *Journal of Applied Psychology, 61,* 348–353.

Guilford, J. P. (1967). *The nature of human intelligence.* New York: McGraw-Hill.

Hadamard, J. (1945). *The psychology of invention in the mathematical field.* Princeton, NJ: Princeton University Press.

Helmholtz, H. von (1898). An autobiographical sketch. In *Popular lectures on scientific subjects, second series* (E. Atkinson, Trans.) (pp. 266–291). New York: Longmans, Green.

Holland, J. (1975). *Natural and artificial systems*. Ann Arbor: University of Michigan Press.

Holland, J. H. (1992). Genetic algorithms. *Scientific American, 267*(1), 66–72.

Hull, D. L. (2001). In search of epistemological warrant. In C. Heyes, & D. L. Hull (Eds.), *Selection theory and social construction: The evolutionary naturalistic epistemology of Donald T. Campbell* (pp. 155–167). Albany: State University of New York Press.

James, W. (1880, October). Great men, great thoughts, and the environment. *Atlantic Monthly, 46*, 441–459.

Jeppesen, L. B., & Lakhani, K. R. (2010). Marginality and problem-solving effectiveness in broadcast search. *Organization Science, 21*, 1016–1033.

Jevons, W. S. (1900). *The principles of science: A treatise on logic and scientific method* (2nd ed.). London: Macmillan. (Original work published 1877).

Kantorovich, A. (1993). *Scientific discovery: Logic and tinkering*. Albany: State University of New York Press.

Kim, K. (2001). Nested hierarchies of vicarious selectors. In C. Heyes & D. L. Hull (Eds.), *Selection theory and social construction: The evolutionary naturalistic epistemology of Donald T. Campbell* (pp. 101–118). Albany: State University of New York Press.

Klahr, D. (2000). *Exploring science: The cognition and development of discovery processes*. Cambridge, MA: MIT Press.

Klahr, D., & Simon, H. A. (1999). Studies of scientific creativity: Complementary approaches and convergent findings. *Psychological Bulletin, 125*, 524–543.

Ko, Y., & Kim, J. (2008). Scientific geniuses' psychopathology as a moderator in the relation between creative contribution types and eminence. *Creativity Research Journal, 20*, 251–261.

Koza, J. R. (1992). *Genetic programming: On the programming of computers by means of natural selection*. Cambridge, MA: MIT Press.

Koza, J. R. (1994). *Genetic programming II: Automatic discovery of reusable programs*. Cambridge, MA: MIT Press.

Koza, J. R., Bennett, F. H., III., Andre, D., & Keane, M. A. (1999). *Genetic programming III: Darwinian invention and problem solving*. San Francisco, CA: Morgan Kaufmann.

Kronfeldner, M. E. (2010). Darwinian "blind" hypothesis formation revisited. *Synthese, 175*, 193–218. doi: 10.1007/s11229-009-9498-8

Kuhn, T. S. (1970). *The structure of scientific revolutions* (2nd ed.). Chicago: University of Chicago Press.

Langley, P., & Jones, R. (1988). A computational model of scientific insight. In R. J. Sternberg (Ed.), *The nature of creativity: Contemporary psychological perspectives* (pp. 177–201). New York: Cambridge University Press.

Langley, P., Simon, H. A., Bradshaw, G. L., & Zythow, J. M. (1987). *Scientific discovery*. Cambridge, MA: MIT Press.

Lenski, R. E., & Mittler, J. E. (1993). The directed mutation controversy and neo-Darwinism. *Science, 259*, 188–194.

Leung, A. K., Maddux, W. W., Galinsky, A. D., & Chiu, C. (2008). Multicultural experience enhances creativity: The when and how. *American Psychologist, 63*, 169–181.

Mach, E. (1896, January). On the part played by accident in invention and discovery. *Monist, 6*, 161–175.

Maier, N. R. (1931). Reasoning in humans: II. The solution of a problem and its appearance in consciousness. *Journal of Comparative and Physiological Psychology, 12*, 181–194.

Mandler, G. (1995). Origins and consequences of novelty. In S. M. Smith, T. B. Ward, & R. A. Finke (Eds.), *The creative cognition approach* (pp. 9–25). Cambridge, MA: MIT Press.

Martindale, C. (1990). *The clockwork muse: The predictability of artistic styles*. New York: Basic Books.

Mednick, S. A. (1962). The associative basis of the creative process. *Psychological Review, 69,* 220–232.

Mendelsohn, G. A. (1976). Associative and attentional processes in creative performance. *Journal of Personality, 44,* 341–369.

Newell, A., Shaw, J. C., & Simon, H. A. (1962). The processes of creative thinking. In H. E. Gruber, G. Terrell, & M. Wertheimer (Eds.), *Contemporary approaches to creative thinking* (pp. 63–119). New York: Atherton Press.

Newell, A., & Simon, H. A. (1972). *Human problem solving*. Englewood Cliffs, NJ: Prentice-Hall.

Nickles, T. (2003). Evolutionary models of innovation and the Meno problem. In L. V. Shavinina (Ed.), *The international handbook on innovation* (pp. 54–78). New York: Elsevier Science.

Perkins, D. N. (1994). Creativity: Beyond the Darwinian paradigm. In M. A. Boden (Ed.), *Dimensions of creativity* (pp. 119–142). Cambridge, MA: MIT Press.

Peterson, J. B., & Carson, S. (2000). Latent inhibition and openness to experience in a high-achieving student population. *Personality and Individual Differences, 28,* 323–332.

Poincaré, H. (1921). *The foundations of science: Science and hypothesis, the value of science, science and method* (G. B. Halstead, Trans.). New York: Science Press.

Proctor, R. A. (1993). Computer stimulated associations. *Creativity Research Journal, 6,* 391–400.

Rothenberg, A. (1986). Artistic creation as stimulated by superimposed versus combined-composite visual images. *Journal of Personality and Social Psychology, 50,* 370–381.

Rothenberg, A. (1990). *Creativity and madness: New findings and old stereotypes*. Baltimore, MD: Johns Hopkins University Press.

Russ, S. W. (1999). An evolutionary model for creativity: Does it fit? *Psychological Inquiry, 10,* 359–361.

Schooler, J. W., & Dougal, S. (1999). Why creativity is not like the proverbial typing monkey. *Psychological Inquiry, 10,* 351–356.

Seifert, C. M., Meyer, D. E., Davidson, N., Patalano, A. L., & Yaniv, I. (1995). Demystification of cognitive insight: Opportunistic assimilation and the prepared-mind perspective. In R. J. Sternberg & J. E. Davidson (Eds.), *The nature of insight* (pp. 65–124). Cambridge, MA: MIT Press.

Shrager, J., & Langley, P. (Eds.). (1990). *Computational models of scientific discovery and theory formation*. San Mateo, CA: Morgan Kaufmann.

Simon, H. A. (1983). Discovery, invention, and development: Human creative thinking. *Proceedings of the National Academy of Sciences, 80,* 4569–4571.

Simonton, D. K. (1984). Is the marginality effect all that marginal? *Social Studies of Science, 14,* 621–622.

Simonton, D. K. (1988). *Scientific genius: A psychology of science*. Cambridge: Cambridge University Press.

Simonton, D. K. (1993). Genius and chance: A Darwinian perspective. In J. Brockman (Ed.), *Creativity: The reality club IV* (pp. 176–201). New York: Simon & Schuster.

Simonton, D. K. (1995). Foresight in insight? A Darwinian answer. In R. J. Sternberg & J. E. Davidson (Eds.), *The nature of insight* (pp. 465–494). Cambridge, MA: MIT Press.

Simonton, D. K. (1998). Donald Campbell's model of the creative process: Creativity as blind variation and selective retention. *Journal of Creative Behavior, 32,* 153–158.

Simonton, D. K. (1999a). The continued evolution of creative Darwinism. *Psychological Inquiry, 10,* 362–367.

Simonton, D. K. (1999b). Creativity as blind variation and selective retention: Is the creative process Darwinian? *Psychological Inquiry, 10,* 309–328.

Simonton, D. K. (1999c). *Origins of genius: Darwinian perspectives on creativity.* New York: Oxford University Press.

Simonton, D. K. (2000). Creativity: Cognitive, developmental, personal, and social aspects. *American Psychologist, 55,* 151–158.

Simonton, D. K. (2003). Scientific creativity as constrained stochastic behavior: The integration of product, process, and person perspectives. *Psychological Bulletin, 129,* 475–494.

Simonton, D. K. (2004). *Creativity in science: Chance, logic, genius, and zeitgeist.* Cambridge: Cambridge University Press.

Simonton, D. K. (2005). Darwin as straw man: Dasgupta's (2004) evaluation of creativity as a Darwinian process. *Creativity Research Journal, 17,* 299–208.

Simonton, D. K. (2007a). The creative imagination in Picasso's *Guernica* sketches: Monotonic improvements or nonmonotonic variants? *Creativity Research Journal, 19,* 329–344.

Simonton, D. K. (2007b). Picasso's *Guernica* creativity as a Darwinian process: Definitions, clarifications, misconceptions, and applications. *Creativity Research Journal, 19,* 381–384.

Simonton, D. K. (2009a). Creativity as a Darwinian phenomenon: The blind-variation and selective-retention model. In M. Krausz, D. Dutton, & K. Bardsley (Eds.), *The idea of creativity* (2nd ed., pp. 63–81). Leiden, Netherlands: Brill.

Simonton, D. K. (2009b). Varieties of (scientific) creativity: A hierarchical model of disposition, development, and achievement. *Perspectives on Psychological Science, 4,* 441–452.

Simonton, D. K. (2010a). Creative thought as blind-variation and selective-retention: Combinatorial models of exceptional creativity. *Physics of Life Reviews, 7,* 156–179.

Simonton, D. K. (2010b). Reply to comments on "Creative thought as blind-variation and selective-retention: Combinatorial models of exceptional creativity." *Physics of Life Reviews, 7,* 190–194.

Simonton, D. K., & Damian, R. I. (in press). Creativity. In D. Reisberg (Ed.), *Oxford handbook of cognitive psychology.* New York: Oxford University Press.

Sobel, R. S., & Rothenberg, A. (1980). Artistic creation as stimulated by superimposed versus separated visual images. *Journal of Personality and Social Psychology, 39,* 953–961.

Sober, E. (1992). Models of cultural evolution. In E. Sober (Ed.), *Conceptual issues in evolutionary biology* (2nd ed., pp. 477–492). Cambridge, MA: MIT Press.

Staw, B. M. (1990). An evolutionary approach to creativity and innovations. In M. A. West & J. L. Farr (Eds.), *Innovation and creativity at work: Psychological and organizational strategies* (pp. 287–308). New York: John Wiley.

Stein, E., & Lipton, P. (1989). Where guesses come from: Evolutionary epistemology and the anomaly of guided vision. *Biology & Philosophy, 4,* 33–56.

Sternberg, R. J. (1998). Cognitive mechanisms in human creativity: Is variation blind or sighted? *Journal of Creative Behavior, 32,* 159–176.

Sternberg, R. J. (1999). Darwinian creativity as a conventional religious faith. *Psychological Inquiry, 10,* 357–359.

Thagard, P. (1988). *Computational philosophy of science*. Cambridge, MA: MIT Press.

Toulmin, S. (1972). *Human understanding: The collective use and evolution of concepts*. Princeton, NJ: Princeton University Press.

Tweney, R. D. (1989). A framework for the cognitive psychology of science. In B. Gholson, W. R. Shadish Jr., R. A. Neimeyer, & A. C. Houts (Eds.), *The psychology of science: Contributions to metascience* (pp. 342–366). Cambridge: Cambridge University Press.

Wallas, G. (1926). *The art of thought*. New York: Harcourt, Brace.

Wan, W. W. N., & Chiu, C.-Y. (2002). Effects of novel conceptual combination on creativity. *Journal of Creative Behavior, 36*, 227–240.

Watson, J. D. (1968). *The double helix: A personal account of the discovery of the structure of DNA*. New York: Atheneum.

Weisberg, R. W. (2000). An edifice built on sand? [Review of the book *Origins of genius*, D. K. Simonton]. *PsycCRITIQUES, 45*, 589–593.

Weisberg, R. W., & Hass, R. (2007). We are all partly right: Comment on Simonton. *Creativity Research Journal, 19*, 345–360.

16 Creative Combination of Representations

Scientific Discovery and Technological Invention

Paul Thagard

INTRODUCTION

Human creativity operates in many domains, including scientific discovery, technological invention, artistic imagination, and social innovation. What are the cognitive processes that produce these creative results? Are there psychological mechanisms common to such diverse products of creativity as Darwin's theory of evolution, Edison's lightbulb, van Gogh's paintings of sunflowers, and Bismarck's introduction of old-age pensions? This chapter will develop and evaluate the *combinatorial conjecture* that all creativity, including scientific discovery and technological invention, results from combinations of mental representations.

The combinatorial conjecture has been proposed or assumed by many authors, but the evidence presented for it has been restricted to a few examples (e.g., Boden, 2004; Finke, Ward, & Smith, 1992; Koestler, 1967; Mednick, 1962; Poincaré, 1921; Thagard, 1988, 1997). This chapter gives a more thorough evaluation of the conjecture by seeing whether it applies to 100 important cases of scientific discovery and to 100 important cases of technological invention. The primary result of examination of these cases is support for the combinatorial conjecture: No counterexamples were found. But the study of 200 creative episodes was interesting in other ways, and this chapter will report a collection of findings about the nature of the representations and processes used. These findings concern the role of visual and other kinds of representations and the extent to which discoveries and inventions were accidental, analogical, and observational or theoretical.

Before getting into the historical studies, I will provide a theoretical perspective on creative conceptual combination by reviewing a new neurocomputational account that provides an explanation of how neural representations can be combined. This account accommodates visual and other nonverbal kinds of representations and therefore is capable of applying to a wide range of creative episodes. I then describe the results of study 1, which looks at 100 examples of scientific discovery, and study 2, concerning 100 examples of technological invention. These large samples confirm the combinatorial conjecture, whose claim, however, is nontrivial, as I will show by considering theoretical objections to it from the extreme embodiment perspective that thinking and hence creativity are not representational and computational. I will argue that these objections are unwarranted and that the combinatorial conjecture remains highly plausible for scientific discovery and technological invention.

NEURAL THEORY OF CONCEPTUAL COMBINATION

If creativity is to be explained as a combination of mental representations, we need a rigorous scientific account of the nature of representations and the processes that combine them. Thagard and Stewart (2011) use a neurocomputational model to show how representations construed as brain processes can be combined. This section will sketch the basic assumptions of this model without attempting to give mathematical details or general justification.

From the perspective of current work in theoretical neuroscience, concepts and other representations are patterns of firing activity in neural populations (Dayan and Abbott, 2001; Eliasmith and Anderson, 2003; O'Reilly and Munakata, 2000; Thagard, 2010a). Hence conceptual combination needs to be understood as a process of putting together new patterns of firing activity from old ones. This approach has the potential of being far more flexible than previous psychological accounts of conceptual combination that have been restricted to verbal representations. There is growing evidence that concepts are neural representations that encode information in various modalities, including verbally but also encompassing the results of sensory processes such as vision (Barsalou, Simmons, Barbey, & Wilson, 2003). In contrast to the verbal and mathematical data structures that have traditionally been used in cognitive science, neural representations are adept at capturing visual and other kinds of encodings.

There are currently two main theoretical approaches in cognitive science to the problem of binding multiple representations together. The most prominent is synchrony, through which different neural representations are bound together by virtue of their temporal coordination (Hummel & Holyoak, 2003; Shastri, 1999). Thagard and Stewart (2011) follow Eliasmith (2005, forthcoming) in employing a different method called *convolution*, which is a mathematical technique for braiding structures such as waves and vectors together. Convolution was originally developed for applications to waves in electrical engineering, but Plate (2003) showed how it could be adapted to provide an account of how vectors of numbers corresponding to the firing rates of neurons could be combined into larger structures without losing crucial structural information (see Eliasmith & Thagard, 2001, for an introduction). Then Eliasmith (2005, forthcoming) showed how convolution can be performed by populations of spiking neurons. The contribution of Thagard and Stewart (2011) is to apply such mechanisms to the Aha! experience that results when combination of neural representations is sufficiently novel that it generates an emotional reaction.

Accordingly, the combinatorial conjecture can be fleshed out as follows: All creativity results from convolution-based combination of mental representations consisting of patterns of firing in neural populations. The historical studies to be described next do not serve to evaluate these neurocomputational claims, but directly address the underlying assumption that representation combination is the fundamental mechanism of creativity in various domains.

STUDY 1: SCIENTIFIC DISCOVERY

Case studies in the history, philosophy, and psychology of science have usefully looked in detail at select examples of advances in science and technology (e.g., Gorman, Tweney, Gooding, & Kincannon, 2005). However, generalizations about the nature of scientific discovery need a more systematic look at a large number of episodes. Accordingly, I conducted an analysis of the cases described in a book called *100 Greatest Science Discoveries of All Time* (Haven, 2007). The author, Kendall Haven, is a reputable science journalist with a background as a research scientist and many publications. For my purposes, there is no need to defend the claim that these are exactly the "100 greatest," only that they are undeniably a large collection of very important discoveries, from the law of the lever to the human genome. Most crucial for a serious test of the combinatorial conjecture, the examples were not chosen by me and so were not biased by motivation to confirm rather than refute it.

For each of the 100 discoveries, it was possible to identify concepts whose combination contributed to the discovery. The first example discussed by Haven is Archimedes' discovery of the principle of the lever, that the weights pushing down on each side of a lever are proportional to lengths of the board on each side of the balance point. This principle is a newly created proposition, which I take to be a mental representation analogous to a sentence. Philosophers sometimes talk of propositions as abstract meanings of sentences, but there is no reason to believe in the existence of such abstract entities, so I will employ the cognitive science idea of a proposition as a mental representation carried out by neural processes. The new proposition about levers is clearly the result of combining other mental representations in the form of concepts such as *weight, push, side,* and *proportional.* Hence Archimedes discovery of the principle of the lever and Haven's 99 other examples all confirm the combinatorial conjecture. The spreadsheet containing my analysis of the 100 is available on request.

The example that presented the biggest potential challenge to the combinatorial conjecture was the 1938 discovery in South Africa of a coelacanth, from a species that was thought to have been extinct for over 80 million years. The curator of a local museum came across a novel fish and sent it to a biologist who recognized it from fossil records. My first impression was that this discovery was a simple perceptual recognition that did not amount to the generation of any new representations. On reflection, however, it became clear that what made this discovery creative was recognition of the existence of coelacanths that are currently alive. The criteria for creativity, as suggested by Boden (2004), are that a development be novel, surprising, and important. Merely finding a coelacanth fossil would not satisfy any of these criteria, but a live specimen was indeed surprising and important. Thus the creative discovery in this case is not just the recognition of a coelacanth, but the proposition that living coelacanths currently exist. This proposition requires the combination of concepts such as *living* and *coelacanth*, and hence serves to confirm rather than refute the combinatorial conjecture. Similarly, the serendipitous discovery of penicillin might be erroneously construed as simply a matter of perception, but what made Alexander Fleming's discovery novel, surprising, and important was his more complex recognition that mold was killing bacteria, producing the key conceptual combination *bacteria-killing mold.* Serendipity and conceptual combination go together.

Note, however, that the coelacanth discovery did not require the generation of any new concepts, as the concept *coelacanth* was already familiar to evolutionary biologists. I was surprised to find that I could identify original new concepts (indicated by newly coined words) in only 60 of the 100 cases. This is probably an undercount that could be increased by more detailed study of the cases, but there still seem to be many cases where a major scientific discovery was made without introducing novel, permanent concepts. Combination of concepts into new propositions, but not permanent new concepts, include Copernicus on the earth rotating around the sun, Galileo on falling objects, and Boyle's law of gases. It is important to recognize that the combinatorial conjecture is true concerning mental representations in the form of propositions, but would be false if it were interpreted as a claim about creativity requiring the generation of new, permanent concepts.

In looking at Haven's sample of scientific discoveries, I was interested in what kinds of mental representations were used in discovery generation. Obviously, all 100 involved verbal representations that have been used to communicate them to others, but I conjectured that visual and other kinds of mental representations that encode information in nonverbal formats might also be relevant. (For a review of different kinds of mental representation, see Thagard, 2005.) Mathematical representations are a subset of verbal representations that use numbers and/or equations. There are many scientific discoveries in which mathematics was important, ranging from Archimedes' law of the lever to Einstein's generation of $E=mc^2$. I identified 46 of the 100 discoveries as involving mathematical representations, although some digging could probably increase this number. Mathematical representations were much more common in physics than in biology and medicine, where many discoveries such as the existence of cells were qualitative. For tables summarizing the results described in this section and comparing them for those of technological invention, see the next section.

What role do nonverbal representations play in scientific creativity? Visual representations akin to pictures seemed to me important in at least 41 of the discoveries, ranging from Copernicus picturing the earth going around the sun to Robert Bakker's imagining the activities of warm-blooded dinosaurs. Some of these were more obviously visual than others as seen from their pictorial presentations, for example in Andreas Vesalius's revisionary drawings of human anatomy and Robert Hooke's drawings of cells viewed through a microscope. I found only five examples, however, where nonverbal, nonvisual representations seemed to play an important role in creative thinking, although more detailed historical analysis may well turn up more. I speculate that kinesthetic representations contributed to Archimedes' discoveries about levers and buoyancy and possibly to Galileo's discoveries about falling objects. Touch seems to have been relevant to Benjamin Franklin's discoveries about lightning and electricity because he felt sparks, and to Count Rumford's ideas about heat from friction. Sound definitely contributed to Christian Doppler's thinking about shifting frequencies. Otherwise, scientific thinking seems to have operated well just with visual and verbal (including mathematical) representations. The next section reports that nonverbal representations are much more important in technological invention.

Analysis of this relatively large sample of scientific discoveries provided an opportunity to examine questions independent of the combinatorial conjecture. I was curious how many of the discoveries were based in large part on accidents, that is events that were not the result of an intentional plan of investigation. Based on Haven's brief accounts and my own knowledge of historical events, I estimated that around a quarter of the discoveries had a substantial accidental component. For example, Galileo was not looking for moons of Jupiter with his telescope, van Leeuwenhoek was not seeking microbes with his microscope, and Roentgen was very surprised to encounter X-rays. Hence serendipity is an important part of scientific discovery, but the majority of cases seem to result from intentional problem solving. For a discussion of many cases of serendipity in scientific discovery, see Roberts (1989) and Meyers (2007).

Most unintended discoveries are observational, but serendipity can also be a feature of theoretical research. Heisenberg's uncertainty principle was an unanticipated consequence of his mathematical explorations, and Lorenz was surprised to find large outcomes from tiny changes in the starting conditions of his computational model of atmospheric storms.

Analogy has often been recognized as an important creative cognitive process, and many important examples have been identified (Holyoak & Thagard, 1995, Chapter 8). I found a significant analogical component in 14 of Haven's samples, with many cases that have not been analyzed in the philosophical or psychological literatures on analogy. It is quite possible that more examples of analogy could be found through more detailed historical analysis of the cases, but I doubt that would change the conclusion that analogy is an important but by no means exclusive mechanism for scientific creativity. Table 16.1 summarizes 14 cases of analogical discovery, noting the source analog that generated ideas about the target domain leading to a discovery. Table 16.1 also indicates the representational modalities used in addition to the ubiquitous verbal one. It is interesting that visual representations seem to be important in a greater proportion of analogical discoveries (11/14) than in discoveries in general (41/100).

Dunbar (1995) distinguished between *local* analogies that operate within a single domain and *long-distance* analogies that cross domains. All but two of the important analogical discoveries are long-distance, requiring a major mental leap across domains. Particularly interesting are three cases, indicated with an asterisk in "long,*" where the analogy served to unite previously disparate domains. Before Newton, projectile and

Table 16.1 Scientific discoveries based on analogy. See text for explanation of the asterisk in "long*"

DISCOVERY	*TARGET*	*SOURCE*	*MODALITY*	*DISTANCE*
Living cells	Living cells	Monk cells	Visual	Long
Gravity	Planetary motion	Projectile motion	Visual, mathematical, kinesthetic	Long*
Fossils	Sharks teeth	Stone teeth	Visual	Long
Life	Hierarchy	Tree	Visual	Long
Lightning	Lightning	Spark	Visual, heat	Long*
Vaccination	Smallpox	Cowpox	Visual	Local
Ultraviolet light	Ultraviolet	Infrared	Mathematical	Local
Electromagnetism	Magnetism	Electricity	Visual, mathematical	Long*
Evolution	Natural selection	Malthusian competition	Mathematical	Long
Periodic table	Elements	Piano scale	Visual	Long
Relativity	Gravity	Elevator	Visual, mathematical	Long
Fault lines	Rock layers	Rubber bands	Visual	Long
Earth mantle	Earth	Egg	Visual	Long
Quantum theory	Electrons	Crystals	Mathematical	Long

planetary motion were distinct domains, but Newtonian mechanics unified them in a common physical framework. Similarly, before Franklin, sparks and lightning were different kinds of things, and before Faraday electricity and magnetism were unconnected. In these three cases, analogical thinking brought about an important kind of conceptual change in the nature of domains that were changed through new unifying theories, so that a long-distance analogy turned into a more local one.

Philosophers often debate about the relative importance of theory and observation in science, so I coded the 100 examples for whether the discoveries were primarily:

Observational, based on perception using human senses;
Instrumental, based on observations using instruments; or

Theoretical, requiring hypotheses that go beyond the results of sensory and instrumental observation.

The results are interesting, with 70 of the discoveries theoretical, 18 observational, and 12 instrumental. Examples of discoveries made with unaided observations include Davy's discovery of anesthesia, Mendel's findings about heredity, and Fleming's discovery of penicillin. Instruments important for making observations not possible with ordinary human perception include the telescope (Galileo), microscope (Pasteur), and spectograph (Hubble).

Finally, I was interested in the extent to which scientific discoveries depended on previous technological advances and the extent to which science led to subsequent advances in technology. My counts are preliminary and should be viewed only as approximate minimums to be made more precise by more thorough historical analysis, but they are nevertheless interesting. I found 37 of the discoveries as depending in important ways on technological advances. This was obviously much more than the 12 that were directly based on instrumental observations: Many of the theoretical discoveries arose from observations that required new technologies, for example, Davy's electrochemical ideas resulting from the availability of batteries. I identified 21 scientific discoveries that in turn led to technological advances, such as Franklin's discoveries about electricity enabling him to invent the lightning rod. Thorough historical research would undoubtedly generate more examples.

Thus the study of 100 examples of scientific discovery was useful for much more than just testing the combinatorial conjecture. It served to clarify the differences between generating new propositions and generating new concepts, with only the former occurring in all cases of scientific discovery. Verbal representations seem to be universal in discoveries, but are complemented in many cases by mathematical and visual ones. Other sensory representations did not seem to be very important for scientific discovery. For lack of data, I have not addressed the role of another kind of nonverbal representation that is important for human thinking—emotion (see Thagard, 2006, 2010a). I would conjecture that every one of the 100 discoveries generated a strong, positive emotional response in the discoverer, just like Archimedes' famous Eureka! moment when he discovered the principle of buoyancy. Thagard and Stewart (2011) give an account of the neural processes that might generate such treasured moments.

This study has also generated interesting data about the extent to which discovery is accidental, analogical, theoretical, observational, and dependent on instruments. These data must be viewed as highly tentative, since they are based on brief accounts of one author and background knowledge of one interpreter. I hope they will serve to stimulate further systematic research on a larger sample of scientific discoveries that are examined in much more depth. Let us now compare scientific discovery with a similar survey of cases of scientific invention.

STUDY 2: TECHNOLOGICAL INVENTION

Previous investigation based on a few examples has suggested that invention of new technologies involves the same basic set of cognitive processes of scientific discovery (Saunders and Thagard, 2005; Thagard and Croft, 1999). Both invention and discovery require basic cognitive processes such as problem solving, analogical inference, and concept generation. But a study of a large number of inventions also turned up some interesting cognitive differences.

For my sample of inventions, I used *100 Greatest Inventions of All Time* by an experienced nonfiction writer, Tom Philbin (2003). As with my discovery sample, the "greatest" assertion should not be taken too seriously, but there is no question that Philbin identified many very important inventions, ranging from the wheel (#1) to the video recorder (#100). Unlike Haven, Philbin ranks the creations 1–100 in order of importance, but it would be hard to defend his entire ordering. A few examples appear on both lists: anesthesia, X-rays, and the transistor. As with the 100 discoveries, the analysis of 100 inventions should be viewed as highly provisional, since it may depend on both Philbin's and my idiosyncrasies. Still, preliminary data may help to point the way to future studies that are broader and deeper.

Tables 16.2–16.5 summarize the similarities and differences found between discovery and invention. As Table 16.2 shows, all inventions, like all discoveries, involved verbal representations. This unanimity may be an artifact of the need for people (including the creators as well as commentators such as Philbin) to use language to report their discoveries to others, but inspection suggests language may well have played a role in all cases. For example, the invention of the wheel plausibly had a substantial nonverbal component owing to visual and kinesthetic representations of crucial ingredient concepts such as *log* and *rolling*, but these concepts have verbal representations as well.

It is striking in Table 16.1 that visual representations seem to be much more common in the sample of inventions than in the sample of discoveries. The reason for this difference is probably that most inventions are *things* that people can see: the wheel, lightbulb, computer, and so on. Hence people naturally have visual representations of them. For the same reason, inventions are more susceptible to other nonverbal representations such as touch, heat, kinesthesia, and sound. Mathematical representations were much less common for discoveries than for inventions, 26 rather than 46. This discrepancy may reflect the fact that Philbin had many more early examples than Haven, including 10 inventions before the Christian era in contrast to only 1 discovery that early. Cases such as the plow, sail, and bow and arrow are clearly important inventions, but the frequency of mathematical representations

Table 16.2 Kinds of representation used in scientific discovery and technological invention

	Verbal representation	*Visual representation*	*Mathematical representation*	*Other kind of representation*
Scientific discovery	100	41	46	5
Technological invention	100	87	26	48

Table 16.3 Kinds of novelty in scientific discovery and technological invention

	New propositions	*New concepts*	*Accidental*	*Analogical*
Scientific discovery	100	60	26	14
Technological invention	100	100	5	12

would have increased if Philbin had included many more recent science-dependent examples.

Table 16.3 summarizes aspects of novelty in discovery versus invention. All invention, like all discovery, generates new propositions, minimally of the form: This device serves to perform that function. Unlike discoveries, many of which do not introduce new concepts, all the inventions involved the introduction of new concepts. The difference again arises from the fact that all inventions are things, and the selected kinds were clearly important enough to warrant naming. Inventions seem to have occurred much less accidentally than discoveries, 5 versus 26, since inventions are usually the result of an intentional effort to solve some identified problem. Nevertheless, there are a few interesting cases of inventions such as anesthesia and X-ray machines that arose accidentally: Davy was not looking for a way to kill pain,

Table 16.4 Analogies in technological inventions

INVENTION	*TARGET*	*SOURCE*	*MODALITY*	*DISTANCE*
Printing press	Printing	Olive press	Visual, kinesthetic	Long
Telephone	Telephone	Ear	Visual, sound	Long
Paper	Bark paper	Hemp paper	Visual, touch	Local
Airplane	Airplane	Bird	Visual	Long*
Stethoscope	Stethoscope	Wood	Sound	Long
Microscope	Microscope	Eyeglasses	Visual	Local
Braille	Braille	Previous dots	Touch	Local
Incubator	Baby incubator	Chick hatchery	Visual, heat	Long*
Cotton gin	Cotton machine	Hand movements	Visual, kinesthetic	Local
Windmill	Windmill	Sail	Visual	Long
Washing machine	Machine	Hand movements	Visual, kinesthetic	Local
Oil derrick	Oil	Gallows	Visual	Long

Table 16.5 Theory and observation in discovery and invention

	Theoretical	*Observational*	*Instrumental*
Scientific discovery	70	18	12
Technological invention	24	76	0

and Roentgen was not looking for a way to examine bones. I was struck, however, by the highly incremental nature of invention, with many new technologies being part of a whole series of improvements in attempts to accomplish some task such as building a better lightbulb. According to Philbin's descriptions, 71 of the inventions were incremental in this way, whereas the scientific discoveries seemed to involve more dramatic leaps.

It would be interesting to determine by more detailed historical examination whether these incremental examinations can be viewed as cases of analogical inference, using past inadequate inventions as source analogs to develop new, improved targets. That finding would increase dramatically the occurrence of analogies in invention, which at 12% is a bit less than the 14% for discovery. Table 16.4 displays the analogies identified for technological invention. As with discovery, analogy seems to have been an important cognitive process in invention, but is far from being universal. All the inventive analogies plausibly have a nonverbal component. There are proportionately more local analogies than I found in discovery (5/12 vs. 2/14). Two of the long-distance analogies indicated by an asterisk were originally cross-domain but redefined the nature of domains so that we can now view them as the same domain. For example, before the Wright brothers used what they knew about birds to inform airplane construction, birds and flying machines were different kinds of objects, but now are unified under the general theory and practice of aerodynamics.

Finally, Table 16.5 shows an interesting difference between the theoretical and observational status of discoveries and inventions. I counted 76 of the inventions as observational in that they were made using ordinary human senses, with only 24 requiring theoretical leaps beyond observation. Many of these theory-based inventions were electrical devices such as the telephone. Instruments for measuring the effects of theoretical entities were undoubtedly important in many of these inventions, but none seemed to be based just on instrumental observations without a large theoretical contribution.

Comparison of discovery and invention raises interesting questions about the relation of science to technology. I found 37 cases of scientific discoveries that depended on preceding technological developments such as the telescope and spectograph. Moreover, at least 21 of the discoveries led to new technologies such as radio and the atomic bomb. Surprisingly, looking at inventions I only identified 9 of Philbin's cases as depending on prior scientific discoveries, and only 3 as generating new scientific discoveries. Perhaps these low numbers result from Philbin's assessment of "greatest" in terms of everyday usage rather than scientific importance.

A sample of 20th-century inventions would probably display a much stronger interconnection of technology and science more in accord with the findings from Haven's 100 discoveries.

OBJECTIONS TO COMBINATION

My survey of 200 examples of scientific discovery and technological invention did not turn up any counterexamples to the combinatorial conjecture, but there are other ways in which that conjecture might turn out to be false. This section considers three: abstraction, mutation from single representations, and (most radically) creativity that does not at all rely on mental representations. I have already argued that serendipitous perception in cases such as the coelacanth and penicillin are actually cases that confirm rather than refute the combinatorial conjecture.

Welling (2007) discusses four creative processes: application of existing knowledge, analogy, combination of concepts, and abstraction. The first three of these clearly involve combinations of representations, but what about the fourth, abstraction? According to Welling (2007, p. 170):

> The mental *process* of abstraction may be defined as: the discovery of any structure, regularity, pattern or organization that is present in a number of different perceptions that can be either physical or mental in nature. From this detection results the *product* abstraction: a conceptual entity, which defines the relationship between the elements it refers to on a lower, more concrete, level of abstraction.

To illustrate abstraction, Welling uses Piaget's example of children learning the concept of *weight* by abstraction of experiences of objects that are heavy and light. He speculates that children and other learners use Gestalt principles of perceptual organization such as grouping and closure.

Without a more detailed model of how abstraction works, it is difficult to assess whether it constitutes a challenge to the combinatorial conjecture. It does seem, however, that abstraction requires the combination of perceptual representations, for example the physical, kinesthetic sensations involved in assessing an object as heavy or light. I conjecture that when children learn the abstraction *weight* they put together a combination of verbal and nonverbal representations of experiences of heavy objects and light objects. If there are any cases of abstraction that are not combinatorial in this fashion, I expect that they are not particularly creative according to Boden's criteria of being new, surprising, and valuable.

Another kind of possible counterexample to the combinatorial conjecture would be if creativity arose from a kind of mutation in a single concept, analogous to the way that mutations occur in genes. Although various authors have attempted to exploit an analogy between genetic mutation and concept generation (e.g., Dawkins, 1976), I think this analogy is feeble from a cognitive perspective (Thagard, 1988, Chapter 6). Thinking is much more structured and constrained than biological evolution. In particular, no one has ever identified an interesting case of creativity arising from a random alteration in a single concept analogous to mutation in a single gene. Conceptual combinations occur in a much more focused way in the context of problem solving and hence are a much more plausible mechanism of creativity than single-concept mutations. See the Appendix for further analysis of why discovery and invention do not result from blind variations.

My analyses and arguments in defense of the combinatorial conjecture may suggest to the reader that the claim is true but trivial, making no substantive assertion. Response to the charge of triviality can be made first by pointing to the neurocomputational theory of combination that proposes a detailed neural mechanism for combining representations using convolution (Thagard & Stewart, 2011). This theory shows that the conjecture can be fleshed out into a specific claim about the cognitive and emotional processes that underlie human creativity. Second, the nontriviality of the combinatorial conjecture is evident from serious theories that deny a theoretical role for representation altogether. If there are no mental representations, then creativity is obviously not the result of combining them.

Denial of mental representations was a hallmark of behaviorism, which dominated American psychology until the cognitive revolution of the 1950s. It has been revived in a movement espousing "radical embodied cognitive science" that draws on a combination of Heideggerian philosophy, Gibsonian psychology, and dynamic systems theory to propose an alternative to the dominant computational-representational view of thinking (see, e.g., Chemero, 2009; Clark, 1997; Dreyfus, 2007; Thompson, 2007; Warren, 2006). If radical embodiment is true, then creativity does not require combining representations at all: It can be "action-first" rather than "thought-first" (Carruthers, 2011). Then the combinatorial conjecture would be false.

These extreme embodiment claims need to be distinguished from more moderate ones made by researchers such as Gibbs (2006) and Barsalou et al. (2003), who maintain that the kinds of representations and computations performed by the brain are closely tied to bodily processes such as perception and emotion. My account of neural representation is sufficiently broad to encompass a wide range of perceptual modalities, all of which can be understood as patterns of activation in populations of neurons. The body also plays a large role in my account of emotion (Thagard, 2010a; Thagard and Aubie, 2008). Hence I am happy to endorse a *moderate* embodiment thesis that acknowledges the importance of perceptual and other physiological processes (Thagard, 2010b). This moderate thesis is fully compatible with the combinatorial conjecture as long as cognition is not mistakenly restricted to only language-like representations.

So why does cognitive science, and particularly the theory of creativity, need representations? The answer is that postulation of various kinds of representations currently provides the best available explanation of many kinds of human thinking, including perception, inference, learning, problem solving, and language use (see e.g. Anderson, 2010; Smith & Kosslyn, 2007; Thagard, 2005). Proponents of the extreme embodiment thesis have barely scratched the surface in matching the explanatory successes of the computational-representational approach. Indeed, it can be argued that even the basic problem of motor control is too complex to understand without postulating representations and computations (Thagard, 2010b; Todorov & Jordan, 2002; Wolpert & Ghahramani, 2000). Abilities such as grasping objects require building complex mental models to predict the effects of various kinds of muscular operations.

More specifically, no one has a clue how to use pure embodiment to explain creative developments in science and technology. Discoveries such as relativity theory and inventions such as the telephone require the full range of human representational capacities, from verbal and mathematical to more obviously embodied representations such as vision and sound. Humans are indeed embodied dynamic systems embedded in their environments, but our success in those environments depends heavily on our ability to represent them mentally and to perform computations on those representations. Hence the embodied aspect of much of mental cognition does not refute the combinatorial conjecture, although the claims of radical embodiment do serve to show that the conjecture is a substantive one about human creativity.

According to Arthur (2009), new technologies arise as combinations of other technologies, and he even talks about how "technologies modify themselves over their lifetime" (p. 87) and describes technology as "self-producing" (p. 170). Obviously, however, past technologies have not had the capability to actually combine or modify themselves, although in the future more intelligent machines may do so (Lipson & Pollock, 2000). Rather, new technologies from wheels to iPads have resulted from the combination of human mental *representations* of previous technologies. Technological creativity is a physical process of interaction with the world, and a social process of interaction with other people. But it is also a psychological process carried out by brains that are capable of computationally modifying representations through such mechanisms as visualization, conceptual combination, analogy, and inference in general.

CONCLUSION

The two studies in this chapter have found support for the combinatorial conjecture in 200 examples of discovery and invention, but do not address whether it holds in other domains of human creativity. It should not be too hard to apply the conjecture to social innovation, which concerns the creation of new organizations, institutions, and practices that benefit human society. Innovations such as democratic government, public education, pension plans, universal health care, and international governance have contributed greatly to the quality of human lives, and I expect to show that all of these resulted in part from the combination of representations. My guess is that social innovation will turn out to be more like technological invention than like scientific discovery, except for a reduced contribution of representations that are visual.

More difficult will be assessment of the applicability of the combinatorial conjecture to the great many examples of creativity resulting from artistic imagination, including poems, plays, novels, films, music, dance, and architecture. Examination of even a few examples from these categories will require attention to many kinds of representation beyond the verbal, such as emotion in poetry, vision in film, sound in music, and kinesthesia in dance and sculpture. I expect that scrutiny of such examples will serve not just to confirm the combinatorial conjecture but also to flesh it out with greater understanding of the kinds of representations and processes that contribute to human creativity. Also needed is a general theory of how newly generated representations are evaluated for their coherence with other representations and overall value.

Although my analysis of 200 examples has been highly provisional and needs to be supplemented by a deeper and broader study, it has helped to characterize representational aspects of creativity along such dimensions as mode of representation, role of accident and analogy, and relative contribution of theory and observation. I would like to see the development of an *Atlas of Creative Science and Technology*, which would contain not only historical descriptions of great discoveries and inventions, but also their assessment with respect to the kinds of cognitive factors identified in this chapter. Creativity is combination of representations, but there is much more to be learned about the nature of these representations and the cognitive processes that produce them.

ACKNOWLEDGMENTS

This research is supported by the Natural Sciences and Engineering Research Council of Canada. I am grateful to Robert Proctor and Daniel Saunders for helpful suggestions.

REFERENCES

Anderson, J. R. (2010). *Cognitive psychology and its implications* (7th ed.). New York: Worth.

Arthur, W. B. (2009). *The nature of technology: What it is and how it evolves*. New York: Free Press.

Barsalou, L. W., Simmons, W. K., Barbey, A. K., & Wilson, C. D. (2003). Grounding conceptual knowledge in modality-specific systems. *Trends in Cognitive Sciences, 7*, 84–91.

Boden, M. (2004). *The creative mind: Myths and mechanisms* (2nd ed.). London: Routledge.

Carruthers, P. (2011). Creative action in mind. *Philosophical Psychology, 24*, 437–461.

Chemero, A. (2009). *Radical embodied cognitive science*. Cambridge, MA: MIT Press.

Clark, A. (1997). *Being there: Putting brain, body, and world together again*. Cambridge, MA: MIT Press.

Dawkins, R. (1976). *The selfish gene*. New York: Oxford University Press.

Dayan, P., & Abbott, L. F. (2001). *Theoretical neuroscience: Computational and mathematical modeling of neural systems*. Cambridge, MA: MIT Press.

Dreyfus, H. L. (2007). Why Heideggerian AI failed and how fixing it would require making it more Heideggerian. *Philosophical Psychology, 20*, 247–268.

Dunbar, K. (1995). How scientists really reason: Scientific reasoning in real-world laboratories. In R. J. Sternberg & J. Davidson (Eds.), *Mechanisms of insight* (pp. 365–395). Cambridge, MA: MIT Press.

Eliasmith, C. (2005). Cognition with neurons: A large-scale, biologically realistic model of the Wason task. In B. Bara, L. Barasalou, & M. Bucciarelli (Eds.), *Proceedings of the 27th Annual Conference of the Cognitive Science Society* (pp. 624–629). Mahwah, NJ: Lawrence Erlbaum.

Eliasmith, C. (forthcoming). *How to build a brain*. Oxford: Oxford University Press.

Eliasmith, C., & Anderson, C. H. (2003). *Neural engineering: Computation, representation and dynamics in neurobiological systems*. Cambridge, MA: MIT Press.

Eliasmith, C., & Thagard, P. (2001). Integrating structure and meaning: A distributed model of analogical mapping. *Cognitive Science, 25*, 245–286.

Finke, R., Ward, T. B., & Smith, S. M. (1992). *Creative cognition: Theory, research and applications*. Cambridge, MA: MIT Press / Bradford Books.

Gibbs, R. W. (2006). *Embodiment and cognitive science*. Cambridge: Cambridge University Press.

Gorman, M. E., Tweney, R. D., Gooding, D. C., & Kincannon, A. P. (Eds.). (2005). *Scientific and technological thinking*. Mawah, NJ: Lawrence Erlbaum.

Haven, K. (2007). *100 greatest science discoveries of all time*. Westport, CT: Libraries Unlimited.

Holyoak, K. J., & Thagard, P. (1995). *Mental leaps: Analogy in creative thought*. Cambridge, MA: MIT Press/Bradford Books.

Hummel, J. E., & Holyoak, K., J. (2003). A symbolic-connectionist theory of relational inference and generalization. *Psychological Review, 110*, 220–264.

Koestler, A. (1967). *The act of creation*. New York: Dell.

Lipson, H., & Pollock, J. B. (2000). Automatic design and manufacture of robotic lifeforms. *Nature, 406*, 974–978.

Mednick, S. A. (1962). The associate basis of the creative process. *Psychological Review, 69*, 220–232.

Meyers, M. A. (2007). *Happy accidents: Serendipity in modern medical breakthroughs*. New York: Arcade.

O'Reilly, R. C., & Munakata, Y. (2000). *Computational explorations in cognitive neuroscience*. Cambridge, MA: MIT Press.

Philbin, T. (2003). *The 100 greatest inventions of all time*. New York: Citadel Press.

Plate, T. (2003). *Holographic reduced representations*. Stanford, CA: CSLI.

Poincaré, H. (1921). *The foundations of science* (G. Halsted, Trans.). New York: Science Press.

Roberts, R. M. (1989). *Serendipity: Accidental discoveries in science*. New York: John Wiley.

Saunders, D., & Thagard, P. (2005). Creativity in computer science. In J. C. Kaufman & J. Baer (Eds.), *Creativity across domains: Faces of the muse* (pp. 153–167). Mahwah, NJ: Lawrence Erlbaum.

Shastri, L. (1999). Advances in SHRUTI—A neurally motivated model of relational knowledge representation and rapid inference using temporal synchrony. *Applied Intelligence, 11*, 79–108.

Simonton, D. K. (2010). Creative thought as blind-variation and selective-retention: Combinatorial models of exceptional creativity. *Physics of life reviews, 7*, 156–179.

Smith, E. E., & Kosslyn, S. M. (2007). *Cognitive psychology: Mind and brain*. Upper Saddle River, NJ: Pearson Prentice Hall.

Thagard, P. (1988). *Computational philosophy of science*. Cambridge, MA: MIT Press.

Thagard, P. (1997). Coherent and creative conceptual combinations. In T. B. Ward, S. M. Smith, & J. Viad (Eds.), *Creative thought: An investigation of conceptual structures and processes* (pp. 129–141). Washington, D C: American Psychological Association.

Thagard, P. (2005). *Mind: Introduction to cognitive science* (2nd ed.). Cambridge, MA: MIT Press.

Thagard, P. (2006). *Hot thought: Mechanisms and applications of emotional cognition*. Cambridge, MA: MIT Press.

Thagard, P. (2010a). *The brain and the meaning of life*. Princeton, NJ: Princeton University Press.
Thagard, P. (2010b). How brains make mental models. In L. Magnani, W. Carnielli, & C. Pizzi (Eds.), *Model-based reasoning in science and technology. Abduction, logic, and computational discovery* (pp. 447–461). Berlin: Springer.
Thagard, P., & Aubie, B. (2008). Emotional consciousness: A neural model of how cognitive appraisal and somatic perception interact to produce qualitative experience. *Consciousness and Cognition, 17,* 811–834.
Thagard, P., & Croft, D. (1999). Scientific discovery and technological innovation: Ulcers, dinosaur extinction, and the programming langage Java. In L. Magnani, P. Nersessian, & P. Thagard (Eds.), *Model-based reasoning in scientific discovery* (pp. 125–137). New York: Plenum.
Thagard, P., & Stewart, T. C. (2011). The Aha! experience: Creativity through emergent binding in neural networks. *Cognitive Science, 35,* 1–33.
Thompson, E. (2007). *Mind in life: Biology, phenomenology, and the science of mind*. Cambridge, MA: Harvard University Press.
Todorov, E., & Jordan, M. I. (2002). Optimal feedback control as a theory of motor coordination. *Nature Neuroscience, 5,* 1226–1235.
Warren, W. H. (2006). The dynamics of perception and action. *Psychological Review, 113,* 358–389.
Welling, H. (2007). Four mental operations in creative cognition: The importance of abstraction. *Creativity Research Journal, 19,* 163–177.
Wolpert, D. M., & Ghahramani, Z. (2000). Computational principles of movement neuroscience. *Nature Neuroscience, 3,* 1212–1217.

CHAPTER 16 APPENDIX: BLIND VARIATION

Simonton (2010, Chapter 15, this volume) has attempted to revive the idea of blind variation in creativity, but I think his mathematical analysis does not go to the heart of the matter. Here is an alternative.

Let V be the set of variants that can arise in a set of structures such as genes, mental representations, machines, and so forth. V will be very large, but not infinite, as biological and physical entities are finite. The variants can be numbered $V_1 \ldots V_k$, with V_i indicating some specific variant.

Then we can define G_i as the generation of variant V_i, and U_i as the utility of V_i. I propose that a variant is *blind* if its generation is independent of its utility, that is, the probability of generation given nonzero utility is the same as the probability of its generation if it were useless:

V_i is blind iff $P(G_i/U_i > 0) = P(G_i/U_i = 0)$.

Then we can say that a process of variation is overall-blind if every variant in it is blind. Genetic mutation is overall-blind, but scientific discovery, technological invention, and other forms of human creativity are overwhelmingly not, because psychological processes of problem solving and representation generation focus thinking toward variants that are more useful than random ones.

A quantitative approach has the advantage that we can talk about degree of blindedness, which is the extent to which variants in a process are blind:

Blindedness = # of blind actual variants/# of actual variants.

My survey of 100 scientific discoveries and 100 inventions suggests that the blindedness of these processes is near 0, although the difficulty of assessing the relevant probabilities and utilities makes it hard to say. Many discoveries (but hardly any inventions) have an unintentional component, but even in these cases it seems that more useful variants are more likely to be generated than useless ones. For example, Galileo never intended to find the moons of Jupiter when he turned his new telescope on the heavens, but his interests, background knowledge, and cognitive processes made it more probable that he would generate the representation "Jupiter has moons" than some utterly useless representation such as "Rome has toes." Hence discovery is not blind, and biological evolution is a poor model for scientific discovery and other kinds of creativity.

17 On the Unreasonable Reasonableness of Mathematical Physics

A Cognitive View

Ryan D. Tweney

INTRODUCTION

The use of mathematical representations is pervasive in all of science, and it is most pervasive in physics. Walk into the office of a physicist and you will likely see a blackboard or whiteboard filled with equations. Open a physics journal, and the appearance will be similar; most articles will have equations and expressions scattered among the text, and sometimes the math will have more lines than the text. For the nonphysicist, this is intimidating—physics looks arcane, and while math is used in all the sciences, it dominates physics as in no other field. Yet, for the physicist, the math seems natural, like a first language, and what seems to the outsider to be abstract and lifeless, a domain of cold symbols, to the physicist is fraught with meaning. How can cognitive science understand this difference? What does the physicist know that the outsider does not? Or, to put the question in cognitive terms, what processes are involved in the fluent use of mathematics in physics, how does it work, and how is it learned? There is a great deal of implicit cognition going on. Can we characterize it?

There is, of course, a simple explanation, which works at one level. We could argue, and we would surely be right, that the physicist has *learned* mathematics and that part of the acquisition of expertise in physics is precisely expertise in the use of mathematics. A few existing studies of physics expertise bear on the issue, for example, the classic study by Chi, Feltovich, and Glaser (1981) which compared novices (graduate students in physics) with experts (faculty members in physics). Using a variety of methods centered on the two groups' understanding of elementary (college) problems in physics, Chi et al. showed that there were key differences in how novices and experts organized their understanding of the problems. For novices, surface features loomed large (e.g., they categorized together all the inclined plane problems, all the rotating mass problems, and so on), whereas for experts, deep features (e.g., problems involving conservation of angular momentum, problems involving kinetic energy, and so on) were used to organize problems. While both experts and novices had many common elements in their knowledge base, the organization of that knowledge differed. Further, even when experts and novices were equally successful in solving problems, their approach differed: Experts in physics tended to work forward in solving problems, from the general characteristics of the problem as given to the specific means of solving it. By contrast, novices tended to work backward, that is, from a statement of the solution state to the starting point (Larkin, McDermott, Simon, & Simon, 1980). Most to the point for the present chapter, the graduate students tended to write down *more* equations than the faculty members, seeking an algorithmic solution apparent in the equations. Experts used their mathematical expressions more sparingly but also more meaningfully, using mathematics as a representation rather than as a possible path to an algorithmic solution. While studies like that of Chi et al. have been important in understanding the nature of expertise in general (Ericsson, 2006), they still do not speak to the central issue of the present study: What is it about *mathematics* that is so powerful and so pervasive in physics and what exactly will students learn as they become expert physicists?

In spite of the ubiquity of mathematics in science, the relative neglect of the subject in the cognitive disciplines is puzzling, especially given the huge amount of attention given to elementary math—arithmetic, algebra, and geometry. Thus, the recent *Handbook of Mathematical Cognition* (Campbell, 2005) has exactly one chapter devoted to higher math, that by Núñez and Lakoff (2005) (about which I say more below). In one of the few cognitive studies to address the use of higher mathematics, Kurz-Milcke (1998, 2004) gave a fluid flow problem that required a differential equation for its solution to three faculty members, a physicist, a physical chemist, and a mathematician. Think-aloud protocols revealed that each of the three had to reconstruct the necessary mathematics from "first principles," in effect, reinventing the calculus. For the mathematician, this involved an approach via symbolic considerations of the use of limits, but for the physicist and physical chemist, model-construction was paramount. There is an important clue here to the way in which mathematics works in physics; the suggestion is that mathematics allows a particular kind of model-based reasoning.

Recent cognitive analyses have included a central emphasis on the role of models in the practices of science (e.g., Brewer, 2003, Chapter 13 of this volume; Clement, 2008; and the papers in Magnani & Nersessian, 2002). Nersessian characterizes a model "as a representation of a system with interactive parts and with representations of those interactions" (Nersessian, 2008, p. 12) and notes that model-based reasoning can be conducted via computational, conceptual, physical, or mathematical models. Still, there is a lack of attention to the overwhelming importance of mathematics in such models.

Here, I develop a framework to address part of this gap. The framework is based on a reconsideration of mathematical representations as metaphors and on an argument that metaphor is the appropriate way to approach the issue. My strategy is to develop an extension of the metaphorical analysis of pure mathematics developed by Lakoff and Núñez (2000), adapting it to the specific use of mathematics in physics. To establish the argument I use a cognitive-historical analysis of an important case study, the electromagnetic theory developed by James Clerk Maxwell, now epitomized within physics as "Maxwell's Equations." The framework used in the analysis, and the larger project of which it is a part (see, e.g., Tweney, 2009, 2011) extends cognitive approaches that have been successfully applied in recent years to understand scientific practices.[1]

The choice of this particular subject stems from the fact that British physics underwent a fundamental transition in the 1830s, from a primarily experimental "natural philosophy" to a mathematized theoretical physics.[2] During the 18th and most of the 19th centuries, French and British mathematical physics differed greatly. The French tradition emphasized pure mathematics, with physical applications being derivative from the mathematics. British mathematical physics, especially in Scotland where Maxwell received his first training, was skeptical about the French approach and emphasized instead the primacy of physical intuition and geometrical approaches to physics. From the 1830s, Cambridge University, where Maxwell studied from 1850 to 1854, was firmly within the British tradition (see Harman, 1998; Olson, 1975).

An important aspect of this transition in the nature of mathematical physics was a result of Maxwell's mathematization of Michael Faraday's field theory (Garber, 1999, esp. Chapter 7). As I argue here, what Maxwell achieved was in part a cognitive shift in the way mathematical expressions represented physical concepts. Whereas Faraday had used a kind of "intuitive mathematics," a dynamic geometry of fields and their forces, Maxwell converted this into a mathematically represented theory of fields. We can understand the bases of this shift by considering the way in which Maxwell's mathematics instantiated Faraday's concepts, and allowed for their further extension.

Understanding this transition opens new questions for cognitive science and extends the scope of recent developments in embodied, situated, and distributed cognition (see, for example, the chapters in Robbins & Aydede, 2009). A central claim of these approaches is that cognition cannot be understood as a disembodied, disconnected set of abstract processes, but instead must always include attention to the "hand-eye-brain" relations present in cognitive actions. Chandrasekharan and Osbeck (2010; see also Osbeck, Nersessian, Malone, & Newstetter, 2010; Osbeck & Nersessian, Chapter 5, this volume) provide an interesting discussion of the similarities among these three recent developments in cognitive science, arguing that all

[1] The argument is also an extension of my prior work on Michael Faraday's research practices (e.g., Tweney, 1992a, 2006) and of my work on the cognitive underpinnings of scientific thinking (e.g., Ippolito & Tweney, 1995; Tweney, 2001, 2004).

[2] Historians of science have written extensively about Maxwell's role in this transition (e.g., Buchwald, 1985; Darrigol, 2000; Garber, 1999; Harman, 1998; Nersessian, 1984; Siegel, 1991; Simpson, 2010; Smith & Wise, 1989; Wise, 1982).

three can be understood as reflections of the "common coding" of perception, action, and imagination of movement. Similarly, Barsalou (2010) argues that "grounded cognition" bridges all three approaches. In this chapter, I hope to show how a similar approach can speak to the role of mathematics in physics. "Hand-eye-brain" are as involved in this domain as in any other.

I begin with a caveat. For the nonphysicist, Maxwell seems impossibly hard, in large part because of the profusion of mathematical expressions and equations. To fully understand his work is to first learn a good deal of mathematics; calculus, advanced calculus, and vector analysis in particular. For this reason, Maxwell's electromagnetic field theory (in a "modernized" form—see below) is usually taught only to physics majors, and only after several years of instruction in mathematics.[3] But in light of the analysis I present in this chapter, this need not be the case. I hope to convince the reader that, with sufficient attention to the *underpinnings* of mathematics, as used by Maxwell, the task is not so formidable as it seems. The present analysis, by presenting a framework for understanding Maxwell via a cognitive grounding in embodied practices, can open the "black box" of Maxwell's equations to audiences not versed in physics. In the present chapter, I use the modern version of Maxwell's equations, introducing the reader to a mode of analysis that permits the "nonexpert" to understand some of the cognitive bases of Maxwell's innovative approach, without requiring prerequisite knowledge of advanced mathematics or physics.[4]

METAPHOR AND ANALOGY

The role of analogy in science has been a frequent topic in the cognitive psychology of science, for example, the classic study of analogies between electricity and fluid flow by Gentner and Gentner (1983), and Dunbar's (1999) account of analogy use in molecular biology laboratories. Of particular relevance to the present chapter, Nersessian (2002, 2008) has closely examined Maxwell's use of analogy from a cognitive perspective, with a special focus on the three papers leading up to his 1873 *Treatise on Electricity and Magnetism*.

The distinction between metaphor and analogy is not often drawn explicitly in psychology, analogy being the preferred term for most existing research. Gentner, Bowdle, Wolff, and Boronat (2001) examined the way metaphor is reducible to analogy in cognitive terms, proposing a resolution via structure mapping theory

[3] An interesting exception occurs at the two campuses of St. John's College in Santa Fe, New Mexico, and Annapolis, Maryland. As part of a nonelective "great books" curriculum, Faraday and Maxwell are read, in the original, by all students in the junior year. My own conversations with students at the two campuses suggest that some acquire a deep understanding of Maxwell and nearly all develop a deep appreciation for the role of mathematics in his science. While beyond the scope of this chapter, the example opens many questions about the nature of science education and the implicit presuppositions which the higher education establishment brings to the task of curricular design (Tweney, 2011).

[4] Collins and Evans (2007) argue that "interactional expertise," by which they mean the ability to master the language of the true expert by an engaged outsider lacking the practical competence of the true expert, "provides a bridge between the rest of us and full-blown physically engaged experts" (p. 77). In this sense, one claim of the present paper is that interactional expertise is easier to acquire, in the case of mathematical physics, than has been thought.

(Gentner, 1983; Gentner & Bowdle, 2008; Woolf & Gentner, 2011; see also Utsumi, 2011, for a recent overview of cognitive approaches to metaphor). For the present project, I distinguish the two by using *metaphor* to stand for usages that are tacit and undiscussed comparisons between a source and target, even when these have been learned (perhaps with great difficulty and much conscious effort). Once learned, these become automatic and generally undiscussed relations between sources and targets; in our case, between the purely mathematical expression and the physical entity or process that it represents. There is of course a large gray area in which either term could be used, and, obviously, my distinction does not capture all of the ways in which metaphor and analogy have been seen to differ. Thus, metaphors may emphasize the surface features of a comparison, while analogies tend to center on deep structural relations (Gibbs, 2008, contains chapters that address the nature of metaphor in a variety of ways).

By contrast to the usual approach, I use *analogy* to include only *explicit* comparisons between a source and target, in which the mapping from source to target is deliberately identified and discussed. Clement (2008), summarizing his extensive research on model construction in science and science education, has argued for a multiplicity of kinds of analogy used in scientific reasoning, ranging from simple exemplars (e.g., various exemplars of acceleration in differing problems given to students) to analogies as "proto-models" (e.g., billiard balls as a starting point for a formal thermodynamic model). Explicit analogies in all of these senses are prevalent in Maxwell's work and have been the subject of much prior scholarship (see, for example, Hesse, 1973; and the studies of Nersessian, cited above). They are important aspects of a full understanding, but I am concentrating here upon the metaphorical underpinnings of the analogies; the *implicit* bases which Maxwell takes for granted.[5]

My analysis of the metaphors underlying Maxwell's approach is derived from the approach of Lakoff and Núñez (2000; see also Núñez, 2000; Núñez & Lakoff, 2005, for a brief outline), who argue that even the most "abstract" of mathematical formulations are grounded in basic cognitive embodiments via the use of metaphor. They begin with the basic arithmetical operations and expressions; for example, they relate basic arithmetic to the elementary cognitive operations of collecting objects, arguing that "Object collection" as source is mapped onto "Arithmetic" as target. Thus, "Collections of objects of the same size" are mapped onto "Numbers," "Putting collections together" onto "Addition," and "Taking a smaller collection from a larger one" onto "Subtraction" (Lakoff & Núñez, 2000, p. 55). In this way, numbers become "grounded" via the metaphors of object collection, and arithmetic itself can then become the source for further extensions to new target domains. "Grounding metaphors" in their analysis are linked directly to sensorimotor experience (as in the examples given), and these are extended via other conceptual metaphors and conceptual blends; "The more indirect the grounding in experience, the more 'abstract' the mathematics is" (p. 102).

[5] Of course, the mathematics used in physics is not always implicit, since explicit mathematical arguments do occur, even in Maxwell. The focus here, though, is what Maxwell takes for granted on the part of his reader, and it is this that is needed to penetrate the otherwise arcane equations.

Lakoff and Núñez used two separable kinds of mappings; first, a mapping between a source domain and a target domain (linking, for example, numbers as source to functions as target, p. 386), then, second, "blends" to extend the scope of two or more domains (compare Turner, 2001). For example, circles in the Euclidean plane (source) can be mapped to the Cartesian plane (target), forming the "Unit Circle Blend" (p. 388). Such blends can participate in new mappings to function as new metaphors. For example, the Unit Circle Blend can be mapped onto the trigonometric functions, to create what they call the "Trigonometry Metaphor" (p. 393).

The use of grounding metaphors and conceptual blending culminates in what they call an "idea analysis" of mathematics. Working from basic arithmetic, through algebra and geometry, to the "Basic Metaphor of Infinity" (which becomes the grounding metaphor for irrational numbers, limit concepts, differentiation, and much more), Lakoff and Núñez conclude their book by explicating Euler's formula:

$$e^{i\pi} = -1.$$

Rather than offering a proof, which Leonhard Euler first did in the 18th century, Lakoff and Núñez (2000, Part 6) develop the *meaning* of the formula by considering its separate components in terms of the underlying metaphors and blends. Thus, beginning with trigonometry, itself grounded on analytic geometry, the "Unit Circle Blend," and "The Recurrence Is Circularity" metaphor, they proceed by developing the meaning of the irrational number e and what it means to raise it to a power. The imaginary number i is developed as a rotation in the complex plane (with real numbers on the x-axis and complex numbers on the y-axis). If we put these together with other blends, too numerous to review here, the resulting idea analysis, which they call "Euler's Blend," incorporates a series of blends involving power series, Taylor's series, and the exponential complex plane, together with more elementary operations, to produce an account of the meaning of Euler's formula. The approach is very general; for example, Núñez (2005) has extended the analysis, showing its applicability to the historical case of the development of transfinite cardinal numbers by Georg Cantor.

Lakoff and Núñez (2000) were concerned only with the analysis of mathematics as such; they did not consider the way mathematics is used in physical science as representations of physical processes. In the present chapter, I sketch such an application. Some of the metaphors and blends used by Lakoff and Núñez will be used, but others will be developed (in particular, they did not discuss vector mathematics, which are central to my account). In the remainder of the chapter, I first discuss the use of representational mathematics in physics, then sketch a metaphorical analysis of the mathematical representation of fields via vector calculus, showing how the concepts of scalar and vector fields can be grounded in basic physical concepts and intuitive conceptions of motion and force.[6] I then apply the analysis to the case of

[6] Krieger (1992) presented an analysis with some similarity to the one given here. He uses social metaphors (division of labor and the factory, clockworks and calculus, kinship relations, and so on) to describe central concepts, particles and fields especially. Still, his focus was on communicating the concepts of physics, rather than their mathematical representation, and his metaphors were explanatory, rather than foundational.

Maxwell (first sketching the historical background for Maxwell's "Faraday-like" use of mathematics in physics). The result is a cognitive account of Maxwell's representations in terms of their meaning as physical representations.

REPRESENTATIONAL MATHEMATICS

Many of the representational uses of mathematics in science involve a kind of shorthand for visual representation. Thus, the equation for the distance covered by a falling body, $s = \frac{1}{2} gt^2$, where *s* is distance, *t* is time, and *g* is the gravitational acceleration, can be plotted on Cartesian coordinates as a parabola. Mathematical expressions can be used in this way to represent changing phenomena; for example, a point moving along a parabola can be seen as a kind of "movie" of the kinetics of a falling object. Calculus extends this even further, allowing representations of an entire family of possible "movies," each of which can be instantiated when appropriate boundary conditions are specified.

Still, translating an equation into a visual representation does not capture all of the representational uses of mathematics in physics. Consider the following anecdote told by Jacob Bronowski about John von Neumann:

> When I worked with him during the war, we once faced a problem together, and he said to me at once, "Oh no, no, you are not seeing it. Your kind of visualizing mind is not right for seeing this. Think of it abstractly. What is happening on this photograph of an explosion is that the first differential coefficient vanishes identically, and that is why what becomes visible is the trace of the second differential coefficient." (Bronowski, 1973, pp. 433–434)

Two things stand out in von Neumann's statement: first, that the relevant "seeing" is "not right for a visualizing mind" and, second, that one must see phenomena "abstractly." Visualization aside, however, it is interesting to note that differentials *can* capture nonvisual phenomena that *can* in some cases be directly experienced. For example, a car speedometer indicating a steady 60 miles per hour captures a first derivative (velocity or speed is the first derivative of distance covered with respect to time). If the car speeds up, the twisting motion of the speedometer captures the acceleration (the second derivative of the distance). A person in the car cannot "feel" a constant speed but can feel an acceleration as a push backward. This is detection of the second derivative—and, to extend the case, a sudden "jerk," say, if the car is hit from behind, is a felt experience of the third derivative.

For the case at hand, Maxwell's theory of the electromagnetic field, the mathematics is based on similar directly experienced phenomena, sometimes visual and sometimes (like the speedometer case) nonvisual and closely allied to muscular and kinesthetic sensations. As I argue below, "Maxwell's Equations" (the set of four vector differential equations that are now taken to summarize his theory, Figure 17.8) capture more than simply the visual representation of electrodynamics. They also capture important nonvisual properties of the field—forces, tensions, compressions, and energy relations. Cognitively, these have an important sensorimotoric aspect that renders them not "abstract"; in fact, they are closely tied to the embodied representations of physics knowledge. Further, Maxwell's equations are a compact and efficient representation of the properties of an electromagnetic *field*. In this sense, they capture the properties of an entire dynamic system of relationships; to learn physics "the Maxwell way" is to learn to think from *wholes* to *parts*, rather than the reverse (Simpson, 2005, 2010). The cognitive implications of such radical conceptual change are an important focus of the present project.

Maxwell's equations are notoriously difficult, even for advanced students (Dyson, 1999). More generally, as Gingras (2001) noted, the mathematization of physics after the 17th century served to exclude actors who otherwise might have been central figures. Similarly, the exclusion of a "math-averse" public from an understanding of physics generally is doubtless one reason why Maxwell, while well known to physicists, is relatively unappreciated by the general public. By using the extensive groundings of his ideas in perceptual and motoric intuitions, and his lucid use of analogy, Maxwell's main ideas can be presented and the general appreciation of mathematical physics can be enhanced.

REPRESENTING FIELDS WITH MATHEMATICS

The notion of "electromagnetic field" has an intuitive character in popular writings about science, a kind of invisible "something" that fills space and carries such things as radio waves, cellphone messages, and the like. Within physics, fields are central, one of the basic concepts, along with "particles," of modern understanding. To gain an appreciation of what the physical concept of field means, I begin with scalars and vectors. Scalars are magnitudes, that is, numbers representing something like temperatures or lengths. They express the size or amount of something. Vectors are *directional* quantities that express the magnitude and the direction of something. Often represented by arrows, pointing in a given direction and with length representing the magnitude, vectors can be used to represent physical quantities like forces and motions.

Now consider a thin plate of metal which is heated near the middle (Figure 17.1). It is easy to imagine a temperature at each point of the surface, and easy to imagine these temperatures varying continuously from one part of the sheet to another. The temperature of the metal constitutes a *scalar field*. Note that it is infinitely divisible and continuous; we can use the "Basic Metaphor of Infinity" to conceptualize

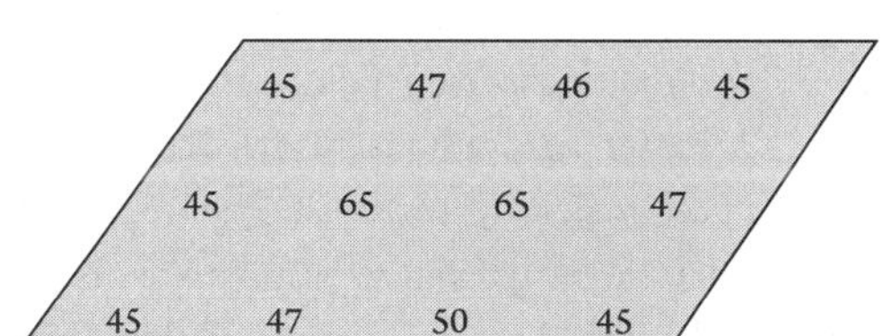

Figure 17.1 A scalar field, representing the distribution of temperature across a metal plate heated in the center.

a scalar field, without difficulty, and we can understand that an infinity of temperature scalars covers the diagram.

Next imagine a modern "wind farm," an array of windmills generating electricity. Such windmills are made to orient themselves facing the wind, which means that all of the devices in one farm are likely to be facing nearly the same way. They rotate faster as the wind speed increases, so the speed of rotation is a measure of the magnitude of the wind speed. Our imagined wind farm is a *vector field*, and can be represented as in Figure 17.2 by an array of arrows.

Again, we have to imagine an infinity, in this case an infinite number of arrows, but note that the prevailing direction at the top and bottom of the field is from the northwest to the southeast, but that there is a high-speed gust blowing through the center from west to east.

As informal geometric conceptions, scalar and vector fields are easy enough. How do we operate on them as mathematical entities? For the scalar field, this is easy. Thus, if told that the diagram was wrong, and that the temperatures in the diagram were exactly half of what they should be, it is an easy matter to simply multiply each of the numbers by 2. Our grounding metaphors for such numerical operations are well established and need no further argument to be clear; we can map operations on single numbers onto an array of numbers. But learning how to operate on vectors, let alone vector fields, requires learning some new metaphors.

As an example, we can define addition and subtraction of vectors, but now the "object collection" metaphor of normal addition is no longer adequate. To add two vectors, **A** and **B**, we, in effect, place them head-to-tail, as shown in Figure 17.3.

If **A** and **B** represent two forces on an object (say one is gravity and the other is the force from a rocket engine), then **A** + **B** represents the resultant of the two forces, the direction and the magnitude of the force when the two components are acting together. Note that this is not like ordinary addition; you cannot add the magnitudes (that is, the scalar part) of the two vectors and get the magnitude of their sum. Instead, you get the vector sum by placing the two vectors head-to-tail, as shown in the figure—the vector sum depends on the angle between the vectors. Still, vector addition obeys the properties of ordinary (scalar) addition; we can define a unit vector (corresponding to the scalar unit, the number 1); the operation is commutative and associative; and so on. Note that if the two vectors both point in the same direction, the vector sum reduces to the ordinary arithmetical sum.

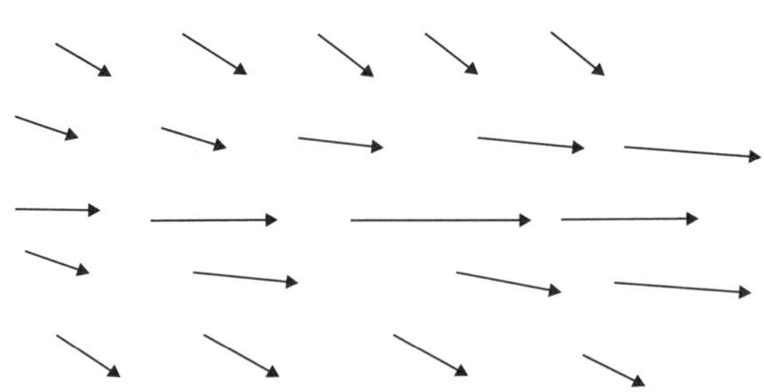

Figure 17.2 A vector field, representing a flow from northwest to southeast, but with a strong gust from the west in the center of the field.

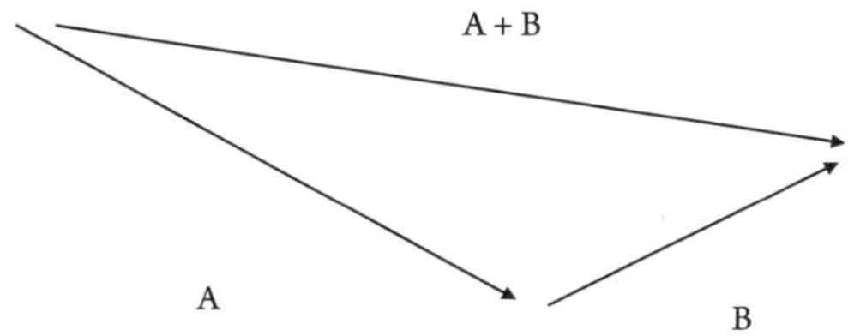

Figure 17.3 The addition of two vectors, A and B.

When it comes to multiplication, things are more complex; there are *three* kinds of vector multiplication. First, we can multiply a vector by a scalar, which simply involves multiplying (in the usual way) the magnitude of the vector by the scalar. This is equivalent to stretching the vector. Two vectors can be multiplied in one of two ways. One, the "dot product," gives a scalar as the result and depends on the angle between the vectors, and the other, the "cross product," gives a vector at right angles to the plane of the two vectors. We need not follow the mathematics of these to make a general point, however; mathematical operations on vectors can be defined; they can be rooted in basic metaphors; and these can be used to represent physical aspects of things like forces and velocities, which are directional, as well as possessing a specific size or magnitude.

Like scalars, vectors can be seen as changing quantities, and there is a differential calculus of vector change, just as there is of scalar change. Thus, consider a simple algebraic equation from elementary physics, $s = 1/2gt^2$, where s is distance fallen, t is time, and g is the gravitational acceleration at the earth's surface. In elementary calculus, we learn that the derivative of this function is just gt and that this represents the rate of change of the function. The first function, $1/2gt^2$, gives the distance traveled by a falling body from rest, and the derivative, gt, gives the rate at which the distance is changing (that is, distance per unit of time), in other words, the velocity of the object. The distance increases as the square of the time, while the velocity increases linearly with time.

Finding the rates of change of vector functions is more complex even than defining the operations of addition and multiplication, and I will not develop the details here. However, we can get a rough sense of how it works by considering, first, one kind of vector rate of change, a concept that can be applied to an entire scalar field.

Consider a function that is probably familiar to most psychologists (even given that they use it only indirectly via tables of its values). The function for a bivariate normal distribution, assuming no correlation between the variables x and y, with means of 0 for both and standard deviations of 1 for both, is the following

$$z = \frac{1}{2Pi} e^{-x^2 - y^2}.$$

The equation gives the probability density, z, and has the familiar bell shape shown in Figure 17.4.

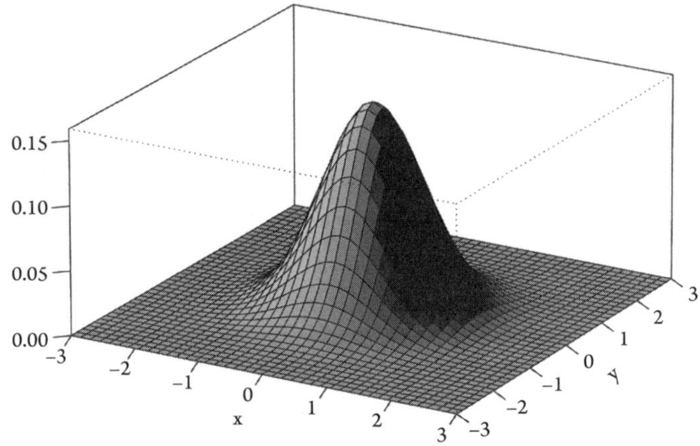

Figure 17.4 A bivariate normal distribution, with no correlation between x and y. The mean is 0 and the standard deviation is 1 for both variables.

This kind of visualization is familiar to most psychologists; we have the grounding metaphors and conceptual blends needed to see the relation between the equation, the tabled values of the bivariate normal distribution, and the three-dimensional figure. It is also easy to imagine this as represented by a scalar field, like the heated metal sheet. We can imagine a plane figure with higher numbers at the center and smaller numbers, approaching zero near the edges. We can also imagine this as a representation of a physical hill, and that can help with conceptualizing the rate of change of the function. At the bottom of the hill or at the very top, there is almost no change—we could place a marble at the bottom or (very carefully!) at the top and it wouldn't roll. But anywhere else, and the marble will roll downhill. In fact, the marble will pick its own direction; it will always roll downhill in the direction of the fastest rate of change (assuming a smooth hill, of course). Further, depending upon where it starts rolling, it will have a greater or a lesser force acting on it; zero at the top or bottom, highest somewhere in the middle. In effect, we could represent the hill by a vector field, in which vectors always point in the direction of greatest change (the direction the marble will take) and with magnitudes corresponding to the resultant force on the marble at each point.

Such a vector field is actually not hard to compute and is known as the *gradient* in vector analysis. For the function shown, we can compute it using partial derivatives, and a graphical representation (looking down on the hill, from above) looks as shown in Figure 17.5.

Note that the vectors point away from the top of the hill, centered on the top, and that their size changes in accord with the steepness of the hill at each point, vanishingly small at the top, increasing to a maximum in the middle, decreasing as the bottom is approached.

In the above remarks, I've asked the reader to use metaphors (wind, temperature, marbles rolling down a hill, etc.) to conceptualize the mathematical operations involving fields. In effect, the reader has participated in something like an idea

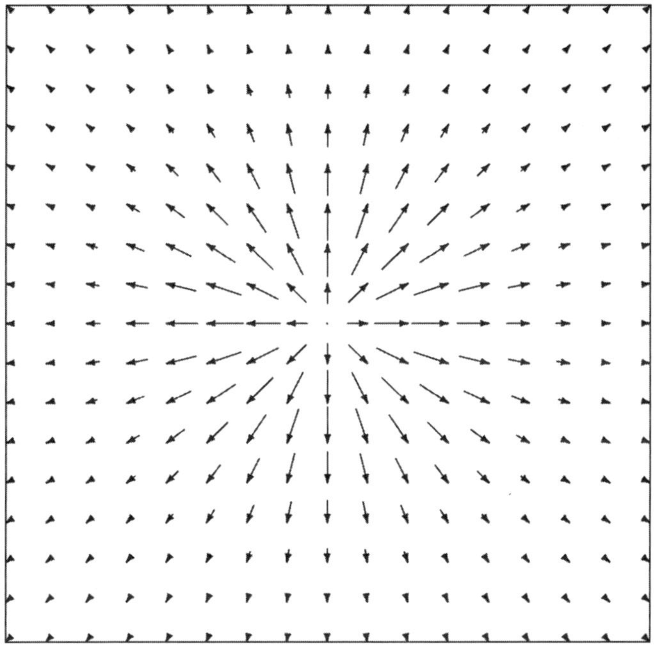

Figure 17.5 A vector field formed from the function shown in Figure 17.4 by taking the *gradient* of the function at each point. Each vector shows the direction and magnitude of steepest descent at the point represented by the tail of the vector.

analysis of the meaning of the mathematics. The basic metaphors are summarized in Table 17.1.

The gradient operator develops a vector field from a scalar field. There are two additional differential operators that operate on vector fields (they are especially central in a Maxwellian representation of the electromagnetic field). The *divergence* of a vector **A** (written as div **A** or by using the symbol ∇, read "del" as in $\nabla \bullet$ **A**, "del

Table 17.1 Basic metaphors for vectors and scalars

Source domain *Physical processes & basic math*	*Target domain* *Mathematical operators*
Velocity, force, etc.	Vectors
Numbers on a line	Numbers on a plane (scalar field)
Wind farms	Vectors on a surface (vector fields)
Rate of change (single variable)	Derivative
Rate of change (two variables)	Partial derivative
Release of marble on a hill	Gradient as steepest descent (a vector)
Steepest descent	Gradient field of vectors

dot **A**") gives a scalar at each point of the vector field. The curl (written as curl **A**, **or as** $\nabla \times \mathbf{A}$, read "del cross **A**") gives a vector at each point. To understand these, I will use a fluid flow metaphor (Skilling, 1942)[7]. The scalar result of the divergence operator is related to the amount of flow passing through the field, and the vector result of the curl operator gives a measure of the direction and magnitude of the "rotational tendency" of the flow at a point. Consider a faucet running into a sink and represent the water flow as a vector field. If the sink drain is open, there is as much water flowing out as is flowing in. In this case, the divergence is zero. If we put a stopper in the sink, more water flows in than out, and the divergence is nonzero. As with the gradient, the actual operator is a partial differential equation, and we skip the details here. The divergence operator is defined for every infinitesimal volume of the field as a whole, giving a scalar result, a net "flow" in and out, for each volume.

The curl operator is harder to intuit. Consider a stream of water flowing with equal velocity in a straight channel. If we immersed a small paddlewheel in the flow (holding the axle just above the flow), there would be no tendency for the wheel to rotate—as much pressure would be on the left of the wheel as on the right. In this case, the curl would be zero. Now suppose the stream flows more slowly near the banks than near the center (say, because of frictional forces). Now the wheel would turn, because there would be less pressure on the side nearest the bank. In this case, the curl would be nonzero, and would be a vector showing the direction of greatest pressure. (Once again we are actually working with a partial differential equation and skipping the details.) Notice that we can have a rotational tendency in a vector field, even when (as in the example) all of the flow is linear; a nonzero curl does not necessarily mean that the flow itself is curved. Notice also that the divergence and the curl can express the nonvisual "feel" of a vector field. For example, imagine holding your hand in the streams of water; different pressures might be exerted on different parts of the hand if there is nonzero curl.

At this point, we have a simplified metaphorical analysis of the conceptual analysis underlying the curl and divergence operators, shown in Table 17.2. The mappings were only partially developed, of course, since the actual mathematics used to develop the concepts has not been presented.

Table 17.2 Basic metaphors for the divergence and curl operators

Source domain *Physical processes*	*Target domain* *Mathematical operators*
Flow into and out of a field	Divergence (a scalar)
"Rotational" tendency of a field	Curl (a vector)

[7] My discussion of the grad, curl, and div as flow metaphors is derived from Skilling's book. Those interested in learning about Maxwell's equations in a modern context will find this an accessible text that assumes only elementary calculus as background.

Equipped with some metaphoric understanding of the relevant mathematics, we are now in a position to see the relation to Maxwell's retranslation of Faraday's field theory. As did Maxwell, the above account relies heavily upon sensorimotor intuitions that are already present and can be taken without further detailing of the relation between source and target: the reason why I've preferred the term "metaphor" to the term "analogy" for the mappings between source and target. In particular, metaphors are *implicit*, while analogies are *explicit*. Of course, Maxwell and his audience had a much more detailed knowledge of the mathematics involved—my analysis is merely a sketch, but it will enable an overview of some of the basic processes. To begin, some background in the history of the science involved is necessary.

BACKGROUND

By 1820, a great deal was known about electricity and magnetism, but there was no known relationship between the two. In that year, the Danish scientist Hans Christian Oersted discovered that an electric current produced by a voltaic battery had an associated magnetic influence. Thus, passing a current through a wire made a magnetized needle near the wire "jump around." It was soon found that passing a current through a coil of wire made the coil act as if it were a common magnet with opposing poles, an "electromagnet." The discovery inspired a search by many scientists for the reciprocal relationship; if electricity could produce magnetism, could the reverse effect be found as well?

The question was answered in 1831 by Michael Faraday (1791–1867). He discovered that an electric current could be generated by the magnetic influences of a nearby current-carrying coil when it was switched on or off (but not when the current was steadily on in the first coil). A moving bar magnet (but not a stationary one) could also induce a current in a nearby coil. The discovery was the beginning of a long series of experiments by Faraday, in the course of which he developed the first true field theory in the history of modern physics (Einstein & Infeld, 1938).

Across a series of thousands of experiments, published primarily in 30 articles in the *Philosophical Transactions of the Royal Society* (29 of which are collected in the three volumes of his *Experimental Researches in Electricity*, 1839–1855), Faraday gradually developed his theory, culminating in an account of the "lines of force" that he took to be the ultimate physical reality, even, he supposed, more fundamental than matter itself (Gooding, 1981; Tweney, 1992a; see also Fisher, 2001, for an accessible introduction to the text of Faraday's *Experimental Researches*). Along the way, he made many discoveries—electromagnetic induction, the distinction between paramagnetic and diamagnetic substances, the optical rotation of polarized light by a magnetic field, and others. The discoveries posed a challenge to theorists, yet Faraday's own theoretical accounts failed to satisfy his contemporaries (except, of course, Maxwell).

If you sprinkle iron filings on a sheet of paper laid over a bar magnet, the filings will arrange themselves in characteristic patterns (Figure 17.6, left).

From an early date, Faraday conceptualized the magnetic field of a magnet in terms of invisible lines of force (Figure 17.6, right). These were initially taken

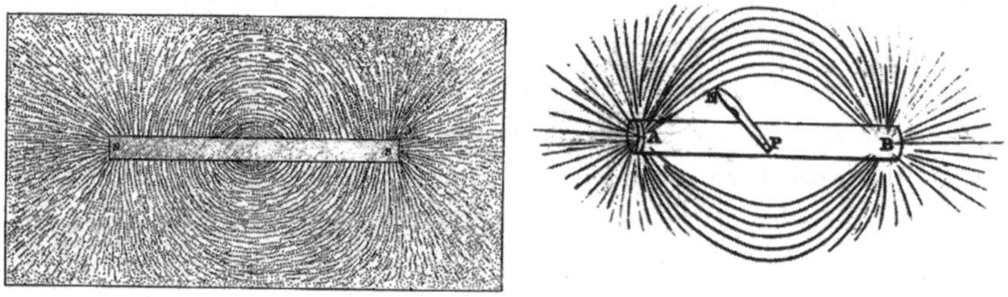

Figure 17.6 [Left] The pattern of iron filings around a bar magnet (Faraday, 1852a, Plate 1). [Right] Faraday's depiction of the lines of force around a bar magnet. The knife blade, which is right-handed, was used to describe the directionality of the lines of force (Faraday, 1832, Plate 1).

merely as a heuristic device to represent the force on a particle (such as an iron filing), but Faraday went further than this. Across two decades of research, he sought to show that the lines of force were in fact real entities—invisible and nonmaterial but real. In the case of magnets, the lines were closed curves (i.e., they reentered the magnet and were present even within the substance of the magnet itself), they filled all of space, and they were possessed of determinate physical properties. Electric lines of force were similar, streaming outward from a negative electric charge and inward toward a positive charge. Unlike magnetic lines, however, electric lines always began and ended at a charge (or a neutral ground) (Figure 17.7).

James Clerk Maxwell (1831–1879) claimed repeatedly that his own work was founded upon that of Faraday, that he was "following the method of Faraday," and that the approach taken by Faraday, while not mathematical in the usual sense (there are no formal equations in any of Faraday's works), was nonetheless "intuitively mathematical" (Maxwell, 1873a; see also Gooding, 1992; James 1998; Tweney, 2009). For Maxwell, his equations of electrodynamics (Figure 17.8, in modern form) reflected an underlying theory that was closely similar to Faraday's. There were some conceptual differences: Whereas Faraday argued for physically real lines of force (Faraday, 1852b), a view that does not require an ether, Maxwell posited an ether filling all space and matter. Faraday's lines of force became, for Maxwell, properties of the energy relationships in the ether, akin to the stresses and strains within

Figure 17.7 Faraday's depiction of the lines of electric force around two test charges, positive on the left and negative on the right. Note the use of arrows to depict direction, although magnitude is not explicitly represented here (Faraday, *Diary* entry for August 4, 1836, in Martin, 1933, Vol. 3, p. 84).

an ordinary solid. The representation developed by Maxwell thus synthesized the visual and the nonvisual aspects of Faraday's representations.

Faraday (1846b) had speculated that light was a transversal vibration in the lines of force. That there was an intimate relation between magnetism, electricity, and light had been suggested by Faraday's experimental discovery that the plane of polarization of a light beam was rotated when a magnetic field was present in a dense optical medium (Faraday, 1846a). Maxwell (1864, p. 580) extended Faraday's speculation by showing that the equations of the field implied that the velocity of propagation of an electromagnetic beam would be identical to that of the velocity of light: "The agreement of the results [for the calculated and experimentally determined velocity of light] seems to show that light and magnetism are affections of the same substance, and that light is an electromagnetic disturbance propagated through the field according to electromagnetic laws." This leads of course to a prediction, one famously confirmed by Heinrich Hertz in 1888. He directly produced electromagnetic waves using a spark gap and detected an induced current in a coil placed some meters away—the first production and detection of what became known as radio waves. Note that this prediction represents a use of the hypothetico-deductive method. Maxwell himself did not frame this as a prediction, however, and instead devoted experimental efforts to refining the measurement of electric and magnetic properties, seeking a closer match between the estimated speed of light and the measured speed. Maxwell's thinking here is an instance of pattern-finding, rather than the logical prediction of the consequences of a hypothesis.

The transition from Faraday's experimentally based theory of the lines of force to Maxwell's mathematically expressed version offers a rich set of exemplars for a cognitive understanding of the use of mathematics as a representational medium. It is, as noted, a transition of immense importance for the history of modern physics. Secondly, and more to the point of the present chapter, because it was a novel take on the theoretical understanding of electricity and magnetism, Maxwell was very explicit about what he was doing and why he was doing it. Thus, his many statements about the way in which mathematics should be used in physics offer themselves as important cognitive clues to the wider issue of how it is that mathematics attains its "unreasonable effectiveness."

Maxwell characterized Faraday as possessing a "dynamic geometry," one which led to the unique way in which Faraday integrated experimental results and the formal development of field theory. In his extensions of Faraday, Maxwell insisted repeatedly that the proper use of mathematics in physics was *representational*. For example, speaking of his use of Lagrangian formulations of dynamics, he stated, "We . . . avail ourselves of the labours of the mathematicians, and retranslate their results from the language of the calculus into the language of dynamics, so that our words may call up the mental image, not of some algebraical process, but of some property of moving bodies" (Maxwell 1873/1891, Vol. 2, p. 200).

MAXWELL'S "FARADIC" THEORY

To see how mathematical expressions can come to represent physical processes in electricity and magnetism, consider the modern formulation of Maxwell's field theory, as represented by the four equations shown in Figure 17.8.[8]

In what sense do these represent the electromagnetic field theory that Maxwell inherited from Faraday? Note first that all four are actually differential equations involving vectors. Three vectors are involved, **B**, the magnetic field intensity at a point, **E**, the electric field intensity at a point, and **J**, the current strength at a point. The left hand side of each equation gives either the divergence or the curl of a vector quantity (**B** or **E**), using the standard "del" notation, in which $\nabla \bullet \mathbf{B}$ is the divergence and $\nabla \times \mathbf{B}$ the curl of a vector field. The del operators expand into partial differential equations. Thus, the first equation (when expanded) gives the divergence of a magnetic field at a point,

$$\nabla \bullet \mathbf{B} = 0.$$

In words, the divergence of a magnetic field at any point is always zero; as much magnetic flux is leaving the point as is entering. In Faraday's terms, this is equivalent to saying that as many field lines of force are entering a given area as are leaving it. Put another way, magnetic field lines are always closed curves.

The same is not true for the electric field, as the second equation indicates:

$$\nabla \bullet \mathbf{E} = \rho / \varepsilon_0 .$$

In words, there is net flow out of a volume that depends upon the charge within, ρ, and a constant, ε_0, the "permittivity of space." In Faraday's terms, lines of electric force flow out of an area containing an electric charge.

The remaining two equations capture the relation between the electric and magnetic fields. The third expresses the curl of an electric field associated with a changing magnetic field:

$$\nabla \times \mathbf{E} = -\partial \mathbf{B} / \partial t .$$

In words, the curl of the electric field (its "rotational" character at a point) is a function of the rate of change of the magnetic field, **B**, with respect to time, *t* (the expression on the right is the partial derivative of magnetic field intensity with respect to time; it is a *partial* derivative because the field may also be changing in space, i.e., from place to place). This is equivalent to Faraday's 1831 discovery that an electric field is created by a changing magnetic field, but not a stationary one, and to his

[8] "Maxwell's Equations" in their modern form (as in Figure 17.8) do not occur in Maxwell, the nearest approach being his summary in Part 4 of the *Treatise* (1873/1891, Vol. 2, pp. 258–259). The modern form is due primarily to Oliver Heaviside (Hunt, 1991). The differences are more than merely notational, however. For Maxwell, as noted, the field is characterized in part by stresses and strains in an ether, aspects not included in the modern interpretation of his equations.

$$\nabla \bullet \mathbf{B} = 0$$

$$\nabla \bullet \mathbf{E} = \frac{\rho}{\varepsilon_0}$$

$$\nabla \times \mathbf{E} = \frac{\partial \mathbf{B}}{\partial t}$$

$$\nabla \times \mathbf{B} = \mu_0 \mathbf{J} + \mu_0 \varepsilon_0 \frac{\partial \mathbf{E}}{\partial t}$$

Figure 17.8 Maxwell's Equations in modern form (see text for explanation).

claim that one must cut magnetic lines of force to induce a current or an electric field.

Finally, the fourth equation represents the curl of the magnetic field associated with either a current or a changing electric field (or both):

$$\nabla \times \mathbf{B} = \mu_0 \mathbf{J} + \mu_0 \varepsilon_0 \, {}^{\partial \mathbf{E}}\!/_{\partial t} \quad .$$

Here, the curl of the magnetic field is a function of a nearby current (given by the first term) and/or a changing electric field (the second term), each multiplied by suitable constants (μ_0 is the "magnetic susceptibility" of space). This equation represents Oersted's 1820 discovery that a magnetic field can be created by a current.

In each case, note that I have mapped particular properties of the electric or magnetic field as source onto a target mathematical expression. The mappings are summarized in Table 17.3.

Up to this point, I have sketched an analysis of Maxwell's equations using a modern formulation of the four equations, and relating each to a set of underlying metaphors. It is only a sketch, because the full details (the partial differential equations and such) have not been developed—it is what someone learns in a second- or third-year undergraduate physics course. In the next section, I elaborate on some of the details by which Maxwell led his readers to his formulations.

Table 17.3 Conceptual blends for fields

Source domain *Physical processes*	*Target domain* *Mathematical operators*
Flow into and out of a magnetic field	Divergence: $\nabla \bullet \mathbf{B} = 0$
Flow into and out of an electric field	Divergence: $\nabla \bullet \mathbf{E} = {}^{\rho}\!/_{\varepsilon_0}$
"Rotational" tendency of an electric field	Curl: $\nabla \times \mathbf{E} = -{}^{\partial \mathbf{B}}\!/_{\partial t}$
"Rotational" tendency of a magnetic field	Curl: $\nabla \times \mathbf{B} = \mu_0 \mathbf{J} + \mu_0 \varepsilon_0 \, {}^{\partial \mathbf{E}}\!/_{\partial t}$

This also will be a sketch, since a full analysis would take many pages, but it will illustrate the main points that are open to cognitive understanding.

WHAT MAXWELL DID

Maxwell developed his theory of the electromagnetic field, initially in three papers and then in his 1873 *Treatise on Electricity and Magnetism*. These four works can be summarized as follows:

1. "On Faraday's Lines of Force" (1855–1856). Presented a hydrodynamic model of the field, treating lines of force by analogy to thin *tubes* of force carrying a moving fluid. Applied to field strength, potential difference, electric current, but not to electrodynamics.
2. "On Physical Lines of Force" (1861–1862). Developed a "vortex" model of the magnetic field, with electric current as movement of idler wheels between rotating magnetic vortices. Applied to (e.g.) electromagnetic induction. First derivation of the speed of light.
3. "A Dynamical Theory of the Electromagnetic Field" (1864). Elaborated a mathematical model using 20 equations (very roughly corresponding to the modern "Maxwell's Equations," above). Based on abstracted mechanical phenomena and energy considerations.
4. The *Treatise* (1873b). Comprehensive theory of electricity, magnetism, and the electromagnetic field, based heavily on Lagrangian formulations of energy and introducing novel vector-analytic expressions (Crowe, 1967) for forces and fluxes. Also includes extensive treatment of experimental procedures, measurement issues, and extensive discussions of heuristics and cognitive concerns, as well as the overall methods of mathematical physics.

As the above makes clear, analogy played an important—and complex—role in Maxwell's thought.[9] He began his study of electricity and magnetism by first centering on Faraday's account of the lines of force. In "Faraday's Lines of Force," Maxwell used a mechanical representation of the field as an incompressible fluid flowing through thin tubes. In this case, the lines of force (recast by Maxwell as "tubes of force") were analogized to the flow lines of the fluid. The 1861–1862 paper, "Physical Lines of Force," used a mechanical analogy, one in which "molecular vortices" were analogous to the lines of force. Maxwell's mechanical analogy included a set of imagined "idler wheels" rather like ball bearings, which rotated between the vortices and represented electric current. But this paper also pushed the theory further in that the vortices were offered as having *explanatory* power. That is, following

[9] Daniel Siegel (1991) has traced the development of Maxwell's physical concepts across the three papers, paying special attention to the rhetorical character of the individual works. A more "cognitive" account (and one more accessible, to those without the mathematics and physics) is given by Nersessian (1984, 2008). See also Simpson (2005), who outlines the rhetorical tropes of Maxwell's approach. Lambert (2011) has recently described the historical context for Maxwell's use of analogy.

William Thomson (later Lord Kelvin),[10] Maxwell was beginning to assert the reality of the analogy as representing, in effect, what was "really there." On this account, the vortices were actual entities within a material ether, and Maxwell even tried (unsuccessfully) to demonstrate the reality of the vortices experimentally. In the 1864 paper "Dynamical Theory of the Electromagnetic Field," he dropped the idea that the vortices were *actual* entities and instead conceptualized the field in terms of the dynamical properties of a material ether—stresses and strains that distributed kinetic and potential energy throughout the field.[11]

Maxwell's final version of the theory was developed in his *Treatise* (1873/1891, 1873b). The overall theory is similar to that in the 1864 paper (Siegel, 1991), although the argument is arranged very differently, and the book (unlike the three papers just discussed) covers a variety of topics, not necessarily in what might seem the most logical order. Still, there is an underlying unity, a *rhetoric* (Siegel, 1991; Simpson, 2005, 2010), one in which advocacy of the field theory is a central concern. Toward this end, Maxwell's *Treatise* covers much ground that seems, at first sight, to be surprising. Long sections cover the way in which mathematics ought to be used, the role of analogy in science, the nature of the visualizations used to picture the field. It is, to be sure, a difficult work, but for cognitive scientists it is filled with interesting material, precisely because Maxwell elaborates not just his theory of electricity and magnetism, but an outline of his novel approach to the use of mathematics in physics. In effect, as Simpson notes (2005, p. xvii), Maxwell "appeared to be expressing a new way of looking at the natural world; and to this end, style and literary concerns would be central concerns to him" (see also Darrigol, 2000, p. 166). Maxwell argued convincingly and at length that mathematics can be used in a way that reflects the physical theory that is being represented.

Particular emphasis is given in the *Treatise* to Lagrangian formulations (Harman, 1998; Maxwell, 1873/1891, Vol. 2, pp. 199–210; Simpson, 2010). Lagrange had developed an approach to mechanics using an energy equation, $L = T - U$, where T and U are the kinetic energy and the potential energy, respectively. The Lagrangian, L, is defined at every point in the space of a mechanical system. This formulation was originally developed by Lagrange because, for some problems, it is calculationally more convenient. Maxwell had a different aim, however: The Lagrangian formulation implied a different physics than the Newtonian one, a physics in which

[10] Smith and Wise (1989, esp. pp. 453–463 and 469) describe the differences between Thomson and Maxwell on the role of such analogies. Thomson, in the end, did not accept Maxwell's final version of the electromagnetic theory, since it included unobservable entities taken as real. For Thomson, this violated the need for direct sensory verification of hypothetical entities, as expressed in his 1884 Baltimore lectures (Thomson, 1884/1987).

[11] Buchwald (1985) argued that Maxwell's use of the term "dynamical" reflects a distancing from the mechanical model, and that, by focusing only upon the relations between kinetic and potential energy, Maxwell left open the exact nature of the mechanisms that produced those relations. Maxwell himself was quite explicit that he was doing this; see, especially, his comments in the *Treatise* (1873/1891, Vol. 2, pp. 199–210) and Maxwell (1879). In effect, he was doing something far more radical than merely "distancing" his account from a mechanical one. For Maxwell, *all* natural systems, even mechanical ones, could be construed in terms of energy relationships, and it was exactly these that constituted the fundamental relationships, rather than the material components.

the energy relationships *at each point between the masses* were defined. It was thus a *field-like* representation and did not invoke action-at-a-distance. This change in representation is perhaps the most central of Maxwell's many re-representations, and one that had a substantial impact on physics.

Much existing literature on Maxwell centers on the exact role of analogy in his work. For Maxwell, the proper use of analogy was essential in science, and he commented frequently upon it in his papers, noting both uses and misuses. Mary Hesse (1973) argued that Maxwell's rejection of the exclusive use of hypothetico-deductive methods and of strictly abstract mathematical methods was supplanted by what he considered to be a middle way in which physical analogy "allows the mind at every step to lay hold of a clear physical conception without being committed to any theory" (Maxwell, 1855–1856, p. 156). In fact, in all three papers and in the *Treatise* there are multiple methods of argument, including deductive and purely mathematical arguments, as well as a variety of types of analogy (see also Cat, 2001).

But what of the metaphorical underpinnings of his approach? In making the distinction between metaphor and analogy at the beginning of this chapter, I noted that the term "metaphor" is more appropriate than the term "analogy" for those instances in which the metaphor will be taken for granted. Thus Maxwell, in the first of the four works listed above, has to argue for the value of treating lines of force as tubes carrying an incompressible fluid, but he does not need to argue for the representation of the tubes in terms of mathematically expressed dynamical relations derived from theories of fluid flow. Instead, he assumes this knowledge on the part of his reader. His approach is more clear in the fourth work, the *Treatise* itself. In this work, he is acting as an advocate for a particular, Faraday-like, view of electric and magnetic fields. He develops some of the metaphorical and conceptual blends that he uses—in fact, the first chapter of the *Treatise* is devoted to these preliminaries. Among other things, he there discusses the vector operations of gradient, divergence (which he called "convergence," with a reversal of sign), and the curl (which he called "curl" in the first edition of 1873, but changed to "rotation" for the second, 1881, edition).

Today, it is a commonplace in nearly all texts of electromagnetism to use the curl, div, and grad language to discuss the nature of the field.[12] Ironically, although these concepts have their origin in Maxwell, he used the terms themselves only occasionally within the *Treatise*, preferring, instead, to write out the partial differential equations whenever needed. Still, it is clear that he was relying upon the metaphorical groundings of these equations. He did not rely exclusively upon such groundings; there are formal ("abstract") derivations in the text, along with extensive discussion of instruments, measurement techniques, and experimental tests of his claims. But it is not possible to see the structure of his arguments without a grasp of the metaphorical underpinnings along lines similar to those I have sketched above.

[12] Still, there are major differences in the modern conception, which no longer uses the "embodied" aspect of Maxwell's theory. There are no stresses, strains, or tensions in the modern conception, since there is no medium (no ether) for their manifestation.

We are now in a position to see what exactly is similar between Faraday's scientific practices and "the method of Faraday" as used by Maxwell. Several aspects of Faraday's approach are especially relevant. First, his style of experimentation was primarily *exploratory* rather than *confirmatory* (Steinle, 2003). That is, his experimentation was not primarily hypothetico-deductive—he did not, on the whole, engage in the explicit testing of hypotheses. Instead, he used experiment in a back-and-forth manner to create the meanings of the phenomena he was observing; for him, experiments were a mode of refining successive construals to make more satisfactory concepts (Gooding, 1990). Rather than proceeding from a theory to a predicted perceptual observation, most of Faraday's work proceeds from the perceptual to the conceptual (Ippolito & Tweney, 1995; Tweney, 1992b). Maxwell, similarly, did not primarily approach the development of his mathematical expressions by derivation. Instead, he developed the explanatory concepts of electric and magnetic phenomena by successive refinement of the mathematical representations and by integrating them with experimental phenomena. Just as Faraday did not rely upon the logical derivation of predictions from hypotheses, Maxwell did not rely upon the mathematical derivation of expressions from previous expressions.

Second, Faraday was, from an early date, uncomfortable with the notion that action-at-a-distance theories (like Newtonian gravity) were correct—he felt throughout his career that "something" had to be in the space between interacting bodies. The idea fit well with his notion of a *force-centered* universe in which forces were primary and the phenomena of the universe (including matter) were manifestations of force, rather than force being a manifestation of matter (Cantor, 1991; Nersessian, 1985; Tweney, 1985, 1992a). Maxwell shared this view, even as he relied upon the notion of an ether, rather than lines of force, and on energy, rather than force, as the underlying "stuff" of the physical universe.

An interesting correspondence between Maxwell and Faraday illuminates the relation between the two men's views of the role of mathematics in physics. Maxwell wrote, "Now as far as I know you are the first person in whom the idea of bodies acting at a distance by throwing the surrounding medium into a state of constraint has arisen, as a principle to be actually believed in" (Maxwell to Faraday, November 9, 1857, in James, 2008, p. 301). Faraday's response, while acknowledging his lack of knowledge of mathematics, included the following: "When a mathematician engaged in investigating physical actions and results has arrived at his own conclusions, may they not be expressed in common language as fully, clearly, and definitely as in mathematical formulae? If so would it not be a great boon to such as I to express them so?—translating them out of their hieroglyphics, that we might also work upon them by experiment. I think it must be so, because I have always found that you could convey to me a perfectly clear idea of your conclusions... so clear in character that I can think and work from them" (Faraday to Maxwell, November 13, 1857, in James, 2008, pp. 305–306).

Several points of cognitive interest can be seen in this exchange. Note first the reference in Maxwell's letter to Faraday's view of a "medium" in a "state of constraint." This is, of course, a clear statement of the essential field-like character of both men's theories. Beyond this, however, note that the conception of the field as a state of "constraint" is sensorimotor in character. For both Maxwell and Faraday, the field was conceptualized in embodied, although not necessarily visual, terms. Second, note Faraday's claim that Maxwell could "convey to me a perfectly clear idea," that is, that in spite of the "mathematical hieroglyphics," Maxwell's approach was one that could be translated into Faraday's more qualitative, nonsymbolic, form. That such translation could be done verbally supports the claim of this chapter that the metaphoric grounding of Maxwell's physics was different than those of the "mathematician investigating physical actions," different in a fashion that resembled Faraday's and that Maxwell himself emphasized again and again in his *Treatise*. Thus, by unraveling the groundings of the mathematical approach, we can see more clearly the sense in which Maxwell was following "the method of Faraday." For both men, physics was an effort at representing the physical world in terms that could be related specifically to much simpler conceptual understandings, one shared by all scientists, indeed, all people. What differs is the specific tools used to achieve that grounding, not the overall goal of the approach and not its cognitive underpinnings.

Finally, note that Faraday is asking Maxwell for something he can "think and work upon." He is not asking for mere verbal understanding, but for one that he can "get his hands on," for an understanding that involves his entire *embodied* approach to nature (see also Cat, 2011). For Faraday the mode of doing this was experiment; for Maxwell, it included also mathematical representation. For both, hand, eye, and brain were dynamic wholes that worked in concert, just as the electromagnetic field was a space-filling whole, rather than a sum-of-parts.

DISCUSSION

The physicist Eugene Wigner (1960) once commented on "the unreasonable effectiveness of mathematics in the natural sciences." He offered no final answer as to why this was the case, concluding instead: "The miracle of the appropriateness of the language of mathematics is a wonderful gift which we neither understand nor deserve" (1960, p. 14). My analysis does not solve the puzzle as far as the ontological objects of physics or mathematics are concerned, but it does shed light on the cognitive nature of their relationship. The metaphorical grounding of mathematical expressions, and their extensions via conceptual blends (which in turn can serve as grounding metaphors for subsequent expressions), hold the key to understanding mathematical representations in physics. The way in which this works has been argued here primarily via analysis of a simplified version of representations which had their origin in Faraday's and Maxwell's writings. My argument is a "plausibility" argument, a demonstration of the possibility of applying similar analytic tools to the understanding of the fine-scale details of Maxwell's actual arguments (a project I am currently working on), and of their extension to other cases. Aspects of the argument point toward the kind of cognitive understanding that will be possible in seeking to answer the question of the "unreasonable effectiveness" of mathematics, as Wigner (1960) put it.

That effectiveness depends upon the effectiveness of mathematical expressions as *representations*. The term "representation" has a variety of meanings in cognitive science. In the domain of scientific thinking, it is obviously central; theories and hypotheses, even speculations, are meant to be representations of a reality, and the practices of science are largely devoted to the construction, testing, and refining of those representations. Of course, from a philosophical point of view, the work does not resolve the issue of whether or not "realism" (the view that scientific theories reflect an *actual* reality) is an appropriate ontological stance, but it does shed light on what practicing scientists actually do when doing science. Much has been learned about the way scientists think about representations in experimentation (e.g., Gooding, 1990; Klahr, 2000; Kulkarni & Simon, 1988; Tweney, 2006), in theory construction (e.g., Nersessian, 2008), and in theory change (e.g., Andersen, Barker, & Chen, 2006; Chi, 1992). Yet, as noted earlier, almost nothing has been said about how scientists think using *mathematical* representation. I hope to have convinced the reader that an approach through metaphors and conceptual blends has much to offer the understanding of mathematics as a representational medium in physics, just as Lakoff and Núñez (2000) have shown its value for mathematics as such.

While the approach developed here to understanding mathematical physics is complementary to those used in the understanding of mathematics, several tasks remain. First, it is necessary to integrate the metaphors and blends like those used above with accounts of the role of experiments and data, the empirical nitty-gritty of science (Tweney, Doherty, & Mynatt, 1981). A similar machinery could be used, since the grounding metaphors could be substituted by grounding experiments (for example), showing how mappings can be made from the empirical to the mathematical and vice versa. The arguments used to assemble experimental data into a theoretical account of phenomena rely on both explicit analogies and implicit groundings. Thus, while Faraday used no mathematical representations at all, his "intuitive mathematics" (as Maxwell called it) *and* his experimental thinking alike relied on blends and mappings. In this sense, his theory resembles that of Maxwell precisely because he used similar means to develop and articulate some of the same concepts, albeit with different tools. What differs between Faraday and Maxwell (specific content aside) is the *mode* of developing representations, not the processes that underlie those developments. This is not to say that the two representations are equivalent, rather, they are complementary, but it is possible for Maxwell to say that he is "following the method of Faraday," even when he fills the page with mathematics, because the path from one representation to another (i.e., from one step in an argument to another) is based on cognitively equivalent processes.

It is frequently argued that cognition must be considered with attention to its *embodiment*. In extending the metaphorical conception of mathematics developed by Lakoff and Núñez (2000), Núñez (2008) recorded the gestures used by mathematics professors in upper division courses. The gestures accompanied "enactments" of mathematical concepts; for example, a sequence of real numbers that "oscillate" around a fixed value were accompanied by a back-and-forth hand motion. According to Núñez, such gestures betray the underlying grounding metaphors of the seemingly abstract formalisms and emphasize the dynamic character of the concepts. The analysis is reminiscent of the gestures studied by Clement (2008), gestures produced by physical scientists attempting to solve a difficult "spring problem." For both mathematicians and physicists, gestures capture the underlying metaphors that ground the more abstract verbal and mathematical expressions and are, in this sense, constitutive of the dynamic thinking involving representations. We have no record of Maxwell's gestures, but it has been noted (Cat, 2001, 2011) that there is a "muscular" sense to many of Maxwell's illustrations. For example, in setting up a convention for the handedness of a rotation about an axis, the "right-hand rule," Maxwell used "the muscles of the arm when we turn the upper side of the right hand outwards, and at the same time thrust the hand forwards, [which] will impress the right-handed screw motion on the memory more firmly than any verbal definition" (Maxwell, 1873/1891, Vol. 1, p. 25n). Here again we find a "hand-eye-brain" aspect at the foundation of Maxwell's thinking.

There has been much attention to the role of diagrammatic representations in cognition, and such representations have an important link to the metaphoric argument presented here (Gooding, 2005). Thus, Qin and Simon (1995) presented undergraduates with copies of the first part of Einstein's 1905 paper on special relativity, asking them to understand the text. The analysis focused upon the diagrams sketched by subjects as they worked through the algebraic bases of the argument. These diagrams and the verbal protocols gathered at the same time were interpreted in terms of the mental models developed by the subjects and the changes in these models as they worked through the text. Qin and Simon concluded that the mental models alone were not sufficient for understanding, but that they had to be accompanied by appropriate mental images which could be manipulated by the subjects (whether in the "mind's eye" or on paper), and the images could be used to derive both the qualitative aspects of the situation and the equations that described it quantitatively. The present analysis complements and extends their point: Thinking in physics is more than simply derivation and calculation, and the diagrams used in science function, in this sense, in a way similar to the gestures recorded by Clement and by Núñez. In both cases, they betray the underlying *metaphoricity* of the representations. In this sense, understanding the role of mathematics in physics is part of the goal of understanding how gestures and diagrams are used in physics.

Finally, note that the present analysis contributes to the understanding of the role of implicit cognition in scientific thinking. By tying mathematical representation to groundings in metaphor and conceptual blends, the account lends substance to the idea that mathematical physics is not abstract, not foreign to the embodied cognition of everyday thinking and problem solving, and not removed from cognitive analysis. The case study opens new possibilities for understanding mathematical representations and suggests that cognitive science can explore and understand those chalkboard equations in our physicist friend's office.

ACKNOWLEDGMENTS

This chapter is based on two talks, one at the Purdue meeting and one at the Berkeley meeting of the International Society for the Psychology of Science and Technology. I am grateful to the audiences of both meetings for questions and discussion.

John Clement helped to clarify many of the ideas. Yanlong Sun helped with the figures. A draft version of the paper was read and extensively commented upon by the editors of this book and by Howard Fisher, Frank James, Nancy Nersessian, Kenneth Shemberg, and Thomas Simpson, and their comments have substantially improved it. The ultimate origins of the paper date back to many conversations with many people, but I must single out the late David Gooding, who started me on this line of thought, and Elke Kurz-Milcke who helped it to continue.

REFERENCES

Andersen, H., Barker, P., & Chen, X. (2006). *The cognitive structure of scientific revolutions*. Cambridge: Cambridge University Press.

Barsalou, L. W. (2010). Grounded cognition: Past, present, future. *Topics in Cognitive Science, 2*, 716–724.

Brewer, W. F. (2003). Mental models. In L. Nadel (Ed.), *Encyclopedia of cognitive science* (Vol. 3, pp. 1–6). London: Nature.

Bronowski, J. (1973). *The ascent of man*. Boston: Little, Brown.

Buchwald, J. Z. (1985). *From Maxwell to microphysics: Aspects of electromagnetic theory in the last quarter of the nineteenth century*. Chicago: University of Chicago Press.

Campbell, J. I. D. (Ed.). (2005). *Handbook of mathematical cognition*. New York: Psychology Press.

Cantor, G. (1991). *Michael Faraday: Sandemanian and scientist. A study of science and religion in the Nineteenth Century*. New York: St. Martin's Press.

Cat, J. (2001). On understanding: Maxwell on the methods of illustration and scientific metaphor. *Studies in History and Philosophy of Modern Physics, 32*, 395–441.

Cat, J. (2012). Into the 'regions of physical and metaphysical chaos': Maxwell's scientific metaphysics and natural philosophy of action (agency, determinacy and necessity from theology, moral philosophy and history to mathematics, theory and experiment). *Studies in History and Philosophy of Science*, 43, 91–104.

Chandrasekharan, S., & Osbeck, L. (2010). Rethinking situatedness: Environment structure in the time of the common code. *Theory & Psychology, 20*, 171–207.

Chi, M. T. H. (1992). Conceptual change within and across ontological categories: Examples from learning and discovery in science. In R. N. Giere (Ed.), *Minnesota studies in the philosophy of science: Vol. 15. Cognitive models of science* (pp. 129–186). Minneapolis: University of Minnesota Press.

Chi, M. T. H., Feltovich, P. J., & Glaser, R. (1981). Categorization and representation of physics problems by experts and novices. *Cognitive Science, 5*, 121–152.

Clement, J. J. (2008). *Creative model construction in scientists and students: Imagery, analogy, and mental simulation*. Dordrecht, Netherlands: Springer.

Collins, H. M., & Evans, R. (2007). *Rethinking expertise*. Chicago: University of Chicago Press.

Crowe, M. J. (1967). *A history of vector analysis: The evolution of the idea of a vectorial system*. South Bend, IN: Notre Dame University Press.

Darrigol, O. (2000). *Electrodynamics from Ampère to Einstein*. Oxford: Oxford University Press.

Dunbar, K. (1999). How scientists build models: In Vivo science as a window on the scientific mind. In L. Magnani, N. J. Nersessian, & P. Thagard (Eds.), *Model-based reasoning in scientific discovery* (pp. 85–100). New York: Kluwer Academic.

Dyson, F. J. (1999). Why is Maxwell's theory so hard to understand? In James Clerk Maxwell Foundation, *James Clerk Maxwell Commemorative Booklet* (pp. 9–14). Edinburgh: James Clerk Maxwell Foundation.

Einstein, A., & Infeld, L. (1938). *The evolution of physics: From early concepts to relativity and quanta*. New York: Simon & Schuster.

Ericsson, K. A. (2006). The influence of experience and deliberate practice on the development of superior expert performance. In K. A. Ericsson, N. Charness, P. J. Feltovich, & R. R. Hoffman (Eds.), *The Cambridge handbook of expertise and expert performance* (pp. 683–704). Cambridge: Cambridge University Press.

Faraday, M. (1832). Experimental researches in electricity, 1st series. On the induction of electric currents. Reprinted in M. Faraday (Ed.). (1839). *Experimental researches in electricity* (Vol. 1, pp. 1–41). London: Taylor & Francis.

Faraday, M. (Ed.). (1839–1855). *Experimental researches in electricity* (3 vols.). London: Taylor & Francis.

Faraday, M. (1846a). Experimental researches in electricity, 19th series. On the magnetization of light and the illumination of magnetic lines of force. Reprinted in M. Faraday (Ed.). (1855). *Experimental researches in electricity* (Vol. 3, pp. 1–26). London: Taylor & Francis.

Faraday, M. (1846b). Thoughts on ray vibrations. Reprinted in M. Faraday (Ed.). (1855). *Experimental researches in electricity* (Vol. 3, pp. 447–452). London: Taylor & Francis.

Faraday, M. (1852a). Experimental researches in electricity, 29th series. On the employment of the induced magneto-electric current as a test and measure of magnetic force. Reprinted in M. Faraday (Ed.). (1855). *Experimental researches in electricity* (Vol. 3, pp. 371–401). London: Taylor & Francis.

Faraday, M. (1852b). On the physical character of the lines of magnetic force. Reprinted in M. Faraday (Ed.). (1855). *Experimental researches in electricity* (Vol. 3, pp. 407–437). London: Taylor & Francis.

Fisher, H. J. (2001). *Faraday's experimental researches in electricity: Guide to a first reading.* Santa Fe, NM: Green Lion Press.

Garber, E. (1999). *The language of physics: The calculus and the development of theoretical physics in Europe, 1750–1914.* Boston: Birkhäuser.

Gentner, D. (1983). Structure-mapping: A theoretical framework for analogy. *Cognitive Science, 7,* 155–170.

Gentner, D., & Bowdle, B. F. (2008). Metaphor as structure mapping. In R. W. Gibbs Jr. (Ed.), *The Cambridge handbook of metaphor and thought* (pp. 109–128). Cambridge: Cambridge University Press.

Gentner, D., Bowdle, B. F., Wolff, P., & Boronat, C. (2001). Metaphor is like analogy. In D. Gentner, K. Holyoak, & B. N. Kokinov (Eds.), *The analogical mind: Perspectives from cognitive science* (pp. 199–253). Cambridge, MA: MIT Press.

Gentner, D., & Gentner, D. R. (1983). Flowing waters or teeming crowds: Mental models of electricity. In D. Gentner & A. L. Stevens (Eds.), *Mental models* (pp. 99–130). Hillsdale, NJ: Lawrence Erlbaum.

Gibbs, R. W. Jr. (Ed.). (2008). *The Cambridge handbook of metaphor and thought.* Cambridge: Cambridge University Press.

Gingras, Y. (2001). What did mathematics do to physics? *History of Science, 39,* 383–416.

Gooding, D. C. (1981). Final steps to the field theory: Faraday's study of magnetic phenomena. *Historical Studies in the Physical Sciences, 11,* 231–275.

Gooding, D. C. (1990). *Experiment and the making of meaning: Human agency in scientific observation and experiment.* Dordrecht, Netherlands: Kluwer Academic.

Gooding, D. C. (1992). Mathematics and method in Faraday's experiments. *Physis, 29* (New Series), 121–147.

Gooding, D. C. (2005). Seeing the forest for the trees: Visualization, cognition, and scientific inference. In M. E. Gorman, R. D. Tweney, D. C. Gooding, & A. P. Kincannon (Eds.), *Scientific and technological thinking* (pp. 173–218). Mahwah, NJ: Lawrence Erlbaum.

Harman, P. M. (1998). *The natural philosophy of James Clerk Maxwell.* Cambridge: Cambridge University Press.

Hesse, M. (1973). Logic of discovery in Maxwell's electromagnetic theory. In R. N. Giere & R. S. Westfall (Eds.), *Foundations of scientific method: The nineteenth century* (pp. 86–114). Bloomington: Indiana University Press.

Hunt, B. R. (1991). *The Maxwellians.* Ithaca, NY: Cornell University Press.

Ippolito, M. F., & Tweney, R. D. (1995). The inception of insight. In R. J. Sternberg & J. E. Davidson (Eds.), *The nature of insight* (pp. 433–462). Cambridge, MA: MIT Press.

James, F. A. J. L. (1998). Faraday, Maxwell, and field theory. In F. A. J. L. James (Ed.), *Semaphores to short waves* (pp. 71–84). London: Royal Society for the Encouragement of Arts and Manufactures.

James, Frank A. J. L. (Ed.). (2008). *The correspondence of Michael Faraday, Vol. 5:1855–1860.* London: Institution of Engineering and Technology.

Klahr, D. (2000). *Exploring science: The cognition and development of discovery processes.* Cambridge, MA: MIT Press.

Krieger, M. H. (1992). *Doing physics: How physicists take hold of the world.* Bloomington: Indiana University Press.

Kulkarni, D., & Simon, H. A. (1988). The processes of scientific discovery: The strategy of experimentation. *Cognitive Science, 12,* 139–176.

Kurz, E. M. (1998). Representation, agency, and disciplinarity: Calculus experts at work. In M. A. Gernsbacher & S. J. Derry (Eds.), *Proceedings of the 20th Annual Conference of the Cognitive Science Society* (pp. 585–590). Mahwah, NJ: Lawrence Erlbaum.

Kurz-Milcke, E. (2004). The authority of representations. In E. Kurz-Milcke & G. Gigerenzer (Eds.), *Experts in science and society* (pp. 281–302). New York: Kluwer Academic/Plenum.

Lakoff, G., & Núñez, R. E. (2000). *Where mathematics comes from: How the embodied mind brings mathematics into being.* New York: Basic Books.

Lambert, K. (2011). The uses of analogy: James Clerk Maxwell's "On Faraday's lines of force" and early Victorian analogical argument. *British Journal for the History of Science, 44,* 61–88.

Larkin, J. H., McDermott, J., Simon, D. P., & Simon, H. A. (1980). Models of competence in solving physics problems. *Cognitive Science, 4,* 317–345.

Magnani, L., & Nersessian, N. J. (Eds.). (2002). *Model-based reasoning: Science, technology, values.* New York: Kluwer Academic/Plenum Publishers.

Martin, T. (Ed.). (1933). *Faraday's diary: Being the various philosophical notes of experimental investigation made by Michael Faraday during the years 1820–1862,* Vol. 3. London: G. Bell.

Maxwell, J. C. (1855–1856). On Faraday's lines of force. Reprinted in W. D. Niven (Ed.). (1890). *The scientific papers of James Clerk Maxwell* (Vol. 1, pp. 155–229). Cambridge: Cambridge University Press.

Maxwell, J. C. (1861–1862). On physical lines of force. Reprinted in W. D. Niven (Ed.). (1890). *The scientific papers of James Clerk Maxwell* (Vol. 1, pp. 451–513). Cambridge: Cambridge University Press.

Maxwell, J. C. (1864). A dynamical theory of the electromagnetic field. Reprinted in W. D. Niven (Ed.). (1890). *The scientific papers of James Clerk Maxwell* (Vol. 1, pp. 526–597). Cambridge: Cambridge University Press.

Maxwell, J. C. (1873a). Faraday. Reprinted in W. D. Niven (Ed.). (1890). *The scientific papers of James Clerk Maxwell* (Vol. 2, pp. 355–360). Cambridge: Cambridge University Press.

Maxwell, J. C. (1873b). *A treatise on electricity and magnetism* (2 vols.). Oxford, UK: Clarendon Press.

Maxwell, J. C. (1873/1881).*A treatise on electricity and magnetism* (2 vols., 2nd ed., W. D. Niven, Ed.). Oxford, UK: Clarendon Press.

Maxwell, J. C. (1873/1891). *A treatise on electricity and magnetism* (2 vols., 3rd ed., J. J. Thompson, Rev.). Oxford, UK: Clarendon Press.

Maxwell, J. C. (1879). Thomson and Tait's "Natural philosophy" [Review]. Reprinted in W. D. Niven (Ed.). (1890). *The scientific papers of James Clerk Maxwell* (Vol. 2, pp. 776–785). Cambridge: Cambridge University Press.

Nersessian, N. J. (1984). *Faraday to Einstein: Constructing meaning in scientific theories.* Dordrecht, Netherlands: Nijhoff.

Nersessian, N. J. (1985). Faraday's field concept. In D. Gooding & F. A. J. L. James (Eds.), *Faraday rediscovered: Essays on the life and work of Michael Faraday, 1791–1867* (pp. 175–188). Basingstoke, UK: Macmillan.

Nersessian, N. J. (2002). Maxwell and the "method of physical analogy": Model-based reasoning, generic abstraction, and conceptual change. In D. B. Malament (Ed.), *Reading natural philosophy: Essays in the history and philosophy of science and mathematics* (pp. 129–166).Chicago: Open Court.

Nersessian, N. J. (2005). Interpreting scientific and engineering practices: Integrating the cognitive, social, and cultural dimensions. In M. E. Gorman, R. D. Tweney, D. C. Gooding, & A. P. Kincannon (Eds.), *Scientific and technological thinking* (pp. 17–56). Mahwah, NJ: Lawrence Erlbaum.

Nersessian, N. J. (2008). *Creating scientific concepts.* Cambridge, MA: MIT Press.

Núñez, R. E. (2000). Conceptual metaphor and the embodied mind: What makes mathematics possible? In F. Hallyn (Ed.), *Metaphor and analogy in the sciences* (pp. 125–146). Dordrecht, Netherlands: Kluwer Academic.

Núñez, R. E. (2005). Creating mathematical infinites: Metaphor, blending, and the beauty of transfinite cardinals. *Journal of Pragmatics, 37,* 1717–1741.

Núñez, R. E. (2008). Conceptual metaphor, human cognition, and the nature of mathematics. In R. W. Gibbs Jr. (Ed.), *The Cambridge handbook of metaphor and thought* (pp. 339–362). Cambridge: Cambridge University Press.

Núñez, R. E., & Lakoff, G. (2005). The cognitive foundations of mathematics: The role of conceptual metaphor. In J. I. D. Campbell (Ed.), *Handbook of mathematical cognition* (pp. 109–124). New York: Psychology Press.

Olson, R. (1975). *Scottish philosophy and British physics, 1750–1880: A study in the foundations of the Victorian scientific style.* Princeton, NJ: Princeton University Press.

Osbeck, L. M., Nersessian, N. J., Malone, K. R., & Newstetter, W. C. (2010). *Science as psychology: Sense-making and identity in science practice.* Cambridge: Cambridge University Press.

Qin, Y., & Simon, H. A. (1995). Imagery and mental models in problem solving. In J. Glasgow, B. Chandrasekaran, & N. H. Narayanan (Eds.), *Diagrammatic reasoning: Cognitive and computational perspectives* (pp. 403–434). Menlo Park, CA: AAAI Press/MIT Press.

Robbins, P., & Aydede, M. (Eds.). (2009). *The Cambridge handbook of situated cognition.* Cambridge: Cambridge University Press.

Siegel, D. M. (1991). *Innovation in Maxwell's electromagnetic theory: Molecular vortices, displacement current, and light.* Cambridge: Cambridge University Press.

Simpson, T. K. (2005). *Figures of thought: A literary appreciation of Maxwell's Treatise on electricity and magnetism.* Santa Fe, NM: Green Lion Press.

Simpson, T. K. (2010). *Maxwell's mathematical rhetoric: Rethinking the Treatise on electricity and magnetism.* Santa Fe, NM: Green Lion Press.

Skilling, H. H. (1942). *Fundamentals of electric waves.* New York: John Wiley.

Smith, C., & Wise, M. N. (1989). *Energy & empire: A biographical study of Lord Kelvin.* Cambridge: Cambridge University Press.

Steinle, F. (2003). The practice of studying practice: Analyzing research records of Ampère and Faraday. In F. L. Holmes, J. Renn, J., & H.-J. Rheinberger (Eds.), *Reworking the bench: Research notebooks in the history of science* (pp. 93–118). (Published as *Archimedes*, Vol. 7). Dordrecht, Netherlands: Kluwer Academic.

Thomson, W. (1884). *Notes of lectures on molecular dynamics and the wave theory of light.* Reprinted; R. Kargon & P. Achinstein (Eds.). (1987). *Kelvin's Baltimore lectures and modern theoretical physics.* Cambridge, MA: MIT Press.

Turner, M. (2001). *Cognitive dimensions of social science.* Oxford: Oxford University Press.

Tweney, R. D. (1985). Faraday's discovery of induction: A cognitive approach. In D. Gooding & F. A. J. L. James (Eds.), *Faraday rediscovered: Essays on the life and work of Michael Faraday, 1791–1867* (pp. 189–210). London: Macmillan.

Tweney, R. D. (1992a). Inventing the field: Michael Faraday and the creative "engineering" of electromagnetic field theory. In R. J. Weber & D. N. Perkins (Eds.), *Inventive minds: Creativity in technology* (pp. 31–47). Oxford: Oxford University Press.

Tweney, R. D. (1992b). Stopping Time: Faraday and the scientific creation of perceptual order. *Physis, 29,* 149–164.

Tweney, R. D. (2001). Scientific thinking: A cognitive-historical approach. In K. Crowley, C. D. Schunn, & T. Okada (Eds.), *Designing for science: Implications from everyday, classroom, and professional settings* (pp. 141–173). Mahwah, NJ: Lawrence Erlbaum.

Tweney, R. D. (2004). Replication and the experimental ethnography of science. *Journal of Cognition & Culture, 4,* 731–758.

Tweney, R. D. (2006). Discovering discovery: How Faraday found the first metallic colloid. *Perspectives on Science, 14*(1), 97–121.

Tweney, R. D. (2009). Mathematical representations in science: A cognitive-historical case study. *Topics in Cognitive Science, 1,* 758–776.

Tweney, R. D. (2011). Representing the electromagnetic field: How Maxwell's mathematics empowered Faraday's field theory. *Science & Education, 20,* 687–700.

Tweney, R. D., Doherty, M. E., & Mynatt, C. R. (Eds.). (1981). *On scientific thinking.* New York: Columbia University Press.

Utsumi, A. (2011). Computational exploration of metaphor comprehension processes using a semantic space model. *Cognitive Science, 35,* 251–296.

Wigner, E. P. (1960). The unreasonable effectiveness of mathematics in the natural sciences. *Communications on Pure and Applied Mathematics, 13,* 1–14.

Wise, M. N. (1982). The Maxwell literature and British dynamical theory. *Historical Studies in the Physical Sciences, 13,* 175–205.

Woolf, P., & Gentner, D. (2011). Structure-mapping in metaphor comprehension. *Cognitive Science, 35,* 1456–1488.

18 Digging into Implicit/Explicit States and Processes

The Case of Cognitive/Social Process Interaction in Scientific Groups

Susannah B. F. Paletz and
Christian D. Schunn

PROCESSES IN COGNITIVE/SOCIAL PSYCHOLOGY OF SCIENCE

To what extent are key psychological processes of science implicit or explicit? This central question has significant theoretical and methodological implications for how these processes can be studied, which include the extent to which practitioner self-reflection and autobiographies can be taken as serious data points. In general, the psychology of science sits within two broader fields, the cognitive science of science (Dunbar, 1995; Holyoak & Thagard, 1995; Nersessian, 2009) and social studies of science (Knorr-Cetina, 1999), both of which include history of science and philosophy of science. Each of these perspectives relies to some extent on self-reflection and self-report methodologies.

Within psychology, there is a long theoretical and empirical history regarding the implicit versus explicit nature of psychological processes. This line of research has not always been called implicit versus explicit, but can be identified under studies of tacit versus explicit and unconscious versus conscious processing. Within cognitive psychology, there is now considerable agreement that some cognitive content (e.g., knowledge, beliefs, goals) is explicit, which includes aspects that are verbalizable, accessible to consciousness, and open to cognitive control (Ericsson & Simon, 1993). On the other hand, some cognitive content is implicit, being nonverbalizable, not accessible to consciousness, and not open to cognitive control (Reder & Schunn, 1996). Different amounts of practice with tasks is thought to influence automaticity and the use of working memory for manipulating intermediate products, which then will influence implicit/explicit status (Ericsson & Simon, 1993; Schunn, Lovett, & Reder, 2001). However, there are also some processes that in general are not open to access, such as priming (Schunn & Dunbar, 1996).

Within social psychology, there is a somewhat different view. In general, self-reports of processes and explanations for behaviors are held in great suspicion, such as when individuals are asked to self-report socially undesirable attitudes (Nisbett & Wilson, 1977). The rise of the implicit association test (IAT) to measure stereotypes and attitudes hidden from conscious thought, as well as the distinction between implicit and explicit attitudes, have their philosophical roots in this suspicion. Even so, a moderate correlation exists between implicit (automatic) and explicit (controlled) attitudes (Nosek, 2005). At the same time, surveys are the most common form of data collected, and the study of the actual behavior is rare in social psychology (Baumeister, Vohs, & Funder, 2007). In social psychology, the *content* of constructs is typically conceived of as explicit (e.g., attitudes, beliefs, role classifications). The *processes* are typically conceived of as implicit (e.g., social comparison, dissonance, priming, schema formation), although some content is implicit (e.g., implicit theories of creativity, Paletz & Peng, 2008; Paletz, Peng, & Li, 2011), and some processes have overt behavioral components (e.g., the success of various persuasion techniques).

Why Psychology of Science Needs Cognitive and Social Psychology Integration

In response to the high-level question about whether key processes in science are explicit or implicit, a simple first cut would divide the processes into cognitive and social psychological processes, and then apply the relevant frameworks and methods from those literatures. But what about processes that sit on the boundary between cognitive and social psychology? Whose frameworks and methods should take priority in those cases?

We claim that scientific activity is inherently cognitive: Scientific activity is a process by which information is transformed through complex problem-solving and decision-making component processes. But it is also inherently social—a process that engages groups of individuals frequently working together collaboratively or in opposition, involving interaction and/or social perception. Further, we claim that the social/cognitive divide is a somewhat arbitrary sociological distinction of communities as found in graduate training programs, journals, and conferences, rather than a real ontological distinction with clear dividing lines based on underlying processes. For example, motivation and shared mental models are psychological constructs that are important to the psychology of science and are partially owned by both the cognitive and social psychology literatures.

The potential interconnections between primarily social and cognitive processes further complicate the implicit/explicit issue in the psychology of science. Suppose, for example, that analogical reasoning, a process intellectually owned primarily by cognitive psychologists, had an influence on intragroup conflict, a process intellectually owned primarily by social and organizational psychologists. Would the influence process follow the frameworks of cognitive psychologists or that of social psychologists?

To come to a more complete understanding of the psychology of science and in order to address the explicit/implicit distinction, integration between social and cognitive psychology will become important. In any real-world situation, both kinds of processes will be in play, with many processes actually sitting at the boundary of cognitive and social. In addition, there can be interactions among cognitive and social variables. Thus, making performance predictions will require understanding both components and their interactions. The current chapter is an example of work in this integration space.

A SOCIAL/COGNITIVE MODEL OF SCIENTIFIC COLLABORATIVE PROCESSES

We begin with a description of a particular model that integrates social and cognitive processes to produce a deeper understanding of the effects of multidisciplinary knowledge on science team innovation. These input and output variables have been studied by both cognitive and social psychologists but with very different intermediate process variables. The model includes a range of processes to illustrate a number of substantive ways in which cognitive and social variables can come together.

Our particular model (described in greater depth in Paletz & Schunn, 2010) goes from external team structures to social processes, which in turn influence cognitive processes, which in turn influence team innovative outcomes. Thus, we assume no direct links from team structures to cognitive processes or from team processes to team outcomes. Furthermore, we assume that cognitive processes do not influence social processes directly, though we note such relationships are possible.

In a later section in this chapter, we describe some empirical work that reexamines these assumptions of the model with the test case of one particular social/cognitive connection, specifically that of conflict and analogy. This particular case also acts as a test of the implicit versus explicit nature of the interconnections between social and cognitive processes in science.

Convergent and Divergent Thinking in Scientific Creativity

The most commonly studied aspect of creativity in general is divergent thinking. Divergent thinking includes, among other factors, fluency (the generation of many ideas), originality (ideas that are novel compared to prior solutions), and elaboration (the depth with which each idea is explored;Torrance, 1988). However, divergent thinking alone is an incomplete view of creativity (Amabile, 1996; Hocevar & Bachelor, 1989). Research on scientific creativity must take into account the *quality* of ideas (Cropley, 2006). In Paletz and Schunn (2010), we argued that divergent thinking results in new ideas, but that there must also be convergent thinking in which good ideas are selected, adapted, and refined from the generated set (e.g., Cropley, 2006). Furthermore, we suggested in our model that different processes and relationships are important for divergent and convergent thinking. The connections we described in our models are likely the strongest relationships between the variables of interest, based on the available literature. These models did not preclude the possibility of other relationships but simply focused on those most likely to have an effect.

Our social/cognitive models addressed both divergent and convergent processes and showed both potential causal pathways. We did not seek to reduce social factors to cognitive factors, although cognitive variables may mediate the relationship between certain social factors and innovation; rather, we sought to understand whether cognitive factors previously associated with innovation were also influenced by team factors. In addition, by integrating social and cognitive explanations, we may understand why interventions aimed at the cognitive level in team settings may sometimes produce unexpected outcomes. In other words, this approach enables us to gain a deeper understanding of the interactive influence of social and cognitive factors in scientific creativity.

Cognitive Processes of Scientific Creativity

Many cognitive processes have been implicated in science creativity. Our model focused on three variables that have been most commonly studied: information search, analogy, and information evaluation. Information search is a systematic search through possible solutions, often directed by simple techniques (e.g., simplicity first, more available first). Analogy is the process of drawing abstract relational comparisons between the current problem and prior problem/solutions to generate solutions for the current problem. Analogy has been studied in depth in science (Dunbar, 1997; Gentner et al., 1997). Information search and analogy are primarily associated with divergent thinking: They lead to novel ideas, a wider variety of ideas, and more ideas. They also help problem solvers overcome mental blocks or fixation on one particular solution (Christensen & Schunn, 2005). Information evaluation is a convergent thinking cognitive process by which problem solvers examine the value of and select among the various generated ideas to determine which ones best solve the task at hand. Along the explicit/implicit continuum, all three of these are explicit cognitive process, insofar as their results (e.g., the analogy, the query, the selection) can be perceived by the thinker and by a potential audience. However, the causal cognitive processes behind these three are likely to be implicit, involving knowledge structure activation, association, and comparisons. Surprisingly, of these three simple cognitive processes commonly associated with creativity in the cognitive literature, only information search has been discussed in the social psychology creativity literature (Nemeth & Rogers, 1996). This gap is particularly surprising because commonly discussed social variables underlying scientific creativity are likely connected to analogy and evaluation cognitive processes as well.

Social Processes of Scientific Creativity

The team creativity literature has a rich and varied history (see Ilgen, Hollenbeck, Johnson, & Jundt, 2005; Kozlowski & Ilgen, 2006 for reviews). Brainstorming is often the focus of team creativity researchers (Thompson & Choi, 2006), but it is a subset of what scientists do: They also expound, problem solve, argue, criticize, test, and get off-topic. Most conceptual models of teamwork highlight the broader organizational context, the task demands, team processes, and/or team outputs, generally in the input-process-output type of model (Kozlowski & Ilgen, 2006; McGrath, 1984). One weakness of input-process-output models is that they often do not examine individual cognitive processes. Our model took as illustrative examples two team structures of likely importance to both creative outcomes in science and to creative cognitive processes: knowledge diversity and member roles (Saunders & Ahuja, 2006). Team processes in the model that were hypothesized to be influenced by these team structures included communication, task conflict, and shared mental models.

Diversity in teams, including knowledge domain diversity, has historically resulted in mixed effects on teamwork: It is often positively associated with divergent thinking but can also hinder convergent thinking via increasing unproductive conflict and making it difficult to achieve a shared vision (Gebert, Boerner, & Kearney, 2006; Kurtzberg & Amabile, 2000–2001; Mannix & Neale, 2005; van Knippenberg & Schippers, 2007). Team members' knowledge diversity may be implicit or explicit, depending on team metaknowledge about members' expertise. Formal roles are a type of team structure encompassing the behavioral norms and scripts surrounding individuals' roles. In our model, we proposed that formal roles help guide team processes and manage the contradictory effects of knowledge diversity. The three team process variables were interrelated with each other and the structure variables. Groups where individuals do not share unique information have poorer decision making (Stasser & Titus, 1985), so broad participation and communication of information are necessary to harness knowledge diversity. Knowledge diversity is thought to lead to task conflict, which is commonly argued to lead to better team performance (Mannix & Neale, 2005, although see De Dreu & Weingart, 2003). Task conflict, as opposed to relationship and process conflict, involves disagreements about the work being done by the group (Jehn & Chatman, 2000). Social and organizational psychologists have also discussed the importance of a shared vision to team innovation (West, 1990). Shared mental models are the similarity and overlap of schemas regarding tasks, teamwork, and norms held by group members. As with cognitive processes, social processes are explicit when they are obviously experienced (e.g., long-lasting task conflict), but not when they lie outside of conscious thought (e.g., many mental models and the degree to which they are shared).

Social/Cognitive Model—Divergent Path

On the surface, the relevant social factors for divergent and convergent elements seem to be in opposition. Multidisciplinary teams are particularly illustrative of this contradiction: They both hold the promise of broad solutions and a great deal of difficulty (Derry, Schunn, & Gernsbacher, 2005). Attempts to test and describe the effects of diversity have resulted in weak or inconsistent results (Mannix & Neale, 2005; van Knippenberg, De Dreu, & Homan, 2004), suggesting that moderated and mediated models are necessary (van Knippenberg et al., 2004; van Knippenberg & Schippers, 2007). Our model argued that formal roles are a key factor of this moderation. Figure 18.1 presents the divergent path of our model. The rest of this section will explain the empirical and theoretical origins of each link in the divergent path; the next section justifies the convergent path (Figure 18.2).

Generally, diversity is theorized to be positively associated with team innovation via the team having access to a broader range of information, perspectives, and opinions (Mannix & Neale, 2005; van Knippenberg & Schippers, 2007). Thus, diversity would be particularly likely to enhance team performance when the team works on a task that involves information processing and requires innovative solutions (van Knippenberg et al., 2004). A diversity of opinions also increases perceived novelty and complex, integrative thinking (Antonio et al., 2004). Furthermore, according to theory by Nijstad and Stroebe (2006), the more knowledge structures available and accessible to members of the group, the more ideas that group can generate. One important caveat is that these different knowledge structures must be communicated between members of the group. For a team to take advantage of knowledge diversity, team members must communicate relevant and unshared information. Therefore, broad participation is considered both a likely outcome of and a necessary condition for knowledge or disciplinary diversity to be innovative.

Informational diversity may increase performance via disagreements about task-relevant ideas (Jehn, Northcraft, & Neale, 1999; Pelled, Eisenhardt, & Xin, 1999). Task conflict focuses on the work done by the group; process conflict relates to plans and delegation; and relationship conflict revolves around issues of interpersonal incompatibility and off-task social issues (Jehn & Chatman, 2000). Following a much-expressed but contentious opinion in organizational and social psychology

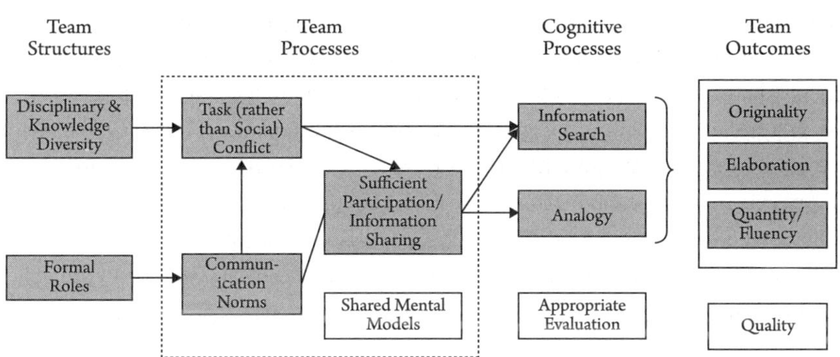

Figure 18.1 Hypothesized social/cognitive pathways of team divergent thinking. Reprinted from Paletz, S. B. F., & Schunn, C. (2010). A social/cognitive framework of multidisciplinary team innovation. *Topics in Cognitive Science, 2*, 73–95, Figure 2. Permission granted by the Cognitive Science Society.

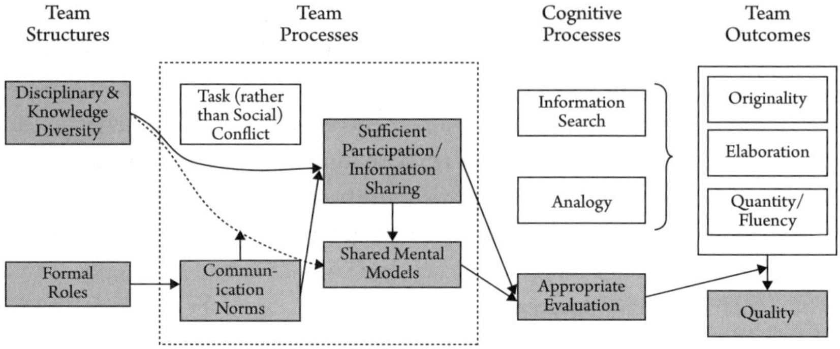

Figure 18.2 Hypothesized social/cognitive pathways of team convergent thinking. Reprinted from Paletz, S. B. F., & Schunn, C. (2010). A social-cognitive framework of multidisciplinary team innovation. *Topics in Cognitive Science, 2*, 73–95, Figure 2. Permission granted by the Cognitive Science Society.

(De Dreu & Weingart, 2003), we hypothesized that task conflict, but neither process nor relationship conflict (Jehn, 1997), would be an important positive factor in the effect of functional diversity on team performance. Arguments about the positive effects of task conflict are supported by a lengthy literature showing systematic, positive effects of minority opinion dissent on others' cognitions (Hinsz, Tindale, & Vollrath, 1997). Dissent of any kind has been found to increase discussion intensity (Schulz-Hardt, Brodbeck, Mojzisch, Kerschreiter, & Frey, 2006). In particular, a minority opinion dissenter promotes more original solutions (Nemeth & Kwan, 1985) via a more thorough search for information by other members of the group (Nemeth & Rogers, 1996).

A second structure we hypothesized to be important is formal roles. Formal roles, if set up correctly with associated communication norms and added task structure, should influence the relationship between knowledge diversity and task conflict, heightening knowledge diversity's positive effect on task conflict and lessening the potential relationship conflict. In particular, formal roles make explicit the implicit differences in background knowledge between team members. Communication norms in this case are expectations and rules regarding communication behaviors stemming from formal roles, such as respecting expert opinion and tolerating expert-driven task conflict via formal communication structures (e.g., turn taking, asking for concurrence). Formal roles can make implicit communication norms explicit and privilege unique information. For example, minority dissent has a stronger positive impact if the source of the minority opinion can claim expertise (Hart, Stasson, & Karau, 1999). In the context of knowledge diversity, formal roles that reify and manage domain expertise may help encourage productive task conflict without the associated disruptive relationship conflict. In addition, the communication norms made salient by formal roles could encourage broader participation and information sharing. Individuals within a team would encourage sharing of unique information if the norm is that individuals within certain roles are both respected in those roles as experts and expected to share their expertise. When experts are assigned (i.e., formalized, made explicit) rather than assumed, the tendency in groups to withhold unshared information can be significantly overcome (Stewart & Stasser, 1995).

We hypothesized that the cognitive process of analogies might be involved in the connection between task conflict and scientific discoveries. Knowledge diversity has been associated with an improved ability to generate useful analogies. For example, mixed-background microbiology laboratories have been found to be more successful than homogeneous microbiology labs: The mixed-background labs brought a broader set of analogies to bear, and were thus better able to solve problems (Dunbar 1995, 1997). We hypothesized that task conflict, in the context of knowledge diversity, would lead each individual to actively consider different perspectives, likely through the activation of a broader range of domains (Nijstad & Stroebe, 2006). The most difficult step to develop useful analogies is often retrieving the best analogy from memory. Once the right case is retrieved, the analogical mapping and productive inference is relatively easy (Forbus, Gentner, & Law, 1995). Thus, thinking about different perspectives should prompt different knowledge within the individual. Furthermore, the previously discussed participation/information sharing might cue additional knowledge from memory across individuals. Therefore, conflict via information sharing could create opportunities to bring a broader set of useful analogies to bear.

In sum, we hypothesized that knowledge diversity under formal roles would lead to task rather than relationship conflict and increased breadth of participation, which would then lead to information search and analogies, which would increase the originality, quantity, and richness (elaboration) of solutions (Paletz & Schunn, 2010). Some of the underlying mechanisms behind this model involved making explicit what was implicit, such as conflicting knowledge and unique information.

Social/Cognitive Model—Convergent Path

Group members must also select the best of their creative ideas (De Dreu & West, 2001). Because diverse groups may have more difficulty in creating shared mental models, there have been recent calls to examine shared mental models in diversity research (van Knippenberg & Schippers, 2007). Shared mental models are "dynamic, simplified, cognitive representations of reality that team members use to describe, explain, and predict events" (Burke, Stagl, Salas, Peirce, & Kendall, 2006, p. 1199). Team members use shared mental models to integrate and interpret new information about the team, task, and the problem situation (Burke et al., 2006). Therefore, we hypothesized that knowledge diversity would have a negative impact on shared mental models, resulting in more representational gaps (Cronin & Weingart, 2007). Cropley (2006) described the importance of convergent thinking in effective creativity, discussing it as originating primarily via established knowledge, which defines, guides, and shapes what is considered creative. This established knowledge includes not only explicit information about judgment criteria, but also implicit knowledge such as heuristics, norms, and unspoken assumptions about a field. These are all, in fact, shared mental models. Knowledge diversity offers a broader base of information to be shared, but also offers conflicting and contradictory norms and assumptions regarding what might be considered optimal solutions.

In the model, we further hypothesized that shared mental models in turn positively influence evaluative cognitive processes, which are useful for idea quality. Evaluation is defined here as reasoning about constraints and tradeoffs on the scientific discovery problem at hand (e.g., between parsimony and fit to data, or between ease of explanation and inclusion of underlying mechanisms). Original ideas, richly elaborated ideas, and many ideas all indirectly lead to better ideas, on average. Evaluation, however, is critical to select among the various generated ideas to choose the better ideas.

Shared mental models help teams to function (Burke et al., 2006; Kozlowski & Ilgen, 2006) and are considered necessary to timely problem solving (Jones, Stevens, & Fischer, 2000; Orasanu, 1995). Shared mental models shape how individuals interpret their environment and include norms regarding how they should react to it (Salas, Stagl, & Burke, 2004). Promoting shared mental models is often considered vital for increasing team effectiveness (Kozlowski & Ilgen, 2006). We hypothesized that shared mental models create high-level common ground upon which fine-grained details could be effectively co-constructed across the team. On the one hand, this shared understanding can allow team members to accurately evaluate the various choices on a subtask as well as provide useful evaluative feedback to others regarding their subtasks. On the other hand, when shared mental models are implicit, it can be more difficult to tell what parts are genuinely shared versus dissimilar or nonoverlapping. Such unshared elements, particularly with the details of perceptions of the team's problem, can cause difficulties for a team (Cronin & Weingart, 2007). On the other hand, if enough of a shared language exists to facilitate teamwork (via a "trading zone"), the team may yet succeed (Gorman, 2008).

We hypothesized that formal roles would influence the effects of knowledge diversity on convergent thinking processes (Paletz & Schunn, 2010). For example, shared conceptions of team roles can improve coordination and are thus vital to performance (Kozlowski & Ilgen, 2006). In addition, the act of assigning experts is associated with increasing the group's tendency to share unique information (Stasser, Stewart, & Wittenbaum, 1995; Stasser & Titus, 1985), suggesting that formal roles, because of communication norms regarding those roles, may make mental models easier to share via broader participation and communication. In fact, we hypothesized that formal roles could serve to moderate the negative relationship between knowledge diversity and shared mental models such that when formal roles exist, the negative relationship would no longer be significant. In effect, the tendency for formal roles to make explicit team members' assumptions about others' knowledge and how to interact with them can help a team achieve more shared mental models.

In sum, in this model, we hypothesized that knowledge diversity would negatively impact shared mental models (Paletz & Schunn, 2010). However, formal roles would increase the commonality and accuracy of shared mental models via communication norms (i.e., regarding participation and listening) and would therefore mitigate the negative relationship between knowledge diversity and shared mental models. Shared mental models and broad participation would then increase the likelihood that the cognitive process of evaluation would be appropriate, leading to high-quality scientific discovery outputs.

Implicit versus Explicit Social/Cognitive Connections

Within this larger model, we hypothesized many different connections between social and cognitive processes (Paletz & Schunn, 2010). Each of these connections might have implicit or explicit elements. Scientists may explicitly choose a cognitive strategy in response to a social situation, or they may simply implicitly react with a cognitive behavior in response to a social situation without knowing why they are doing so. Here, explicit can refer to either the selection process, which is sometimes called metacognitive control of the choice, or simply to the triggering environment and selected action itself, which is sometimes called metacognitive awareness (Reder & Schunn, 1996).

In the empirical section that follows, we examine in depth just one of the social/cognitive connections from the larger model, that between conflict and analogy. We select these two because they are each particularly richly described processes within disparate subfields in the psychology of science. We examine whether there are empirical connections, how the connections are raised by the individuals (and hence more likely to be explicit processes), and whether they are simply statistical regularities of behavior (and hence more likely to be implicit processes).

AN OBSERVATIONAL STUDY OF SCIENTIST TEAMS IN ACTION

Overview of Study

We summarize some results of an observational study of the interplay between conflict and analogy in one particular context (Paletz, Schunn, & Kim, 2012). The study involved coding videos for conflict and analogy events as displayed in natural, informal, task-related conversations and then statistically examining lagged regressions for evidence of an influence of one variable on the other, followed by a qualitative examination of the relationship within statistically significant connections.

Context

The Mars Exploration Rover (MER) mission's goals were to determine whether Mars ever had liquid surface water by collecting data using two rovers on opposite sides of Mars. Water on Mars is a necessary precursor to life having developed or existed on Mars. Prior to the MER team's discoveries, the hypothesis that water historically existed on Mars was the minority view; after the first 90 days of their mission, it became a foregone conclusion. In addition to this major discovery, the mission produced large general improvements in scientific understanding of the geology, geochemistry, and atmospheric conditions on Mars. Overall, MER was considered one of the most successful scientific missions by NASA in the last several decades. This success was not just due to the robotics and logistical planning before the mission, but was in large part because of the continual, creative adaptation to challenges during those first 90 days and beyond. The greater MER science team had over 100 members and was inherently multidisciplinary. In the first 90 days, they operated in a colocated, real-time problem-solving manner, often in a large room with workstation clusters for each subteam.

Sampling and Coding Method

Video cameras were moved around the larger science room for a subset of the days and captured hundreds of hours of individual work and informal conversations involving groups of various sizes. From the larger video collection, a subset of informal (unstructured) but task-relevant conversations totaling over 11 hours were selected for transcription and analysis based on the audibleness of conversations. The transcripts were segmented into thought statements and then independently coded for presence or absence of both analogies and microconflicts. Analogies were identified purely from the transcripts. Disagreements among coders regarding conflicts were verified from the video because tone and body gestures were very important for identifying real moments of conflict. Conflict was identified not simply because the speaker took a controversial viewpoint, but because the speaker was disagreeing with something said previously (see Paletz, Schunn, & Kim, 2011, for the microconflict coding method). Then each analogy and microconflict event was further coded for types of analogy and types of conflict to see whether the connections depended upon the subtypes (see Paletz, Schunn, & Kim, 2011, for details regarding conflict, and Paletz, Schunn, & Kim, 2012, for details on analogy and the study).

One key dimension of analogy was analogical distance: within domain (where the target and source of the analogy were from the same domain, i.e., within the domain of geology on Mars), within discipline (where the target and the source of the analogy were within the same discipline, such as geosciences or organizational processes in general), or outside discipline (e.g., the target was a geological formation and the source was from general knowledge, such as a food). Dunbar (1997) argues that outside-discipline analogies are not used for discoveries per se but rather simply serve an explanatory function.

One key dimension of conflict was whether it involved conflicts about the task (scientific data analysis or planning involving the rovers), human work processes, or relationships. Task conflict is often argued to be potentially useful whereas process and relationship conflict are generally thought to be harmful (Jehn, 1997, but also see De Dreu & Weingart, 2003).

Analysis Methods

The initial step was to look for statistical associations in the temporal relationships between conflict and analogy. This step was accomplished by dividing the larger transcripts into blocks of 25 lines (utterances/thought statements), with this segmentation conducted in two different ways. First the blocks were created centered around analogy moments, with sequential blocks of 25 lines before and following each analogy and continuing throughout the video clip transcript. A given block could possibly be shorter than 25 lines if a video clip transcript ended or another analogy began. Second, the blocks were created centered around conflict moments, with sequential blocks of 25 lines before and following each microconflict event. Then time series logistic regressions of lag 1 were computed. For analogy to conflict, the regressions were conducted using the analogy-centered blocks and tested whether the absence or presence of an analogy in block n predicted the absence or presence of a microconflict in block $n+1$. For conflict to analogy, the regressions were conducted using the conflict-centered blocks and tested whether the absence or presence of a microconflict in block n predicted the absence or presence of an analogy in block $n+1$. In the case of the analogy-centered analyses, these regressions were actually hierarchical logistic regressions (of blocks nested within conversational clips) because there was significant clip-level variance in the dependent variable. Simple logistic regression was used in the case of conflict-centered blocks because there was not significant clip-level variance in the dependent variable. Additional statistical analyses examined subtypes, and then qualitative analyses examined the cases captured by the statistically significant subtype connections.

RESULTS

Main Variable Connections

Analogies overall were not a significant predictor of subsequent conflict overall (Paletz, Schunn, & Kim, 2012). However, within-domain analogies led to a twofold increase in the occurrence of conflict in the subsequent block, whereas within-discipline and outside-discipline analogies did not have this effect. Furthermore, the effect was localized to particular kinds of conflict. In particular, within-domain analogies were associated with a twofold increase in scientific data-analysis science conflicts and work process conflicts, but no change in scientific rover and experiment planning conflicts. Table 18.1 presents examples of within-domain analogies followed closely by conflict.

Examining the reverse relationship, again the overall relationship of conflict to analogy was not statistically significant, but a more specific relationship was (Paletz, Schunn, & Kim, 2012). In particular, process conflicts were associated with a more than twofold increase in the occurrence of subsequent within-discipline analogies. Note that here the relationship involves a different type of analogy; thus, we can rule out simple third-variable context explanations in which a third-variable C led to both more conflict and analogy because it was a different kind of analogy implicated in the two different directional effects. Table 18.2 presents examples of process conflict followed closely by within-discipline analogies.

Data on Explicit versus Implicit Nature of Connections

In our study, we hypothesized that within-discipline analogies would result from more conflicts because they would play a direct role in resolving those conflicts;

Table 18.1 Examples of within-domain analogies (in italics) leading to conflict (most key words underlined)

Within-domain analogies leading to science conflict	
S2:[a]	You don't see it there. That *looks like everything we've been seeing so far in transition.*
S1:	But it's different. It's been there for a long time in a windy spot.
S1:	...Do an IDD <instrument reading> on it, get the Mössbauer <another instrument>.
S2:	Yeah, it's probably not going to look *like these little spheroids,* it's going to look different. And so, let's see what it looks like. ...
S2:	Inside the crater, then that would be a good place to do it Because we can work <missing words>
S1:	Now why do you think that that's the case? [said in a challenging manner]
S2:	Well, you're more likely to be able to tease out mineralogy <missing words>
S1:	But why do you think there are courser grains inside the crater?
S2:	Um, I'm just surmising, um
S4:	Well, you do see chunks—
S2:	The wind may have done this, this kind of the lag deposit inside here the sand is so coarse it's coarser than I would have expected.
S1:	Well background sand is fine, fine sand, The particles are pretty <missing word>
S3:	But clearly larger particles are up on the slopes outside the crater, even some cobbles.
S1:	It may be that that's the background is the fine ground up basalt sand.
Within-domain analogies leading to process conflict	
S2:	Okay, so you just do the... It's *almost like the context meeting presentation* in the downlink.
S1:	That's right, except I have abbreviated the context meeting because that ten minutes is all we have, we've got to move.
S2:	...I think the name of it, don't you mean a "pancam," "pancam can" because panoramic is more of a generic term for looking across the surface... Well, I think it's pancam.
S1:	Yeah but we can't say that *Like we say navigation camera* or...
S2:	Oh I see, okay. Well, if you... Go ahead and say what you want. I would guess panorama or pan-cam, pan-camera, but I don't know. I, we were told to use pancam in everything we write.

[a] S2, S1, etc. are speaker numbers. The first speaker in the clip is S1, etc. S2 in one clip may not be S2 in another clip.

Table 18.2 Examples of process conflicts (most key words underlined) leading to within-discipline analogies (in italics)

S5[a]:	Put the little thumbnails on there Cause I can look at them without opening the report…
S4:	Yeah, that's true, but, well, what I'm saying [missing words as loud laughter from offscreen drowns out S4] I mean, if I just have, if it's a Word doc, It just has a generic icon. So you have to <missing words> But if you see that it's over a couple hundred kilobytes then you know there's <missing words> like the SOWG documentarian <missing word> to yesterday that's 600 <missing words>
S5:	Umm, I see.
S4:	Yeah, but thank you for that because I hadn't thought about it. Somebody who was just reading the text wouldn't have a clue.
S5:	See, <missing words>
S4:	Nah, no, I mean I can say Rock, you know, and I usually, I try to do that because in the illustration it says "see attachment."
S5:	Wow, I mean I, <missing words>
S4:	[Laughs] Nah, I'm glad you told me that.
S5:	<missing words> …
S4:	In fact, <missing words> We made a picture of, we made a picture of the first soil targets. But I guess that's gone now in this morning's report, probably because it didn't have <missing words>
S5:	<missing words>
S4:	Yeah, if you'd have seen the, um, Pancam <missing words> higher resolution <missing words>
S5:	Yeah…
S4:	*See, it looked little, it looked like little SEM images*, you know? [Note: SEM images are scanning electron microscope images.]
S5:	I saw them.
S1:	But then—exactly, but the point is why then— It sounded like you have a bias towards the RATting versus getting [person's name] in a position where he can really analyze some rocks with a Mini-Tes.
S3:	Hey, let me make a suggestion—

(*continued*)

Table 18.2 (Continued)

S2:	It's not my bias. It's one that—it's one that you're hearing from several people.
S1:	Well, it's just what [second person's name] said.
S2:	Yeah, that's my point.
S1:	But is that in general?
S2:	And [third person] agrees with him.
S1:	I don't know that he does.
S2:	Well, maybe he's trying to be—
S3:	We were poo-pooing [fourth person's] super resolution, But maybe that's what you want to do to get a—something like the white rock.
S1:	That's exactly why I brought it up in SOWG yesterday, because it seems to me that if we don't go to this white rock, which is the one I know [fifth person] was thinking about, if we don't ever go to that white rock that's the best we'll do.
S3:	That way it's not a—it's not just an experiment— It's an important experiment.
S2:	Well, look, I've walked around in Nevada. And there's hydrothermal systems out there that were active a million years ago <missing words>. As a matter of fact, *gold prospectors walked over that terrain all through the 19th century and only when they realized that those things were sitting on top of gold deposits did they* <missing words>. The point is: you walk out there, you see low white rocks that are part of these quartz, you know super <missing words> deposits, And you can go another mile and not see another one.
S3:	I thought of something, The likelihood that we landed on one is pretty slim.
S2:	Well, yeah, sure you can get into that discussion, but—
S5:	Yes.
S1:	But we have to get into that though, because that's the sniper mentality. Not applied to engineering, but applied to science.

[a] S2, S1, etc. are speaker numbers. The first speaker in the clip is S1, etc. S2 in one clip may not be S2 in another clip.

similarly, the within-domain analogies would be a direct cause of the science (data analysis) and work process conflicts (Paletz, Schunn, & Kim, 2012). Qualitative unpacking of the specific pairs revealed that these explanations were not generally the case. First, the analogies were not usually aiding in the immediate resolution of preceding conflicts. We theorized in the discussion that the process conflicts might have sparked the subsequent within-discipline analogies via more indirect, implicit mechanisms, such as subtle types of social challenges and knowledge structure activation.

The analogy-to-conflict relationship was similarly complex. When we unpacked the 11 within-domain analogy/microconflict event pairs, the within-domain analogies were sparking 45% of the science and work process conflicts, but 55% were only sparked indirectly or in some other way related to the conflicts. These results suggest that the main relationships between within-domain analogies and the science and work process conflicts were somewhat more likely to be indirect and implicit rather than direct and explicit. The within-domain analogies preceding process and certain types of science conflicts could have been due to representational gaps in shared mental models regarding the domain of science on Mars. Given that the MER team was inherently multidisciplinary, the within-domain analogies may have been both assumed to be the most shared and, in fact, no more shared than within-discipline analogies. Both of these explanations appeal to underlying, hidden, implicit sociocognitive processes, rather than explicit, direct sociocognitive processes.

DISCUSSION OF THE EXPLICIT VERSUS IMPLICIT PROCESSES QUESTION

Both our model and empirical study contained explicit and implicit elements. The content of shared mental models can be either implicit or explicit. Similarly, most of the social and cognitive processes in our model are explicit when they are experienced and noted, but are implicit when they are, essentially, underlying knowledge structures, representational gaps, norms, and differences in background knowledge. In qualitatively examining analogy-to-conflict and conflict-to-analogy pairs, we found both that subtype was important and that the relationship between the two must involve implicit processes. Most of our hypothesizing about the utility of formal roles was because of their power to reify and *make explicit* what would normally be implicit: communication norms, differences in unique knowledge between team members, and the need for broad participation.

Our empirical study extended our original model in that, instead of simply drawing a connection from social and cognitive processes, we examined possible links from a cognitive process (analogy) to a social one (conflict). Both conflict and analogy can be explicit: Our method relied on both conflict and analogy being represented in conversation and interaction. That said, the method had strengths and weaknesses. On the one hand, the results were based on an observational study, without experimental controls, that could be specific to that setting. In coding interaction, we lacked data on both self-reports (explicit, known content) and internal, implicit processes. On the other hand, by coding communication and interaction, we examined what people actually said and did, even if those actions were extremely brief and may have occurred without conscious thought or later memory. In fact, brief interactions can elude even the most perceptive participant (Gottman & Notarius, 2000), suggesting that coded interactions may be both or either implicit and explicit. Furthermore, these interactions were not staged, forced, or subject to demand characteristics outside of the real-world situation of mission planning and scientific analysis.

CONCLUSION

Within psychology, different constructs are often identified as implicit, explicit, or both. While the content of attitudes, beliefs, and knowledge is often considered explicit, content can also be implicit (e.g., a familiar face, Reder et al., 2000; stereotypes, implicit theories of creativity). Similarly, processes are often considered automatic, but not all escape conscious awareness (e.g., the incrementally changing results of persuasion or power plays). Many processes require explicit goals to launch the process (Anderson, Bothell, Byrne, & LeBiere, 2004) or explicit goal structures to form new processes (Crowley, Shrager, & Siegler, 1997). In this chapter, we reviewed a pair of complex social/cognitive models of team innovation, drawing on both divergent and convergent thinking processes. In both pathways, most variables could be implicit or explicit. Because some of the variables sit between social and cognitive worlds, it will be important to negotiate a shared meaning of implicit/explicit across the cognitive and social psychology communities in conducting empirical work to establish the implicit/explicit status of each variable.

One particularly novel element of this work was to consider new connections between social and cognitive variables, and in this chapter we considered the nature of those connections. Overall, connections between variables may typically be implicit, either as part of processes in general or through implicit associations and priming in particular. However, some of the moderating variables (i.e., formal roles) in our model were suggested to have an effect because of their power to increase metaknowledge, that is, make explicit and known underlying, implicit factors. Overall, in examining the interaction between important variables from two such different subfields, we can better understand the nature of problem-solving conversations such as the implicit/explicit nature of thinking and, thus, creativity as it occurs in real time.

REFERENCES

Amabile, T. M. (1996). *Creativity in context: Update to the social psychology of creativity.* Boulder, CO: Westview Press.

Anderson, J. R., Bothell, D., Byrne, M., & LeBiere, C. (2004). An integrated theory of the mind. *Psychological Review, 111,* 1036–1060.

Antonio, A. L., Chang, M. J., Hakuta, K., Kenny, D., Levin, S., & Milem, J. (2004). Effects of racial diversity on complex thinking in college students. *Psychological Science, 15,* 507–514.

Baumeister, R. F., Vohs, K. D., & Funder, D. C. (2007). Psychology as the science of self-reports and finger movements: What happened to actual behavior? *Perspectives on Psychological Science, 2,* 396–403.

Burke, C. S., Stagl, K. C., Salas, E., Peirce, L., & Kendall, D. (2006). Understanding team adaptation: A conceptual analysis and model. *Journal of Applied Psychology, 91,* 1189–1207.

Christensen, B. T., & Schunn, C. D. (2005). Spontaneous access and analogical incubation effects. *Creativity Research Journal, 17*(2–3), 207–220.

Cronin, M. A., & Weingart, L. R. (2007). Representational gaps, information processing, and conflict in functionally diverse teams. *Academy of Management Review, 32,* 761–773.

Cropley, A. (2006). In praise of convergent thinking. *Creativity Research Journal, 18,* 391–404.

Crowley, K., Shrager, J., & Siegler, R. S. (1997). Strategy discovery as a competitive negotiation between metacognitive and associative knowledge. *Developmental Review, 17,* 462–489.

De Dreu, C. K. W., & Weingart, L. R. (2003). Task versus relationship conflict, team performance, and team member satisfaction: A meta-analysis. *Journal of Applied Psychology, 88,* 741–749.

De Dreu, C. K. W., & West, M. A. (2001). Minority dissent and team innovation: The importance of participation in decision making. *Journal of Applied Psychology, 86,* 1191–1201.

Derry, S. J., Schunn, C. D., & Gernsbacher, M. A. (Eds.). (2005). *Interdisciplinary collaboration: An emerging cognitive science.* Mahwah, NJ: Lawrence Erlbaum.

Dunbar, K. (1995). How scientists really reason: Scientific reasoning in real-world laboratories. In R. J. Sternberg & J. E. Davidson (Eds.), *The nature of insight* (pp. 365–395). Cambridge, MA: MIT Press.

Dunbar, K. (1997). How scientists think: On-line creativity and conceptual change in science. In T. B. Ward & S. M. Smith (Eds.), *Creative thought: An investigation of conceptual structures and processes* (pp. 461–493). Washington, DC: American Psychological Association.

Ericsson, K. A., & Simon, H. A. (1993). *Protocol analysis: Verbal reports as data* (2nd ed.). Cambridge, MA: MIT Press.

Forbus, K. D., Gentner, D., & Law, K. (1995). MAC/FAC: A model of similarity-based retrieval. *Cognitive Science, 19,* 141–205.

Gebert, D., Boerner, S., & Kearney, E. (2006). Cross-functionality and innovation in new product development teams: A dilemmatic structure and its consequences for the management of diversity. *European Journal of Work and Organizational Psychology, 15,* 431–458.

Gentner, D., Brem, S., Ferguson, R. W., Markman, A. B., Levidow, B. B., Wolff, P., & Forbus, K. D. (1997). Analogical reasoning and conceptual change: A case study of Johannes Kepler. *Journal of the Learning Sciences, 6,* 3–40.

Gorman, M. E. (2008). Scientific and technological expertise. *Journal of Psychology of Science and Technology, 1,* 23–31.

Gottman, J. M., & Notarius, C. I. (2000). Decade review: Observing marital interaction. *Journal of Marriage and the Family, 62,* 927–947.

Hart, J. W., Stasson, M. F., & Karau, S. J. (1999). Effects of source expertise and physical distance on minority influence. *Group Dynamics: Theory, Research, and Practice, 3,* 81–92.

Hinsz, V. B., Tindale, R. S., & Vollrath, D. A. (1997). The emerging conceptualization of groups as information processors. *Psychological Bulletin, 121,* 43–64.

Hocevar, D., & Bachelor, P. (1989). A taxonomy and critique of measurements used in the study of creativity. In J. A. Glover, R. R. Ronning, & C. R. Reynolds (Eds.), *Handbook of creativity* (pp. 53–75). New York: Plenum Press.

Holyoak, K. J., & Thagard, P. (1995). *Mental leaps: Analogy in creative thought.* Cambridge, MA: MIT Press.

Ilgen, D. R., Hollenbeck, J. R., Johnson, M., & Jundt, D. (2005). Teams in organizations: From input-process-output models to IMOI models. *Annual Review of Psychology, 56,* 517–543.

Jehn, K. A. (1997). A qualitative analysis of conflict types and dimensions in organizational groups. *Administrative Science Quarterly, 42,* 530–557.

Jehn, K. A., & Chatman, J. A. (2000). The influence of proportional and perceptual conflict composition on team performance. *International Journal of Conflict Management, 11,* 56–73.

Jehn, K. A., Northcraft, G. B., & Neale, M. A. (1999). Why differences make a difference: A field study of diversity, conflict, and performance in workgroups. *Administrative Science Quarterly, 44,* 741–763.

Jones, R. G., Stevens, M. J., & Fischer, D. L. (2000). Selection in team contexts. In J. F. Kehoe (Ed.), *Managing selection in changing organizations: Human resource strategies* (pp. 210–241). San Francisco, CA: Jossey-Bass.

Knorr-Cetina, K. (1999). *Epistemic cultures: How the sciences make knowledge.* Cambridge, MA: Harvard University Press.

Kozlowski, S. W. J., & Ilgen, D. R. (2006). Enhancing the effectiveness of work groups and teams. *Psychological Science in the Public Interest, 7,* 77–124.

Kurtzberg, T. R., & Amabile, T. M. (2000–2001). From Guilford to creative synergy: Opening the black box of team-level creativity. *Creativity Research Journal, 13,* 285–294.

Mannix, E., & Neale, M. A. (2005). What differences make a difference? The promise and reality of diverse teams in organizations. *Psychological Science in the Public Interest, 6*(2), 31–55.

McGrath, J. E. (1984). *Groups: Interaction and performance.* Englewood Cliffs, NJ: Prentice-Hall.

Nemeth, C. J., & Kwan, J. (1985). Originality of word associations as a function of majority vs. minority influence processes. *Social Psychology Quarterly, 48,* 277–282.

Nemeth, C. J., & Rogers, J. (1996). Dissent and the search for information. *British Journal of Social Psychology, 35,* 67–76.

Nersessian, N. J. (2009). How do engineering scientists think? Model-based simulation in biomedical engineering laboratories. *Topics in Cognitive Science, 1,* 730–757.

Nijstad, B. A., & Stroebe, W. (2006). How the group affects the mind: A cognitive model of idea generation in groups. *Personality and Social Psychology Review, 10,* 186–213.

Nisbett, R., & Wilson, T. (1977). Telling more than we can know: Verbal reports on mental processes. *Psychological Review, 84,* 31–259.

Nosek, B. A. (2005). Moderators of the relationship between implicit and explicit evaluation. *Journal of Experimental Psychology: General, 134,* 565–584.

Orasanu, J. (1995). *Evaluating team situation awareness through communication.* Paper presented at the International Conference on Experimental Analysis and Measurement of Situation Awareness. Daytona Beach, Florida.

Paletz, S. B. F., & Peng, K. (2008). Implicit theories of creativity across cultures: Novelty and appropriateness in two product domains. *Journal of Cross-Cultural Psychology, 39,* 286–302.

Paletz, S. B. F., Peng, K., & Li, S. (2011). In the world or in the head: Internal and external implicit theories of creativity. *Creativity Research Journal, 23,* 83–98.

Paletz, S. B. F., & Schunn, C. D. (2010). A social-cognitive framework of multidisciplinary team innovation. *Topics in Cognitive Science, 2,* 73–95.

Paletz, S. B. F., Schunn, C. D., & Kim, K. H. (2011). Conflict under the microscope: Micro-conflicts in naturalistic team discussions. *Negotiation and Conflict Management Research, 4,* 314–351.

Paletz, S. B. F., Schunn, C. D., & Kim, K. H. (2012). *The interplay of conflict and analogy in multidisciplinary teams.* Manuscript under review.

Pelled, L. H., Eisenhardt, K. M., & Xin, K. R. (1999). Exploring the black box: An analysis of work group diversity, conflict, and performance. *Administrative Science Quarterly, 44,* 1–28.

Reder, L. M., Nhouyvanisvong, A., Schunn, C. D., Ayers, M. S., Angstadt, P., & Hiraki, K. (2000). A mechanistic account of the mirror effect for word frequency: A computational model of remember-know judgments in a continuous recognition paradigm. *Journal of Experimental Psychology-Learning Memory and Cognition, 26,* 294–320.

Reder, L. M., & Schunn, C. D. (1996). Metacognition does not imply awareness: Strategy choice is governed by implicit learning and memory. In L. M. Reder (Ed.), *Implicit memory and metacognition* (pp. 45–78). Mahwah, NJ: Lawrence Erlbaum.

Salas, E., Stagl, K. C., & Burke, C. S. (2004). 25 years of team effectiveness in organizations: Research themes and emerging needs. In C. L. Cooper & E. T. Robertson (Eds.), *International review of industrial and organizational psychology* (Vol. 19, pp. 47–91). Chichester, UK: John Wiley.

Saunders, C. S., & Ahuja, M. K. (2006). Are all distributed teams the same? Differentiating between temporary and ongoing distributed teams. *Small Group Research, 37,* 662–700.

Schulz-Hardt, S., Brodbeck, F. C., Mojzisch, A., Kerschreiter, R., & Frey, D. D. (2006). Group decision making in hidden profile situations: Dissent as a facilitator for decision quality. *Journal of Personality and Social Psychology, 91,* 1080–1093.

Schunn, C. D., & Dunbar, K. (1996). Priming, analogy, and awareness in complex reasoning. *Memory & Cognition, 24,* 271–284.

Schunn, C. D., Lovett, M. C., & Reder, L. M. (2001). Awareness and working memory in strategy adaptivity. *Memory & Cognition, 29,* 254–266.

Stasser, G., Stewart, D. D., & Wittenbaum, G. M. (1995). Expert roles and information exchange during discussion: The importance of knowing who knows what. *Journal of Experimental Social Psychology, 31,* 244–265.

Stasser, G., & Titus, W. (1985). Pooling of unshared information in group decision making: Biased information sampling during group discussion. *Journal of Personality and Social Psychology, 48,* 1467–1478.

Stewart, D. D., & Stasser, G. (1995). Expert role assignment and information sampling during collective recall and decision making. *Journal of Personality and Social Psychology, 69,* 619–628.

Thompson, L., & Choi, H. S. (2006). *Creativity and innovation in organizational teams.* Mahwah, NJ: Lawrence Erlbaum.

Torrance, E. P. (1988). The nature of creativity as manifest in its testing. In R. J. Sternberg (Ed.), *The nature of creativity: Contemporary psychological perspectives* (pp. 43–75). Cambridge: Cambridge University Press.

van Knippenberg, D., De Dreu, C. K. W., & Homan, A. C. (2004). Work group diversity and group performance: An integrated model and research agenda. *Journal of Applied Psychology, 89,* 1008–1022.

van Knippenberg, D., & Schippers, M. C. (2007). Work group diversity. *Annual Review of Psychology, 58,* 515–541.

West, M. A. (1990). The social psychology of innovation in groups. In M. A. West & J. L. Farr (Eds.), *Innovation and creativity at work: Psychological and organizational strategies.* Chichester, UK: John Wiley.

PART SIX
UNCONVENTIONAL PERSPECTIVES ON THE CONDUCT OF SCIENCE

19 Implicit Ontological Reasoning

The Problems of Dualism in Psychological Science

Brent D. Slife, Jeffrey S. Reber, and James E. Faulconer

This chapter describes a theoretical meta-analysis in which the literature of science itself, especially psychological science, provides the data that ground its examination. Specifically, the chapter focuses on one portion of the implicit scientific reasoning or assumptions that psychologists use in conducting their scientific investigations—their ontological reasoning.

Few psychologists are likely to be familiar with the notion of ontological reasoning, even though they engage in it every day. They are more likely to be aware of their epistemological reasoning, where their assumptions about what can be known pervade their method decisions. We chose to focus on ontological reasoning for two reasons. First, ontological assumptions are typically viewed as even *more* fundamental than epistemological assumptions, and thus are considered more fundamental in the implicit reasoning of scientists (e.g., Richardson, Fowers, & Guignon, 1999). Epistemology may concern the nature of knowing, but ontology concerns the nature of what is. Ontological reasoning is reasoning done in terms of particular ontological assumptions that are usually implicit. Second, the ontological reasoning we identify in psychology's research methods has been highly criticized. In fact, we outline what seems to be a growing consensus, across a diverse set of contexts, that this particular ontology is deeply problematic. Yet, because many psychologists are generally unaware of it, they embrace this ontology as if it were problem-free.

As our analysis will show, part of the reason for this lack of awareness is that current method practices seem to be the "only game in town." That is to say, psychologists are so familiar with one form of ontological reasoning in their methods that they have taken it for granted. One way to combat this overfamiliarity is to introduce an alternative form of reasoning, and thus a methodological point of comparison. Indeed, one of the reasons psychologists' consciousness has been raised about epistemologies is that they are now aware of *rival* epistemologies in qualitative methods. Consequently, the latter portion of this chapter describes a highly successful program of research that uses alternative ontological reasoning. We hope that this contrasting research program helps psychologists not only consider other options, but also gain greater awareness of their current ontological reasoning.

THE NATURE OF ONTOLOGICAL ASSUMPTIONS

Identifying ontological assumptions in our discipline requires knowing more about them. Philosophers have not always been consistent in the way they have used the term "ontology," but here we mean assumptions about the nature of what is. The question, what is fundamentally real? is an ontological question, as is the question, how are the fundamental things of reality in relation? In psychology, for instance, what is mind? is a pivotal question (Kalat, 1992; Schwartz & Begley, 2002, p. 35), along with related issues, such as, is there a mind different from the brain? And is there an objective realm that is independent of a subjective realm? These latter questions raise the most fundamental issues of ontology: Is there only one fundamental reality (ontological monism), or are there two (ontological dualism)? We could, of course, consider the possibility of *many* fundamental realities or ontological pluralism, but historically monism and dualism have, for one reason or another, been the more compelling alternatives (Viney & King, 2003).

The pervasive influence of dualism, especially in Western culture, is widely acknowledged (Searle, 1997). For almost 300 years, as Solomon (1988) chronicles, each succeeding major thinker has recognized the problem of dualism and tried to offer a solution, unsuccessfully, at least until philosophers such as Heidegger (1926/2010) offered nondualistic accounts. Descartes, whom many consider the Father of modern philosophy, is credited with conceptualizing a version of dualism that continues to be pervasive in scholarly and popular culture (Damasio, 1994; Noë, 2010). Consider John Searle's (1996) depiction of this pervasiveness:

> It's a striking fact that we continue to pose and answer questions in terms that Descartes really popularized, in terms that he would have found completely comfortable and completely familiar to him. And in particular, this distinction between the mental and the physical, the mind and the body, spiritual and the material, and the idea that all of reality must divide in those two categories.

Essentially, from Descartes's perspective, the mind, which he takes to be identical to subjectivity, not only works differently from the body or the objective world, but is also *independent* of the objective realm in many important and real ways. In other words, Cartesian dualism presumes that the subjective and objective are two fundamentally distinct realities, each with its own ontic or reality status.

Descartes supposed that this dualism is a matter of the mind on one side and the body on the other, but his mind-body assumption is unnecessary to dualism. All that dualism requires is the presupposition that there are, in some sense, two distinct realities to be dealt with. Mind-body dualism does make this assumption, but it is not the only form of two distinct realities. Indeed, dualism has hung on as long as it has in the modern world *because* it is not the same as mind-body dualism. As we will see, dispensing with mind-body dualism does not mean that we have dispensed with dualism itself. Actually, subjectivity and objectivity are the most common ways of thinking of two ontological realities in the modern world, and we can understand the world in those terms without assuming mind-body dualism (Jones, 2010; Searle, 2004; Solomon, 1988). According to subject-object dualism, the subjective world consists of our opinions, feelings, and meanings, whereas the objective world consists of the objects and laws of the world. These two "worlds" are generally considered to have distinctly different qualities. The meanings of the subjective realm are thought to be more changeable and value-laden, whereas the objects and laws of the objective realm are viewed as less changeable and less value-laden. Physical laws, for example, are typically considered unchangeable and ultimately free of values.

These differences in quality do not mean that the two realms are not thought to interact. Dualists do consider them to interact, but from different origins, meaning that the subjective and objective worlds are ontologically independent of one another. Objects, such as a ring made of gold, can be *given* meaning from our subjective realm and thereby become a wedding ring, so the dualistic logic goes, but this meaning does not come from the ring's objective properties. Of course, a particular subjective meaning or idea about the objective world can be *grounded* in objective reality. As we will see, this grounding is considered one of the functions of a dualistic science, objectively grounding the subjective theories of the scientist. Paraphrasing Kant's famous couplet, theory without data is empty; data without theory are blind. Still, these meanings and theories are not actually *in* the objective world; they are subjective explanations or hypotheses *about* the objective world that can change as we obtain new information from the objective realm. From the dualistic perspective, the unchangeability of the objective world is generally thought to be more fundamental and truthful than the changeable meanings of the subjective world.

If this approach to ontology sounds familiar, it is probably because this ontological dualism is often viewed as *the* primary interpretive framework of traditional quantitative research. As Donald Jones (2010) puts it in his treatise on dualism, our Western method practices "take a particular [Cartesian] picture of the world as a starting point or foundation for everything that is done within those practices" (p. 6). In one of the best discussions of the problem of dualism, John Searle (2004, Chapters 2–4; see also 1996, 1997), argues that the problem can be understood by recognizing that consciousness has only a first person ontology, yet scientific analysis of the brain requires a third person ontology. According to Searle, traditional researchers consider themselves in the first person, as they subjectively view their personal world and even their research purposes, but they cast their results in the third person. As we will describe, this dualistic interpretive framework includes many mainstream psychological understandings of research. In fact, as we will attempt to show, these research practices are some of the more formalized manifestations of this dualistic ontology.

THE REJECTION OF DUALISM

"So what?" many psychologists might ask. Why should research psychologists care about these ontological issues, let alone the domination of a particular ontology in their research practices? Our answer is a simple one: They should care because these ontological issues pervade their scientific reasoning and methodological practices in a number of potentially *problematic* ways. As we will show, many psychologists have recognized this problematic status in the mind-body problem, perhaps most conspicuously in neuroscience. However, philosophers too, as we will document, have argued that dualism (mind-body or otherwise) simply does not work. Our argument is that psychologists should recognize the validity of these philosophical and neuroscientific conclusions and question not only mind-body dualism specifically, but also ontological dualism generally.

Psychologists are probably most familiar with the critiques of mind-body dualism in neuroscience. As Garza and Fisher Smith (2009) put it, "dualism... [is] a specter that haunts the conceptual and theoretical landscape in psychology, neurology, and cognitive neuroscience" (p. 520). Neuroscientist Antonio Damasio (1994) is, of course, noted for his pointed critique of Descartes's separation of mind and body, which he refers to as a significant and often unrecognized "error" that has misled many cognitive scientists and neuroscientists (see pp. 248–252). Noë (2010) is also clear that the last 25 years have led a growing number of neuroscientists to abandon the Cartesian dualism of mind and body for "an embodied, situated approach to mind" (p. 186) in which "we are dynamically coupled with the world, not separate from it" (p. 181). Rand and Llardi (2005) echo the same conclusion: "To the degree that a scientist subscribes to the still-widespread Western belief in mind-body dualism..., his or her ability to investigate the relationship between mental events and brain events may be compromised" (p. 9).

We submit that the arguments against mind-body dualism apply equally to dualism more broadly understood, that is, to ontological dualism. Along with Taylor (1995) and Solomon (1988), we argue that the ontological categories of mind-body dualism continue "beyond the demise of dualism in the contemporary demand for a neutral, objectifying science of human life and action" (Taylor 1995, pp. 5–7). Consider how this broader dualism is itself rejected by a qualitatively different discipline—philosophy. If this rejection is true, it would be particularly impressive because philosophers are famous, or rather infamous, for not agreeing on much of anything. Indeed, this lack of agreement is part of the reason that psychologists, and perhaps psychologists of science, have declined to include philosophical issues in the mainstream of our discipline: They have presumed that philosophical issues are intractable, an ideological whirlpool that ultimately provides no help for practical matters. How striking, then, would it be if a consensus existed among normally quarrelsome philosophers that dualism of any variety is wrong?

Consider, for example, two philosophers across the divide of Continental and Analytical philosophies, perhaps two of the more subsuming camps of professional philosophy. Ludwig Wittgenstein (1981), who spans both camps depending on the work you read, views dualism as "one of the most dangerous ideas" of Western philosophy (p. 104). John Searle, who typically disagrees with everything Wittgensteinian, nevertheless agrees on the pernicious influence of dualism. He makes clear in several books that the most significant intellectual problems of our contemporary world are the consequences of Cartesian dualism (Searle, 1996, 1997, 2004). As Taylor (1995) has chronicled in some detail, the last four hundred years of philosophy have been an ongoing struggle to escape from the problems created by Cartesian dualism. Indeed, he declares the whole dualistic approach to knowledge "to be mistaken" (Taylor, 1995, p. 12). Passmore (1970) may best sum up this philosophical consensus about dualism in any of its forms: "The rejection of dualism is indeed one of the few points on which almost all the creative philosophers of modern times have agreed" (p. 38).

Other fields have become similarly resistant to this problematic ontology, from biology to religion. For example, it may be of interest to note how many theologians have disputed dualism recently. Theologians are often viewed as advocates of dualism, because it supposedly helps them to give supernatural meanings some credence. Perhaps it is surprising, then, that many prominent religions and theologians avoid or directly oppose dualism. For example, one of the Four Noble Truths of Buddhism is that suffering stems from the creation of illusory dualisms, especially the creation of an independent reality such as objectivity. Islam also resists this type of dualism, opting to move away from divisions between self and world (Nelson, 2009, pp. 366–367). Christianity has historically championed some types of dualism, but as James Nelson (2009) explains, recent scholarship in Christianity explicitly rejects dualism and emphasizes the "unitive state" of life (p. 90).

The Basic Problem of Dualism

So what is going on? Why are so many writers across so many divergent disciplines and modes of inquiry so adamant about the rejection of dualism? Why is dualism, in all its forms, viewed as so profoundly problematic? These questions can, of course, thrust us into deep waters, but let us note in passing a few of the "in principle" problems that are at the heart of ontological dualism. The first, and perhaps the best-known, issue with dualism has been variously termed the *problem of interaction* (Griffin, 2000; Leahey, 1992; Viney & King, 2003). This problem involves the difficulties of explaining how two fundamentally different realities (e.g., subjective and objective, mind and body) can be unified in any way, whether that means how they interact or work together, or whether it means how a human being can understand them as a unity. This problem of interaction is one of the primary reasons that many neuroscientists have abandoned mind-body dualism, even though the problem of the unity of mind and brain continues (Noë, 2010).

As a second in-principle problem for dualism, the ultimate subjectivity of experience highlights the question of *where to draw the line* between subjective and objective realities (Griffin, 2000; Slife & Richardson, 2009; Viney & King, 2003). If we can never get outside our subjective experience of objective reality—that is, if we only know the objective world through our subjective experience—then how do we really know where our perceptions end and the external reality begins? Indeed, what are our grounds for believing in an objective reality at all? This problem in drawing the line between the two realities highlights or compounds a third problem of dualism: How do we correct our subjective meanings (Griffin, 2000; Viney & King, 2003)? As any dualist knows, "objective" observations are subject to problematic "subjective" influences, such as selective attention, demand characteristics, and confirmation bias (Nickerson, 1998). If objective observations fall prey to these types of supposedly subjective influences, how can we correct our subjective theories about objective reality? How can we measure the influence of our subjectivity on perceptions of an external objective world that cannot be directly known in the first place?

The bottom line here is not only that these "in-principle" problems of ontological dualism are difficult to overcome, but also that no historical remediation in these disciplines, from neuroscience to philosophy, has seemed to provide sufficient traction to solve these dualistic difficulties. Of course, there is a much longer story in these disciplinary efforts, including interactionism, parallelism, and emergentism, but the short story is that none of these efforts has mitigated dualism's difficulties. These efforts at remediation have not diminished how much dualism's problems continue to dominate Western intellectual culture, as treatises such as Solomon (1988), Taylor (1995), Searle (2004), and Jones (2010) describe.

THE IMPLICATIONS OF DUALISM FOR EPISTEMOLOGY

Does ontological dualism also dominate the logic of mainstream psychological research? To answer this question, we need to reflect on at least three of the epistemological implications of these "in-principle" ontological problems that bear directly upon psychological research. First, the notion that subjective reality is separate from objective reality suggests the possibility that the subjective might interfere with us gaining knowledge of the objective. Specifically, subjective experiences such as values, preferences, beliefs, and interpretations could distort an objective rendering of the world. In fact, there is a sense in which there is no such thing as "objective experience" because our experience of the objective world is always through our subjective "lens," and thus is viewed by ontological dualists as a representation of that world *by* the person rather than an appearing of the world *to* the person.

As a second epistemological implication of dualism, consider how observation is typically viewed as the closest subjective experience we have to the objective world. Observation is still part of our experience, and thus is ultimately subjective from the dualist perspective. Still, it is at least a portion of our experience that deals with the external world, a portion in which careful method can attempt to control for subjective distortion. Thus, according to dualism, observation is scientifically superior to our experiences with our subjective thoughts and feelings. Consequently, empiricism, and its emphasis on sensory experiences, is usually the chosen epistemology of the dualist.

Given that experiences of the external world can be distorted by subjective factors, such as biases and values, the third epistemological implication of ontological dualism is that it implies an ongoing methodological task: Dualists must constantly seek correction of their subjective representations of reality. This correction requires rigorous attempts both to eliminate the distortion of subjective biases and to seek the correspondence between subjective or intersubjective representations (theories) and objective reality (data) (Fishman, 1999; Stiles, 2009). (It also requires ignoring the epistemological question of how such correspondence is possible.) Gauging this corrective correspondence is frequently considered the province of the scientific method—the logic for gathering systematic, bias-minimized observations.

THE DUALISM OF PSYCHOLOGICAL RESEARCH

How might these epistemological implications of a dualistic ontology manifest themselves in psychologists' ways of conceptualizing and conducting research—what we view in this book as the implicit reasoning of scientists? Consider, in response to this question, our "Top Ten" methodological manifestations or practices of this reasoning. As you will see, this list does not require a *personal* subscription to ontological dualism (much less mind-body dualism) for the researcher to engage in dualistic method practices. In other words, no personal, formal, or conscious assumptions are necessary for one to reason in terms of ontological dualism. All that is required is to follow conventional training in psychological methods, as the quotes from the various method texts (cited below) evidence.

1. *Theory versus data.* Consider, first, our usual way of conceptualizing and dividing methodological inquiry. On one side of method is theory, which reflects the psychology researchers' subjectively agreed-upon ideas, models, and assumptions about the objective world. On the other side of method are data, which supposedly map onto the properties, quantities, and regularities of the objective world. Consistent with the first epistemological implication discussed above, theory and data are often thought to have the characteristics of each of the two separate realms of dualism. In a widely adopted methods text, Schweigert (2006) describes the characteristics of theory and data this way:

> To avoid being swept away by either unfounded speculations or biased perceptions, scientists tie their beliefs to concrete, observable, physical evidence that both independent observers and skeptics can double-check. Scientists look for *independent* evidence of their claim: objective evidence that does not depend on the scientist's theory or personal viewpoint. (p. 2)

Schweigert's point is logical from a dualistic perspective. Data could not be the "test" of theories, in the conventional sense, if the data were originally dependent on such theories. In other words, the issue for the dualist is not whether theory and data interact, but whether they originate from separate sources *when* they interact.

2. *Data as objective.* As noted in the second epistemological implication, sensory observation is considered to be closer to the objective world than subjective thoughts and feelings or subjectively agreed-upon ideas, like theories. Consequently, data are thought to provide the grounding for theory because they are collected through systematic, standardized, and repeatable empirical observations, and thus—in themselves—are supposedly uninterpreted indications of the value-free world of objects. Data collected in this manner, as Dyer (2006) put it in his methods text, "ground theories in reality" (p. 13). This is the reason, according to methodologists Mitchell and Jolley (2007), that empirically derived data are often referred to as "objective evidence" (p. 2). The data are the facts about objects beyond our subjective experience. However, if they are to have meaning to subjective humans, they must be interpreted using the subjectivity of theory. In this sense, empirical science begins with the difference between the subjective and objective in order to bring them together in an orderly way in the intersubjective (i.e., theoretical) interpretation of objective data.
3. *Avoiding subjective bias.* As close to objectivity as empirical data may be, the dualist views them as potentially susceptible to subjective interference. In Dyer's (2006) words, "The information acquired by the senses always requires some degree of interpretation by the person whose senses they are. Sensory information cannot, therefore, 'speak for itself' but always requires a reasoned act of interpretation by the observer" (p. 4). Given that subjective distortion is possible, it is imperative that the objective researcher attempt to avoid any bias in his or her methods of data collection and interpretation (Mitchell & Jolley, 2007). Indeed, any indication of subjective biases in the data is a sure sign of problems with the researcher's objectivity, because the objective world is itself presumed to be free of subjectivity, and thus bias-free. As Schweigert (2006) tells us, "the scientific" approach is adopted so that the results of the research will be meaningful, unambiguous, and uncontaminated by the biases of either the participants or the researcher" (p. 2). In this sense, dualism manifests itself in the clear injunction to strive to separate the subjective (biases) from the objective (information) for the sake of valid knowledge.
4. *Correcting for subjectivity.* If, as the first epistemological implication indicates, all scientists experience the world (including their data) subjectively, then psychology researchers must be constantly on the lookout for subjective distortions, and as the third epistemological implication enjoins, must constantly take corrective actions to remove those distortions. Indeed, this is the primary value of the scientific method; it is the systematic corrector of problematic subjective factors in the light of supposedly value-free, objective data. This correction supposedly ensures a clear and relatively undistorted correspondence between the objective and subjective worlds. As Stiles (2009) writes, "Researchers creatively modify their theories by (abductively) adding to them or altering them so that they correspond to accumulating observations" (p. 1).
5. *The objectivity of scientific method.* Of course, this corrective function would need to extend to the logic of the scientific method itself, which must also be as free of values as possible. If the scientific method were itself biased and value-laden,

it would be systematically biased, and thus yield to subjectivity. It would not be objective, and thus not an accurate rendering of the objective world. True to form, and despite much analysis on the value-ladenness or interpretive nature of the scientific method (e.g., Gadamer, 1993; Griffin, 2000; Kuhn, 1970; Slife & Williams, 1995), method texts almost never identify these values and interpretations. Instead, the dualistic presumption is that the scientific method, with the logic of its procedures, is a relatively fair and unbiased means of mapping the world. For the psychological researcher, according to Mitchell and Jolley (2007), the scientific method is "our most objective way of knowing" (p. 16).

6. *The objectivity of researchers.* The presumed corrective objectivity of this dualist ontology is frequently thought to ensure the objectivity of the community of researchers. As Marczyk, DeMatteo, and Festinger (2005) stated it, "Scientific knowledge is not based on the opinions, feelings, or intuition of the scientist. Instead, scientific knowledge is based on objective data that were reliably obtained in the context of a carefully designed research study" (p. 4). This is the reason no biases, personal or professional, are typically admitted or reported in research publications. Indeed, these reports are nearly always written in third person, as if the researcher is absent or irrelevant to the gathering or interpreting of the data. Objective data and subjective biases are two different realms of thought and experience, and their separation must be maintained.
7. *Theory as universal propositions.* Despite the subjectivity of theories, from the dualist perspective the best psychological theories are nearly always framed in terms of postulated, universal propositions or principles. This is the case primarily because the natural objective world is assumed to consist of universal and unchangeable laws (Heiman, 1999). Thus, the truly superior theories in psychology are those that provide mechanistic, and more recently, computational models of human thoughts, feelings, and actions that are thought to parallel the lawfully determined and mathematically derivable processes of human life (Wyschogrod, 2002). The closer the parallel between the universal model and the objective world, the greater the confidence psychologists can have that their theories represent the universal principles of the world with as little subjectivity as possible.
8. *The replication of findings.* The presumed presence of these objective laws implies that research findings need to be replicable. Laws, and thus objective findings that reflect those laws, should occur across contexts, and be repeatable in this sense. Ray (2006), for example, reflects this assumption in his methods text: "It must be possible for different people in different places and at different times using a similar method to produce the same results" (p. 7). Supposedly, this repeatability occurs only in dealing with the objective realm, hence the perceived need for reliability, standardization, and stability (Cohen & Swerdlik, 2005).
9. *The need for prediction.* Replication provides a consistency to research findings that strengthens researchers' confidence that their subjective hypotheses accurately reflect the natural laws acting in the objective world (Heiman, 1999, p. 21). It also leads to an increased confidence in the power of subjective hypotheses to predict the behavior of objects in the objective world under specified conditions.

Such predictions can only be made probabilistically, of course, because they are always hampered by the inescapable subjectivity of our experience (Okasha, 2002). Researchers can never be sure their predictions are undistorted and actually realized in the objective world. It is for this dualistic reason that researchers avoid terms like "certainty" and "proof" in their findings and predictions.

10. *The enhancement of objectivity.* The reduction of subjective biases, and thus the increase in probability, is generally understood as accuracy in gauging the objective realm. Because, given the assumption of dualism, the objective realm is presumed to be independent of the subjective realm, the best way to gauge the objective is through the elimination of the subjective (biases). This elimination is achieved, in part, by developing observational devices and instruments that enhance sensory observation and standardize measurement (Cohen & Swerdlik, 2005). The reduction in bias, and thus in dualistic error, increases researcher confidence that the observations are as close to the objective world as possible.

We could add to this list of dualistic conceptions and practices, but we think the point has been made: Ontological dualism is central to contemporary psychological research. Indeed, the dualism of psychological theory and practice is like water for fish—so dominant and so pervasive in the discipline that many mainstream psychologists might ask: How else could they be? In other words, many psychologists may not know of another way to think about investigation.

THE DOMINANCE OF DUALISM

The obvious questions about the overwhelming dominance of a potentially problematic conceptualization are: Why does dualism have such swaying and staying power in psychological research? and Why don't at least some researchers see its difficulties and search for an alternative? Like all complex issues, the answers to these questions are themselves complex, with a myriad of factors contributing to this dominance. Space constraints prohibit the full development of all the factors here. Clearly, as Taylor (1995) and Searle (1997) describe, there are numerous social and institutional supports for the dualistic status quo. Clearly, also, there has been a long and complex history of dualism, particularly during the rise of science and especially when psychologists took their cues on the identity of science from early manifestations of the natural sciences.

We believe that many psychologists would defend their use of dualistic methods with a twofold rationale: (1) it works, and (2) it is the "only game in town." Actually, we think these two defenses are part of the same issue: Psychologists really do not have an alternative nondualist approach, either to explore or to compare to the current approach. Claims that current method practices are successful are shallow in this regard because they have not been compared to nondualist forms of method reasoning. Historic rivals for this reasoning, such as introspectionism, have been dualist themselves, merely favoring the subjective side of the dualism (Viney & King, 2003). Moreover, the radical behaviorist's and the neuroscientist's historic dislike of dualistic theory has rarely transferred to their methods, which partake of the same method practices just reviewed (Packer, 2010).

The assumption seems to be that the denial of mind-body dualism implies the denial of dualism generally, but, as we have described, one can deny one type of dualism while embracing another. The dominance of dualism is so pervasive in psychology that many qualitative, nondualist approaches are often conducted in a dualist manner and judged according to dualist criteria (e.g., Packer, 2010; Slife & Richardson, 2009; Stiles, 2009). For example, qualitative findings are often viewed as too subjective to be objective knowledge (Shek, Tang, & Han, 2010). Given this dominance in the methodology of the discipline, how would we know if the "only game in town" truly works? Surely, we wouldn't judge its working solely by our ability to play the game successfully. We would need other "games"—other logics of investigation—with which to compare it.

Unfortunately, the need for a nondualist methodology does not make it magically materialize, especially when dualism is so institutionalized. As this chapter argues, psychologists are just now beginning to discern the dualism of their methods. Other steps are required to even begin to reform our procedures, such as seeing this dualism as problematic, seeing where it makes its appearances, and then desiring alternatives. But desiring alternatives is not sufficient either, especially if psychologists assume that viable nondualist alternatives are not possible. Thus, an important first task in even identifying this dualistic reasoning of psychological scientists is to identify nondualistic methods in *any* sphere of inquiry. Without successful alternatives, dualistic methods really are the "only game in town", and nondualism is a convenient and perhaps interesting philosophical point, but a practical methodological fiction.

As it happens, there are many individuals working to develop nondualistic modes of inquiry, or what is sometimes called *interpretive approaches to inquiry*. Martin Packer (2010), Paul Ricoeur (1965, 1981, 1991, 1992, 2004), Charles Taylor (1971), Hans-Georg Gadamer (1993), and Frank Richardson (with Fowers & Guignon, 1999) are just a few of the persons who are, in our view, providing the most promising conceptual bases for this new approach to psychological investigation. However, a summary of these efforts is beyond the scope of this chapter, and the work of these thinkers remains largely theoretical. Consequently, it may be more instructive to offer a practical, real-life example of a fully fledged and currently working program of nondualist research. We think many psychologists may be pleasantly surprised by how relatively easy it is to understand and deploy such a program.

A NONDUALIST ALTERNATIVE: CONSUMERS UNION

Consumers Union (CU) is a nonprofit organization founded in 1936 to inform consumers about the quality and safety of the products they use. Their more recognizable publication, *Consumer Reports*, is CU's primary outlet for providing this information to consumers. CU has a long history of a widely praised research program for gauging the quality of particular consumer goods, including cars, which will be our exemplar of a nondualist approach to inquiry (Slife & Richardson, 2009).

We should note at the outset that the CU project can be interpreted, we believe erroneously, in a dualistic manner. CU, for example, regularly characterizes its car ratings as "unbiased." This characterization would seem, at first blush, to fit with a dualist approach to car investigations. The dualist would want investigators to strive to eliminate as many subjective biases as possible, so that only the objective truth of the cars would remain (method practice #3). For this reason, CU investigators should seek to eliminate CU's own values and biases, among other subjective factors, because subjective values and biases would presumably distort the investigators' "corrective experiences" of the cars themselves in their pristine objectivity (method practice #4).

A closer look at CU's car rating program, however, reveals that this elimination is not what they do. When CU claims to be unbiased, the notion of bias refers to the possibility of undue influence rather than subjective values. Specifically, CU takes no revenue or samples from carmakers because it fears that consumers will wonder whether its ratings are influenced by these kinds of financial considerations. CU's fears in this regard could also be the result of its awareness of prominent dualist interpretations of its ratings, and thus consumer anxieties about subjective biases.

Still, CU does not attempt to eliminate or even minimize its supposedly subjective values in formulating its assessment of cars. Like other rating organizations, CU, in formulating the criteria for its rating system, presumes a general moral framework, which includes values like justice, fairness, and consumer protection (see Consumers Union, 2010). In rating cars specifically, CU values factors like reliability, drivability, and owner satisfaction, but it puts the greatest premium on safety. CU's car ratings, in this sense, are partially constituted by CU's biases and values. We say "partially" because the cars themselves are an important part of the ratings; the construction and functioning of the cars influence what kind of ratings can be reasonably made and whether the cars get high or low ratings.

In this sense, the data produced by CU's investigation of the cars is neither objective nor subjective. CU's car ratings—as helpful as they may be for understanding cars—are not a bias-free description or map of the cars as they exist in their objective relations with other objects in the objective world. Neither are they merely a report of the subjective meanings that researchers have imposed on the world they investigated. Rather, CU's car ratings are explicitly value-laden ratings of the cars' quality and performance. Another moral framework with a different set of values would lead to a *completely different* set of car ratings. For example, if another rating organization valued drivability over safety, some cars ranked at the bottom of the CU's safety-first list might now appear at the top of the other organization's ratings. Two different rating organizations, perhaps CU and a racing association, could even evaluate the *exact* same set of cars but reach an entirely different set of car rankings, just because different values guided the ratings. This means that values are a vital constituent of ratings; ratings do not exist without them. In terms familiar to the dualist, they are inextricably "subjective objectivities"; the subjective and the objective have never been and can never be separated. In nondualist terms, they are *meanings*, expressions of a human historical and cultural situatedness.

DUALISM AND THE TRUTH OF CAR RATINGS

The *methodological dualist* might understandably challenge this contention of the inseparability of subjectivity and objectivity in meanings, and the exploration of this challenge will help us to clarify why so many disciplines have abandoned a dualist approach to inquiry. The general thrust of a dualist approach to car ratings would be to formulate methods that attempt to move the ratings to greater objectivity (method practice #2). In other words, the main goal would be to move *closer* to the supposedly objective world of the cars and *farther* from the presumably subjective world of beliefs and values (method practices #1–#3). The problem with this dualistic approach is that there is never a time in which the ratings are closer or farther from this subjective world, because the ratings require the values even to exist.

The methodological dualist might say that the truth is approximated when our subjective beliefs *correspond* to the objective world (method practice #4). The problem, again, is that no such truthful ratings exist in the objective world; that is, there are no objective, value-free ratings for CU's ratings to correspond to. The cars are necessary to the ratings, of course, but they are never sufficient alone to account for the ratings. The cars exist in the *ratings* only insofar as the cars are revealed by the values of safety and drivability. (Recall how different values would mean different ratings.) Indeed, the values lead to certain measures, such as crash tests, that help to form the ratings. From this perspective, the truth of the car *ratings*, as distinguished at this point from the cars themselves, can never be objective or unbiased in the dualist sense that they pertain *only* or even *more* to this objective world, or that they are determined by value-free judgments. Even financial independence from the carmakers does not make the ratings any less dependent on the value-laden criteria used to rate the cars. Financial independence may help us to put more trust in the *criteria* or values used as well as the tests that embody those values. However, this type of independence does not make the ratings any less value-laden or bring us any closer to a value-free, objective description of the world.

A methodological dualist might attempt to avoid the ratings altogether and try to access the different subjective and objective worlds that supposedly form the truth of the ratings (method practice #5). This tack assumes that the values and cars themselves, the supposed subjectivities and objectivities of this example, would each need to be evaluated for their proximity or correspondence to the objective world. But how can the dualist evaluate how true to the objective world CU's values are? If these values are inherently subjective, and thus unrelated to the objective world by definition of their separation from this world, then dualists cannot evaluate the truthfulness of this subjectivity because, in principle, it can never be closer to or farther from objective reality. This is part of the reason that objective science doesn't attempt to evaluate moral systems: They are thought not to be part of the objective world. Even positive psychology, in this sense, only gathers data *about* moral systems; it does not and cannot evaluate their truthfulness empirically (Slife & Richardson, 2008). For the methodological dualist, then, the objective facts are one thing, and human valuings are something completely separate.

"Wait a moment," says the *non*dualist. "Are CU's values really subjective in a strict sense? Are they really independent of the presumed objectivity of the cars themselves, as the dualist would hold? What could the specific value of 'drivability' mean, except in its relation to the objects called 'cars'?" This CU value, from a nondualist perspective, *comes from* as much as it is *applied to* the world of cars. The value of safety, though not unique to cars, is nevertheless related to our experience with fast-moving, death-dealing modes of transportation. In other words, CU's particular values *matter* because of the nature of the cars themselves and our relation to them. In this sense, it is difficult to understand CU's values as subjective in the conventional dualist sense of *un*related to the cars as objects. These values have a status similar to the ratings from the outlook of a nondualist: They are neither subjective nor objective, but a unitary, interpreted reality or meaning.

What about the cars themselves? From a dualist's perspective, the cars make up the objective, material things to which everything truthful about the ratings should correspond (method practice #2). This perspective implies that the cars are unrelated in some original sense to the values CU uses to evaluate them. "But is this strictly true?" asks the *non*dualist again. As mentioned, carmakers clearly attempt to manufacture cars that fit the values of car-buyers. Weren't cars, and all their components, invented and shaped, in fact, because of the values of the people who desired them? If this is true—that cars themselves cannot be fully understood except in relation to values—then even the so-called objective portion of the car ratings is not objective in the methodological dualist's sense.

At this point, the nondualist could conclude that the car ratings cannot be dualistically understood because they are meanings "all the way down" (Held, 2007, p. 283). That is, all aspects of the ratings are expressions of cultural and historical embeddedness, with neither the supposedly subjective nor the supposedly objective aspects of these meanings separable from one another. This nondualist position, sometimes referred to as *hermeneutic realism* or *ontological relationality*, assumes that the world consists of relational meanings rather than isolatable objects (Bernstein, 1983; Fishman, 1999; Messer, Sass, & Woolfolk, 1988; Packer & Addison, 1989; Richardson et al., 1999; Slife, 2004).

Some readers may infer from this conclusion that we are arguing for a kind of idealism; that is, the only real things are ideas, or at least the only things we can investigate are ideas. That inference comes from assuming that scientific investigation has the character that the methodological dualist assumes—the existence and separability of objective facts and subjective values. With that assumption, anyone who argues against the existence of objective facts, as we have done, appears to be arguing for subjectivity, an abandonment of the real from a dualist's perspective (i.e., the objective). In other words, "real" and "objective" have been so confounded in dualism that an argument against the latter is perceived to be an argument against the former.

But nothing could be further from the truth. Interpretive or hermeneutic realism *is* realism rather than idealism. For example, CU's measurement of "braking distance" and "miles per gallon" is real to the hermeneuticist, but it is not real in the objective sense; hermeneutics is a different ontology (a different understanding of what is real). Braking distance and miles per gallon are meanings, not objects, in which case they cannot be understood except in relation to us in our historical and cultural situatedness. Still, *this relationship does not imply that these meanings have to be subjectivities or unreal,* except from a dualist perspective. Nor does hermeneutics assume that reality is ideal. Rather, it supposes that human valuings are inseparable from scientific investigation. Without these valuings, for instance, we would not care about braking distance or gas mileage, and we would not measure them, or even invent ways to measure them.

Methodological dualists might admit that the selection of such factors for measuring cars reflects the values of people, but they might reject the notion that the *measurement* of these factors is itself value-laden (method practice #10). Here, we could certainly refer to historic battles concerning the standards for measuring braking distance and gas mileage, which were all, ultimately, battles over particular values. However, we suspect methodological dualists are more concerned about measurement in the sense of the scientific method (method practice #5 above). In other words, they might concede that the *discovery* context of science is value-laden (i.e., the generation of hypotheses), but dispute the value-ladenness of the *justification* context of science (i.e., the actual testing of the hypotheses). Given our limited space, we can only refer the reader to the number of articles and books that make clear the values inherent in the justification context of the scientific method (Bernstein, 1983; Fishman, 1999; Gadamer, 1993; Griffin, 2000; Heidegger, 1926/2010 Messer et al., 1988; Packer, 2010; Ricoeur, 1981; Taylor, 1995). Indeed, the first author (Slife, 2008) took part in a special issue of the journal *Counseling and Values* (2008), where many authors explicated the values involved in counseling research, including those involved in the context of justification.

The point here is that such valuings are not irrelevant at *any* point in the process of investigation. There are not two ways of understanding the world, the objective and the subjective; there is only one, from the hermeneutic perspective. Psychological researchers must choose the topics of investigation (some of which have been mistakenly described as "objective" and others as "subjective"), choose the methodology or philosophy of science (e.g., positivism, hermeneutics), choose the procedures or methods from that philosophy which best investigate the topic at hand, choose the best approach to analyzing the data generated (which have their own assumptions), and then choose how to best interpret the findings. Values are involved at every step. The only question is whether they are acknowledged, monitored, and accounted for in the process of knowledge advancement. If this nondualist position is correct, then dualist notions of truth are not correct and can only yield the truth coincidentally. This is, in fact, the conclusion of the many disciplines we cited previously.

Nondualism and Psychology

What about psychology in this regard? Could a nondualist method, like that of CU, really work as a viable alternative for psychological research? Not only have several scholars answered this question affirmatively, but many also believe that psychology's successes are actually *already* due to nondualist practices that have been unwittingly smuggled into the dualist story of psychological method (Slife & Richardson, 2009). For example, the instruments and measures used in psychological research could be understood as value-laden ratings designed to study those aspects of human beings that the discipline views as morally significant. Consider the many self-esteem instruments used in various psychology studies and therapeutic assessments

(Mruk, 2006). For the dualist, the data collected using these instruments map some feature of the objective human world about which subjective theories have been formed.

For the nondualist, on the other hand, self-esteem is much like the value of safety for the CU. It is part of psychology's moral framework and has long been valued as an important contributor to health and well-being, two core values of the discipline. Recall that a value-free, objective evaluation of a car (in the dualist sense) would never reveal or disclose the safety of the car. Conventional scientific instruments and scientific observations, such as through a microscope or telescope, might tell us many things about cars, but they could *never* disclose a car's safety or drivability. To disclose these meanings, one must devise procedures for evaluation that embody these values, such as crash tests. In other words, *the values themselves are necessary for the car to disclose this meaning*; values and cars cannot be understood apart from one another in arriving at a meaningful truth. As Heidegger (1926/2010) put it, disclosure takes place *within* our concerns and involvements with the world, not *outside* them.

Similarly, an objective, value-free study of persons would never disclose their self-esteem because there is no self-esteem out there in the objective world to which the subjective idea corresponds. Rather, self-esteem, like safety, is a meaning that is no less real but cannot be understood apart from the values of health and well-being and is disclosed only by procedures for evaluation that embody these values. Consequently, study procedures are designed and instruments are developed that tend to focus researchers on participants' self-esteem in relation to obesity, stress, heart disease, and many other issues that reflect the core values of the discipline (Mruk, 2006). In this sense, self-esteem, as well as the values of psychology and the broader society that inform and are informed by self-esteem, cannot be understood apart from each other.

A second implication of CU's nondualistic method for psychology is that the logic and design of the scientific method are themselves inseparable from the values of the scientists who practice them. As described previously (method practice #5), the logic of science for the dualist is presumed to be relatively free of values and thus capable of providing a relatively undistorted tool for mapping the objective reality of the world. However, this presumption is held in spite of, rather than because of, scholarship on psychology's research methodology. As enumerated above, many analyses of this methodology have noted the abundance of preinvestigatory assumptions and values that are implicit in our research methods (Bernstein, 1983; Fishman, 1999; Gadamer, 1993; Griffin, 2000; Heidegger, 19626/2010 Messer et al., 1988; Packer, 2010; Ricoeur, 1981; Taylor, 1995).

For the nondualist, science is viewed from the outset as a value-laden set of procedures designed to test for the values scientists endorse. The question is not whether scientific research is value-laden, but which values are and ought to be at play. Just as the CU designs its crash tests, consumer surveys, and driving evaluations to illuminate the meanings of car safety, consumer protection, and drivability, so too psychologists develop their methods, procedures, hypotheses, and analyses to bring to light the values they find important for human beings, such as predictability, generalizability, and reliability. As suggested, this could mean that psychology's methods are *already* value-laden, through and through, and thus *already* nondualist in nature, despite their dualist depiction and description in method texts and psychology's journal articles.

What then is the problem? The problem is that psychologists teach these methods and conduct their studies as if the 10 dualistic practices we described are correct. Instead of identifying their assumptions and values, and monitoring their effects on studies, budding and veteran investigators assume they can and should eliminate or reduce them. This method practice means that one of the most vital ingredients in any study from a nondualist perspective, its root values, is not only left unexamined, but also thought to be in need of elimination. This is just one of the 10 dualistic method practices that a nondualist would challenge. There is little question that the other nine are just as impactful, with their continuing uncritical exercise just as consequential for our understanding of investigations.

Implications for the Psychology of Science

What does all this mean for the psychology of science (as apart from psychology more generally)? The ontological issues we have described could be understood as more philosophy than psychology, but are they? We contend that they are just as much about implicit scientific reasoning as they are about philosophical frameworks. As we have shown, dualism and nondualism inform the logic of practical method decisions just as much as they inform philosophy. Why not identify them as some of the assumptions of scientists' methods and permit them to yield their insights into the minds of scientists when conducting these methods? As psychologists of science, we cannot avoid the philosophical when important issues such as dualistic reasoning are at least partly conceptual rather than empirical. We realize that most psychologists of science have been trained in empirical methods only, and these methods are what are familiar, but many psychologists have been doing conceptual analyses for years (e.g., Slife, Reber, & Richardson, 2005).

If we limit the psychology of science to only one type of investigation, such as empiricism, we run the risk of overlooking key components of our subject matter. Such limits could be analogous to the proverbial drunk who is searching for his car keys where the light is better, though he dropped them somewhere else. In other words, just because we have a light, in the sense of our empirical methods, does not mean that everything we need to know in the psychology of science is discoverable with that method, especially if we endorse a deeply problematic logic of method like dualism. And why wouldn't we welcome the grounding of methods on alternative assumptions? The historical case is that we have endorsed dualism and empiricism for philosophical reasons, not because we have systematically compared them to nondualist and nonempirical methods and found them to be more effective. The effectiveness of CU's project should be a cautionary tale here. Why not acknowledge our *philosophy* of science and allow it to help us with our *psychology* of science?

Some psychologists of science, such as Gregory Feist (2006), agree that "it is of utmost importance to understand the history of science, the philosophy of science, and the sociology of science" (p. 33). But if these disciplines are seen only as allied studies of science, alongside the psychological study of science, there will be little appreciation for the important philosophical, historical, and sociological issues and insights that bear upon our *own* scientific study of scientists' thoughts and behaviors. Indeed, shouldn't we consider the bootstrap problem involved in using a method of reasoning to examine a scientist engaged in the same method of reasoning? What is the impact of using empirical and dualistic methods to examine the act of scientists using empirical and dualistic methods? What is ignored? Shouldn't we consider alternatives in such a study? When is the philosophy that guides the reasoning of the empirical scientist itself discussed and explored for the insights it can yield in understanding the psychology of science?

We are not advocating the rejection of dualistic methods in the psychology of science, or psychology for that matter. We are advocating a type of methodological pluralism where the pros and cons of particular types of method reasoning are evaluated in the light of particular investigative questions. Although we cannot develop this pluralism fully here (see Slife & Gantt, 1999; Slife & Melling, in press), our main point is that we should develop alternative methodological resources, so that we have not only other method options at our disposal, but also greater awareness of our implicit methodological reasoning. Just like a carpenter, we could evaluate the advantages and disadvantages of all the tools in our tool chest and use the best mode of inquiry for the job at hand.

REFERENCES

Bernstein, R. J. (1983). *Beyond objectivism and relativism: Science, hermeneutics, and praxis.* Philadelphia: University of Pennsylvania Press.

Cohen, R. J., & Swerdlik, M. E. (2005). *Psychological testing and assessment: An introduction to tests and measurement.* New York: McGraw Hill.

Consumers Union. (2010). Our mission. Retrieved July 20, 2010, from http://www.consumerreports.org/cro/aboutus/mission/overview/index.htm

Damasio, A. R. (1994). *Descartes' error: Emotion, reason, and the human brain.* New York: Putnam.

Dyer, C. (2006). *Research in psychology: A practical guide to methods and statistics.* Malden, MA: Blackwell.

Feist, G. J. (2006). *The psychology of science and the origins of the scientific mind.* New Haven, CT: Yale University Press.

Fishman, D. B. (1999). *The case for pragmatic psychology.* New York: New York University Press.

Gadamer, H. G. (1993). *Truth and method* (2nd ed.). (J. Weinsheimer & D. G. Marshall, Trans.). New York: Continuum.

Garza, G., & Fisher Smith, A. (2009). Beyond neurobiological reductionism: Recovering the intentional and expressive body. *Theory & Psychology, 19,* 519–544.

Griffin, D. R. (2000). *Religion and scientific naturalism: Overcoming the conflicts.* Albany: State University of New York Press.

Heidegger, M. (1926/2010). *Being and time* (J. Stambaugh, Trans., D. J. Schmidt, Rev.). Albany: State University of New York Press.

Heiman, G. W. (1999). *Research methods in psychology* (2nd ed.). Boston: Houghton Mifflin.

Held, B. S. (2007). *The interpretive turn: The search for truth and agency in theoretical and philosophical psychology.* Washington, DC: American Psychological Association.

Jones, D. (2010). *Changing pictures of social science theory and practice: A Wittgensteinian approach to human minds and experience.* Unpublished doctoral dissertation, University of Texas, Austin.

Kalat, J. (1992). *Biological psychology.* Belmont, CA: Thomson Wadsworth.

Kuhn, T. S. (1970). *The structure of scientific revolutions.* Chicago: Chicago University Press.

Leahey, T. H. (1992). *A history of modern psychology.* Englewood Cliffs, NJ: Prentice-Hall.

Marczyk, G., DeMatteo, D., & Festinger, D. (2005). *Essentials of research design and methodology.* Hoboken, NJ: John Wiley.

Messer, S. B., Sass, L. A., & Woolfolk, R. L. (1988). *Hermeneutics and psychological theory: Interpretive perspectives on personality, psychotherapy, and psychopathology.* New Brunswick, NJ: Rutgers University Press.

Mitchell, M. L., & Jolley, J. M. (2007). *Research design explained* (6th ed.). Belmont, CA: Thomson Wadsworth.

Mruk, C. (2006). *Self-Esteem research, theory, and practice: Toward a positive psychology of self-esteem* (3rd ed.). New York: Springer.

Nelson, J. (2009). *Psychology, religion, and spirituality.* New York: Springer.

Nickerson, R. S. (1998). Confirmation bias: A ubiquitous phenomenon in many guises. *Review of General Psychology, 2,* 175–220.

Noë, A. (2010). *Out of our heads. Why you are not your brain, and other lessons from the biology of consciousness.* New York: Hill & Wang.

Okasha, S. (2002). *Philosophy of science: A very short introduction.* Oxford: Oxford University Press.

Packer, M. J. (2010). *The science of qualitative research.* Cambridge, MA: Cambridge University Press.

Packer, M. J., & Addison, R. B. (Eds.). (1989). *Entering the circle: Hermeneutic investigation in psychology.* Albany: State University of New York Press.

Passmore, J. (1970). *Philosophical reasoning* (2nd ed.). London: G. Duckworth.

Rand, K. L. & Llardi, S. S. (2005). Toward a consilient science of psychology. *Journal of Clinical Psychology, 61,* 7–20.

Ray, W. J. (2006). *Methods: Toward a science of behavior and experience* (8th ed.). Belmont, CA: Thomson Wadsworth.

Richardson, F. C., Fowers, B. J., & Guignon, C. B. (1999). *Re-envisioning psychology: Moral dimensions of theory and practice.* San Francisco, CA: Jossey-Bass.

Ricoeur, P. (1965). *Freud and philosophy: An essay on interpretation* (D. Savage, Trans.). New Haven: CT: Yale University Press.

Ricoeur, P. (1981). *Hermeneutics and the human sciences.* Cambridge University Press.

Ricoeur, P. (1991). *From text to action: Essays in hermeneutics II* (K. Blamey & J. B. Thompson, Trans.). Evanston, IL: Northwestern University Press.

Ricoeur, P. (1992). *Oneself as another* (K. Blamey, Trans.). Chicago: Chicago University Press.

Ricoeur, P. (2004). *The course of recognition* (D. Pellauer, Trans.). Cambridge, MA: Harvard University Press.

Schwartz, J. M., & Begley, S. (2002). *The mind and the brain: Neuroplasticity and the power of mental force.* New York: HarperCollins.

Schweigert, W. A. (2006). *Research methods in psychology: A handbook* (2nd ed.). Long Grove, IL: Waveland Press.

Searle, J. R. (1996). *The philosophy of mind* [Cassette Recording]. Cantilly, VA: Teaching Company.

Searle, J. R. (1997). *The mystery of consciousness*. New York: New York Review of Books.

Searle, J. R. (2004). *Mind: A brief introduction*. New York: Oxford University Press.

Shek, D., Tang, V., & Han, X. (2010). Evaluation of evaluation studies using qualitative research methods in the social work literature (1990–2003): Evidence that constitutes a wake-up call. *Research on Social Work Practice,* 180–194. Retrieved from http://rsw.sagepub.com/content/15/3/180

Slife, B. D. (2004). Taking practice seriously: Toward a relational ontology. *Journal of Theoretical and Philosophical Psychology, 24*(2), 157–178.

Slife, B. D. (2008). A primer of the values implicit in counseling research. *Counseling and Values, 53*(1), 8–21.

Slife, B. D., & Gantt, E. (1999). Methodological pluralism: A framework for psychotherapy research. *Journal of Clinical Psychology, 55*(12), 1–13.

Slife, B. D., & Melling, B. (in press). Method decisions: Quantitative and qualitative inquiry in the study of religious phenomena. *Pastoral Psychology.*

Slife, B. D., Reber, J., & Richardson, F. (2005). *Critical thinking about psychology: Hidden assumptions and plausible alternatives.* Washington, DC: American Psychological Association.

Slife, B. D., & Richardson, F. C. (2008). Problematic ontological underpinnings of positive psychology: A strong relational alternative. *Theory & Psychology, 18,* 699–723.

Slife, B. D., & Richardson, M. (2009). Evaluating the philosophies of theory-building in case studies. *Pragmatic Case Studies in Psychotherapy, 5*(3), 108–125.

Slife, B. D., & Williams, R. N. (1995). *What's behind the research? Discovering hidden assumptions in the behavioral sciences.* Thousand Oaks, CA: Sage.

Solomon, R. (1988). *Continental philosophy since 1750: The rise and fall of the self.* Oxford: Oxford University Press.

Stiles, W. B. (2009). Logical operations in theory-building case studies. *Pragmatic Case Studies in Psychotherapy, 5*(3), 9–22.

Taylor, C. (1971). Interpretation and the sciences of man. *The Review of Metaphysics, 25*(1), 3–51.

Taylor, C. (1995). *Philosophical arguments.* Cambridge, MA: Harvard University Press.

Viney, W., & King, D. B. (2003). *A history of psychology: Ideas and context.* Boston: Allyn & Bacon.

Wittgenstein, L. (1981). *Zettel.* Oxford, UK: Basil Blackwell.

Wyschogrod, E. (2002). Pythagorean bodies and the body of altruism. In S. Post, L. Underwood, J. Schloss, & W. Hurlbut (Eds.), *Altruism and altruistic love: Science, philosophy, and religion in dialog* (pp. 29–39). Oxford: Oxford University Press.

20 Notre Trahison des Clercs

Implicit Aspirations—Explicit Exploitations

P.A. Hancock

'Let us make no mistake: higher education is a business.' (Taylor, 2010)

PREAMBLE

The pursuit of knowledge is conceived of as being a good thing. It is often argued that our increased understanding of the world leads to a better life in terms of both people's material well-being and their purportedly, concomitant moral development. The implicit assumption is that superior understanding also leads to "progress," which supports a vision of a better future for all humankind and indeed potentially all of life itself on our planet. The primary engine that fuels this progress is science. As the formalized method of systematic advance in understanding, science has been remarkably successful in its search to comprehend the world around us. Technology, the operational handmaiden of science, has been equally successful in providing the material manifestations of the discoveries of science. In consequence, we collectively and continually proclaim that we are better off now than at any time in the whole existence of our species.

For many of the people reading this work, this statement is true. For most of the people not reading this work, this statement is much less true. Indeed, for some it is an outright falsehood. The benefits, which have accrued from science and technology, are not equally distributed. While few have much, many have little. In recent times, fewer have even more and many more now have nothing (and now sometimes even less than nothing if they are born into the negative equity of debt). The dictations of such patterns of distribution are essentially political in nature. The scientist or the creator of the knowledge that supports such a pattern of distribution has traditionally had little, if anything, formally to do with the process by which the "good" he or she produces is subsequently disbursed. Indeed, in many visions of science, this outflow mechanism is explicitly considered to be beyond the purview of the scientists themselves. Here, I argue that it is not. Neither theoretically, practically, nor especially morally, can science, nor scientists themselves, abdicate from this responsibility (and see Harris, 2010). Further, I argue that society is always an ongoing experiment in which the psychological sciences absolutely play the central role in how we see, understand, and organize ourselves. Thus, scientific psychologists are at the very center of this crucial issue. In what follows, I look to clothe these arguments through reference to my own area of applied psychology, namely, Human Factors and Ergonomics (HF/E). However, to begin let us consider the following quote:

Here is the decisive pair of questions: Am I apart from the universe? That is, whenever I look I am looking as through a peephole upon an unfolding universe. Or am I a part of the universe? That is, whenever I act, I am changing myself and the universe as well. Whenever I reflect on these two alternatives, I am surprised again and again at the depth of the abyss that separates the fundamentally different worlds that can be created by such choices: either to see myself as a citizen of an independent universe, whose regularities, rules, and customs I may eventually discover, or to see myself as a participant of a conspiracy, whose customs, rules, and regulations we are now inventing. Whenever I speak to those who have made their decision to be either discoverers or inventors, I am impressed again and again by the fact that neither of them realizes that they have ever made that decision.

I have a dear friend who grew up in Marrakesh. The house of his family stood on the street that divides the Jewish and Arabic quarters. As a boy he played with all the others, listened to what they thought and said, and learned of their fundamentally different views. When I asked him once, "Who was right?" he said, "They are both right." "But this cannot be," I argued from an Aristotelian platform. "Only one of them can have the truth!" "The problem is not truth," he answered. "The problem is trust" (von Foerster, 1990).

INTRODUCTION

Surely, we are the good guys? Aren't scientists, at least in our own implicit view, not only clad in the stereotypical white laboratory coat, but also dressed in the metaphorical white hats of the old-time western heroes? We are on the side of the angels, are we not? Indeed, this is how we most often implicitly choose to see ourselves and often to explicitly eulogize ourselves (see Grunberg et al., 2010). But we also realize, along with anyone who has seen the various film versions of Mary Shelley's novel *Frankenstein*, that this is not how part, if not much, of the rest of society sees us.[1] However, our implicit assumption

[1] There has always been a part of the American psyche, for example, that, if it is not manifestly anti-intellectual, at least expresses deep and abiding reservations about the putative "excesses" of knowledge. Perhaps this is well epitomized by the observation of Garrison Keillor, the doyen of down-home, when he noted concerning his own invented part of Middle America: "Lake Wobegon, where smart doesn't count for so much. A minister has to be able to read a clock. At noon, it's time to go home and turn up the pot roast and get the peas out of the freezer." *Lake Wobegon Days* (1985)

as scientists, and indeed often the manifest expectation of society, remains that science is a good thing. Although there has traditionally been, and nominally still persists, a purported separation of science from ultimate purpose (see Gould, 1997; Moray, 1993, but see also Snow, 1960), nevertheless, most scientists implicitly conceive of their search for knowledge and understanding as a morally positive enterprise. Indeed, to repeat the point, we see science explicitly as supporting the public "good." Yet, is this necessarily so, and more to the point in our present difficult times, is this actually so?

In this chapter, I want to explore these assumptions. I do this in order to ask very pointed questions about the responsibility of the academy in relation to the way in which the knowledge that it creates is subsequently employed in and by society. To begin, however, I want to point to a foundational philosophical dilemma that may address some origins of the tensions with respect to the role of scientists and their place within society. The dilemma is one that faced the philosopher St. Augustine now some 1,600 years ago (Augustine, 400, 410). It is a conundrum that also concerned the cyberneticist Norbert Wiener in the middle of the last century (Wiener, 1954, 1964). It is, indeed, a question that still persists and pervades. Briefly, this is the contrast between the so-called Manichaean and what became to be known as the Augustinian view of the world. These are philosophical positions that concern the foundation of cause and whether the Universe is run by many, one, or no gods. At its very heart, this argument is about the nature of our ignorance.

My dear brothers, never forget, when you hear the progress of enlightenment vaunted that the devil's best trick is to persuade you that he doesn't exist!
(From "The Generous Gambler" by Charles Pierre Baudelaire, February 7, 1864)

AUGUSTINIAN AND MANICHAEAN PERSPECTIVES

Through the long lens of history there are many possible visions of our world. One of the primary differences between the approach embraced by the sciences and alternative perspectives on human development, such as religion, lies in the attribution of causation. In a direct way, this inquiry devolves to the simple question of how much intelligence one is willing to concede to the Universe. If one is willing to see the world as saturated with causal entities (both benevolent and malevolent) who, in their various incarnations, are constantly exerting their positive and negative influences upon you as an individual, then you tend to see the outside world as powerful and controlling. This latter, polytheistic-based perspective tends to downplay personal responsibility and emphasize the vagaries of fate and chance in the course of human events. This worldview is certainly one that has permeated the distant past of Western civilization, and echoes of this perspective still resonate in our own times. Such a worldview even more strongly underpins a number of contemporary, non-Western cultures in which multiple deities are still seen as being in control today.

It might perhaps, however, be considered a somewhat hopeful sign that the number of putative "gods" has tended to contract over human history[2]. Later

[2] As noted, this is largely a Western-based view of historical progress, since other cultures do not view time in terms of progress anyway.

forms of social explanation began to search for and emphasize unity in this apparent causal diversity (Hancock, 2002). A most interesting example of this form of progress was the persistent search by Johannes Kepler (1571–1630) for the mathematical elegance in the description of the orbits of the planets (the wandering stars). It was Kepler's belief in the perfection of God that sustained him in his search through Tycho Brahe's empirical data, which allowed him to eventually identify the elliptical nature of planetary orbits, a discovery that led to Newton's own insights (see Hancock, 2005; Koestler, 1959/1986). What had formerly been a confusion of retrograde motion was now laid bare as a single, comprehensible mechanism. It was from similar forms of simplification and unification that the origins of monotheism derived, in which one single god (or still persistent multiple faces of that single god) began to predominate. But even within these monotheistic structures there remained the necessity to account for the adverse events in existence.[3] Thus, like other forms of evolving belief, Judeo-Christianity was forced to invoke the notion of a "devil" to represent the agency of undesired outcome: in more simplistic terms, the agent of "evil." As recent speeches on the "axis of evil" very evidently show, we have still yet to progress beyond this notion. The idea of forces of light opposed by forces of darkness is one that transcends any single formalized dogma. However, for Christianity the idea of equality between these two respective forces represented an early source of heresy: the heresy of Manichaeism (see Bevan, 1930; Runciman, 1947).

In actuality, the history of Manichaeism itself[4] as a rival to Christianity is less important to my present discourse than is St. Augustine's rejection of it. For it is in part in this rejection of the notion of equivalent causal forces in opposition to each other that our present foundation of the scientific enterprise is laid. Thus, modern science sees the world (and the Universe in which it is framed) as complex but not malign. In science, we explicitly see Nature as something to be explored, not necessarily as something to be feared. Nature possesses certain regularities and principles, and it is the belief of science and scientists that we are able to discover these regularities. The occasional, insightful "shafts of light" that illuminate these regularities are hailed as blooms of science, and their articulators are lauded as the heroes of science accordingly. In science, we do not see Nature as a force that actively (i.e., teleologically) adopts a purposive intention to deceive us. Indeed, can we as experimental psychologists even seriously contemplate a situation in which we set up an experiment, for example, using an fMRI system to explore a certain form of memory loss only to have the machine's electrons themselves plotting to mislead us? Today, it is almost impossible for us to envisage a "God" of the "EEG machine" whose malevolent purpose is to perpetrate some fraud upon our collected data, or even pose a direct threat to us as natural explorers.[5] As scientists, we deal with complex issues,

[3] The schizophrenic God of the Old and New Testaments is an example of this inherent paradox of distressing and tragic events nominally originating from an all-powerful and all-benevolent Deity.

[4] See for example: http://www.newadvent.org/cathen/09591a.htm (Retrieved 1-11-11).

[5] There is, of course, one way in which Nature does menace, and that is through the threat of species extinction. As explored most famously by Malthus, it is our fertility that, in the end, acts as a natural threat to our own survival. I shall not, however, explore the intrinsic teleology of this implicit threat in the present work.

but the nature of Nature is not to oppose our efforts in respect of that exploration. Failure to understand the secrets of Nature is what we attribute to our own failure, not to the perfidious character of the phenomena themselves.[6]

Yet we know that this fundamental assumption about the foundation of our science is not true of people themselves. People, the primary interest of the science of psychology, constantly act to oppose and to deceive (Hancock, 2012). In fact, we have a set of codifications, which look to regulate such forms of social interaction. We call it law. But as John Senders (1998) has pointed out: "What is lawgical is not always logical." By this he means that law can, and indeed regularly does, choose to reject what we understand to be our best state of scientific understanding. This is not simply the rejection of conceptions such as the repudiation of Darwin's theory of the evolution of species by those of a fundamentalist religious persuasion. No, the former rejection is actually much more profound and disturbing. It is epitomized in Richard Feynman's personal observations on the *Challenger* accident when he penned the immortal phrase: "For a successful technology, reality must take precedence over public relations, for nature cannot be fooled" (Feynman, 1986). At its heart, this means that for our eventual survival as a species there are necessary occasions where science (the empirically based study of reality) must take precedence over law (the codification of human interaction). Thus, material reality must, on certain explicit occasions, supersede human perception, however strong and persuasive we might believe that perception to be. The fundamental issue is that law has no precedent to set the force of this maxim in place, which is why science must take its formal place in the process of the exercise of power (see highlighted box below).

ON THE CREATION OF A SUPREME ACADEMY

Preamble

The Constitution of the United States of America was created as an experiment in social organization. Formulated in part in reaction to a single centralized power, the Constitution was never intended to be a fixed and unalterable entity. It was conceived and should remain a growing and evolving system that seeks the betterment of all toward an envisaged state of ever-increasing social welfare. But the times have changed and both we, the citizens, and our Constitution must change with them. Some two and a quarter centuries ago, an individual could aspire to the status of polymath, having command of a broad understanding in science, law, and religion as well opportunity and facility in applying this knowledge to the everyday process of government. The aspiration was valid because there was proof of the existence of individuals who

[6] It is really very interesting and very telling that the purported "shyness" effect associated with failed experiments on phenomena like ESP invokes this very form of Manichaean opposition. The "shyness" effect supposedly occurs when an individual seeks to explicitly test such parapsychological characteristics. When the skeptical individual retires from the scene and less stringent tests are permitted, lo the effect reappears.

had achieved this enviable status. With the explosion of knowledge in the intervening years, it is now not possible for any single individual to fulfill this role. It was the failure of those who formulated the original Constitution to understand this proliferation of technology, with the concomitant constraints on personal knowledge and wisdom, which has led us to our present predicament.

The Issue

Briefly stated, the world is driven by science and its applications through its more public face, expressed through technological innovation. The speed of development in any one single area of this technology can challenge even the most facile, domain-specific expert. Thus, progress in discrete areas of science easily outdistances the comprehension of any intelligent nonspecialist. This is potentially very problematic since our democratically appointed representatives are required, more and more often, to pronounce on matters in which science and technology play an ever-increasing role. How are such individuals, however well-meaning, to grasp and apply knowledge which, for them, continually spirals out of control? Typically, like many in upper levels of command hierarchies, they seek out advice from those considered authoritative. However, for those steeped in the world of law in which subjective opinion is the very essence of interaction, how are they to determine the validity of one opinion over another? Nevertheless, science never votes democratically. Our understanding of reality is a social construct, but the underlying reality is not amenable to manipulation by opinion alone. The founding fathers of the United States encoded law because it was the strictures of law itself that they knew to fear. Law was power. But today technology is power.

The system of checks and balances was created to depreciate the legal opportunity for any single individual to dominate power. Today, we persist with this increasingly archaic structure, laboring under the continuing illusion that the source of power has not itself been the subject of strident metamorphosis. A new member of the triumvirate of executive, legislative, and judicial has emerged. That new member is science, and it is insufficient now that science stands by, sometimes as a mute or even gagged adviser, when the reality at hand will not suffer the illusion of human subjective control.

The Solution

One salient solution to this conundrum is before us, if we have the present courage to grasp it. We must add to the present triumvirate a fourth member of the system of checks and balances. We must create a "Supreme Academy" with powers equivalent to and commensurate with the other existing members of the system of "checks and balances." Those with a natural (and understandable) tendency toward the resistance of change may well object that current organs of government such as the National Academy, or even a resurrected version of the, now sadly missed, Office of Technology Assessment (OTA) could achieve the objective of scientific advisement. Unfortunately, our modern experience has evidently demonstrated that any such system fails to engender an enforceable strategy especially when well-meaning but ill-informed individuals can readily ignore the implications of objective evidence or potentially interfere with the process of scientific research itself.[7] It is most certainly true that the creation and internal constitution of any such Supreme Academy will have to be carefully formulated, as well as the extent of its powers and its interaction with its sister institutions. But absent this foundation, we are approaching an epoch in which irreversible and potentially fatal ill-informed decisions may well be taken by those ill-equipped to understand the scientific ramifications of their deliberations. The stakes are presently high and getting higher. For the sake of our mutual survival, we cannot afford to omit science from the enforceable equation of government.

The Supreme Academy

Science is not egalitarian. Much as it goes against our national aspirations, Nature does not reveal her secrets on an equal basis. We are each given our respective talents and each permitted to express them as best we can. Science, as a social pursuit, is a search for excellence. In its social form, scientists are acutely aware of the leaders in their own respective fields. Despite the natural controversy that accompanies any human activity, there is a remarkable consensus across communities as to who is and who is not a member of the leading cadre. Many of these latter individuals are personally motivated solely toward their science and are either unmotivated or uninterested by greater social issues or the application of their science to the problems of society. In contrast, there is a growing body of scientists whose focus is on the resolution of these immediate human concerns.

Similarly, there is a growing consensus that the disparate nature of discrete scientific disciplines can inhibit our greater collective efforts toward wisdom, as compared to simply more data accumulation (Wilson, 1998). Thus, there is a body of individuals qualified to compose a Supreme Academy. Again the precise constitution of such a body is beyond the purview of this necessarily limited chapter. This is simply to assert not only that such a body should be created, but also that it is evidently feasible that it can be so created. While this proposal may well appear quite drastic to those of a more traditional perspective, we must always remember that our Constitution is less than five generations old, and the pace and scale of our modern innovations such as the World Wide Web attest to the vast social effects that are now almost daily evident from the advance of technology. As Jacob Bronowski so clearly articulated, our advance in technology has created the modern-day capacity to destroy not just one single individual or even groups or particular cultures. Modern technology permits us to extinguish our whole species. When Bertrand Russell envisaged the dissolution of the bipartite superpower structure, he hoped for an American hegemony as the last, best hope for all of humankind. To fulfill such hope we have to acknowledge the impact that science and technology exerts on everyday life around the world. It must, therefore, have a full and untrammeled voice in our process of government. Not to do so will most probably doom us all.

[7] http://www.centerforinquiry.net/advocacy/protecting_scientific_integrity/

From these observations, one central problem for our present times is now revealed. Scientists implicitly adhere to a set of maxims that derive our scientific knowledge of the world from an Augustinian challenge. However, the product of that understanding is let loose into a human world dominated by Manichaean-type conflicts. In the pragmatism of our times we have seen the putative supremacy of human law over natural law. This has been marked by the apparent way in which Manichaean-based victory dominates over the aspiration of the use of Augustinian-based knowledge. In general terms, scientists, and most pointedly the product of their collective efforts, prove naïve and vulnerable in the battle between the forces of light and the forces of darkness. Indeed, science is considered a weapon in such a human social battlefield, and the battle that is raging is not with ignorance but between ourselves. In very general terms, the product of science goes naked and unprotected into this ever-raging human turmoil. Let us proceed to illustrate this conundrum using a specific instance drawn from HF/E.

> The factory of the future could be designed to control the workforce. This is an old story. For example, automation and computer-controlled machine tools were developed in the public sector for a long period, then finally handed over to private industry. Within the state sector the technology was designed in a specific way: to de-skill workers and enhance management control. That choice was not inherent in the technology and does not appear to have been more profitable. But it is a powerful weapon in a class war. (Chomsky, 2010, p. 88)

HUMAN FACTORS AND ERGONOMICS

What I have written so far is in the nature of a polemic exhortation, a form of diatribe that begins to berate our collective scientific enterprise somewhat gently over its greater responsibilities. However, it is without essential specification. To remedy that shortfall I must, albeit briefly, move from the general case to the more particular. In such circumstances, I will focus on the area of science that I know best, which is Human Factors and Ergonomics (HF/E). Sadly, unlike many other fields of contemporary psychology such as memory, attention, decision making, and so forth, I have to explain, even to some of my specialized colleagues, exactly what these research and application disciplines are (and see Hancock, 2012; Proctor & Vu, 2010). HF/E are centrally concerned with how humans work with technology. In general, HF/E draws on base knowledge from a number of differing disciplines, primarily from experimental psychology and applied industrial engineering and system design (see Proctor & Van Zandt, 1994). The notion of optimizing human-machine interaction and thus mitigating or completely excising episodes of system failure are apparently laudable moral aims that predominate in the discipline. HF/E also looks to improve the overall quality of human life though its collective actions (see Hancock, 2009; Hancock & Drury, 2011). As such, HF/E represents a fundamental battleground for the issues I have raised. Indeed, it is most probably the case that my own concern with HF/E has served to motivate my interest in this issue (see Hancock, 2009). In seeking to influence the design and operations of many, if not most, complex societal support technologies, HF/E is closer to the issue of science application than perhaps many areas of psychology, although of course many aspects of psychology have a very substantive impact in the real world (e.g., see Loftus, 1996).

One of the most interesting trends in modern HF/E is the way it has progressed from the consideration of one individual and one machine (e.g., one single person and his or her PC) to a much greater emphasis on teams and their actions. There has been, one might say, a natural progression from even this latter stance toward a focus on "teams of teams." This sequential broadening of concern is echoed in references in general to "a system of systems" and more recently a "culture of cultures." What this line of progress epitomizes is the ever-greater focus on wide-scale sociotechnical systems (and see also Hoffman & Millitello, 2008; Hollnagel, 2009; Reason, 2008). This natural line of progress leads us eventually to the consideration of the "system of all sociotechnical systems," or more simply, world society upon which the present chapter seeks to open direct, HF/E-related discourse. However, before considering the whole world, let us consider HF/E issues associated with a more discrete, but critical, technical case study.

Here, I want to explore only one, but nevertheless very pointed, contemporary example that encapsulates the issues I have sought to raise. This concerns the remote operation of unmanned aerial combat vehicles (UCAVs). There are any number of HF/E issues associated with the operation of these new weapon platforms. In particular, aerospace engineers have been seeking to create such "aircraft" that can stay aloft for extended periods (in the order of multiple days or even weeks). These operational platforms (it is hard to think of them as traditional aircraft) require only intermittent human supervision as to mission goals and the execution of specific tasks. The actual "flight" of the vehicle is becoming more and more automated. Only high-level forms of command (e.g., circle this location) are communicated to the UCAV which is then autonomously responsible for executing the specifics of such flight parameters in terms of manipulations of control surfaces. In essence, the supervisor provides a general goal, and the machine seeks to fulfill it.

For critical operations, such as weapons release, the operator is still very much at the forefront of momentary control. However, despite some stringent rules of engagement, the relative success of these forms of technology in their present operational mission is estimated to be very low.[8] The empirical question is what responsibility those in HF/E (and the associated sciences involved in the creation of such weapons) take in the false positives in this situation (where the latter is a good representation of the jargon to obscure the fact of unjustifiable murder; and see Sinclair, Siemieniuch, & Henshaw, 2010). Should we, in HF/E be directly associated in improving this ratio and thus reducing purported collateral damage (i.e., unjustified murders)? Initially, this seems almost a laudable goal to make such weapons more effective. After all, who would argue that it is immoral to reduce the frequency of the

[8] Of course, the actual level of operational success is a closely guarded military secret. However, Chomsky (2010, p. 239), using information from the *New York Times*, quotes David Kilcullen (a suggestive name indeed in this context), a former counterinsurgency adviser to General David Petraeus, who says that the success rate (kill ratio of terrorists to civilians) is approximately 2%.

death of innocent people? But, regardless of its efficiency, a weapon is still a weapon, and it is both a political and scientific question that harks back to Jeremy Bentham, as to whether such a vector for a technological design adds to the sum total of the quality of human life (Hancock & Drury, 2011).

Here is an explicit instance of advanced science being used as an instrument of Manichaean conflict. The two emergent questions are first, can science and by extension design, be so configured that the actions of such an entity (e.g., weapon) are morally supportable? And second, if this former aim cannot be achieved, what can science do about the organization of the largest form of sociotechnical system, (i.e., human society itself) so that the former aim comes within the bounds of possible (not to say probable) existence (Harris, 2010)? It would be nice and comforting to make some simplistic form of assertion that a major sea change could achieve such a transformation at a stroke. However, since society is fundamentally palimpsestual by nature, this will necessarily not be so. What is required is an iterative and evolutionary process in which the basic tenets of morality (see Hobbes, 1651/2010) are engineered into technology—by design (and see Hancock, Pepe, & Murphy, 2005). As is evident from the immediate fracturing of Isaac Asimov's "laws of robotics" by the first operational robots, a mere statement of advocacy is evidently insufficient. This requires a necessary sequence of actions, not exhortations (which of course makes each of these latter sets of statements somewhat self-contradictory in nature).

Can one engineer morality by design? The simple and simplistic answer is yes; it is conceptually feasible to do so. However, this precept appears to be almost diametrically opposed to an economic structure that is predicated upon the ascendency of profit above all other considerations. In its fullest essence, designed and engineered morality necessarily suboptimizes profit. The issue devolves to whether one sees human beings as a means to the ends of profit in a profit-driven system, or whether one sees human beings themselves as the ends of that selfsame system. This, of course, is a political question (Marx, 1867); however, I would submit that within the greater purview of science for which I am advocating, it is also a scientific question (and see Hancock, 2009; Harris, 2010). I would go one rather drastic step further along this path. If such issues do not become fundamental scientific concerns, I believe it harbingers the end of our species. Clear symptoms of this potential end are already appearing. Although this may seem a somewhat doom-laden perspective, it does not make it an irrational one. Thus, texts such as the present one are well meaning and well intentioned, but they are actually part of our ongoing, modern "treason" if they do not serve to impress upon contemporary scientists the crucial importance of the issue and the necessity to alter both their explicit scientific and pedagogical mission accordingly.

You can't blame a man for making a bullet if that bullet is later used to kill a saint.
I knew that, or I would if I'd taken a minute to think. (Howard, 2010, p. 77)

AN EXCURSION INTO TECHNICALITIES

One of the primary issues with the current state of development of human society concerns the manner of the discourse through which purported consensus is derived. In advocating for a more science-based approach, we have to contrast any such offering with the present state of affairs. If one examines the nature of political discourse, and especially pronouncements made for the media, as opposed to nonpublic expressions of opinion, they are very largely devoid of information. Here, I am using information in its formal terms as the quantitative and quantifiable reduction of uncertainty. Most political statements that derive from those versed in the law, as opposed to individuals versed in science, are almost meaningless. I say this not as a form of insult but rather as a rational examination of the nature of information purportedly being communicated. Some might argue (and there is a potentially pragmatically valid basis to this argument) that in our present climate all such statements necessarily border on the meaningless. Indeed, contemporary politicians seem to get into the most trouble when they make affirmative statements. It is therefore potentially an implicit assumption that in a declining democracy general pronunciations cannot afford to be specific and thus information-rich. In contrast, science explicitly mandates the use of tightly specified theory. For someone to make a profound statement in science it should necessarily be information-rich. In the present chapter, such theory is only touched on briefly but necessarily. The foundation of this theory is the Minkowski space-time description that has been used previously in this general context of large-scale systems research (e.g., Hancock, 2009; Moray & Hancock, 2009).

To begin this brief exploration, let us imagine a population distribution of any human capacity. This capacity can be any talent that humans exhibit. We might imagine that such talent is normally distributed according to the Gaussian description (see Figure 20.1). Of course, it need not necessarily be normally distributed, and the distribution could be a log-linear, a negative binomial, or any of multiple forms. As will become apparent, the precise shape of the distribution does not have a profound effect on the basis of this present argument (although in its practical effect it is highly influential).

Now that we have a single distribution as represented in Figure 20.1, we need to impose a feedback loop such that the most talented end of the distribution exerts an

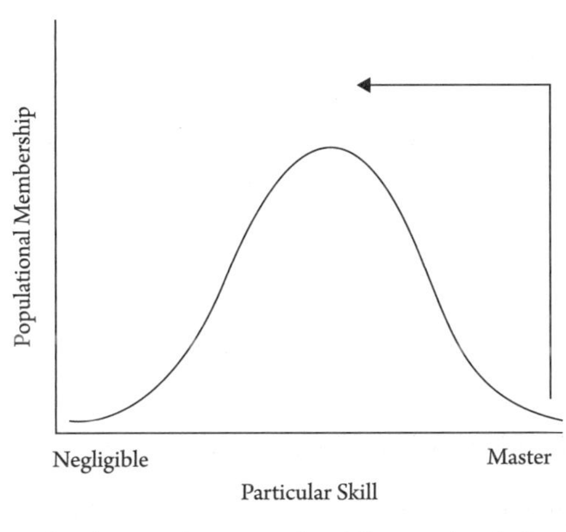

Figure 20.1 Gaussian distribution of any particular skill across the whole population.

influence over the rest of the remaining distribution. In Figure 20.1 this is illustrated by the arrow showing that the cadre of experts (as in the most talented segment, and by talent here I mean the manifest production of skills, not simply the inherent capacity to perform), exert their effect over their less skilled peers (it is of course quite possible and quite common for much of the rest of the distribution to neither know nor care about this one specific skill. In fact is it often the case we are indifferent to much of the rest of the human world). Having imagined one such skill, we now need to elaborate this from a single exemplar into the spectrum of all possible human skills. This step is illustrated for a series of skills in Figure 20.2. The general case for the whole set of human skills is given in Figure 20.3.

Now it is evident that if we have the distributions shown in Figure 20.3, they logically connote the whole of the human population, that is, the total possible set of human skills by the complete distribution of their representations in all human beings. (Of course, I take it to be evident that not everyone possesses all skills, and that the true distribution contains many registered with zero capacity on many specific skills). The circular representation shown in Figure 20.3 is thus the general case of humanity. However, the question that emerges is, what "frame" does this mass of humanity exist in, and how does it navigate through such a representation?

It is here that Minkowski's representation can help us (and see Moray & Hancock, 2009). As reproduced in Figure 20.4, the Minkowski diagram represents a complete description of past, present, and future. The "mass" of humanity, as illustrated in Figure 20.3, can now be seen as a single "entity" navigating in this four-dimensional space-time manifold (the representation of "society" being shown as the semicircle in the absolute future segment of the diagram). The manner in which this unified organism reacts to the constraints and opportunities that surround it is directly contingent upon the feedback derived from each influential segment of society. My central point, on a very general level, is that science (and the scientists who practice it) has traditionally had by far an insufficient influence on the direction that we are collectively taking and, most crucially, that this collective direction is not subject to

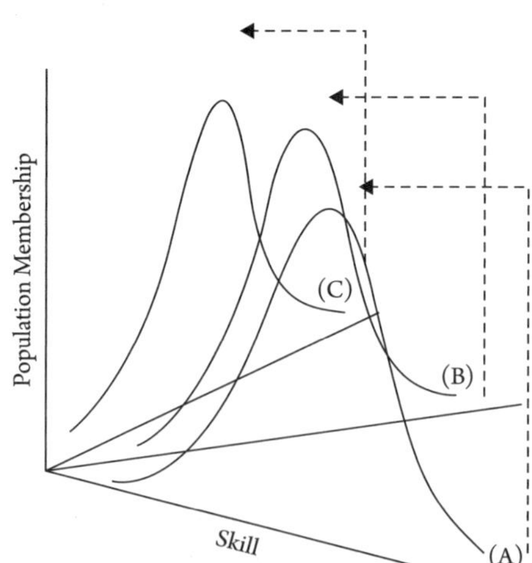

Figure 20.2 Suite of three Gaussian distributions for three specific skills (A), (B), and (C).

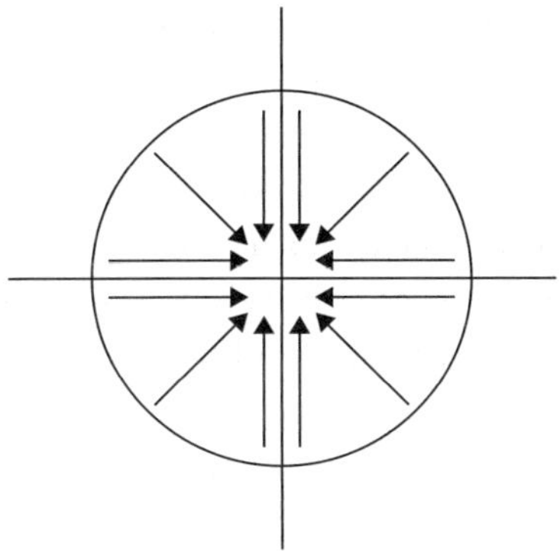

Figure 20.3 Top-down view of the 360-degree representation of all skills. The universal representation of Figure 20.2.

rational "navigation," which I take to be the primary purview of scientific contribution (and see also, Harris, 2010). To not engage our explicit efforts in this crucial action of "direction" but to persist in the implicit assumption that science is still a "good" thing is "notre [contemporary] trahison des clercs."

The strong do as they wish; the weak suffer as they must. (Thucydides)

BENDA'S TREASON AND OUR TREASON

It is important now to articulate explicitly the difference between the present expression of the "treason of the intellectuals" articulated above and that of its first expression by the French philosopher Julien Benda in his original thinking (Benda, 1928). In fact, it is almost diametrically the opposite. For Benda was berating the intellectuals of his time for taking much too intimate and partisan a part in the nature of the political

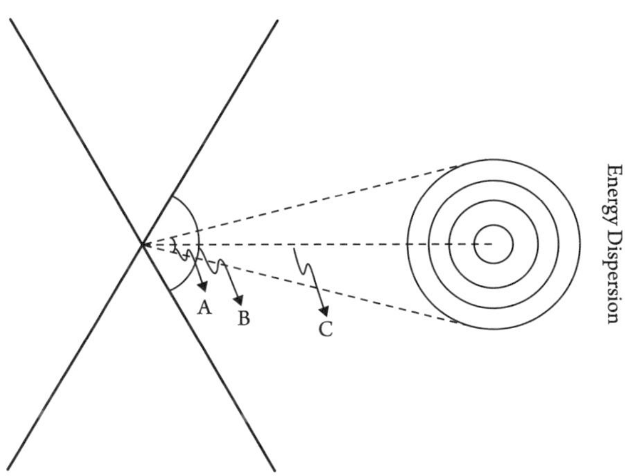

Figure 20.4 Representation of navigation in a Minkowski space-time framework. The collective organism of human society is represented as the semicircle in which the collective feedback of social action can take us toward the destructive levels of future energy dispersion (line C) or generate opportunities to avoid such destructive future consequences (Lines A and B).

milieu of the time; the key word here being "partisan." In its pure essence, I am advocating for a very similar form of treason. However, I am not arguing for the national, racial, cultural, and individualistic separatism epitomized in Benda's treason. Neither am I advocating for Senator Eugene McCarthy's perversion of Benda's conception (and see Schalk, 1971). In actuality, I stand in the same level of abhorrence as Benda in his original work of this impoverished and partisan participation. Rather, I argue that the neglect of the fundamental theory and practice of political and social organization on behalf of the scientist and technologist as an organized human endeavor, has led to our present circumstance.[9] As we look back on the history of psychology itself, we are justified in suggesting that it grew initially as one fledgling branch of the greater concerns for specific questions of philosophy. Through the focus on the individual and especially the modern seduction of the neurosciences, with all its wonderful and involving imaging techniques, it is easy to miss the human being in the wider context. While this might horrify some social psychologists and indeed sociologists whose whole career has been focused at this level of analysis, I think it fair to say that the ill-founded and in retrospect, ill-conceived theories of only a few modern economists have had more influence on modern society than all the actions of all the ergonomists who have ever lived. As has been once remarked, the true purpose of philosophy (and here read psychology) is not to study the world but to change it. That change must be rationally grounded and empirically based (and see also Harris, 2010). Nature never votes Democrat, and the laws of physics can be very unforgiving to a society who truly believes that perception supersedes reality. But understanding such relationships and constraints is the very stuff of modern psychology. Thus, imperatively, research psychologists must be at the forefront of not only the debate but also crucially the action.[10] It will be the ultimate height of absurdity that we discover the last secrets of the human brain as the last human brain is extinguished from existence.

> Failure to confront... [our challenges] seriously may well confirm the speculation of one of the great figures in modern biology, Ernst Mayr, that higher intelligence is an evolutionary error, incapable of survival for more than a passing moment of evolutionary time. (Chomsky, 2010, p. 175)

[9] Of course, I do not seek to protest the absurd here, which denies that individuals who are indeed scientists and technologists do take part in the political process. Indeed, it is certain that they do. What is less certain is the degree to which such individuals partake at the peak of the political process. Even a cursory examination of the credentials of most members of the Senate and the House of Representatives shows a dearth of technical qualifications. There is usually a smattering of ex-physicians. However, ex-physicists and those with advanced engineering degrees are usually notable for their absence. Parenthetically, it is as absurd to imagine that individuals who are scientists and technologists do not individually partake of the political process as it is to imagine that those of strong religious persuasion also do not so partake. The notion of the separation of church and state is thus as problematic an illusion as the topic of the present discourse. In recent times, however, it has become clear that certain religious creeds have become much more effective at organization and subsequent political influence than have science and technology as professional bodies. Corporations that utilize science and technology are, of course, a case apart in our modern world.

[10] Although I would still argue for psychology as the study of the individuals as being central to the issues I have raised, it is very clear that we are not alone in science in bearing this burden. For example, those in general systems theory have expressed very much the same sentiments (and see Ackoff, 2004).

SUMMARY AND CONCLUSION

It might well be thought that, in many ways, the future is far too important to be left to politicians (and see Rueter, 2000). As individuals trained in the law, their underlying assumption is that truth emerges from the conflict of opposing, spoken discourse (at heart they are Manichaean by persuasion).[11] Sadly, Nature very rarely reveals her secrets based on decibels or even rhetoric. Neither the capacity to make loud noise nor the capacity to make clever noise is necessarily linked to the empirical bedrock of reality. Our world shows increasing evidence of more and more symptoms of growing physical problems that cannot be solved by perpetual discourse but only by scientifically guided action. Indeed, some doubt that for issues like global warming, even such rational amelioration strategies can now work. Given this growing urgency, the wider question is whether we can continue to organize ourselves in the social fashion that we have heretofore adopted. In particular, can we afford to have science act in a mere advisory capacity? One HF/E luminary has posited that "good ergonomics is good economics." The central question in our own times is whether good economics (i.e., the manner in which we collectively organize ourselves) is itself good? In a recent, now famous book, Fukuyama (1992) expressed the opinion that we have begun to asymptote on to a collective social structure that promises to ossify into a set way of life. His dramatic title was "the end of history." If we continue in the mode suggested, I would opine something even more drastic that it is "the end of us." The real question is, what are people like you who read this text going to do about it?

I would ask you not to lecture me on morality. I don't take it kindly. Besides, you say "human being" as if it's something special. There are a lot of them about, you know, and few are worth the price of the calcium in their bones. 'Most of us don't measure a person's worth in calcium' she said a little too hotly.'(Howard, 2010, p. 187)

ACKNOWLEDGMENTS

I am grateful to all of the various agencies that have supported my work now for over three decades. I need to make it explicitly clear that the present opinions are my own and do not possibly represent that of any of the agencies that have supported me. I am also grateful to Professor Colin Drury and the editors of this volume for their comments on an earlier version of the present chapter.

REFERENCES

Ackoff, R. L. (2004). *Transforming the systems movement.* Paper presented at the Third International Conference on Systems Thinking in Management. Retrieved November 29, 2010, from http://www.acasa.upenn.edu/RLAConfPaper.pdf

Augustine. (c. 400). *The confessions of St. Augustine,* Warner, R. (1963). The confessions of St. Augustine. New York: Penguin Books.

[11] Whether such individuals truly believe this to be the case is certainly open to exploration and investigation.

Augustine. (c. 410). *The city of God against the Pagans* (see R. W. Dyson, 1998). New York: Cambridge University Press.

Baudelaire, C.P. (1869). The generous gambler. In *Le Spleen de Paris*. New York: New Directions Publishing.

Benda, J. (1928). *The treason of the intellectuals* (R. Aldington, Trans.), with introduction by R. Kimball (2006). Piscataway, NJ: Transaction.

Bevan, A. A. (1930). Manichaeism. In J. Hastings (Ed.), *Encyclopedia of religion and ethics.* London: Charles Scribner.

Chomsky, N. (2010). *Hopes and prospects.* Chicago: Haymarket Books.

Dyson, R. W. (1998). *The city of God against the Pagans.* New York: Cambridge University Press.

Feynman, R. P. (1986). *Volume 2: Appendix F—Personal observations on reliability of shuttle.* http://history.nasa.gov/rogersrep/v2appf.htm

Flach, J. M. (2009). The dynamic of experience: A search for what matters. In L. Norros, H. Koskinen, L. Salo, P. Savioja (Eds.), *Proceedings of the European Conference of Cognitive Ergonomics,* Helsinki, Finland. Technical Research Centre of Finland.

Foerster, H. v. (1990). *Ethics and second-order cybernetics.* Opening address for the International Conference, "Systems and Family Therapy: Ethics, Epistemology, New Methods." Paris, October 4, 1990 (Reprinted in Understanding understanding [pp. 287–304]. New York: Springer-Verlag, 2003) Retrieved September 14, 2011, from http://www.stanford.edu/group/SHR/4-2/text/foerster.html

Fukuyama, F. (1992). *The end of history and the last man.* New York: Free Press.

Gould, S. J. (1997). Nonoverlapping magisteria. *Natural History, 106,* 16–22.

Grunberg, N. E., Baum, A., Erlbaum, L., Glass, D. C., Krantz, D. S., Faraday, M. M.,...Krauss, R. M. (2010). In appreciation: Jerome E. Singer. *Association for Psychological Science, 23* (7), 25–30.

Hancock, P. A. (2002). The time of your life. *Kronoscope,* 2(2), 135–165.

Hancock, P. A. (2005). Time and the privileged observer. *Kronoscope,* 5(2), 176–191.

Hancock, P. A. (2009). *Mind, machine, and morality.* Aldershot, UK: Ashgate.

Hancock, P. A. (2012). Human factors/Ergonomics. In: V. S. Ramachandran, (Ed.), *Encyclopaedia of human behavior* (2nd ed.). Oxford, UK: Elsevier. in press.

Hancock, P.A. (2012). Hoax springs eternal: The psychology of cognitive deception. Manuscript in preparation.

Hancock, P. A., & Drury, C. (2011). Does human actors/ergonomics contribute to the quality of life? Theoretical Issues in Ergonomics Science, 12 (5), 416–426.

Hancock, P. A., Pepe, A., & Murphy, L. L. (2005). Hedonomics: The power of positive and pleasurable ergonomics. *Ergonomics in Design, 13*(1), 8–14.

Harris, S. (2010). *The moral landscape: How science can determine human values.* New York: Free Press.

Hobbes, T. (2010). *Leviathan. Or the matter, forme, and power of a common-wealth ecclesiasticall and civil* (I. Shapiro, Ed.). New Haven, CT: Yale University Press (Original work published 1651).

Hoffman, R., & Millitello, L. G. (2008). *Perspective on modern cognitive task analysis.* New York: Psychology Press.

Hollnagel, E. (2009). *The ETTO principle: Efficiency-thoroughness trade-off.* Chichester, UK: Ashgate.

Howard, J. L. (2010). *Johannes Cabal: The detective.* New York: Doubleday.

Keillor, G. (1985). *Lake Wobegon Days.* New York: Viking.

Koestler, A. (1959/1986). *The sleepwalkers: A history of man's changing vision of the universe.* New York: Penguin.

Loftus, E. F. (1996). *Eyewitness testimony.* Cambridge, MA: Harvard University Press.

Marx, K. (1867). *Das Kapital.* Hamburg, Germany: Verlag von Otto Meissner.

Moray, N. P. (1993). Technosophy and humane factors. *Ergonomics in Design, 1*(4), 33–39.

Moray, N. P., & Hancock, P. A. (2009). Minkowski spaces as models of human-machine communication. *Theoretical Issues in Ergonomic Science, 10*(4), 315–334.

Proctor, R. W., & Van Zandt, T. (1994). *Human factors in simple and complex systems.* Boston: Allyn & Bacon.

Proctor, R. W., & Vu, K.-P. L. (2010). Cumulative knowledge and progress in human factors. *Annual Review of Psychology, 61*, 623–651.

Reason, J. (2008). *The human contribution.* Chichester, UK: Ashgate.

Rueter, T. (2000). *The 267 stupidest things Republicans ever said: The 267 stupidest things Democrats ever said.* New York: Three Rivers Press.

Runciman, S. (1947). *The medieval Manichee: A study of the Christian dualist heresy.* Cambridge: Cambridge University Press.

Schalk, D. L. (1971). La Trahison des clercs 1927 and later. *French Historical Studies, 7*(2), 245–263.

Senders, J. W, (1998). Analysis of an intersection. *Ergonomics in Design, 6*(2), 4–6.

Sinclair, M. A., Siemieniuch .C, & Henshaw, M. J. de C. (2010). "To shoot, or not to shoot: That is the question"; Ensuring effective management of legal and ethical decision making fort autonomous systems (AS). Internal Memo Number 52, Engineering Systems of Systems Research Group Department of Electronic and Electrical Engineering, Loughborough University, Loughborough, UK.

Snow, C. P. (1960). *The two cultures.* Cambridge: Cambridge University Press.

Taylor, M. C. (2010). *Crisis on campus.* New York: Knopf.

Wiener, N. (1954). *The human use of human beings.* Boston: Houghton Mifflin.

Wiener, N. (1964). *God and golem inc.* Boston: MIT Press.

Wilson, E. O. (1998). *Consilience: The unity of knowledge.* New York: Knopf.

2ND PURDUE SYMPOSIUM ON PSYCHOLOGICAL SCIENCES, JUNE 4 AND 5, 2010

Symposium Participants: Front row: Gregory Feist, William Brewer, Michael Gorman, Dean Simonton; Second row: Corinne Zimmerman, Ronald Giere, Robert Proctor; Third row: Alice Eagly, Barabara Koslowski, Lisa Osbeck; Fourth row: Paul Thagard, Kevin Dunbar, E. J. Capaldi; Fifth row: Christian Schunn, Jessica Tracy; Sixth row: Anthony Greenwald, Ryan Tweney. Not pictured: Jan De Houwer, Don Dulany, David Klahr, Brent Slife

Index

Note: Page numbers followed by *t* and *f* refer to tables and figures.

Abductive reasoning, 74, 116–117
Abstract thinking of children, 138–139, 399
Academic disciplines, developmental stages of, 2–3, 17–22, 67–68
Accidental discoveries, 393
Accuracy in problem solving
  automaticity and, 192
  in implicit/explicit problem-solving, 236, 237–239, 241, 243–244, 246
Accuracy motivation vs. defense motivation, 283
Achinstein, P., 47–48
The acting person as unit of analysis. *See also* Scientists
  agency and discovery, 88–93
  data collection and analysis, 97–99
  emotion and mental models, 100–103
  implications for research methods, 94–96
  overview, 86
  the person concept, 88
  research context, 96–97
  research overview, 99–100
  scientific norms as context for, 92–94
  social positioning of, 103–107
Activity theory, 90–91
Actual scientific practices. *See* Inference to the best explanation
Addis, Tom, 62–63
Affective priming, 189–190, 192. *See also* Priming
Agency
  deliberative mental episodes and, 205
  discovery and, 88–93
  distributed cognition and, 3, 3, 82–83
Aggression, 6–7, 353–354
Aha! moments, 382, 390–391, 395
Ahn, W.-K., 318
Alampay, D. A., 298–299
Alexander, R. A., 233
Algorithmic mental processes, 190
Algorithmic methods, 114, 376–377, 380–381, 382–383
Alternative explanations
  confirmation bias and, 124–127, 282–283
  evidential relevance and, 3–4, 120–121, 131–132
  in inference to the best explanation, 116–117
  plausibility of, 115
Amabile, T. M., 269
Ambiguous figures, 291, 291*f*, 298*f*, 297–299
American Psychological Association (APA), 34
Amnesia, 63
Analogical thinking
  conflict in multidisciplinary teams and, 8, 448*t*, 449–450*t*, 445–452
  in group problem solving, 439, 443, 446
  metaphor vs. analogy, 409–412, 419, 426
  scientific discovery/invention and, 393–394, 394*t*, 398*t*
The Analysis of Sensations (Mach), 43
Anderson, J. R., 247
Anderson, K. J., 271
Anderson, M., 94
Andreasen, N., 215

Anomalies
children's handling of, 319
perception and, 292, 300–301
scientific creativity and, 383
strong vs. weak, 3–4, 122–124
Anscombe, G. E. M., 99
Anthropological research methods, 97
Anthropomorphization, 100–102
Anti-empiricism, 20
Archimedes, 375, 391
Aristotle, 42, 91–92
Arthur, W. B., 401
Articles, scientific, 6, 262, 262–264, 318, 324–325
Artifacts and distributed cognition, 79–80, 83, 95–96, 100–102
Artistic creativity, 377, 381, 401–402
Asimov's laws of robotics, 488
Asperger's Syndrome, 26
Assessment methods, 345, 346*f*, 349
Associationism, 49, 49
Associative processes, 195, 210, 212, 256–259, 373–375, 379
Assumptions. *See also* Convergent thinking
auxiliary assumptions and dual process models of cognition, 209–210, 212, 214
ontological assumptions, 459, 460, 469–470, 473–476
in paradigmatic stage of a science, 356, 357
Astronomy
perceptual problems in, 290–291
search grids and blind variation, 368–369
theory ladenness of, 296–297, 302–305, 310–311, 312–313
Atkinson, J., 279
Attention, 124, 309*f*, 312*f*, 308–314, 374–375, 380–381
Attitudes
attitude activation, 189–190
attitude change studies, 124, 130–131
attitudinal selectivity, 272–274
explicit vs. implicit, 437, 452
ingroup and congeniality biases and, 268, 272–274, 283
Augustine, Saint, 481–483
Aural representations, 393
Austin, R. H., 304
Authors
biases of, 324
conflicts of interest, 262
effects of sex of, 270–272, 273–274, 283
Autism, 26
Automaticity. *See also* Implicit processes
cognitive decompositional approach, 184, 185–191
difficulty of examining mental processes, 184, 188–190
features of, 185–187
functional decompositional approach, 184, 191–196
implicit processes as synonymous with, 184–185, 187–188
non-mental terms for features of, 192–193
as not an all-or-none feature, 183–184, 184–185
of thought vs. action, 205, 206
Auxiliary assumptions, 209–210, 212, 214
Awareness
automaticity and, 186, 194–195, 452
implicit/explicit processes and, 62, 229, 328, 445
in mentalistic metatheory, 204, 206, 218–222
mindfulness, 124
thought without, 217

Baars, B. J., 215, 221
Babbage, Charles, 307–308
Bachelard, Gaston, 24–27
Background information
alternative explanations and, 122, 132
confirmation bias and, 124–129
in distinguishing strong/weak anomalies, 122–124
emergence as evidence, 120–121, 124
heuristic use of, 115–116, 132
in idealized models vs. actual practice, 114–116, 117
overview, 112–113
scientific judgment about likely causes and, 118–119
theory and data interdependence with, 3–4, 117–118, 314
BACON (computer program), 369
Bacon, Francis, 327

Bain, Alexander, 42, 49, 364, 371, 379
Balance-congruity principle, 257–258
Balanced identity theory (BIT), 256–259, 257
Balance-scale task experiments. *See also* Dual process models of cognition; Rules
experimental results, 237–238, 241, 243–244
further research needed, 248–249
hypotheses, 236
methodology, 235*f*, 234–235, 236, 236–237, 240, 242–243
research findings, 238–240, 241–242, 244–245
Balthazard, C., 247
Banaji, M. R., 262–263
Bandwagon effect, 320–322
Banks, W. P., 203
Bargh, J. A., 230
Barron, Frank, 27
Barsalou, L. W., 400–401, 409
Baumeister, R. F., 203
Bayesian inference, 132, 212–215
Bechtel, William, 298
Becker, B. J., 271
Behaviorism, 41, 44–46, 202–203, 400
Beliefs. *See also* Theory
background information and changes in, 129–131
in causal hypotheses, 4, 211–212
control of variables and, 158–159, 169, 173–174
networks of, 3–4, 116–117, 119–120, 129, 130–131
prior beliefs, 4, 171
Bell, Alexander Graham, 61–62, 61*f*
Bem, S. L., 279
Benbow, Camilla, 33
Benda, Julien, 491–492
Bentham, Jeremy, 488
Bentler, Peter, 335–336
Ben-Zeev, T., 230
Bergman, Gustav, 46, 219
Berkeley, George, 42
Berkson, W., 50–51
Berry, D. C., 59, 230–231, 233–234
Bettelheim, Bruno, 277
Biases. *See also* Conflicts of interest
children's preconceptions and, 137–138
confirmation bias, 58–59, 76, 124–127, 127–129, 256, 260–262
congeniality bias, 268, 272–274, 283
correspondent inference, 6, 268, 268–270, 277–278
falsification bias, 59
ingroup bias, 6, 268, 270–272, 281–282
methodological corrections for, 267, 465, 466
publication bias, 324
Bickhard, Mark, 88
Biomedical engineering laboratories, 96–97
Biopsychology, 339
Birge, R. T., 320
BIT (balanced identity theory), 256–259, 257
Blind analysis procedures, 328
Blindsight, 220
Blind variation and selective retention (BVSR). *See also* Scientific creativity
blind variants defined, 364–366
criticisms of, 363–364, 377–381, 404–405
dispositional qualities of creative people, 383
explicit blind variation, 368–370
history of, 363–364
implicit blind variation, 370–371
originality, utility and nonobviousness, 7–8, 381–383
partially blind variations, 366–367
processes yielding blind variants, 373–377
selective retention, 7–8, 367–368
variations with properties of blindness, 371–373
Block, N., 221
Blondlot, René, 305–306
Boden, M. A., 370, 375, 381–383, 391–392
Bogen, J., 294
Bookkeeping and bias, 125–126, 127
Boronat, C., 409
Bottom-up/top-down effects (perception). *See* Top-down/bottom-up effects
Bowdle, B. F., 409
Bowers, K. S., 247
Brahe, Tycho, 290–291, 312

The brain. *See also* The mind; Neuroscience
brain imaging, 16–17, 63, 221
as computational system, 80–81
conscious states in, 206
research on thought and behavior, 33
Bransford, J. D., 314
Brentano, F., 205
Brewer, William F., 6, 123, 171, 315, 316, 322, 322, 323
Bridgman, Percy, 92, 98, 107–108, 219
British empiricism, 42, 49
Broadbent, D. E, 59, 230–231, 233–234, 339
Bronowski, Jacob, 412, 485
Brooks, K., 230
Bruce, Robert, 62
Bruner, J. S., 292, 300–301, 339
Buchwald, J. Z., 425n18
Bugelsky, B. R., 298–299
Bühler, Karl, 49–51
Burns, B. D., 232
Bushman, B. J., 354
BVSR. *See* Blind variation and selective retention

Campbell, Donald, 7–8, 51, 363–365, 379, 381
Candolle, Alphonse de, 24
Cantril, H., 325
Capaldi, E. J., 203, 212
Carey, Susan, 76
Carli, Linda, 270–271
Carnap, Rudolph, 45–46, 47, 219
Carnegie-Mellon University, 27–28
Carruthers, P., 213
Causal reasoning. *See also* Experiment design misconceptions of children
background information and changing explanations, 129–131
background information and confirmation bias, 124–129
background information and identifying causal events, 3–4, 118–119
beliefs about non-causal variables, 158–159
brain imaging and, 16–17
as deliberative vs. associative, 210–212
heuristic use of background information in, 115–116, 132
idealized models and formal rules, 113–115, 117
inference to the best explanation, 116–117
as propositional vs. associative, 210–212
reasoning about causal mechanisms, 114, 119–124
scientific/nonscientific perspectives, 481
Chabris, C. F., 309–310, 328
Chadwick, James, 306–307
Chalmers, T. C., 324
Chamberlin, T. C., 260
Chandrasekharan, S., 408–409
Chapman, J. P., 125–126
Chapman, L. J., 125–126
Charon, discovery of, 312–313
Chemistry in Britain (Wotiz and Rodofsky), 216
Chen, Z., 143
Chi, Michelene T. H., 76, 406–407
Children. *See also* Experiment design misconceptions of children
abstract thinking of, 138–139, 399
in developmental psychology, 14–15, 28–29
as "folk" scientists, 14–15
mental models of, 52
preconceptions of, 137–138`
science vs. engineering goals of, 233
theory-ladenness of thinking of, 302, 314–315, 319
Chinn, C. A., 123, 302, 316
Chomsky, Noam, 9, 486, 492
Choynowski, Mieczyslaw, 27
Christy, James, 312–313
Chun, M. M., 308
Churchland, Paul, 76, 201, 213
Citation patterns, 6, 261, 282, 283, 340
Cleeremans, A., 217
Clement, John, 77, 410
Clinical psychology, 336
Clinical psychology of science, 16
Coding of qualitative data, 98–99, 100, 446–447
Coelacanth discovery, 391–392
Cognition. *See also* Dual process models of cognition; Explicit processes; Implicit processes; Mental processes; Thinking
active nature of, 90

computational models of, 52, 62–63, 80–81, 201
distributed cognition, 3, 79–82, 82–83, 94
embodiment theories, 400–401, 408–409, 429–430
as environmentally situated, 94–95
social cognition, 256–259, 268–270
social interaction with, 93
Cognitive decompositional approach to automaticity, 184, 185–191
Cognitive dissonance, 258
Cognitive partnering, 90, 100
Cognitive psychology. *See also specific types of mental representations*
cognitive psychology of science and, 15
computational models of reasoning, 52, 62–63, 80–81
experimental vs. correlational divisions in, 339
influence on philosophy of science, 49–52
as informed by functional approaches, 193–194
mental representations and, 400–401
positioning theory, 104
social psychology integration with, 437–438, 445
standard views on implicit/explicit processes, 200–203, 208, 228–231, 436–437
theory-ladenness of mental processes, 297–302, 308–310, 322, 325
Cognitive study of science. *See also* Neuroscience
distributed cognition concept, 79–82
overview, 76–77
relationship to psychology of science, 78*f*, 77–79
Cognitive styles, 245, 315, 317
A Cognitive Theory of Consciousness (Baars), 215
Cole, Jonathon, 22
Cole, Stephen, 22
Collaborative processes. *See* Groups, scientific; Multidisciplinary teams
Collective cognition, 79–80. *See also* Distributed cognition
Collins, Harry, 60, 64–66, 409n11
Combinatorial conjecture. *See also* Mental representations; Scientific creativity
applicability outside of science/technology, 401–402
counterexamples to, 399–401
overview, 389–391
in scientific discovery, 391–395
in technological innovation, 395–399
Common sense explanations, 6–7, 347–348
Communication, scientific
articles, scientific, 6, 262, 262–264, 318, 324–325
within scientific teams, 441, 440–443, 444
as theory laden, 324–325
Communism/communalism (scientific norm), 259–260
Competitive support, logic of, 212–215
Comprehension and background information, 311
Computational modeling, 62–63
Computational psychology, 200–201
Computational system, human brain as, 80–81
Computer programs, 62–63, 138, 142*t*, 141–150, 175–178, 369–370
Comte, Auguste, 18
Concepts vs. propositions, 392
Conceptual blending, 411, 423*t*
Conceptual change, 77. *See also* Paradigms
Conferences and scientific discipline development, 17–22
Confirmation bias, 58–59, 76, 124–127, 127–129, 256, 260–262
Conflicts and group problem solving, 8, 446, 448*t*, 449–450*t*, 445–452
Conflicts of interest. *See also* Biases
confirmation bias as ethical issue, 256
ethical codes in science, 6, 255, 264, 326–327
evidence of, 260–262
implicit social cognition and balanced identity theory, 256–259
remedies for, 262–264
social norms of science, 259–260
Confounded vs. unconfounded experiments, 4, 123, 139, 143, 150, 156–161, 159–164, 164
Congeniality bias, 268, 272–274, 283

Congruity theory, 257
Conscious/unconscious mental processes. *See also* Dual process models of cognition; Explicit processes; Implicit processes
agency and, 88–89
defocused attention, 374–375, 380–381
in dual process models of cognition, 200–203, 208, 209–210, 228–231
in mentalistic metatheory, 5, 203–207, 208, 222–223
theory ladenness of, 260, 326–327
unconscious implicit thinking, 215–218, 218–222
(un)consciousness as automaticity feature, 186, 192–193, 436–437
Constitution, United States, 483–484
Construct validity, 345–347
Consumers Union, 8, 469–470
Context of discovery vs. justification, 8, 50, 74
Contiguity and association, 42
Contributory vs. interactional expertise, 64–66, 66*f*
Control of variables strategy (CVS), 4, 139*t*, 139, 139–141, 144*f*, 146*f*, 147–148*t*, 141–150, 151*t*. *See also* Scientific method
Convergent thinking, 8, 438–439, 442*f*, 443–445
Convolution of neural representation, 390–391
Copernicus, 381, 392
Correlational vs. experimental research methods. *See* Experimental vs. correlational research methods; Research methods
Correspondent inference, 6, 268, 268–270, 277–278
Counterintuitive explanations, 201, 347–348
Couvalis, G., 329
Covariation, 113–114, 123, 129–130
The Creating Brain: The Neuroscience of Genius (Andreasen), 215
Creating Scientific Concepts (Nersessian), 77
Creativity. *See also* Scientific creativity
artistic creativity, 377, 381, 401–402
Boden's criteria for, 381–383, 391–392
Critique of Pure Reason (Kant), 41
Cronbach, Lee, 6–7, 213, 335–336, 354–355
Cropley, A., 443
Cultural change and technological advances, 20–22
Cultural psychology, 88, 342–343
Curie, Marie, 317
Curie, Pierre, 317
Curl operator (electromagnetic fields), 418*t*, 418
Cvencek, D., 259
CVS. *See* Control of variables strategy

Damasio, Antonio, 462
Darden, Lindley, 76
Darwin, Charles, 274–276, 280, 317–318, 326*f*, 327, 364, 371
Dasgupta, S., 364
Data. *See also* Evidence
data analysis strategies, 343–345
dualism of theory and data, 465
objectivity of, 466
rejection of, 310–311, 313, 320–322
theory ladenness of, 116–117, 294, 315–316, 317–319
David C. Gooding Prize, 33
Davy, Humphrey, 395, 397–398
Dean, D., 159
Deaux, K., 280
Debner, J. A., 221
Decision making, 51–52, 217, 440
Deductive reasoning, 74, 113–114, 421, 426
Defense vs. accuracy motivation, 283
Defocused attention, 374–375, 380–381
De Houwer, Jan, 4–5, 185, 191
Deliberative vs. evocative mental episodes, 203–207, 208, 217–218, 222–223
Demand characteristics, 259, 451–452, 463–464
DeMatteo, D., 467
Democracy, 484
Dennett, D. C., 368
Depersonalization, 88
Descartes, René, 93, 460–461
DeShon, R.P., 233
Developmental psychology, 339
Developmental psychology of science, 14–15, 28–29

Dewey, John, 90–91, 102–103
Diagnostic information, 4, 127–129
Differences, individual, 7, 15, 354
"Different outcome" experimental goals, 168–169
Difficulty of task and problem solving approaches, 5, 231–234, 235*f*, 238*f*, 239, 242*f*, 244*f*, 246–247, 248, 374
Dijksterhuis, A., 217, 230–231, 239
Disconfirmation of hypotheses, 58–59, 113, 124–127, 129–131, 213, 260–262, 371. *See also* Confirmation bias; Hypotheses
Discovery. *See* Scientific discovery
Discovery, context of, 8, 50, 74
Disinterestedness (scientific norm), 259–260, 260–262
Dispositional inferences, 269
Distributed cognition, 3, 79–82, 82–83, 94
Divergence operator (electromagnetic fields), 417–418, 418*t*
Divergent thinking, 8, 374, 438–439, 441*f*, 440–443
Domain knowledge, 4, 137–138, 140, 173–174
Dualism. *See also* Subjective-objective dualism
  alternatives to, 469–470
  defined, 460
  entrenchment in psychological research, 468–469
  epistemological implications, 464–465
  mind-body dualism, 460–461, 462, 463
  nondualist methods in psychology, 473–476
  pervasiveness in science, 460–461
  problems with, 461–464
  in psychological research methods, 465–468
  subjective-objective dualism, 8, 461, 463–464, 465–468
  values and meanings, 470–473
Dual process models of cognition, 194–195. *See also* Balance-scale task experiments; Explicit processes; Implicit processes
  difficulty and problem solving approaches, 5, 231–234, 235*f*, 238*f*, 239, 242*f*, 244*f*, 246–247, 248
  explicit strategy and rule discovery, 245
  generally, 228–229
  implicit/explicit interactions in process control tasks, 231
  implicit strategy and task difficulty, 246
  implicit strategy benefits, 246–247
  in reasoning about causal explanations, 112–113
  in scientific thinking, 229–231
  standard view of, 200–203, 208, 209–210
Duhem-Quine thesis, 212–215
Dulany, Donelson E., 5, 204, 208, 209–210, 213, 219
Dunbar, Kevin
  brain research, 33
  cognitive contributions to philosophy of science, 52
  on implicit/explicit processes in scientific discovery, 229, 230–231, 315
  local vs. long-distance analogies, 393, 409, 446
  as teacher, 31
  in vivo observations of groups, 59, 95
Dweck, C. S., 355
Dyer, C., 466
"A Dynamical Theory of the Electromagnetic Field" (Maxwell), 424–426
Dynamic systems theory, 81

Eagly, Alice H., 6, 280
Ecological validity of experiments, 57–58, 58–59, 234, 247, 301–302, 339
Edison, Thomas, 62, 372
Editors/editorial boards, 262–264, 340–341
Effects
  functional research and implicitness of, 193
  mental processes vs., 192–193
Efficient (automaticity feature), 186–187, 192
Ehrenhaft, Felix, 318–319
Einstein, Albert, 376
Electromagnetic fields. *See also* Faraday, Michael; Maxwell, James
  discovery and description of, 420*f*, 420*f*, 419–421
  Maxwell's theory of, 422–424, 423*f*, 424–426

Electromagnetic fields. (*Cont.*)
metaphors and mathematical representation of, 413–418
non-expert understanding of, 409
Elements of Psychophysics (Fechner), 43
Eliasmith, C., 390–391
Embodiment theories and mental processes, 400–401, 408–409, 429–430
Emergence. *See* Discovery
Emotion, 90–91, 100–103, 107, 354, 379, 390–391, 395, 400
Empirical evidence. *See* Evidence
Empiricism, 20, 42, 49, 74, 475
English Men of Science: Their Nature and Nurture (Galton), 24
Epiphenomenalism, 201, 203
Epistemology
evolutionary epistemology, 51
logical positivism, 41, 45–46, 47, 74
natural epistemology, 18
ontological dualism and, 464–465
psychology relationship to, 42
relativism and theory ladenness, 289–290, 292, 329
scientific method and, 74
Epstein, S., 229
Ericsson, K. A., 215
Erikson, Erik, 277
Ernst, D., 269–270
Essentialism bias, 268, 268–270, 276–279, 282
Ether hypothesis (physics), 420–421, 425, 427
Ethical codes, 6, 255, 264, 326–327. *See also* Morality of science
Ethnographic studies of science, 96, 97
Euler, Leonard, 411
Eureka moments, 382, 390–391, 395
Evans, J. St. B. T., 64–65, 229
Evans, Robert, 200, 409n11
Evidence. *See also* Data
alternative explanations and, 116, 120–121
bottom-up evidence, 312, 316
confirmation bias and, 124–127, 129–130, 256
as objective, 466
in philosophy of science, 18, 289–290
in propositional learning, 211–212
theory and evidential relevance, 3–4, 116, 117–118, 119–120, 131, 325–326
Evil vs. good in science, 8–9, 82–83, 470, 471, 479–480, 487–488
Evocative vs. deliberative mental episodes, 203–207, 208, 217–218, 222–223
Evolution, human, 73–74, 274–276, 276–278, 280
Evolutionary epistemology, 51
Evolutionary theory, 116, 378–379, 378, 399–400, 404–405
Experience in Gestalt psychology, 47
Experimental vs. correlational research methods. *See also* Research methods
integration of, 352–353, 354–355
meta study in social-personality psychology, 340–342
as separate paradigms in psychology, 355–357
social vs. personality researcher use of, 343*f*, 344*f*, 346*f*, 347*f*, 342–349
as split in psychology, 6–7, 337–338*t*, 335–340, 350–351*t*, 349–351, 353
Experiment design misconceptions of children
control of variables strategy (CVS), 4, 139*t*, 139
CVS instruction and data collection, 144*f*, 146*f*, 147–148*t*, 141–150
CVS logic misunderstandings, 159–164
CVS misconceptions overview, 139–141, 151*t*
domain knowledge goal misinterpretations, 140, 173–174
engineering and domain knowledge goal relationship, 174
engineering goal approaches to, 166*t*, 165–171, 174–175
failure to control non-causal variables, 158–159
overextended "fairness" schema, 140, 171–173
overview, 137–138, 175–178
science goal misconceptions, 152–156, 153*t*, 154–155*t*, 164
science vs. engineering experimental goal, 138–139, 152, 151*t*, 164–165, 171, 172*f*, 233

TED computer tutor design, 138, 142*t*, 141–150, 175–178
testing for wrong variable, 156–158
Experiments. *See also* Research methods
confounded vs. unconfounded, 4, 123, 139, 143, 150, 156–161, 159–164, 164
control features, 105–106, 304, 307, 328, 464
ecological validity of, 57–58, 58–59, 234, 247, 301–302, 339
in scientific method, 74
theory ladenness of, 320*f*, 319–322
Experts and expert knowledge
effects on scientific creativity, 377–381
interactional vs. contributory expertise, 64–66, 66*f*
meta-cognitive awareness, 62
novice differences from, 60–61, 76, 406–407
roles in scientific groups and, 442–443, 444
Expert witnesses, 262–264
Explaining Science: A Cognitive Approach (Giere), 75–76
Explanation. *See* Theory
Explicit blind variation, 368–370
Explicit processes. *See also* Dual process models of cognition; Implicit processes
in balance-scale task experiments, 236, 237–238, 239, 241–242, 243–244, 245
benefits/drawbacks of, 229–230, 247, 248–249, 249
consciousness and, 199–200, 215–218
definition, 200
difficult tasks and, 5, 231–234, 248
hypothesis testing and, 234, 244–245, 248–249
interaction with implicit processes, 207, 231
propositional model of learning, 210–212
rule discovery and, 5, 207–210, 237–239, 241–242, 243–244, 245, 248
standard cognitive view of, 5, 200–203, 208, 228–231, 436–437
Ex vivo research methods, 60–61

Fairness notions, 171–173
Falsifiability (philosophy of science), 18, 49, 213
Falsification bias, 59
Faraday, Michael. *See also* Maxwell, James Clerk
analogical thinking, 393–394
differences/similarities to Maxwell, 427–428, 429
intuitive mathematics, 408, 429
research notebooks, 61
on superfluity of ideational variants, 371
theory of electromagnetism development, 420*f*, 419–421
Faulconer, James E., 8
Fechner, Gustav, 43–44, 202
Feest, U., 47
Feist, Gregory, 1–2, 2–3, 229, 475–476
Feltovich, P. J., 406–407
Feminism and psychology of gender, 276–279
Ferguson, M. J., 230
Festinger, L., 258, 467
Feynman, Richard, 31, 320–322, 483
Finite state grammar (FSG) experiments, 207–208
First International Congress of Philosophy, 18
Fisher-Smith, A., 462
Fixation on incorrect rules, 230, 239
Fleming, Alexander, 392, 395
fMRI (functional magnetic resonance imaging), 63, 221
Fodor, Jerry, 295
The Formation of the Scientific Mind (Bachelard), 24–27
Fragmented figures (perception), 299*f*, 299–300
Frank, C. S., 324, 328
Frankenstein (Shelley), 9, 480–481
Franklin, A., 320, 328
Franklin, Benjamin, 393, 393
French, K. E., 271
Freud, Sigmund, 103, 276, 277
Friedan, Betty, 277
Fugelsang, Jonathan, 33
Fukuyama, Francis, 493
Functional-cognitive research framework, 191

Functional decompositional approach to automaticity, 191–196
"Fundamental attribution error," 269

Galileo
discovery of Jupiter's moons, 393, 405
instrument use, 80, 395
mental representations, 392
prediscovery observation of Neptune, 311
theory ladenness of perceptions, 292, 303, 303*f*
thought experiments, 381
Gallagher, S., 203
Galle, Johann Gottfried, 296–297
Galton, Francis, 24
Gardner, Howard, 75
Garza, G., 462
Gatekeepers, 262–264, 341
Geddes, B. W., 233–234, 239
Gender research. *See* Psychology of gender
Genetics of scientific talent, 33–34
Gentner, Dedre, 76, 409
Gentner, Donald R., 409
Gergen, Kenneth, 57
Gestalt psychology, 46–49, 89
Gestures as enactments of mathematical concepts, 429–430
Gholson, Barry, 28
Gibbs, R. W., 400–401
Giere, Ronald N., 3, 51–52, 290
Gigerenzer, G., 222
Gilligan, C., 279
Gingras, Y., 413
Glaser, R., 406–407
Glisky, E. L., 63
Goal-dependent/independent (automaticity feature), 186
Goals and automaticity, 188–189, 192–193
Goal specificity, 232–233
Goffman, E., 104
Gooding, David, 61, 62–63, 67
Good vs. evil in science, 8–9, 82–83, 470, 471, 479–480, 487–488
Gopnik, Alison, 28–29, 52
Gorilla experiments (attention), 310*f*, 309–310
Gorman, Michael, 2, 29, 30, 32, 32–33, 76
Government, science advisement to, 483–485, 493
Grants, research, 2–3, 6, 32, 263–264
Greenwald, Anthony G., 256, 256–259, 261, 262–264
Griffiths, P., 102
Grounded theory research method, 98–99, 466
Groups, scientific. *See also* Multidisciplinary teams
as ameliorating individual bias, 6
analogy and conflict in problem solving, 8, 446, 448*t*, 449–450*t*, 445–452
formal roles in, 440–441, 442–443, 444, 451
human factors studies on, 487
interactional expertise and, 64–66, 66*f*
knowledge diversity and member roles, 440
shared mental models in, 443–445
social/cognitive model, 438–445, 452
in social psychology of science, 16, 83
in vivo research and, 59–60
Gruber, Howard, 76
A Guide to the History of Science (Sarton), 20
Gunstone, R. F., 301–302

Hadamard, J., 215
Hall, G. Stanley, 270, 276–277
Hancock, Peter A., 8–9
Hanson, Norwood Russell, 6, 41, 47–48, 289–290, 290–291
Harris, Sam, 9
Hart, W., 272
Haslam, N., 269–270
Hastorf, A. H., 325
Haugland, H., 201
Haven, Kendall, 391–395
Hayhoe, M. M., 309
Health psychology, 336–339
Heidegger, Martin, 460, 474
Heider, Fritz, 257
Helmholtz, Herman von, 43, 372
Hempel, Carl, 45–46, 213
Hereditary Genius (Galton), 24
Herrnstein, R. J., 117
Herschel, William, 304
Hertz, Heinrich, 421
Hesse, Mary, 426

Hetherington, N. S., 305
Heuristics
in actual scientific practice, 115–116
in discovery computer programs, 369
of novices vs. experts, 60–61
scientific creativity and, 376–377, 380–381
Heyes, Cecilia, 51
Higgins, E. T., 355
Higher order thought theory, 220
Histoire des Sciences et des Savants depuis deux Siècles (Candolle), 24
History of science
development of, 21*t*, 18–20
examples of theory ladenness of attention, 310–314
examples of theory ladenness of experimentation, 320–322
examples of theory ladenness of mental processes, 316–319, 323, 325–326
examples of theory ladenness of perception, 302–308, 394–395
Kuhn's use of, 289–290
methods used in, 20, 74–75
on role of theory in discovery, 296–297, 297
Hoddeson, L., 322
Hogarth, R. M., 231, 239
Hold One Thing (constant) At a Time (HOTAT) strategy, 163, 165
Hollingworth, Leta Stetter, 276
Holmes, Larry, 62–63, 67
Holton, G., 318–319
Horner, M. S., 279
Houts, Arthur, 28
Howard, J. L., 488, 493
Hubble, Edwin, 304–305, 395
Hubble telescope, 80
Hull, David, 18, 46
Human Factors and Ergonomics (HF/E), 486–488
Humans. *See also* The brain; The mind
benefits of HF/E to, 486–487
distributed cognition of, 79–82, 82–83, 94
science as unique to, 13
sex differences (*See* Psychology of gender)
skills of, 489*f*, 490*f*, 491*f*
species extinction/survival, 8–9, 483, 485, 485, 488, 491*f*, 492
Hume, David, 42
Humean indices, 113–115, 115–116
Hunches, tacit knowledge and, 247, 248–249
Hutchins, Ed, 79–80
Hypotheses. *See also* Alternative explanations; Confirmation bias
conflicts of interest and reporting of, 261, 262, 264
disconfirmation of, 58–59, 113, 124–127, 129–131, 213, 260–262, 371
divergent/convergent thinking and, 8, 381, 438–439, 440–445
goal specificity and testing of, 232–233
in mentalistic metatheory, 208, 211–212, 213, 213–214, 219
in scientific method, 74, 113–114, 473
scientist vs. nonscientist approach to, 18
testing of and explicit cognition, 234, 239, 244–245, 248–249
Hypothetico-deductive method, 113–114, 421, 426. *See also* Scientific method

Idealized models of science, 113–118. *See also* Scientific method
Ideas and ideational variants. *See* Blind variation and selective retention; Scientific creativity
Identification stage (scientific discipline development), 17
Identity awareness, 222
Identity (personal) negotiation, 99–100, 103–107
Iguanodon, 316
Images, mental, 42, 45, 49, 430
Imbalance-dissonance principle, 258–259
"Imitation game" experiments, 64–65
Implicit processes. *See also* Automaticity; Blind variation and selective retention; Dual process models of cognition; Explicit processes
as automatic, 4–5, 183
automaticity as not all-or-none, 183–184, 184–185
benefits/drawbacks of, 230–231, 247, 248–249, 249

Implicit processes. (*Cont.*)
cognitive decompositional approach to automaticity, 184, 185–191
definition, 200
difficulties in examining, 5, 184, 188–190, 247
difficult tasks and, 231–234, 246–247
discovery as, 88–93
dual process theories and, 194–195
in finite state grammar experiment, 207–210
functional decompositional approach to automaticity, 4–5, 184, 191–196
in group problem solving, 447–452
interaction with explicit processes, 207, 231
lawfulness, 218
in mentalistic metatheory, 5, 203–207, 222–223
metaphor and mathematical representation, 430
null awareness and, 218
objectivity impairment and, 262–263
standard cognitive views on, 5, 200–203, 208, 228–231, 436–437
unconscious implicit processes, 215–218, 218–222
unconsciousness and, 199–200
Inattentional blindness, 308–309, 309–310, 309*f*
Incubation effect, 216, 375, 380–381
Individual differences, 7, 15, 354
Inductive logic/reasoning, 49, 74
Inference to the best explanation (IBE), 3–4, 116–117, 131
Information evaluation (cognitive process), 439
Information processing, 201
Information-processing load, 127–129, 132
Information search (cognitive process), 439
Ingroup bias, 6, 268, 270–272, 281–282
Inhelder, B., 319
Ink blot experiments, 322, 322*f*
Insight, 89, 202, 230, 267, 375, 482
Institute for the Science of Sciences (Poland), 27
Institute of the History of Science and Technology (Russia), 27
Institutionalization stage (scientific discipline development), 17
Instruments, scientific, 3, 7–8, 79–80, 302–303, 395, 398
Integrative interactionist approaches to psychology, 354–355, 356–357
Intellectual phase locking, 320*f*, 320–322
Intellectuals, moral responsibilities of, 491–492
Intentionality, 88, 92, 186, 187–188, 188
Intentional/unintentional (automaticity feature), 186, 188, 192–193
Interactional expertise, 64–66, 66*f*
Interdisciplinary collaboration. *See* Multidisciplinary teams
Internal consistency, 115, 117
Internal validity, 6–7, 336, 345–347
International Society for the Psychology of Science and Technology (ISPST), 29–30, 30*f*, 31, 33, 67
Interpretation, theory ladenness of, 314–315, 316
Interpretive approaches to inquiry, 469
Interviewing (research method), 61–63, 95, 97–99
Introspectionism, 44, 45, 49
Intuition, sensori-motor, 408–409, 412–413, 419, 429–430
In vitro research methods, 58–59
In vivo research methods, 3, 59–60, 95–96
Isis (journal), 20
Isolation stage (scientific discipline development), 17

Jackendoff, R., 201
Jacoby, L. L., 221
James, William, 44–45, 87, 90, 202, 373
Jastrow, Joseph, 44
Jevons, William, 18, 371–372
Johnson, M. K., 314
Jolley, J. M., 466–467
Jones, Donald, 461
Journal of Psychology of Science and Technology, 29–30, 67
Journals
editors/editorial boards, 262–264, 340–341
explicit description as required by, 62

role in development of scientific disciplines, 17–22, 67
scientific articles, 6, 262, 318, 324–325
Joyce, J., 324
Judgment, scientific. *See* Scientific judgment
Justification, 8, 50, 74, 473

Kant, Immanuel, 41–42, 43, 461
Kanwisher, N., 221
Kaplan, A., 116
Karmiloff-Smith, A., 319
Kaufman, Walter, 317
Keillor, Garrison, 480
Kekada (computer program), 62–63
Kekulé von Stradonitz, F. A., 216
Keller, J., 354
Kendler, Howard, 94
Kepler, Johannes, 290–291, 482
Kihlstrom, J. F., 200, 217–218
Kim, N. S., 318
Kinesthetic mental representations, 393, 396
Kintsch, W., 324
Kirsch, G., 306–307
Klahr, David, 4, 27–28, 31, 52, 141, 143, 164, 229, 315
Klee, R., 329
Klopfer, L. E., 233
Knowledge, scientific. *See also* Experts and expert knowledge; Tacit knowledge
active participation of scientist and, 90
process of acquisition of, 73–74
transfer of, 230–231, 233–234
Kohlberg, L., 279
Koslowski, Barbara, 3–4, 31
Krebs, Hans, 62–63
Kronfeldner, M. E., 363–364
Kuhn, Deanna, 28, 140, 159, 160, 162, 163, 164
Kuhn, Thomas S.
influences on thought of, 24–27
on paradigmatic status of psychology, 355–356
on perception, 48–49
Popper criticism of, 50
psychology of acceptance/rejection of scientific theories, 48–49
role in development of philosophy of science, 18, 41
on scientific creativity, 381
on situatedness of scientific practices, 93
on theory ladenness, 6, 292–293, 300–301, 316, 324–325
use of history of science, 289–290
use of psychology, 48–49
views of individual scientists, 74–75
Külpe, Oswald, 41, 48, 49, 50–51
Kurz-Milcke, E., 407

Laboratory experiments. *See* Ecological validity of experiments; Experiments
Laboratory Life (Latour and Woolgar), 75
Laboratory notebooks, 2, 61*f*, 61–63, 318–319, 327
Lagrange, Joseph-Louis, 425–426
Lakatos, Imre, 75, 356
Lakoff, G., 407, 410–411, 429
Lalande, Jérôme, 311
Langer, E. J., 121
Lassiter, G. D., 217
Latent inhibition, 375, 381
Latour, B., 75, 356
Laudan, Larry, 75
Lave, Jean, 90
Law and public policy, 483, 493
Lawfulness, validity conditions for, 218
Lazarus, R. S., 102
Leaper, C., 271
Learning
associative and propositional theories of, 210–212
behaviorism views of, 44
explicit vs. implicit, 66–67, 200, 207–210
goal-specificity effects on, 232–233
selective, 232
by trial and error, 50–51
Leeper, R., 297–298, 299
Le Gros Clark, Wilfred, 311–312
Le Verrier, Urbain, 296–297
Liberman, Sofia, 29–30, 31
Life of Science (Choynowski), 27
Lipton, P., 212
Literal awareness, 222
Llardi, S. S., 462
Locke, John, 42
Logical behaviorism, 45–46
Logical empiricism, 74
"Logical Machines" (Peirce), 43–44

Logical positivism, 41, 45–46, 47, 74
Logic of competitive support, 210, 212–215
Der logische Aufbau der Welt (*The Logical Structure of the World;* Carnap), 47
Longino, Helen, 87
Longitudinal studies, 33–34, 339
Lubinski, David, 33

Mach, Ernst, 18, 43, 47, 364
Mack, A., 308–309
Mad scientists, 381
Magnetic resonance imaging (MRI), 63, 221
Mahoney, M. J., 318
Maier, N. R., 366, 374
Major, B., 280
The Making of a Scientist (Roe), 27
Malhotra, B. A., 302
Malley, M., 317
Mandler, G., 216
Manichaeism, 481–483
Mantel, Gideon, 316
Mappings
  of metaphor source and target, 411, 417*t*, 418*t*
  structure mapping theory, 409
  of unobservables to observables, 213
Marczyk, G., 467
Marois, R., 308
Marr, D., 190
Mars Exploration Rover mission, 8, 445–452
Marx, Karl, 91–92
Masicampo, E. J., 203
Maslow, Abraham, 27
Masnick, Amy, 27–28, 124, 130
Masterman, M., 356
Matarazzo, Joseph, 17
Material normativity, 93
Mathematical representations. *See also* Maxwell, James
  importance in physics, 8, 406–407, 425, 427–428
  metaphors in, 408, 410–411
  in scientific discovery and invention, 392, 396–397
  sensori-motor intuitions and, 408–409, 412–413, 419, 429–430
  visual representations and, 8, 412–413, 413*f*, 414*f*, 415*f*, 416*f*, 417*f*
Mathematical tools as symbolic artifacts, 79–80
Mathematics, cognitive study of, 407
"Maximum outcome" experimental goals, 165–168
Maxwell, James Clerk
  accessibility to non-experts, 409, 413
  differences/similarities to Faraday, 420–421, 427–428
  historical context, 408
  mathematical representation of sensori-motor intuitions, 412–413, 419, 429–430
  metaphor/analogy use, 409–412, 422–424, 424–426
  theory of electromagnetism, 422–424, 423*f*, 424–426
  use of Faraday's field representations, 61
McClelland, D. C., 279
McDermott, J. J., 44
McGuire, T., 90
Mead, Margaret, 277
Meaning in hermeneutic realism, 472–473, 474–475
Measurements
  theory and instrumental problems, 320–322
Mechanisms, explanatory, 115, 119–120, 123, 129–130
Meehl, Paul, 87, 213
Memory
  attentional processing and, 221
  retrieval as implicit vs. explicit, 187–188
  selective memory, 273
  theory ladenness of, 2, 322–323, 322*f*, 327
  transactive memory, 59–60
  unconscious implicit, 217–218
Mental images, 42, 45, 49, 430
Mentalistic metatheory, 203–207, 208. *See also* Dual process models of cognition; Explicit processes; Implicit processes
Mental models
  Bell's mental model for telephone, 61–62
  conceptual change and, 77
  in developmental psychology of science, 14–15
  model-based reasoning, 77, 407

shared mental models, 60, 66–67, 440, 451
visualization and, 430
Mental processes. *See also* Dual process models of cognition; Explicit processes; Implicit processes; *specific mental processes*
as combination of implicit/explicit, 188
difficulties in studying, 184, 188–190, 212–213, 219
scientific creativity and, 439
Mental representations. *See also* Combinatorial conjecture
in cognitive science, 400–401
neurocomputational model of, 390–391
in scientific discovery and technological invention, 392–393, 396
in sub specie historiae studies, 61–63
theories denying existence of, 400–401
Merton, Robert K., 22, 259–260
Metacognitive awareness, 62, 328, 445
Metaphors
analogy and, 411n13, 409–412, 419
mathematical representations as, 413–418, 417*t*, 428–430
in Maxwell's work, 426
Methodological issues. *See* Research methods
Milgram, Stanley, 57
Mill, John Stuart, 18, 276
Miller, George A., 282, 354
Millikan, Robert A., 318–319
The mind. *See also* The brain
as active vs. passive, 49, 50–51
as computational system, 200–201
distributed cognition and, 81–82
human qualities of science and, 89
ontological definitions of, 460
Mind-body dualism, 460–461, 462, 463. *See also* Dualism
Mindfulness and background information, 124
The Mind's New Science (Gardner), 75
Mishra, P., 322
Mitchell, M. L., 466–467
Model-based reasoning, 77, 407. *See also* Mental models
Moors, Agnes, 4–5, 185
Morality of science, 8–9, 82–83, 470, 471, 479–480, 487–488. *See also* Codes of scientific ethics
Motivation, 90–91, 124–127, 283
Motor transduction systems, 206
MRI (magnetic resonance imaging), 63, 221
Mullins, Nicholas, 17
Multidisciplinary teams. *See also* Groups, scientific
analogical thinking and conflict in, 8, 448*t*, 449–450*t*, 445–452
convergent thinking, 442*f*, 443–445
divergent thinking, 441*f*, 440–443
formal roles and, 107, 440–441, 442–443, 444, 451
interactional expertise and, 64–66, 66*f*
observational study of Mars Rover team, 445–452
social/cognitive model of, 438–445
Murphy, G., 42, 44
Murray, C., 117

Naïve theories, 280, 315
NASA (National Aeronautics and Space Administration), 445–446
National Science Foundation (NSF), 32–33, 66, 67
Natural epistemology, 18
Naturalized philosophy of science, 289–290, 329
Nature-nurture debate, 24, 33–34
Navigation as distributed cognitive process, 79–80
Neimeyer, Robert, 28
Neisser, U., 339
Nelson, James, 463
Nersessian, Nancy J.
on integration problem in psychology of science, 87
mental model research, 62, 77, 407
use of positioning theory, 105
in vivo research, 13, 59, 96–97
Network logic, 212–215
Networks of beliefs, 3–4, 116–117, 119–120, 129, 130–131
Neurath, Otto, 45–46
Neuroscience
challenges to mind-body dualism, 462
computational neuroscience, 80–81

Neuroscience (*Cont.*)
neural models of mental representations, 390–391, 400
neuroscientific psychology of science, 16–17
study of emotion, 354, 395
Newell, A., 376–377, 383
Newstetter, Wendy, 96
Newton, Isaac, 16, 77, 294, 319, 393–394, 482
Nijstad, B. A., 441
Noë, A., 462
Nomothetic-deductive system. *See* Scientific method
Non-scientists
ecological validity of experiments with, 57–58, 58–59, 234, 247, 301–302, 339
hypotheses of, 18
implicit processes, 15
Nordgren, L. F., 217, 230–231, 239
"The Normative Structure of Science" (Merton), 259–260
Norms, scientific. *See* Scientific norms
Notebooks, 2, 61*f*, 61–63, 318–319, 327
Novelty in scientific thinking, 209–210, 397*t*, 397–398
Novice vs. expert problem solving, 60–61, 76, 406–407
Novum Organum (Bacon), 327
N-rays, 305–306
Null awareness, 218–222
Núñez, R. E., 407, 410–411, 429–430

Obedience experiments, 57
Objectivity. *See also* Dualism
confirmation bias and, 256
Merton's norm of disinterestedness and, 259–260, 260–262
subjectivity-objectivity dualism, 8, 461, 463–464, 465–468
Observation
scientific discoveries/inventions based on, 394–395, 398, 398*t*
subjective-objective dualism and, 464, 468
theory ladenness of, 2, 6, 7–8, 301–302, 308–309, 310–311, 394–395, 398
Oersted, Hans Christian, 419
Office of Science and Technology Policy (OSTP), 484
Ogburn, William, 20–22
Ohlsson, S., 230
100 Greatest Inventions of All Time (Philbin), 396
100 Greatest Science Discoveries of All Time (Haven), 391–395
"On Faraday's Lines of Force" (Maxwell), 424–426
"On Physical Lines of Force" (Maxwell), 424–426
On Questions of the Soul (Fechner), 44–45
"On the Subjugation of Women" (Mill), 276
Ontological assumptions, 459, 460, 469–470, 473–476. *See also* Dualism
Ontological relationality, 472–473
Operational definitions, 219
Operationism, behaviorism and, 41, 46
Organized skepticism (scientific norm), 259–260
Osbeck, Lisa M., 3, 105, 408–409
Osgood, C. E., 257
Over, D. E., 229

Pacini, R., 229
Paletz, Susannah B. F., 8
Paradigms. *See also specific experimental paradigms*
paradigmatic status of psychology, 86–87, 94, 355–356
shared mental models as, 60
shifts of, 24–27, 48–49, 74–75, 292, 355–356
in social sciences, 357
Parker, K., 247
Parsimony, 48–49, 115, 213
Patterns of Discovery (Hanson), 47–48, 290–291
Pearson, Karl, 18
Peer review process, 6, 262–264, 318, 324–325
Peirce, Charles Sanders, 42, 43–44
Penicillin discovery, 392
Penzias, Arno Allan, 312
Perception
in Gestalt psychology, 46–49

as theory laden, 6, 47–48, 297–302, 302–308, 329
top-down and bottom-up processes in, 295–296, 307
unconscious implicit perception, 220
Perception and Discovery (Hanson), 47–48, 290–291
The person. *See* The acting person as unit of analysis
Person-situation debate. *See* Situational explanations
Personality psychology. *See* Social-personality research
Personal Knowledge (Polanyi), 88–91
Pettersson, H., 306–307
Philosophy, dualism in, 461–464
Philosophy of science
behaviorism influence on, 44–46
British empiricism influence on, 42
cognitive psychology influence on, 49–52, 75–76
context of discovery vs. justification, 8, 50, 74
development of, 19*t*, 18
Gestalt psychology influence on, 46–49
interdisciplinary approach to, 289–290
psychology influence on, 2, 40–41, 52–53
psychophysics influence on, 43–44
scientific evidence and, 289–290, 329
Physics, mathematical representation in, 8, 406–407, 408, 413, 425, 427–428
Piaget, Jean, 28, 76
Picasso, Pablo, 377
Pickering, Andrew, 87
Piltdown forgery, 311–312
Plausibility, background information and, 115, 116
Pluto, moons of, 312*f*, 312–313
Pocket, S., 203
Poincaré, Henri, 18, 215, 364
Polanyi, M., 52, 88–91, 92–93, 230, 247
Politics
application of science as problem of, 479–480, 486, 488, 493
disregard of scientific information, 488–491
gender research and, 280–281
interference with scientific research, 87, 484–485
scientific participation in, 492n
Popper, Karl, 18, 49–51
Positioning theory, 103–107
Postman, L., 292, 300–301
Power and technology, 484
Practice theory, 92
Pragmatism (philosophy), 42, 74, 90–91
Prediction, 467–468
Prediscovery observations, 310–311
Pre-selection of variations, 367, 382–383
Pretz, Jean E., 5, 194–196
Priming
affective priming, 189–190, 192
associations as primed by stimuli, 375
confirmation bias and, 127
as implicit process, 436–437, 452
transfer of knowledge and, 230
visual perception and, 297–298, 326
Principles of Psychology (James), 44–45
Prior beliefs, 4, 171
Probability and utility in blind variation, 364–366
Problem behavior graphs, 61–62, 64
Problem of interaction (dualism), 463
Problem solving
difficulty of task and approaches to, 231–234, 248
implicit and explicit strategies, 5, 62, 231–234, 247
in scientific groups, 438–445, 448*t*, 449–450*t*, 445–452, 452
unconscious implicit, 216
Procedural knowledge, 139, 160
Process-control tasks
balance scale task paradigm vs., 234, 239, 246
defined, 231
task difficulty and problem solving approaches, 231–234
Process dissociation, 221
Proctor, Robert W., 203, 212
Progress, science role in, 479
Propositional learning, 210–212
Prosopagnosia, 222
Protocol analysis, 64
Psychoanalysis, 201–202

Psychology. *See also specific sub-disciplines of psychology*
British empiricists' views of, 42
defined, 14
functional-cognitive research framework, 191
internal divisions within, 86–87
Kant's views of, 41–42
methodological differences within (*See* Experimental vs. correlational research methods)
ontological assumptions in, 459–460, 465–468, 473–476
the person in, 88
philosophy of science as influenced by, 2, 40–41, 52–53
psychology of science contributions to, 9–10, 34
as quantitative discipline, 43
"Psychology as the Behaviorist Views It" (Watson), 44
Psychology of gender
biases in early research, 274–276
correspondence and essentialism biases, 268, 268–270
feminism and essentialist explanations, 276–279, 278*f*
ingroup and congeniality biases, 6, 268, 270–274, 281–282
overview, 267–268, 278*f*
remedies for bias in, 282–283
situational theories of gender differences, 279–281
The Psychology of Invention in the Mathematical Field (Hadamard), 215
Psychology of science. *See also* Science and technology studies (STS)
cognitive/social psychology integration, 86–87, 437–438
cognitive study of science relationship to, 78*f*, 77–79
conferences, 28*f*, 30*f*
contributions to psychology, 9–10, 34
courses and teaching, 30–32, 67
defined, 14
establishment as a discipline, 2–3, 9, 24–30, 25–26*t*, 67–68
functional decompositional approach and, 194–196
future of, 30, 34
necessity of, 1–2, 13–14
research centers and financial support, 32–33, 67
research methods, 2, 9, 22, 33–34, 58–64, 67
subdisciplines, 14–17
units of analysis in (*See* The acting person as unit of analysis)
The Psychology of Science and the Origins of the Scientific Mind (Feist), 29, 34
Psychophysics, 43–44, 202
Publication bias, 324
Putnam, Hilary, 201

Qin, Y., 430
Qualitative research
coding of data, 98–99, 100, 446–447
data collection and analysis, 97–99
Mars Rover team observational study, 446–447
role in psychology of science, 95, 356–357, 459–460
in sociology of science, 22
subjectivity of, 469
Quine, W. V., 212–215, 289

Rabe-Hesketh, S., 324
Radar sweeps as blind variants, 368
Radical embodiment, 400–401
Raghavan, K., 233
Rand, K. L., 462
Ratio-bias paradigm, 229
Rational argumentation, 18
Rational errors, 230, 239
Rational/social dichotomy, 86–87, 93–94
Ray, W. J., 467–468
Rayas, Javier, 29–30
Reasoning. *See* Dual process models of cognition; Explicit processes; Implicit processes; Thinking
Reber, Arthur, 52, 58, 200, 203, 230
Reber, Jeffrey S., 8
Reese, H. W., 299
Reflexive thinking, 58
Regehr, G., 247
Reitman, D., 324
Relativism, 292, 295, 329, 480

Religion, 463, 481
Remote Associates Test, 374
Replication of experimental results, 467–468
Representational mathematics. *See* Mathematical representations
Representations. *See* Mental representations
Researchers. *See* Scientists
Research grants, 2–3, 6, 32, 263–264
Research methods. *See also* Experimental vs. correlational research methods; Scientific method
  anthropological, 97
  differences in social/personality psychology, 6–7, 345
  ex vivo research methods, 60–61
  grounded theory research method, 98–99, 466
  importance of methodological pluralism, 9, 476
  interviewing, 61–63, 95, 97–99
  nondualist methods, 473–476
  overview, 2, 33–34, 58–64
  self-report methods, 256, 259, 345, 348, 436–437
  subjective-objective dualism in, 464–465
  trial and error methods, 376–377, 380–381
  in vitro research methods, 58–59
  in vivo research methods, 3, 59–60, 95–96
Research notebooks, 2, 61*f*, 61–63, 318–319, 327
Residual decoupling of blind variants, 367
Review of scientific articles, 6, 262–264, 318. *See also* Scientific articles
Reynolds, R. I., 300
Ridgeway, C., 280
Rock, I., 308–309
Rodofsky, S., 216
Roe, Anne, 27
Roediger, H. L., 203
Ross, L. D., 269
Rothschild, L., 269–270
Rouse, J., 92
Rules. *See also* Balance-scale task experiments
  in balance-scale task paradigm, 234–235
  of control of variables strategy, 139
  discovery and confirmation bias, 58–59, 125
  discovery and implicit/explicit thinking, 5, 207–210, 237–239, 241–242, 243–244, 245, 248
  explicit processing and discovery of, 194–196
  fixation on incorrect rules, 239, 241–242
  formal treatment of, 115–116, 117
  in idealized models of science, 113–115
  positioning theory and, 104
  in process-control experiments, 231–232
  seeking and discovery of, 236–237, 244–245
"Rules for the Direction of the Mind" (Descartes), 93
Russell, Bertrand, 44–45, 485
Rutherford, Ernest, 306–307, 317

Salience of experimental data, 231–234, 308, 309–310, 329
"Same outcome" experiment goals, 169–171
Sanders, Gary, 65
Sarton, George, 20
Scalars (physics), 413–418, 413*f*, 417*t*
Scarantino, A., 102
Schachter, D. L., 63
Schauble, L., 138–139, 164–165, 165–167, 168, 233
Schooler, J. W., 230
Schunn, Christian, 8, 27–28, 230–231, 247
Schweigert, W. A., 465–466
Science. *See also* Scientists
  approaches to study of, 3, 73–76
  defined, 14
  development of disciplines within, 2–3, 17–22, 67–68
  as fourth branch of American democracy, 483–485
  idealized models vs. actual practice, 113–118
  morality of, 8–9, 470, 471, 479, 479–480, 480–481, 482–483, 487–488
  non-science vs., 18
  technology relationship to, 398–399

Science and technology studies (STS)
internal divisions within, 86–87
psychology of science relationship to, 57–58, 67, 77–78
support for, 27, 32, 51–52
La Science et L'hypothese (Poincaré), 18
Science instruction, 147–148*t*, 141–150, 176–178
Science Unfettered (McGuire and Tuchanska), 90
Scientific articles, 6, 262, 262–264, 318, 324–325
Scientific creativity. *See also* Blind variation and selective retention; Combinatorial conjecture
1950s interest in, 27
artistic creativity and, 377
convergent and divergent thinking in, 438–439, 441*f*, 442*f*, 443
mental and social processes of, 439–440
mental processes and, 373–377, 402
originality, utility and nonobviousness, 7–8, 381–383, 391–392
social processes of, 440
stochastic combinatorial models of, 369–370, 378
Scientific Creativity (Taylor and Barron), 27
Scientific disciplines, development of, 2–3, 17–22, 67–68
Scientific discovery
agency and, 88–89
artificial intelligence and, 76
combinatorial conjecture and, 7–8, 391–395
computer programs designed for, 369
explicit vs. implicit processes and, 88–89, 229
as human process, 88–93
mental representations in, 391–395
social change and, 20–22
technological advances and, 395
as value laden, 473
Scientific Discovery as Dual Search (SDDS) model, 229
Scientific Discovery: Computational Explorations of the Creative Process (Simon), 76
Scientific Genius: A Psychology of Science (Simonton), 363
Scientific groups. *See* Groups, scientific; Multidisciplinary teams
Scientific instruments, 3, 7–8, 79–80, 302–303, 395, 398.
See also Technological innovation
Scientific judgment. *See also* Biases; Experts and expert knowledge
agency and, 51–52, 90–91
balanced identity theory and, 257*f*, 258
cognitive study of science and, 75
factors affecting theory acceptance/rejection, 48–49
implicitness/explicitness of, 208, 217, 229–230, 230
Scientific method. *See also* Control of variables strategy; Humean indices; Research methods
idealized models vs. actual scientific practice, 113–118
objectivity of, 465, 466–467
philosophical pragmatism and, 74
use in psychology, 22, 57–58, 74
The Scientific Mind (Bachelard), 24–27
Scientific norms
as context for acting person, 92–94, 103, 107–108
ethical norms, 255, 264, 326–327
Merton's prescriptive norms, 259–260
theory ladenness effects and, 327–329
Scientific training, 273, 283, 465–468
Scientific Understanding (de Regt, Leonelli, and Eigner), 52
The Scientist in the Crib: Minds, Brains and How Children Learn (Gopnik et al.), 28–29
Scientists. *See also* The acting person as unit of analysis; Biases; Non-scientists
agency of, 88–93
agent-centered approaches to science and, 82–83
as experiment participants, 294–295
moral responsibilities of, 479–480, 485, 491–492
objectivity of, 262–264, 467
psychopathy and scientific creativity, 381
"Scientists are Human" (Watson), 89
Scientists: Their Psychological World (Eiduson), 27

SDDS (Scientific Discovery as Dual Search) model, 229
Searle, John, 460, 461, 463, 468
Sedgwick, Adam, 325–326
Selectivity processes. *See also* Blind variation and selective retention (BVSR)
  bias in gender studies and, 272–274, 281
  selective learning, 232
  selective retention, 7–8, 367–368
The self, 206, 257–258, 271–272
Self-directed experimentation (SDE), 231
Self-report methods, 256, 259, 345, 348, 436–437
Selz, Otto, 49, 51
Semantic networks, 188–190, 374
Senders, John, 483
Sensations, 43–44, 49, 50–51, 399, 412–413
Sense-making, 99–103
The Senses and the Intellect (Bain), 364
Sensori-motor intuitions, 408–409, 412–413, 419, 429–430
Sensory transduction systems, 206
Serendipity and scientific discovery, 391–392, 393
Sex and Temperament in Three Primitive Cultures (Mead), 277
Sex differences. *See* Psychology of gender
Sex-of-author effects, 270–272, 273–274. *See also* Psychology of gender
Seyle, C., 355
Shadish, William, 28
Shaw, J. C., 376–377
Sheehan, W., 305
Shelley, Mary, 9, 480–481
Shields, Stephanie, 277
Shinoda, H., 309
Shipman, P., 316
Shrager, Jeffrey, 27–28
Shrivastava, A., 309
Siegler, R. S., 150
Sighted variants, 365
Signal detection theory, 221
Siler, Stephanie A., 4
Simon, Herbert A., 27–28, 76, 215, 229, 369, 376–377, 430
Simons, D. J., 309–310, 328
Simonton, Dean, 7–8, 29, 33, 34, 404–405
Simpson, T. K., 425
Single condition experiments, 163–164
Situational explanations
  correspondent inference and, 268
  dispositional explanations vs., 6–7, 268–270, 348–349, 353–354
  essentialist explanations vs., 274–276, 280
  in social psychology, 279–281
  in transactional models of emotion, 102–103
Skinner, B. F., 44, 202–203
Skopeliti, I., 314–315
Slife, Brent D., 8
Smythe, William, 88
Snodgrass, M., 221
Social categories, 269–270, 280
Social Change (Ogburn), 20–22
Social change and technological advances, 20–22
Social cognition, 256–259, 268–270
Social identity theory, 104, 271–272
Social innovation, 401
Social knowledge structure, 257–258, 257*f*
Social-personality research. *See also* Experimental vs. correlational research methods
  experimental vs. correlational methodological split, 335–340, 355–357
  meta social-personality psychology study, 340–342
  personality/social research as merged, 352–353
  personality/social research as separate streams, 349–351, 353–355
  personality/social researcher differences, 342–349
Social psychology. *See also* Social-personality research
  correspondence and essentialism biases, 268, 268–270
  ecological validity crisis, 57–58
  implicit social cognition and balanced identity theory, 256–259
  integration with cognitive psychology, 437–438, 445, 452
  situational theories of gender differences, 279–281
  view of implicit/explicit processes, 437

Social psychology of science, 16, 103–107. *See also* Social-personality research
"Social Psychology of Science Newsletter," 28
Social roles
essentialist vs. situational explanations, 274–276, 280
fundamental attribution error and, 269
in multidisciplinary teams, 440–441, 442–443, 444, 451
positioning theory and, 103–107
Social Studies of Science (journal), 22
Societies, professional, 17–22
Society for the Social Studies of Science (4S), 22, 29
Sociological theories of gender differences, 279–281
Sociology of science
development of, 23*t*, 20–22
views of individual scientists, 75
Soddy, Frederick, 317
Solomon, R., 460, 462
Spatial contiguity, 113–114
Species extinction/survival, 8–9, 483, 485, 485, 488, 492
Speculation, 317
Spence, J. T., 279
Spence, Kenneth, 46, 46
Sperling, G., 221
Spreading activation, 247, 374, 378
Staats, A. W., 87
Standard cognitive view of implicit/explicit processes. *See* Dual process models of cognition; Explicit processes; Implicit processes
Stanley, Julian, 33
Statistical procedures, 343–345, 344*f*
Statistical significance testing, 256
Statistical techniques, 33–34, 256, 343–345, 349
Steinfeld, G. J., 299–300
Steinmetz, J. L., 269
Stevens, S. S., 27
Stevenson, R. J., 233–234, 239
Stewart, T. C., 390–391
Stiles, W. B., 466
Stochastic combinatorial models of creativity, 369–370, 378
Stroebe, W., 441
Structure mapping theory, 409
The Structure of Scientific Revolutions (Kuhn). *See* Kuhn, Thomas S.
STS. *See* Science and technology studies
Stuewer, R. H., 306–307, 328
Subjective-objective dualism, 8, 461, 463–464, 465–468. *See also* Dualism
Subjective values. *See* Values
Subliminal processing, 220–222
Sub specie historiae studies, 61–63
Supreme Academy (of science) proposal, 8–9, 483–485
Swann, W. B., 355
Swiderek, M. R., 130–131
Symbolic representation, 204–205
Synchrony (neural representations), 390
Systematic combinatorial variants, 380
Szaz, T. S., 43

Tacit knowledge
implicit processing and, 52, 247, 248–249
research on, 66–67
shared mental models and, 60
in vitro studies and, 58–59
Tactile representations, 393
Tannenbaum, P. H., 257
Task conflict (scientific groups), 440, 441–442, 443, 446
Taylor, Calvin, 27
Taylor, Charles, 462–463, 468
Teams. *See* Groups, scientific; Multidisciplinary teams
Technological innovation. *See also* Scientific instruments
morality in, 488
power and, 484
scientific discovery and, 7–8, 395, 398–399
social change and, 20–22
TED (Training in Experimental Design) computer tutor, 138, 142*t*, 141–150, 175–178
Temporal contiguity, 113–114
Temporal priority, 113–114
Ter Hark, M., 51
Terman, Lewis, 27
Textbooks as value laden, 324–325
Thagard, Paul, 7–8, 52, 76, 212, 390–391
Theology, dualism in, 463

Theoretical networks, 213
Theory. *See also* Alternative explanations; Confirmation bias; Top-down/bottom-up effects
background information and, 3–4, 117–118, 314
data and, 8, 116, 117–118, 465
evidential relevance and, 119–120, 121–122
as facilitatory and inhibitory, 296–297, 299–300, 302, 322, 328–329
interdependence of evidence/data and, 116–117
logical structure of, 74
naïve theories, 315
observation and, 7–8, 394–395, 398, 398*t*
perception and, 47–48
as top-down knowledge, 294
as universal propositions, 467
Theory ladenness of mental processes
attention, cognitive psychology of, 308–310
attention, history of science of, 310–314
communication, 324–325
controls against, 327–329
experimenting, cognitive psychology of, 319
experimenting, history of science of, 320–322
framework for understanding, 293–297
Kuhn and Hanson on, 290–293
memory, 322–323
multiple processes, 325–326
overview, 6
perception in cognitive psychology, 297–302
perception issues in history of science, 302–308, 326*f*, 394–395
philosophical views of relativism and, 329
scientific thought, cognitive psychology of, 314–316
scientific thought, history of science on, 316–319
top-down processes, (un)conscious nature of, 326–327
Thinking. *See also* Cognition; Scientific creativity
correspondent inference and, 6, 268, 268–270, 277–278
divergent vs. convergent, 8, 374, 438–439, 441*f*, 442*f*, 440–443
relevance of background information and, 124
theory ladenness of, 314–319
thinking out loud, 230
thinking too hard, 239–240, 249
Thomas, J. R., 271
Thomson, J. J., 317
Thomson, William, 425n17, 424–425
Thorndike, E. L., 44–45
Thought experiments, 60, 77, 79
Titchner, Edward B., 202
Tolman, E. C., 91–92
Top-down/bottom-up effects
attention and, 312, 313
perception and, 295–296, 299, 301, 302, 307, 325
researcher unawareness of, 326–327
strengthening bottom-up information, 327–328
thinking and, 316
Toth, E. E., 150, 159
Toulmin, Stephen, 75, 87, 88, 366
Training centers, 17–22, 32–33
Training in Experimental Design (TED) computer tutor, 138, 142*t*, 141–150, 175–178
Transactional models of emotion, 102–103
Transactive memory, 59–60
Transduction systems, 206
Transfer of knowledge and implicit/explicit processes, 209–210, 230–231, 233–234, 244–245, 249
Treason of the intellectuals, 491–492
Treatise on Electricity and Magnetism (Maxwell), 424–426
Trial and error methods, 376–377, 380–381
Tschirgi, J. E., 163, 165
Tuchanska, B., 90
Tweney, Ryan, 8, 15, 31, 58–59, 61, 76
Twin studies on scientific talent, 33–34
2, 4, 6 task, 58–59, 125
Two-strings creativity problem, 366, 374
Tycho Brahe, 290–291

Unconfounded vs. confounded experiments, 4, 123, 139, 143, 150, 156–161, 164

Unconscious implicit thinking, 215–218, 218–222
Unconscious mental processes. *See* Conscious/unconscious mental processes
Unconscious Thought Theory (Dijksterhuis and Nordgren), 230
Uncontrolled/controlled (automaticity feature), 185–186, 192–193
Unintentional discoveries, 393
Unintentional/intentional (automaticity feature), 186, 188, 192–193
United States
  anti-intellectualism in, 480
  Constitution, 483–484
  Supreme Academy (of science) proposal, 483–485
Universalism (scientific norm), 259–260, 467
Unmanned aerial combat vehicles (UCAVs), 487–488
Unobservable phenomena, 184, 188–189, 207–210, 212–213, 219, 425n17
Unusual uses (creativity), 366
Us/them distinctions, 257–259

Validity. *See also* Ecological validity
  internal vs. external, 6–7, 336, 345–347, 347*f*
  validity conditions, 218, 221
Values
  in nondualist research methods, 8, 469–473, 473, 474–475
  textbooks as value laden, 324–325
Van Dijk, T. A., 324
Van Maanen, Adriaan, 304–305
Variables. *See* Control of Variables Strategy
Variation, blind. *See* Blind variation and selective retention
Variation backtracking (blind variation), 372–373
Variation superfluity (blind variation), 371–372
Vectors, 413–418, 414*f*, 415*f*, 416*f*, 417*f*, 417*t*
Vernon, M. D., 48
Vicente, K. J., 323
Vienna Circle (philosophy of science), 18, 40, 45–46, 47
Visual attention. *See* Attention
Visual models/representations
  of inventions, 396
  mathematics relationship to, 8, 412–413, 413*f*, 414*f*, 415*f*, 416*f*, 417*f*
  scientific discovery and, 77, 392, 393–394
Vohs, K. D., 203
Volition, unconscious implicit, 217
Vollmeyer, R., 232
Von Foerster, Heinz, 480
Von Neumann, John, 412
"Voodoo sociology," 22
Vosniadou, S., 171, 314–315
VOTAT (vary one thing at a time) strategy, 62
Vul, E., 221

Wallas, G., 375
Wason, P. C., 58–59
Watson, D. L., 89, 375–376
Watson, John B., 44, 45, 202
The Way Things Are (Bridgman), 98
Weapons, scientific morality and, 487–488
Wegner, D. M., 217
Weick, K. E., 99–100
Weinberg, S., 317
Weiner, Norbert, 481
Weiskrantz, L., 220
Weisstein, Naomi, 277
Welling, H., 399
Wenger, E., 104
Wessely, S., 324
Westrum, Ron, 28
Wettersten, J., 50–51
Whewell, William, 18
White, R. T., 301–302
Whole condition comparisons, 161–162
Wigner, Eugene, 428
Wilson, Robert Woodrow, 312
Wittgenstein, Ludwig, 463
Wolff, P., 409
Wood, R. W., 305–306
Wood, Wendy, 271
Woodward, J., 294
Woolgar, S., 75
Woolley, Helen Thompson, 276
Wotiz, J., 216
Wundt, Wilhelm, 43, 202

Würzburg School (cognitive psychology), 49–51

Yaroshevsky, Mikhail Grigoryevich, 27

Zangwill, O. L., 322

Zimbardo, P., 354

Zimmerman, Corinne, 5, 239

Zuckerman, Harriett, 22

Zur Methodenfrage der Denkpsychologie (*On the Methodological Problem of Cognitive Psychology;* Popper), 49